BRANCHING PROGRAMS AND BINARY DECISION DIAGRAMS

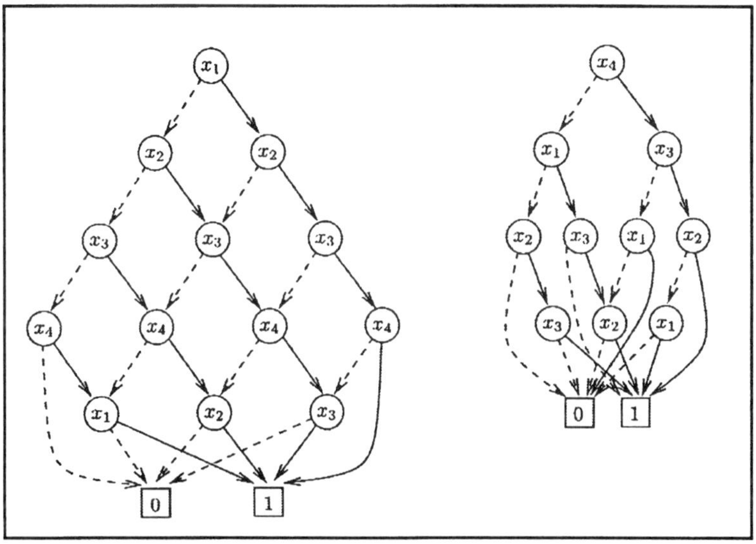

SIAM Monographs on Discrete Mathematics and Applications

The series includes advanced monographs reporting on the most recent theoretical, computational, or applied developments in the field; introductory volumes aimed at mathematicians and other mathematically motivated readers interested in understanding certain areas of pure or applied combinatorics; and graduate textbooks. The volumes are devoted to various areas of discrete mathematics and its applications.

Mathematicians, computer scientists, operations researchers, computationally oriented natural and social scientists, engineers, medical researchers and other practitioners will find the volumes of interest.

Series Volumes

Wegener, I., *Branching Programs and Binary Decision Diagrams: Theory and Applications*
Brandstädt, A., Le, V. B., and Spinrad, J. P., *Graph Classes: A Survey*
McKee, T. A. and McMorris, F. R., *Topics in Intersection Graph Theory*
Grilli di Cortona, P., Manzi, C., Pennisi, A., Ricca, F., and Simeone, B., *Evaluation and Optimization of Electoral Systems*

Branching Programs and Binary Decision Diagrams

Theory and Applications

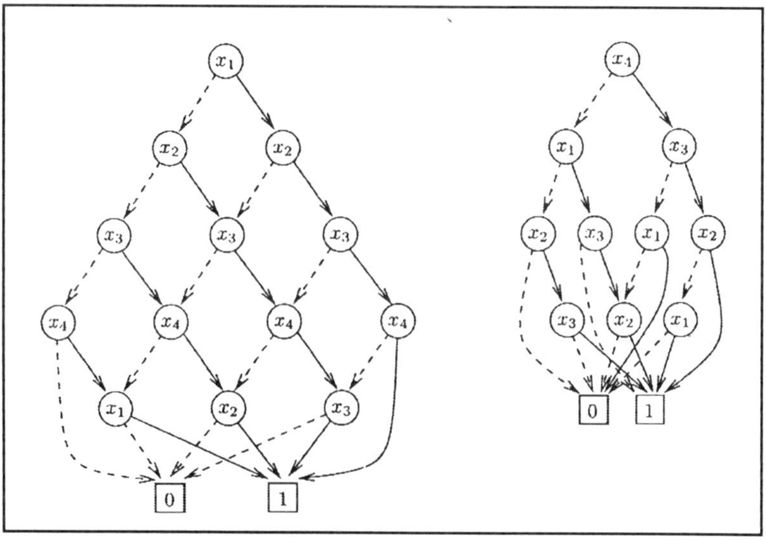

Ingo Wegener

Universität Dortmund
Dortmund, Germany

siam

Society for Industrial and Applied Mathematics   Philadelphia

10 9 8 7 6 5 4 3 2 1

Library of Congress Cataloging-in-Publication Data
Wegener, Ingo.
Branching programs and binary decision diagrams : theory and applications / Ingo Wegener.
p. cm.— (SIAM monographs on discrete mathematics and applications)
Includes bibliographical references and index.
ISBN 0-89871-458-3
1. Branching processes. 2. Decision making—Mathematical models. 3. Computational complexity. I. Title. II. Series.

QA274.76.W44 2000
519.2'34—dc21

00-035749

Contents

Preface

Research on the complexity of Boolean functions has a more than fifty-year-old history. For a long time, research has been focused on circuits and the circuit complexity of Boolean functions. The open problems in circuit complexity are now so hard that not much progress has been made during the past 10 years. However, many new results have been obtained on the communication complexity of Boolean functions and on branching programs (BPs) and binary decision diagrams (BDDs) (which are synonyms) representing Boolean functions. The theory of communication complexity has been described by Kushilevitz and Nisan (1997) and Hromkovič (1997) and now seems to be the right time to present a comprehensive monograph on all aspects of BPs and BDDs including theory and applications.

Theoretical research on BPs has been motivated mainly by relations between their size and the space complexity of Turing machines and, in particular, by the general aim to develop lower bound techniques. Applications of restricted variants of BDDs as data structures or representation types for Boolean functions have led to many new, important, and interesting problems. There is a trade-off between the size of the class of functions representable in polynomial size by some variant of BDDs and the difficulty or complexity of performing operations on the representations. Hence, one has to compare the various representation types. For this reason, upper and lower bound techniques are developed and the representation size of important and selected functions is estimated. Moreover, simulations between the representation types are investigated. For a list of essential operations, one looks for efficient algorithms or proofs showing that they are inherently difficult. The applications of such representations of Boolean functions are manifold: verification; automata theory; model checking; further CAD applications such as synthesis, simulation, test generation, timing analysis, and technology mapping; and applications in optimization algorithms for, e.g., integer programming and network flow, combinatorics, and genetic programming.

The aim of this monograph is to present state-of-the-art knowledge on BPs and BDDs. Before typical applications are discussed, the complexity theoretical and algorithmic properties of various types of BPs and BDDs are investigated.

A theoretically oriented lecture can be based on the chapters and sections 1, 2, 3.1–3.3, 4, 5.1–5.4, 6, 7, 10, 11, and 12 (and perhaps 8 and 9), while a lecture focusing on BDDs as data structures may concentrate on the chapters and sections 1, 2.3–2.4, 3, 4, 5, 6.1, 6.4, 7.1–7.3, 8, 9, 10 without 10.3, 12, 13, 14, and 15. The monograph contains 194 exercises: 66 of them are classified as easy, 87 are of medium difficulty, and 41 are difficult (marked "E," "M," and "D," respectively). Moreover, 29 open problems are presented. Solutions to the exercises (and possibly some open problems) are worked out and presented at

`ls2-www.cs.uni-dortmund.de/monographs/bdd`

where the reader will also find a (we hope short) list of errors and misprints.

I am grateful for the support and affection of many people during this project. Only a few of them can be mentioned here. I have enjoyed researching and writing papers on BPs and BDDs with my coauthors Beate Bollig, Stasys Jukna, Matthias Krause, Martin Löbbing, Sasha Razborov, Martin Sauerhoff, Petr Savický, Olaf Schröer, Detlef Sieling, and Ralph Werchner. The discussions with all of them and many others, in particular, Martin Dietzfelbinger, Jawahar Jain, Noam Nisan, Pavel Pudlák, and Fabio Somenzi, have broadened my way of thinking on BPs and BDDs. My handwritten drafts have been turned into LaTeX style by Oliver Giel, Dominic Heutelbeck, Jörg Pernotzky, Niels Schmitt, and Tanja Zoppa. Beate Bollig, Thomas Hofmeister, Martin Sauerhoff, and Detlef Sieling have read and improved the draft of this monograph. In spite of the many successful applications of BDDs, real-life decisions cannot be based on simple decision diagrams and I am happy to have someone like Christa with whom I share my decisions and who stands by my side in good and difficult times.

Dortmund/Bielefeld, June 1999 Ingo Wegener

Chapter 1

Introduction

1.1 Branching Programs (BPs) and Binary Decision Diagrams (BDDs)

In order to describe and solve a class of concrete problems one has to look for an appropriate abstraction. Functions cover many aspects of connections between different objects. A class of functions consists of *finite functions* if the common domain and the common image of the functions are finite sets. If the elements of the domain and image are encoded by binary strings of fixed length, the resulting functions are called Boolean. The class of *Boolean functions* $f\colon \{0,1\}^n \to \{0,1\}^m$ is denoted by $B_{n,m}$ or by B_n if $m = 1$. The central role of binary encodings in the world of computer science is evident and this implies the central importance of the class of Boolean functions.

Historically, computing has been the main issue of computer science. Circuits or combinational networks and also sequential networks are suitable computation models for Boolean functions. BPs and BDDs are synonyms for the same model. They can be regarded as a computation model as well as a representation or data structure for Boolean functions. They and many of their variants are fundamental in theory and for many applications. This monograph deals with all these aspects. First, we present four equivalent definitions of BPs and BDDs. They reflect the different historical roots of the BP/BDD model. The fact that different objectives lead to the same model emphasizes the importance of this model. It will turn out to be useful that we can refer to the different definitions.

Definition 1.1.1. BP or BDD.

Syntax: A BP or BDD on the variable set $X_n = \{x_1, \ldots, x_n\}$ consists of a directed acyclic graph $G = (V, E)$ whose inner nodes (nonsink nodes) have outdegree 2 and a labeling of the nodes and edges. The inner nodes get labels from X_n and the sinks get labels from $\{0, 1\}$. For each inner node, one of the outgoing edges gets the label 0 (in figures, drawn as a dotted line) and the other one gets the label 1 (drawn as a solid line).

In a BP or BDD on X_n, each node v represents a Boolean function $f_v \in B_n$ defined in the following way.

Semantics 1: The computation of $f_v(a)$, $a \in \{0,1\}^n$, starts at v. At nodes labeled by x_i, the outgoing edge labeled by a_i is chosen. Then $f_v(a)$ is equal to the label of the sink which is reached on the considered path.

Semantics 2: Each input $a \in \{0,1\}^n$ activates all a_i-edges leaving x_i-nodes, $1 \le i \le n$. Then $f_v(a)$ is equal to the label of the final node on the unique *activated path* starting at v.

Semantics 3: A sink s with label b represents the constant function $f_s(a) \equiv b$. Let v_0 and v_1 be the direct successors of an inner node v reached via the 0-edge and 1-edge, resp., and let f_0 and f_1 be the functions represented at v_0 and v_1, respectively. If v is labeled by x_i, then f_v is defined by *Shannon's decomposition rule* $f_v(a) := a_i f_1(a) + \overline{a}_i f_0(a)$ (the product is the Boolean AND, $+$ the Boolean OR, and $\bar{\ }$ the Boolean NOT). This can also be written as $f_v(a) := a_i f_1(a) \oplus \overline{a}_i f_0(a)$ ($\oplus$ is the Boolean EXOR).

It is easy to see that the three definitions of the semantics of a BP or BDD are equivalent. The first one reflects the viewpoint that a BP is a computation model and the main issue is to evaluate the function represented at some specific node efficiently. The second one reflects the more global view that all nodes represent Boolean functions. Both approaches use a top-down view with respect to the underlying directed acyclic graph. The third definition of the semantics suggests a bottom-up approach for the iterative evaluation of the represented Boolean functions. Obviously, the evaluation can also be performed by a top-down recursive approach where the sinks are terminal cases. Shannon's decomposition rule can be written with the Boolean OR or EXOR, since only one of the two terms may equal 1. This rule describes in an obvious way how we can construct a BP representing $f \in B_n$. At each node, we decompose the function that we have to represent for a variable which the function depends on. Functions which do not depend on any variable are constant and can be represented by a sink. We obtain a BP whose underlying graph is a tree with at most $2^n - 1$ inner nodes and 2^n sinks. This proves that each Boolean function can be represented by a BP.

Definition 1.1.2.

(i) The *size* of a BP is equal to the number of its nodes.

(ii) The *depth* of a BP is equal to the length of its longest path.

Usually, the size of a graph is measured by the number of nodes *and* edges. For a BP it is sufficient to consider the number of nodes. Since we can restrict

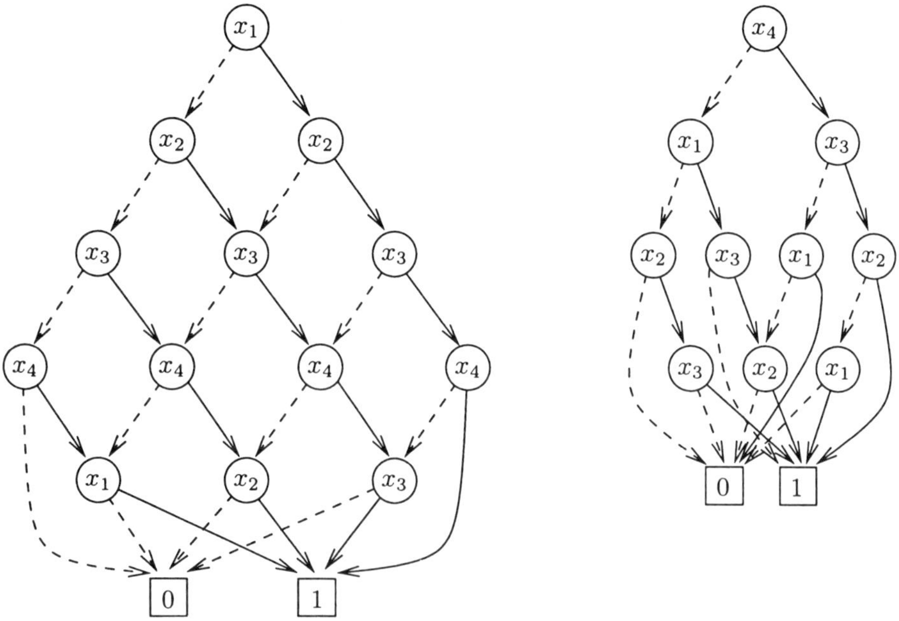

Figure 1.1.1: *BPs for the function HWB_4.*

ourselves to BPs with one 0-sink and one 1-sink and since each BP for a nonconstant function needs both types of sinks, the number of edges of a BP with size s equals $2(s-2)$ for nonconstant functions.

Two examples of BPs are shown in Fig. 1.1.1. Both BPs represent the same function at their sources, called the *hidden weighted bit* function. Since this function will serve as a typical example, we define it formally.

Definition 1.1.3. The hidden weighted bit function $\text{HWB}_n \in B_n$ computes the bit a_{sum} on the input $a = (a_1, \ldots, a_n)$, where $sum := \|a\| := a_1 + \cdots + a_n$ and $a_0 := 0$.

The interesting feature of HWB_n is that the whole input serves as control information *and* as data information. With *sum* we compute the "address" of the bit which has to be computed. The first BP "simulates" this definition of HWB_4. On the first four levels, the actual value of *sum* is "computed." For each of the edges leaving an x_4-node, the value of *sum* is "known." If $sum = 0$, the output is 0, since $a_0 = 0$. If $sum = 4$, then $a_1 = a_2 = a_3 = a_4 = 1$ and the output is $a_4 = 1$. If $sum = i \in \{1, 2, 3\}$, we check the value of a_i. This BP has size 15, depth 5, and follows a simple construction strategy. The second BP has only size 12 and depth 4 but it is more difficult to verify that it represents HWB_4 at its source. For $n = 4$, we can verify this by checking all 16 inputs.

Definition 1.1.4. An *inconsistent path* (or *null chain*) of a BP G is a directed path containing for some variable x_i a 0-edge leaving an x_i-node and a 1-edge leaving an x_i-node.

By definition, the edges of an inconsistent path cannot be activated by one input. On the other hand, it is easy to find an input which activates all edges of a consistent path. If the path contains a b_i-edge leaving an x_i-node, we set $a_i = b_i$. By definition, this does not lead to a contradiction. The bits a_j which are not defined by the above procedure can be defined arbitrarily. The second BP of Fig. 1.1.1 does not contain any inconsistent path and each path from the source to the 1-sink with i tested variables represents 2^{n-i} *satisfying inputs*, i.e., inputs leading to the output 1. The first BP of Fig. 1.1.1 contains inconsistent paths. This causes some difficulties. For example, it is not easy to detect useless nodes. The leftmost x_3-node on level 3 and the leftmost x_4-node on level 4 represent nonconstant functions ($x_1\overline{x}_3x_4+x_1x_3\overline{x}_4+x_2x_3x_4$ and x_1x_4, respectively). If we only have the aim to represent HWB_4 at the source, these nodes can be replaced by the 0-sink, since all paths from the source to the 1-sink using one of these nodes are inconsistent.

As promised we present a fourth definition of BPs and BDDs.

Definition 1.1.5.

(i) An *ite straight line program* is a sequence $I_1, \ldots, I_s$ of instructions of type $I_j = ite(C_j, T_j, E_j)$. The condition C_j, the then-branch T_j, and the else-branch E_j are elements from $\{0, 1, x_1, \ldots, x_n, I_1, \ldots, I_{j-1}\}$. Each instruction I_j represents a Boolean function $f_j \in B_n$ using recursively the semantics $ite(x, y, z) := xy + \overline{x}z$ (if $x = 1$ then y else z).

(ii) A BP or BDD is an ite straight line program where the conditions C_j are variables from $\{x_1, \ldots, x_n\}$ and $T_j, E_j \in \{0, 1, I_1, \ldots, I_{j-1}\}$.

We have to prove that this new definition coincides with Definition 1.1.1. Using a reversed topological ordering of the inner nodes of a BP G (if $(v, w) \in E$, then w has to precede v with respect to a reversed topological ordering), we obtain an ite straight line program for the same Boolean function, where the number of instructions equals the number of inner nodes of G. A node v_k with label x_i, 1-successor v_l, and 0-successor v_m is replaced by $I_k := ite(x_i, I_l, I_m)$, where the sinks are not replaced by instructions but by their labels. If, on the other hand, $I_1, \ldots, I_s$ are the instructions of an ite straight line program fulfilling the conditions of Definition 1.1.5(ii), the corresponding BP graph G gets a 0-sink v_{s+1}, a 1-sink v_{s+2}, and inner nodes $v_1, \ldots, v_s$. The label of v_i, $1 \le i \le s$, equals C_i, the 1-edge leaving v_i leads to v_k if $T_i = I_k$ ($I_{s+1} := 0, I_{s+2} := 1$), and the 0-edge leaving v_i leads to v_l if $E_i = I_l$. The definition of a BP as an ite straight line program reflects the fact that a BP is a special type of "program."

Why do we have two names (and four definitions) for the object of our investigations? The reason is the well-known fact that theoreticians (more precisely, people working in theoretical computer science) and practitioners (more precisely, people working on problems from applications) typically do not talk to each other. So theoreticians have worked on BPs while practitioners used the notion of a BDD. They have worked for some time without any considerable contact and several results have been obtained more than once. Although their research objectives are quite different, their independent developments have led to the same variants of BPs and BDDs. Only since the late 1980s have some theoreticians and some practitioners found out how fruitful and profitable an exchange of ideas and problems is. Nevertheless, we have to live with different names for many objects of interest. In the next two sections, the motivation of theoreticians and practitioners to study BPs and BDDs is described.

1.2 Motivations from Theory

The purpose of complexity theory is to classify problems and functions with respect to their complexity (or "difficulty"). What difficulty means depends on the objective of the investigation.

One of the most important complexity measures is the size of circuits of fan-in 2. The asymptotically optimal upper bound for all Boolean functions $f \in B_n$ has been proved by Shannon (1949). His approach is based on Shannon's decomposition rule, which we have used for the definition of BPs. Hence, the same approach leads to an asymptotically optimal upper bound on the BP size of Boolean functions (see Section 2.2).

The main obstacle in complexity theory is our inability to prove large lower bounds on the complexity of *explicitly defined* functions. For a more sophisticated discussion of the notion "explicitly defined" see Wegener (1987). Here it is sufficient to think of sequences $f = (f_n)$, $f_n \in B_n$, of Boolean functions such that the problem to decide for strings $a \in \{0,1\}^n$, $n \in \mathbb{N}$, whether $f_n(a) = 1$ is contained in NP. In order to develop and strengthen lower bound techniques, one considers restricted computation models. Formulas are circuits whose underlying graph is (after a suitable duplication of the inputs) a tree. BP size lies between circuit size and formula size (see Section 2.1). Hence, lower bounds on the BP size of explicitly defined Boolean functions are of particular interest. Nevertheless, we are unable to prove such bounds. However, the BP model allows a lot of restrictions (see Section 1.5) and for some of them we have powerful lower bound techniques.

The size of BPs is also interesting in itself. Cobham (1966) proves that the logarithm of the minimal size of BPs for a Boolean function is essentially the same as the space complexity of nonuniform Turing machines computing this function (see Section 2.1). Hence, BPs are the best model to study the nonuniform space complexity of problems. Moreover, a well-founded "parallel computation thesis" implies that space complexity and parallel time are closely related (see Wegener (1987)). The complexity class NC^1 consists of all (sequences of) Boolean functions $f = (f_n)$, $f_n \in B_n$, computable by circuits of fan-in 2 and depth $O(\log n)$. It is easy to see that these two conditions imply that the circuit has polynomial size. Hence, NC^1 consists of problems which have efficient parallel solutions. Surprisingly, Barrington (1989) has proved that the class NC^1 also can be described by BPs. Here we consider leveled BPs, i.e., each node belongs to a level and edges lead from one level only to the next level. Then the width of a BP is the maximal number of nodes on one level of the BP. The result of Barrington (1989) states that NC^1 is equal to the class of Boolean functions representable by BPs of polynomial size and width bounded by 5. (For a detailed proof of this result see Wegener (1987).) Altogether, the study of BPs covers many subjects from complexity theory.

1.3 Motivations from Applications

The notion of a decision diagram (DD) (not necessarily binary) is so natural that one cannot date the first use of this type of representation. The famous botanist Carl von Linné (1707–1778) was one of the first to develop a classification scheme for plants. This scheme may be viewed as a DD. The tests at the inner nodes may have more than two outcomes and the number of different labels or names at the sinks is huge. Indeed, typical questionnaires may be understood as DD. Picard (1965) has already written a monograph on this subject. If the number of possible classifications is huge, DDs tend to degenerate to decision trees (DTs). Then it is interesting to find the cheapest possible tests; see Moret (1982) for an overview of these applications of DDs. In applications outside of computer science, tests are usually not binary.

Lee (1959) and Akers (1978a) were the first to use BDDs as representation or data structure for Boolean functions. With BDDs, one can sometimes bound the number of necessary prime implicants in a disjunctive normal form (DNF) for a function. Moreover, BDDs can support the testing and synthesis of circuits. But in these early approaches it is not clear which operations should be supported and which restrictions of BDDs allow efficient algorithms for these operations.

The breakthrough came in the pioneering papers of Bryant (1985, 1986). He recognized the key problem. In order to look for suitable data structures for some kinds of objects, one first has to clarify which operations have to be supported. Bryant presents a list of operations on Boolean functions which are useful for their symbolic manipulation. Then he presents the model of ordered BDDs (OBDDs) which allows efficient algorithms for all these and many more operations. OBDDs are studied in detail in Chapters 3, 4, and 5. In this section, we discuss a list of key operations for data structures for Boolean functions. Examples from applications emphasize the importance of the chosen operations.

First, we draw the reader's attention to an unavoidable time-space trade-off. It is not sufficient to have efficient algorithms to work with the chosen representation type, but the representations also should be as small as possible. The following rule of thumb has turned out to be true. The more compact a representation type is, the harder the operations become. Hence, we have to look for the right compromise.

What does it mean to have small size representations? Since the number of functions $f \in B_n$ equals 2^{2^n}, it holds for any type of representation that the fraction of functions $f \in B_n$ needing at least 2^{n-1} bits for their representation converges exponentially fast to 1 as $n \to \infty$. We need representations of polynomial size and in many applications the polynomial has to be of small degree. The best we can hope for is that the class of Boolean functions (more precisely of sequences of Boolean functions $f = (f_n)$, $f_n \in B_n$) representable in polynomial size is large and contains many "important" functions and many functions we have to deal with in real-life applications. Only the first requirement can be stated in a formal way.

Definition 1.3.1.

(i) For a representation type R for Boolean functions, the complexity class P-R contains all sequences of Boolean functions $f = (f_n)$, $f_n \in B_n$, representable in polynomial size with respect to R.

(ii) A representation type R_1 has at least the same representational power as a representation type R_2 if P-$R_2 \subseteq$ P-R_1.

(iii) R_1 has a larger representational power than R_2 if P-$R_2 \subsetneq$ P-R_1.

The community of researchers and users has to decide which functions have to be considered as "important." It should be obvious that the class of important functions includes storage access functions like the multiplexer function; arithmetic functions like addition, subtraction, multiplication, and division; and symmetric functions. For theoretical reasons we declare the hidden weighted bit function, quadratic functions, some simple functions on matrices, and read-once formulas as important. We do not define these functions here. They are considered in detail in Chapter 4.

It is even more difficult to define the class of functions we have to deal with in "real-life applications." Therefore, the community has reached an agreement on benchmark functions. These functions have fixed input length and most often do not belong to a natural sequence of functions. Hence, the notion of polynomial size is meaningless. We do not report results on benchmark functions. The interested reader may find them in papers we refer to.

Now we present the proposed list of key operations on representations of Boolean functions. Since we consider only graph representations of Boolean functions, we denote the representation of f by G_f.

Evaluation
Input: G_f on n variables, $a \in \{0,1\}^n$.
Output: $f(a)$.

(Binary) Synthesis
Input: G_f, G_g, and a Boolean operation $\otimes$: $\{0,1\}^2 \to \{0,1\}$.
Output: G_h where $h = f \otimes g$.
(In a similar way we may define the synthesis of three or more functions.)

SAT (satisfiability test)
Input: G_f.
Output: Yes if there exists some a where $f(a) = 1$, no otherwise.

SAT-COUNT (satisfiability count)
Input: G_f.
Output: The number of inputs a such that $f(a) = 1$.

Equivalence test
Input: G_f and G_g.
Output: Yes if $f \equiv g$, i.e., $f(a) = g(a)$ for all $a \in \{0,1\}^n$, no otherwise.

Replacement by constants
Input: G_f, $x_i \in X_n$, $c \in \{0,1\}$.
Output: G_g where $g := f_{|x_i=c}$ is the *subfunction* or *cofactor* of f for $x_i = c$, more precisely

$$g(a_1, \ldots, a_n) := f(a_1, \ldots, a_{i-1}, c, a_{i+1}, \ldots, a_n).$$

Replacement by functions
Input: G_f, G_g, $x_i \in X_n$.
Output: G_h where $h := f_{|x_i=g}$, more precisely

$$h(a_1, \ldots, a_n) := f(a_1, \ldots, a_{i-1}, g(a_1, \ldots, a_n), a_{i+1}, \ldots, a_n).$$

Redundancy test
Input: G_f, $x_i \in X_n$.
Output: Yes if f does not essentially depend on x_i, i.e., if

$$f(a_1, \ldots, a_{i-1}, 0, a_{i+1}, \ldots, a_n) = f(a_1, \ldots, a_{i-1}, 1, a_{i+1}, \ldots, a_n)$$

for all $a_j \in \{0,1\}$.

Quantification

Input: G_f, $x_i \in X_n$, a quantifier $Q \in \{\exists, \forall\}$.

Output: G_g. If $Q = \exists$, then $g := (\exists x_i)f := f_{|x_i=0} + f_{|x_i=1}$. If $Q = \forall$, then $g := (\forall x_i)f := f_{|x_i=0} \wedge f_{|x_i=1}$ ($\wedge$ is the Boolean AND).

Minimization

Input: G_f.

Output: G_f^*, a representation of f of the same representation type as G_f which has minimal size for f.

Definition 1.3.2. If the minimal-size representation of each Boolean function f with respect to a given representation type is unique (up to isomorphic representations), the minimal-size representation is called the reduced representation and the operation minimization is called *reduction.*

As an example of reduced representations we mention the representation of regular languages by *deterministic finite automata* (DFAs) (see Hopcroft and Ullman (1979)). It is well known that DFAs of minimal size are unique and that the reduction of DFAs is possible in polynomial time.

The reader may wonder why the listed operations are not independent of each other. We give examples.

1. SAT is the negation of the equivalence test between f and the constant 0.

2. SAT is a special case of SAT-COUNT.

3. The equivalence test can be performed by an EXOR-synthesis of G_f and G_g and the negation of SAT on the resulting G_h.

4. Replacement by constants is a special case of replacement by functions.

5. The result h of a replacement by functions can be written as $h = ite(g, f_{|x_i=1}, f_{|x_i=0})$ and, hence, as the ternary synthesis of g and two functions obtained by replacements by constants. The ternary synthesis can be decomposed to three binary synthesis steps since $h = gf_{|x_i=1} + \bar{g}f_{|x_i=0}$.

6. The redundancy test is an equivalence test for $f_{|x_i=0}$ and $f_{|x_i=1}$.

7. SAT is the test of whether $f(0, \ldots, 0) = 1$ or some variable is not redundant.

8. Quantification is defined as the synthesis of two functions obtained by replacement by constants.

These results can be seen as polynomial-time (Turing) reductions between different problems. They will be used to obtain in an easy way NP-hardness results for some problems from NP-hardness results for other problems (see Garey and Johnson (1979) for an introduction to the NP-completeness theory). On the other hand, the results can be used algorithmically. Nevertheless, it is necessary to consider the operations separately. For example, the most efficient equivalence test is not always an algorithm performing synthesis and a SAT test.

In order to motivate the listed operations, we discuss some typical computer-aided design (CAD) applications in a very general form. These and many more applications are treated in detail in Chapters 13, 14, and 15 (see Bryant (1992) and Clarke and Wing (1996) for surveys on areas of applications).

Symbolic simulation. Before the production of hardware one likes to "work" with the realization. In particular, evaluation is a fundamental operation. The evaluation of a circuit takes linear time with respect to the number of its gates while the evaluation of a BDD can be done in linear time with respect to its depth, which often is much more efficient. Therefore, one transforms given realizations such as circuits into representations such as variants of BDDs. Typically, the inputs have a simple and compact realization. Then the transformation is done gate-by-gate with the operation synthesis. Synthesis steps may cause an avoidable blow-up in the representation size if they are not followed by minimization steps. It is even more desirable to include the minimization in the synthesis procedure. Hence, evaluation, synthesis, and minimization/reduction are fundamental operations. In particular, we apply synthesis and minimization/reduction while transforming circuits into the chosen representation type.

Verification of combinational circuits. The problem is to prove the equivalence of a specification, often an already verified circuit, and a new realization. Verification means a formal proof of correctness and should not be confused with testing or partial simulation. Verification can be done by transforming the specification and the realization (synthesis and minimization/reduction) into the chosen representation type and by an equivalence test for the results. If no better idea is known, equivalence tests are replaced by an EXOR-synthesis and a SAT test. Hence, SAT is well motivated.

Verification of sequential networks. If DFAs can be described by lists of their states and by adjacency lists for the state transition function, the classical algorithms from automata theory (see Hopcroft and Ullman (1979)) are very efficient. But typical sequential networks have more than 10^{20} states. Then another approach is necessary. Using binary encodings we assume that the set of states equals $Q = \{0,1\}^k$ and the input alphabet $\Sigma = \{0,1\}^l$. Also, the state transition function δ can be described as a Boolean function $\delta : Q \times \Sigma \times Q \to \{0,1\}$ where $\delta(q, a, q') = 1$ iff the state transition from q to q' is possible for input a. This also covers nondeterministic automata. The function δ cannot be described by its truth table which is much too large. It has to be given by some logical description which we transform into the chosen representation type. A key problem of verification is to prove that no state in some set F of forbidden states can be reached from one of the initial states in I. Let $f_t(q,q')=1$ iff state q' may be reached from q in at most t steps. Then

$$\begin{aligned} f_1(q,q') &\equiv \exists a \in \Sigma \cup \{\epsilon\} : \delta(q,a,q'), \\ f_{t+1}(q,q') &\equiv \exists q'' \in Q : f_t(q,q'') \wedge f_1(q'',q'), && (*) \\ f_{2t}(q,q') &\equiv \exists q'' \in Q : f_t(q,q'') \wedge f_t(q'',q'). && (**) \end{aligned}$$

We want to compute the set R of states reachable from I. A state is reachable iff it is reachable in at most $|Q|=2^k$ steps. We identify sets of states with their characteristic functions, e.g., $R(q)=1$ iff $q \in R$. Then

$$R(q) \equiv \exists q' : I(q') \wedge f_{2^k}(q',q). \qquad (***)$$

In order to obtain f_{2^k} we apply $(**)$ k times. The approach described in $(*)$ is much too slow. But $(**)$ tends to lead to larger representations (during the process) than $(*)$. We may apply $(*)$ and stop if $f_{t^*} \equiv f_{t^*+1}$. If t^* is small, this approach may be successful and we may replace f_{2^k} by f_{t^*} in $(***)$. Finally, the verification can be performed by a negated SAT test for $R \wedge F$. Altogether, we have used synthesis, minimization/reduction, quantification, and SAT.

Synthesis of circuits. Circuit synthesis uses the principle of abstraction. In intermediate steps, a circuit C_1 gets inputs x and y, and later y is replaced by the outputs of a circuit C_2. If we have transformed C_1 and C_2 into the chosen representation type and if we look for a representation of the synthesized circuit C, this can be done by the operation replacement by functions.

Test pattern generation. A test pattern for a given fault in a circuit C is an input a such that C and its faulty counterpart C' lead to different outputs on a. Again, we have to transform C and C' into the chosen representation type. With SAT-COUNT on $C \oplus C'$ we can compute the number of suitable test patterns and then we need a procedure to compute one satisfying input or a list of satisfying inputs.

1.4 On the Inherent Complexity of Some Problems

In order to describe some interrelations between theory and applications, we prove some easy complexity theoretical results which shed light on the inherent complexity of some problems relevant in applications.

Why do we not use circuits as a representation of Boolean functions? For a circuit of size s evaluation is possible in time $O(s)$; synthesis can be done in constant time if we have direct access to the output gates. The same holds for the replacement problems and quantification. But the other problems are hard for circuits.

Theorem 1.4.1.

(*i*) *SAT is NP-complete for circuits.*

(*ii*) *SAT-COUNT is NP-hard and #P-complete for circuits.*

(*iii*) *The equivalence test is coNP-complete for circuits.*

(*iv*) *The redundancy test is coNP-complete for circuits.*

(*v*) *The minimization problem is NP-hard for circuits.*

Proof.

(i) One may guess an input and verify that it is satisfying in polynomial time. The classical NP-complete problem is SAT for conjunctive normal forms (CNFs) which are special circuits.

(ii) The NP-hardness follows, since SAT is a special case of SAT-COUNT. #P-completeness is (see Garey and Johnson (1979)) the adequate theory for counting problems and SAT-COUNT for circuits is the classical #P-complete problem.

(iii) A problem is coNP-complete iff its complement is NP-complete. In polynomial time, one may guess an input and verify that circuits compute different outputs for this input. Moreover, SAT is the inequivalence test for f and the constant 0.

(iv) In polynomial time, we guess an input and verify that it proves that f essentially depends on x_i. We can check whether $f \not\equiv g$ by checking whether $h := x_{n+1}f \oplus \overline{x}_{n+1}g$ for some new variable x_{n+1} essentially depends on x_{n+1}.

(v) We can solve SAT by minimizing the given circuit and checking whether the minimized circuit does not consist only of the input signal 0. □

Corollary 1.4.2. *If NP $\neq$ P and a representation type admits polynomial-size representations of the variables and polynomial-time algorithms for SAT and synthesis, a sequence of a polynomial number of synthesis steps may lead to representations of nonpolynomial size.*

Proof. Otherwise, we can solve SAT for circuits in polynomial time by constructing a representation of the considered functions from representations of the inputs by gate-by-gate synthesis and then solving SAT on the chosen representation type. □

These complexity theoretical results imply that for representations with an efficient SAT algorithm we can only hope for efficient synthesis algorithms for binary synthesis and for a small number of synthesis steps. For all representations, *clauses*, i.e., disjunctions of literals, can be represented in linear size. Since SAT is NP-complete for CNFs, we have to admit for representations with efficient SAT algorithms that the conjunction of linearly many representations may lead to an explosion of the representation size if NP $\neq$ P.

Later it will turn out that we have difficulties in verifying multipliers. This is disappointing, since multiplication is a fundamental function. Multiplication appears to be easy, since we learn it in primary school. On the other hand, it is one of the hardest functions realized in hardware. We present complexity theoretical arguments for why the verification of multipliers may be hard (Wegener (1994a)).

First, we state the well-known hypothesis that *factorization* is a hard problem. The problem is to identify a number z as prime or to present a proper divisor $d \notin \{1, z\}$. Efficient algorithms for factorization could be used to break most of the public key cryptosystems including the famous Rivest–Shamir–Adleman (RSA) system. Hence, the security of cryptosystems is based on the above-mentioned hypothesis.

For Boolean functions $f\colon \{0,1\}^n \to \{0,1\}^m$, the function $graph(f)\colon \{0,1\}^n \times \{0,1\}^m \to \{0,1\}$ is defined by $graph(f)(x, y) = 1$ iff $f(x) = y$. Here we consider *multgraph*, the graph of multiplication, which tests for three numbers x, y, and z whether $x * y = z$. Those working in complexity theory believe that multiplication is at least as hard as multgraph. Buss (1992) has proved that the functions are equivalent under constant depth reductions. Here we consider *multgraph**, a slight variation which tests whether $x * y = z$, $x \neq 1$ and $y \neq 1$.

Theorem 1.4.3. *If a representation type admits a polynomial-size representation of multgraph*, a polynomial-time algorithm for sequences of replacement by constants operations, and a polynomial-time algorithm for SAT and, possibly, the computation of a satisfying input, factorization can be solved nonuniformly by a polynomial-time algorithm.*

Proof. A nonuniform algorithm (see Garey and Johnson (1979)) may obtain information for free which depends only on the length of the input and not on the input itself. In our case, this information is the polynomial-size representation of multgraph*. In order to factorize z^* we replace in polynomial time the variable vector z with the constant z^*. Afterwards, we check in polynomial time whether multgraph$^*(x, y, z^*)$ is satisfiable. In the positive case, we compute in polynomial time a satisfying input (x^*, y^*, z^*), i.e., we compute proper factors x^* and y^* of z^*. □

Many of the representations we will consider admit the polynomial-time algorithms of the assumptions of Theorem 1.4.3. Then only three possibilities are left:

- The RSA system is broken,
- multiplication has a much smaller representation than multgraph*, or
- multiplication has no polynomial-size representation.

Since the first two possibilities are very unlikely, we have established with complexity theoretical arguments that multipliers are hard to verify.

1.5 Survey

Our aim is to investigate representations of Boolean functions. One task is to classify Boolean functions according to the size of their minimal-size representation of the chosen type. The other task is the description of efficient algorithms for the given list of operations or the proof that these operations do not have efficient algorithms (if NP $\neq$ P). In the final three chapters, we discuss a number of applications and show how the previous results lead to tools for the solution of large-scale problems.

In Chapter 2, we investigate the general model of BPs and BDDs. The size of BPs is compared to the size of circuits and formulas and the tight relationship to the space complexity of nonuniform Turing machines is proved. It is shown that several important operations on BPs are NP-hard problems. Moreover, we are only able to prove polynomial lower bounds of degree "less than 2" for explicitly defined functions. On the other hand, it is shown how efficient BPs for functions like multiplication and threshold functions can be designed. Besides this general model we consider DTs, which are BPs whose underlying graph is a tree. Then all operations can be performed efficiently but this does not mean much. Only very simple functions have DTs of polynomial size. Hence, the merging or recombination of paths is one of the main properties of BPs to make small size possible. For DTs, lower bounds can be proved using the Fourier analysis of Boolean functions. This method works only for trees.

After the consideration of BPs and DTs, the representation of Boolean functions which has found the most extensive applications is introduced in Chapter 3. OBDDs have the restriction that an ordering π on the variable set X_n has to be fixed. On each path in an OBDD each variable occurs only once as a label of a node. If an x_i-node is a successor of an x_j-node, then x_j has to precede x_i with respect to π. It turns out that OBDDs are closely related to DFAs. Then we describe efficient algorithms for the list of operations given in Section 1.3 and for some additional operations. Several OBDD packages are available. Moreover, some simple generalizations of the OBDD model are discussed. If OBDDs become too large to fit into the main memory, the secondary memory has also to be used. This causes a number of difficulties, and methods are presented to decrease the number of page faults. Also investigated is how parallel computers can work on OBDDs.

Knowing that operations on OBDDs can be performed very fast, we consider (in Chapter 4) how large is the class of functions with polynomial or small OBDD size. Communication complexity will play an important role in the investigation of many BDD variants. For OBDDs the lower bound methods are simple. Nevertheless, we introduce the language of communication complexity in this context. The OBDD size of the already mentioned so-called "important" functions is determined. As is often the case, in some situations it is easier not to consider bounds for functions directly but to use reductions. A suitable reduction concept called read-once projections is introduced and applied.

One of the main insights of Chapter 4 is that functions can be represented by small OBDDs only if a good variable ordering is chosen. Chapter 5 is devoted to this variable-ordering problem. It turns out to be hard to find an optimal variable ordering and even to find only a good variable ordering. So heuristics are used to construct variable orderings which are often good enough.

Chapters 3, 4, and 5 show that OBDDs are a representation type one can work with. The most severe limitation is that OBDDs become too large. Hence, the following six chapters present more general BDD variants. In Chapter 6, read-once BPs, which also are called *free* BDDs (FBDDs), are considered. These are BPs where on each path each variable occurs only once as a label of a node. The class of functions representable in polynomial size is larger than for OBDDs. But in order to obtain efficient algorithms for many operations the general FBDD model has to be restricted a little bit. FBDDs are the most general model without inconsistent paths.

Inconsistent paths increase the representational power but make the satisfiability and the equivalence test much harder. So it is natural to ask (in Chapter 7) whether we can be successful with a limited use of inconsistent paths. The model of k-OBDDs admits polynomial-time algorithms for many operations. It is based on a variable ordering which can be used k times. More precisely, a k-OBDD consists of k layers which are OBDDs for a fixed variable ordering. The model of k-indexed BDDs (IBDDs) allows the use of different variable orderings in the different layers. Then problems such as the satisfiability test are NP-complete, even for $k = 2$, and one has to work with heuristic algorithms. The challenge from a complexity theoretical point of view was the development of lower bound techniques for k-BDDs or read-k BPs where on each path each variable is the label of at most k nodes. Such lower bound techniques are now available. The restrictions used in k-OBDDs, k-IBDDs, and k-BDDs are purely syntactic. If we relax these conditions to paths which are consistent, not many results are known.

In Chapter 8, it is investigated whether there are alternatives to Shannon's decomposition rule as semantics of a BDD node. The model of zero-suppressed BDDs (ZBDDs) is based on OBDDs but it treats the variables which are *not* tested on a path in a different way. The Reed-Muller decomposition rule is based on computations within the field $\mathbb{Z}_2$ of two elements. EXOR is the addition and AND the multiplication in this field. This more algebraical viewpoint leads to ordered functional DDs (OFDDs) and a mixture of OBDDs and OFDDs to ordered Kronecker functional DDs (OKFDDs).

We have started our considerations in Section 1.1 with the remark that finite functions can be replaced by Boolean functions after some binary encoding of the inputs and outputs. In Chapter 9, we consider the situation where it is more natural to work directly with functions $f : A^n \to B$, where A and B are finite sets. Using one of the first two semantics of Definition 1.1.1, it is easy to generalize the BDD model. Inner nodes get $a := |A|$ outgoing edges, one for each possible value of the variable which is the label of the node. Moreover, we may have $b := |B|$ sinks labeled by the different values from B. Such representations are known as multivalued DDs (MDDs), algebraic DDs (ADDs), and multiterminal BDDs (MTBDDs). They have several useful applications. A trivial lower bound is the size of the image of the considered function. If this size becomes too large, one has to look for other models. The idea is to introduce further edge labels which control the evaluation. Edge-valued BDDs (EVBDDs) generalize OBDDs and the value on input a is the sum of the new edge labels on the activated path. This may lead to an exponential decrease of the size. The reason is that from a set of s numbers we may obtain exponentially many partial sums. For example, with $2^0, \dots, 2^{s-1}$ we can represent the 2^s numbers in $\{0, \dots, 2^s - 1\}$. Binary moment diagrams (BMDs) generalize OFDDs in the same way and use a more algebraical viewpoint.

In Chapters 10 and 11, two fundamental and successful concepts are combined with the concept of BDDs: nondeterminism and probabilistic methods. It is obvious why nondeterministic BDD models are investigated in complexity theory. Nondeterminism is a powerful theoretical concept leading to the NP-completeness theory. In Chapter 10, it is shown that nondeterministic concepts also may be powerful for applications. But one has to restrict nondeterminism in the right way or choose an appropriate mode of nondeterminism. This is less surprising than one might think. In many applications, people work with nondeterministic finite automata. Moreover, it is not hard to verify that DNFs and nondeterministic DTs are closely related.

Probabilistic methods are presented in Chapter 11. They have turned out to be powerful in almost all areas of computer science. Efficient randomized algorithms can be used if no efficient deterministic algorithms are available. This is the case for the equivalence test of FBDDs. Such a result opens the discussion of whether a probabilistic equivalence test or, more generally, a probabilistic verification may be called verification. Another approach is to replace BDD models with their randomized counterparts. They do not seem to be adequate for applications but many interesting complexity theoretical results have been proved.

The theoretical results are summarized in Chapter 12.

The very first and still perhaps most important applications of BDD representations fall into the area of verification. The distribution of the faulty Pentium processor has emphasized the economical importance of verification.

Computer scientists as producers of hardware and software should feel responsible for their products. Hence, Chapter 13 is devoted to verification of combinational and sequential circuits and to model checking.

Most of the further applications are CAD applications. In Chapter 14, some of them are discussed, among them functional simulation, two-level logic minimization, synthesis and design of circuits, timing analysis, test pattern generation, fault simulation, Boolean unification, computation of synchronizing sequences, and technology mapping.

Finally, in Chapter 15, applications in other disciplines are discussed. Our main emphasis is on combinatorial optimization and counting, e.g., an EVBDD-based integer linear programming (ILP) solver is presented. Very efficient algorithms have been designed and implemented for graph theoretical problems such as network flow and perfect matching. These algorithms work on a description of the underlying graph by adjacency matrices or adjacency lists. They cannot be used if we do not have enough time to enumerate all edges. In this situation, the graph has to be described in a more compact way. This is not possible for random graphs but BDD variants lead to very compact descriptions of very large but somewhat regular graphs. Network flow problems for graphs with more than 10^{36} edges have been solved this way. Furthermore, the problem of counting the number of knight's tours on an 8×8 chessboard has been solved with BDD techniques. In the last section, we discuss how OBDDs can serve as a representation in the new area of genetic programming.

The reader is encouraged to work with the presented methods and results. This can be done by solving the exercises at the end of each chapter. The difficulty of the exercises is estimated by E (easy; should be solvable without much effort after having read the chapter), M (medium; typical exercises solvable with some effort; one has to generalize or transfer methods presented in the chapter), and D (difficult; solvable by being clever or with some amount of effort). Open problems are marked by O.

1.6 Exercises

1.1.E Prove that all definitions of the semantics of a BP or BDD are equivalent.

1.2.M Prove that circuits of fan-in 2 and depth $O(\log n)$ have polynomial size.

1.3.D Prove the easier part of Barrington's theorem. Functions representable by polynomial-size BPs of width 5 can be represented by circuits of fan-in 2 and depth $O(\log n)$.

1.4.E Prove that for any representation type the fraction of $f \in B_n$ which can be represented with at most 2^{n-1} bits tends exponentially fast to 0 as $n \to \infty$.

1.5.M Design a sequence $G = (G_n)$ of BPs representing all $n + 1$ output bits of the addition of two n-bit numbers in size $O(n)$.

1.6.E Generalize the first BP for HWB_n in Fig. 1.1.1 to obtain BPs of quadratic size for HWB.

Chapter 2

BPs and Decision Trees (DTs)

2.1 BPs, Circuits, Formulas, and Space

In order to classify the representational power of BPs we compare them with circuits and formulas.

Definition 2.1.1.

(i) A *circuit* on $X_n = \{x_1, \ldots, x_n\}$ is a B_2 straight line program with inputs from $X_n \cup \{0,1\}$. It is described by a sequence $I_1, \ldots, I_s$ of instructions of type $I_j = (op_j, F_j, S_j)$, where $op_j \in B_2$ is some binary Boolean operation and F_j ($F \mathrel{\widehat{=}}$ first) and S_j ($S \mathrel{\widehat{=}}$ second) are elements from $\{0, 1, x_1, \ldots, x_n, I_1, \ldots, I_{j-1}\}$. The inputs $0, 1, x_1, \ldots, x_n$ represent the corresponding functions. If F_j represents f and S_j represents g, then I_j represents $h := op_j(f, g)$. The size of the circuit is equal to the number s of its instructions.

(ii) A *formula* on $X_n = \{x_1, \ldots, x_n\}$ is a circuit, where each instruction I_k occurs only once as an argument of another instruction I_j.

This unusual definition is chosen to resemble the definition of BPs as ite straight line programs. Using the graphical representation with the inputs $0, 1, x_1, \ldots, x_n$ and inner nodes for the instructions I_j with predecessors F_j and S_j and label op_j, we obtain the usual description of circuits. Formulas are circuits whose underlying graph is a tree if we duplicate the inputs.

Definition 2.1.2. For a Boolean function $f \in B_n$, we denote by $\mathrm{C}(f)$ its *circuit size*, i.e., the size of the smallest circuit representing f, by $\mathrm{BP}(f)$ or $\mathrm{BDD}(f)$ its

BP size, by L(f) its *formula size*, and by L*(f) its formula size if the operation EXOR and its negation are forbidden.

Theorem 2.1.3. *For all* $f \in B_n$ *it holds that*

(*i*) $\mathrm{C}(f) \leq 3 \cdot \mathrm{BP}(f)$ *and*

(*ii*) $\mathrm{BP}(f) \leq \mathrm{L}^*(f) + 3$.

Proof.

(i) This proof is easy, since circuits and BPs are defined as straight line programs. In order to simulate one ite instruction three B_2 instructions are enough, since $ite(x, y, z) = xy + \overline{x}z$.

(ii) We denote the number of leaves in an optimal formula for f by $l = \mathrm{L}'(f) := \mathrm{L}^*(f) + 1$. We describe how we obtain a BP with l inner nodes for the same function. The result follows, since we have to add 2 for the sinks. The simulation is done by induction on l. For $l = 1$, the assertion is obvious, since BPs for a variable x_i need only one inner node. For the induction step, we consider a formula with l leaves. Since EXOR and its negation are forbidden, each Boolean operation can be replaced by AND and perhaps some negations. It is easy to see that $\mathrm{BP}(f) = \mathrm{BP}(\overline{f})$. If we negate the labels of the sinks of a BP, we obtain a BP of the same size for the negated function. Hence, we may assume that $f = g \wedge h$, where g is given by a formula with l_g leaves and h by a formula with l_h leaves. It holds that $l_g + l_h = l$, $l_g < l$, and $l_h < l$. Hence, by the induction hypothesis, g can be represented by a BP G_g with l_g inner nodes and h by a BP G_h with l_h inner nodes. The BP G_f for f is obtained by binary AND synthesis. We start with G_g and replace the 1-sink with the source of G_h. The resulting BP represents f and has $l_g + l_h = l$ inner nodes. □

It is possible and not unlikely that for some function the BP size is exponentially larger than the circuit size. This is due only to the restriction that the conditions in the instructions of a BP have to be variables. Otherwise, it is an easy exercise to simulate circuits. It is also possible that for some function the L* size is superpolynomially larger than the BP size. Since it is known that $\mathrm{L}^*(f) = \mathrm{L}(f)^{O(1)}$, we can conclude from Theorem 2.1.3(ii) that $\mathrm{BP}(f) = \mathrm{L}(f)^{O(1)}$. This result has been improved by Sauerhoff, Wegener, and Werchner (1999).

Theorem 2.1.4. $\mathrm{BP}(f) \leq \alpha(\mathrm{L}(f) + 1)^{\beta} + 2$ *for all Boolean functions where* $\alpha \leq 1.360$ *and* $\beta = \log_4(3 + \sqrt{5}) < 1.195$.

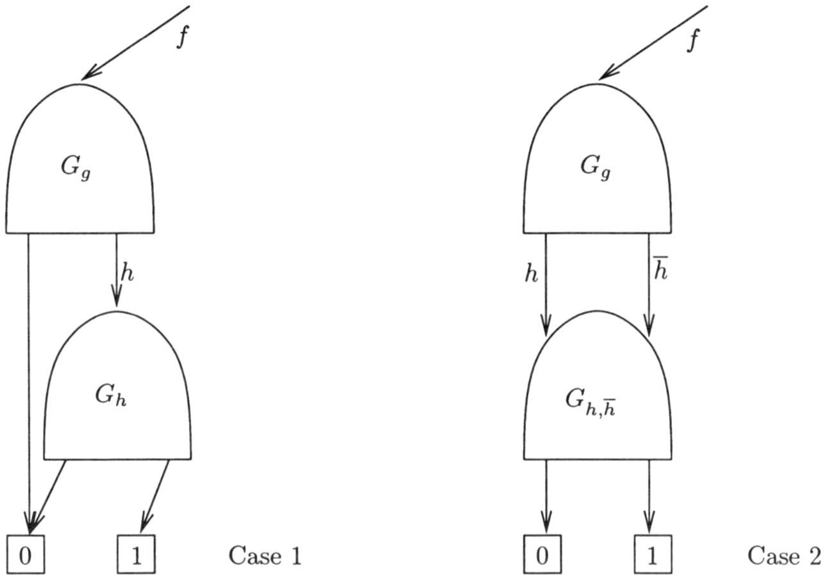

Figure 2.1.1: *BPs for f where $f = g \wedge h$ and $f = g \oplus h$.*

Proof. It is sufficient to prove that we can construct for binary formulas with l leaves a simulating BP with at most αl^β inner nodes. Let $x^1 = x$ and $x^0 = \overline{x}$. Then each binary operation can be described as $(x^a \wedge x^b)^c$ or $x \oplus y \oplus a$ for appropriate $a, b, c \in \{0, 1\}$. Since negations do not change the BP size, we only have to consider the operations AND and EXOR.

We describe a recursive procedure for the construction of the BP for f. By size(f) we denote the number of inner nodes of the constructed BP. The recursion stops for subformulas with one leaf which are replaced by BPs with one inner node, i.e., size(x_i) = 1. Our construction for the two cases $f = g \wedge h$ and $f = g \oplus h$ is shown in Fig. 2.1.1.

Case 1: $f = g \wedge h$.
The BP combines the BP G_g for g and the BP G_h for h. The 1-sink of G_g is replaced by the source of G_h. Obviously, size(f) = size(g) + size(h).

Case 2: $f = g \oplus h$.
Here we combine G_g and a BP $G_{h,\overline{h}}$ with two sources representing h and $\overline{h}$. The 0-sink of G_g is replaced by the h-source of $G_{h,\overline{h}}$ and the 1-sink of G_g by the $\overline{h}$-source of $G_{h,\overline{h}}$. Using the evaluation rule of BPs (Semantics 1 of Definition 1.1.1), it follows that the 1-sink is reached iff $g(a) = 0$ and $h(a) = 1$ or $g(a) = 1$ and $\overline{h}(a) = 1$ which means $h(a) = 0$. Hence, we realize f. We

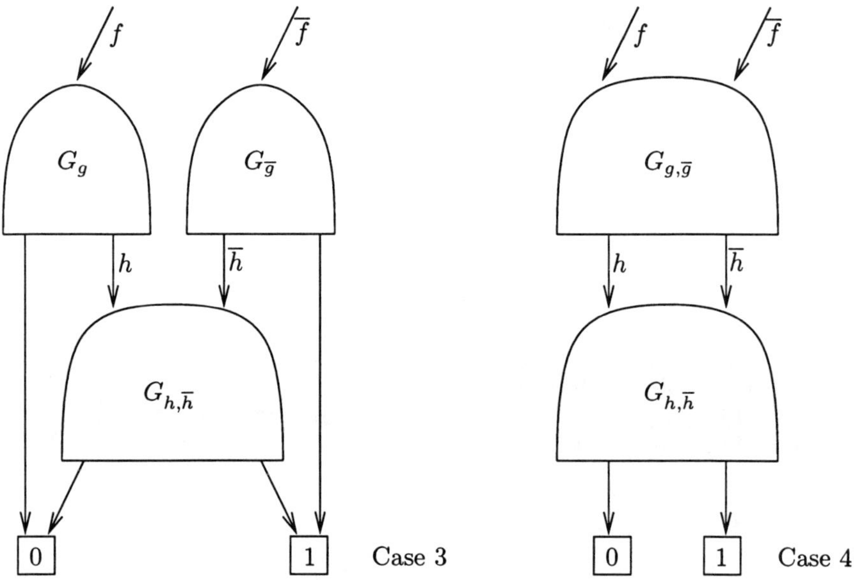

Figure 2.1.2: *BPs for* $(f,\overline{f})$ *where* $f = g \wedge h$ *and* $f = g \oplus h$.

also may interchange the roles of g and h. Obviously, we choose the better alternative. Hence,

$$\text{size}(f) = \min\{\text{size}(g) + \text{size}(h,\overline{h}), \text{size}(h) + \text{size}(g,\overline{g})\}.$$

Now we are faced with a new problem. How do we obtain a BP for the pair $(h,\overline{h})$ and in general for $(f,\overline{f})$? Again, we consider the two cases $f = g \wedge h$ and $f = g \oplus h$, which are illustrated in Fig. 2.1.2. As a terminal case we get $\text{size}(x_i,\overline{x}_i) = 2$.

Case 3: $f = g \wedge h$ and representation of $(f,\overline{f})$.
We use the BPs G_g, $G_{\overline{g}}$, and $G_{h,\overline{h}}$. The BP $G_{\overline{g}}$ is obtained as a copy of G_g with negated sinks. As f-source we choose the source of G_g and as $\overline{f}$-source the source of $G_{\overline{g}}$. The 1-sink of G_g is replaced with the h-source of $G_{h,\overline{h}}$ and the 0-sink of $G_{\overline{g}}$ with the $\overline{h}$-source of $G_{h,\overline{h}}$. It is easy to verify that we represent $(f,\overline{f})$. For the computation of $f = g \wedge h$, we first evaluate g. If $g(a) = 0$, then $f(a) = 0$. If $g(a) = 1$, then $f(a) = h(a)$ and f is evaluated correctly. For the computation of $\overline{f} = \overline{g} + \overline{h}$, we first evaluate $\overline{g}$. If $\overline{g}(a) = 0$, then $\overline{f}(a) = \overline{h}(a)$ and if $\overline{g}(a) = 1$, then $\overline{f}(a) = 1$. Hence, $\overline{f}$ is also evaluated correctly. Again, we may interchange the roles of g and h and choose the better alternative. We obtain

$$\text{size}(f,\overline{f}) = \min\{2 \cdot \text{size}(g) + \text{size}(h,\overline{h}),\ 2 \cdot \text{size}(h) + \text{size}(g,\overline{g})\}.$$

Case 4: $f = g \oplus h$ and representation of $(f, \overline{f})$.
We combine the BPs $G_{g,\overline{g}}$ and $G_{h,\overline{h}}$ as illustrated in Fig. 2.1.2. As f-source we choose the g-source of $G_{g,\overline{g}}$ and as $\overline{f}$-source its $\overline{g}$-source. The 0-sink of $G_{g,\overline{g}}$ is replaced with the h-source of $G_{h,\overline{h}}$ and the 1-sink of $G_{g,\overline{g}}$ with the $\overline{h}$-source of $G_{h,\overline{h}}$. By a simple case inspection, it is shown that we represent $(f, \overline{f})$ and

$$\text{size}(f, \overline{f}) = \text{size}(g, \overline{g}) + \text{size}(h, \overline{h}).$$

We have not obtained any new cases and our construction is complete.

The analysis of this construction is much harder. To prove a sharp upper bound on size(f) we have to bound size(f) and size$(f, \overline{f})$ simultaneously. Both values depend on the structure of the formula (where we have to consider the worst case) and on size(g), size(h), size$(g, \overline{g})$, and size$(h, \overline{h})$. We know only that the sum of the number of leaves of a formula for g and the number of leaves of a formula for h equals l. The tedious and mathematically involved analysis is described by Sauerhoff, Wegener, and Werchner (1999) and leads to the stated result. □

The upper bound of Theorem 2.1.4 can be realized efficiently. We process the nodes of the tree representing the formula from the leaves backward to the root. Then we need constant time at each node to compute size(f') and size$(f', \overline{f'})$ for the function f' represented at this node. Figures 2.1.1 and 2.1.2 show how we can construct the corresponding BPs. The final BP for f can be constructed in linear time with respect to its size. This leads to the following corollary.

Corollary 2.1.5. *Given a formula of size l for f, a BP for f whose size is bounded by $\alpha(l+1)^\beta + 2$ can be constructed in linear time with respect to its size. If only the size of the resulting BP is of interest, the computation can be done in time $O(l)$.*

We do not know whether the bound of Theorem 2.1.4 is optimal. The following example shows that the analysis is at least almost exact.

Definition 2.1.6. The function *alternating tree* AT_k is defined for odd k on $n = 2^k$ variables. $\text{AT}_1(x_1, x_2) := x_1 \oplus x_2$ and

$$\text{AT}_k(u, v, x, y) := (\text{AT}_{k-2}(u) \wedge \text{AT}_{k-2}(v)) \oplus (\text{AT}_{k-2}(x) \wedge \text{AT}_{k-2}(y))$$

for disjoint variable vectors u, v, x, and y of length 2^{k-2}.

Proposition 2.1.7. *The formula size of AT_k equals $n - 1 = 2^k - 1$ and the construction of the proof of Theorem 2.1.4 results in a BP of size $c_1 r^{(k-1)/2} + c_2 s^{(k-1)/2} + 2$, where $c_1 = \frac{1}{10}(15 + 7\sqrt{5})$, $c_2 = \frac{1}{10}(15 - 7\sqrt{5})$, $r = 3 + \sqrt{5}$, $s = 3 - \sqrt{5}$.*

Since $s < 1$, the second term tends to 0 as $k \to \infty$. Since $\beta = \log_4(3+\sqrt{5})$ (see Theorem 2.1.4), $r^{k/2} = (3+\sqrt{5})^{k/2} = n^\beta$. The first term equals $(c_1/r^{1/2})n^\beta \approx 1.340 n^\beta$. This is close to the upper bound $1.360n^\beta$ of Theorem 2.1.4.

Proof of Proposition 2.1.7. The result on the formula size is obvious, since the definition leads to a complete binary formula tree with n leaves and alternating $\wedge$- and $\oplus$-levels. Better formulas are impossible, since the function essentially depends on all its variables and we need for each variable at least one leaf labeled by this variable.

Because of the symmetry in the definition of AT_k, we are able to analyze the size of the resulting BP. Let S_k be the size of the BP for AT_k and T_k be the size of the BP for $(\mathrm{AT}_k, \overline{\mathrm{AT}_k})$. It is easy to check by case inspection that $S_1 = 3$ and $T_1 = 4$. The definition of AT_k can be abbreviated by $f = (f_1 \wedge f_2) \oplus (f_3 \wedge f_4)$, where f_1, f_2, f_3, and f_4 are the same functions on disjoint variable sets. Therefore, $g = h$ in Cases 2 and 3 in the proof of Theorem 2.1.4 and the two terms from which we have to take the minimal one are equal. Hence, by the case inspection in the proof of Theorem 2.1.4,

$$\begin{aligned}\mathrm{size}(f) &= \mathrm{size}(f_1 \wedge f_2) + \mathrm{size}(f_3 \wedge f_4, \overline{f_3 \wedge f_4}) \\ &= \mathrm{size}(f_1) + \mathrm{size}(f_2) + 2 \cdot \mathrm{size}(f_3) + \mathrm{size}(f_4, \overline{f}_4) \quad \text{and} \\ \mathrm{size}(f, \overline{f}) &= \mathrm{size}(f_1 \wedge f_2, \overline{f_1 \wedge f_2}) + \mathrm{size}(f_3 \wedge f_4, \overline{f_3 \wedge f_4}) \\ &= 2 \cdot \mathrm{size}(f_1) + \mathrm{size}(f_2, \overline{f}_2) + 2 \cdot \mathrm{size}(f_3) + \mathrm{size}(f_4, \overline{f}_4).\end{aligned}$$

This leads, since $f_1 = f_2 = f_3 = f_4 = \mathrm{AT}_{k-2}$ and $f = \mathrm{AT}_k$, to

$$\begin{aligned}S_k &= 4S_{k-2} + T_{k-2} \quad \text{and} \\ T_k &= 4S_{k-2} + 2T_{k-2}.\end{aligned}$$

The exact solution of these linear difference equations follows by standard techniques (see Graham, Knuth, and Patashnik (1994)). □

In connection with these investigations, a result of Cleve (1991) should be mentioned. He efficiently simulated formulas with a given depth by BPs. The equivalence of the complexity class NC^1 and the class of functions representable by polynomial-size width-5 BPs (Barrington (1989)) has already been described in Section 1.2.

In Section 1.2, a tight relationship between the BP size and the space complexity of nonuniform Turing machines was also stated. In order to discuss sublinear space complexity we consider Turing machines with three tapes:

- A read-only input tape,
- a write-only output tape, and
- a working tape.

The space of a Turing machine computation is the number of cells visited on the working tape. BPs as well as circuits and formulas are a nonuniform computation model, since we consider only the BP size and not the time to obtain the BP G_n for inputs of length n. In particular, all sequences $f = (f_n)$ of Boolean functions are computable by BPs. Not recursive (not computable) problems do not exist for nonuniform computation models. Hence, a simulation of a sequence $G = (G_n)$ of BPs by ordinary Turing machines is impossible. Therefore, one has defined nonuniform Turing machines.

Definition 2.1.8. A nonuniform Turing machine is a Turing machine equipped with a read-only *oracle* tape. For inputs x of length n, the oracle tape contains an oracle string oracle(n) depending only on the input length and not on the input itself. The space of a nonuniform Turing machine computation is the sum of the number of cells visited on the working tape and the logarithm of the length of oracle(n).

The following result due to Cobham (1966) shows that this definition is appropriate. By $\mathrm{S}_f(n)$, we denote the space complexity of a sequence of Boolean functions $f = (f_n)$, $f_n \in B_n$, with respect to nonuniform Turing machines.

Theorem 2.1.9. *Let* $f = (f_n)$, $f_n \in B_n$.

(*i*) $\mathrm{S}_f(n) = O(\log s_1(n))$ *where* $s_1(n) = \max\{\mathrm{BP}(f_n),\ n\}$.

(*ii*) $\mathrm{BP}(f_n) = 2^{O(s_2(n))}$ *where* $s_2(n) = \max\{\mathrm{S}_f(n),\ \log n\}$.

If f_n essentially depends on all its variables, each BP for f_n contains for $i \in \{1, \dots, n\}$ an x_i-node and $\mathrm{BP}(f_n) \geq n$. We also know that space complexities of size $o(\log n)$ lead to strange properties. Problems which are not very simple need space $\Omega(\log n)$. The typical case is that $\mathrm{BP}(f_n) \geq n$ and $\mathrm{S}_f(n) \geq \log n$. Then Theorem 2.1.9 can be written in brief as $\mathrm{S}_f(n) = \Theta(\log \mathrm{BP}(f_n))$.

Proof of Theorem 2.1.9.

(i) Let $G = (G_n)$ be a sequence of BPs for $f = (f_n)$ of size $\mathrm{BP}(f_n)$. We describe a nonuniform Turing machine evaluating f_n for inputs of length n. As oracle(n) we choose an encoding of G_n. Each instruction of the BP G_n can be encoded by $O(\log n + \log \mathrm{BP}(f_n))$ bits. The length of oracle(n) is $O(\mathrm{BP}(f_n)(\log n + \log \mathrm{BP}(f_n)))$ and its contribution to the space complexity is $O(\log \mathrm{BP}(f_n) + \log\log n) = O(\log s_1(n))$. Then we evaluate G_n by following the path activated by the input. The actual node reached in G_n, more precisely, the encoding of the corresponding instruction, is written on the working tape. Then we look for the value of the variable which is the condition of the instruction. With this value we can find the number of the next node or instruction. The encoding of the instruction which has become actual is copied from the oracle to the working tape. The encoding of the old instruction can be erased. Hence, $O(\log n + \log \mathrm{BP}(f_n)) = O(\log s_1(n))$ cells are sufficient on the working tape.

(ii) Let M be a nonuniform Turing machine evaluating f_n for inputs of length n with space complexity $\mathrm{S}_f(n)$. We bound the number of configurations of M. There are n positions on the input tape, $\mathrm{S}_f(n)$ positions on the working tape, and $l(n)$ positions on the oracle tape, where $l(n)$ is the length of oracle(n). Moreover, if Γ is the alphabet for the working tape, there are $|\Gamma|^{\mathrm{S}_f(n)}$ different words which can be written on the working tape. The number of configurations is bounded by

$$n \cdot \mathrm{S}_f(n) \cdot l(n) \cdot |\Gamma|^{\mathrm{S}_f(n)} = 2^{O(s_2(n))}.$$

Each configuration of M on inputs of length n is simulated by a node of G_n. The function f_n will be represented at the node representing the initial configuration of M. Accepting configurations are 1-sinks and rejecting configurations are 0-sinks. All other configurations c are inner nodes. They are labeled by the input variable x_i read on the input tape if M is in configuration c. The 0-successor is the configuration which is reached if $a_i = 0$ for the actual input a; the 1-successor is defined in an analogous way. By construction, this BP represents f_n with $2^{O(s_2(n))}$ nodes. □

One may also define uniform sequences of circuits $C = (C_n)$ and BPs $G = (G_n)$ (see Wegener (1987)). Loosely speaking they allow the efficient computation of C_n or G_n if n is given. Then a uniform counterpart of Theorem 2.1.9 can be proved. This shows that the important problems on space complexity (lower bounds, or the famous linear bounded automaton (LBA) problem of whether the class of context-sensitive languages equals deterministic linear space) can also be posed in the framework of BPs.

2.2 Lower Bound Techniques for BPs

As already mentioned in Chapter 1, we can only prove small polynomial lower bounds on the BP size of explicitly defined functions. But, by simple counting arguments, we can prove lower bounds for most of the Boolean functions. By Theorem 2.1.3(i) and a well-known result on the asymptotic circuit size, we obtain that the BP size is at least $2^n/(3n)$ for all but an exponentially small fraction of the Boolean functions $f \in B_n$. In a direct way we obtain a better result.

Lemma 2.2.1. *At most $B(s,n) := sn^s(s+1)^{2s}/s!$ functions $f \in B_n$ can be computed by BPs of size s.*

Proof. An upper bound of $sn^s((s+2)!)^2$ is easy to obtain using the definition of a BP as a straight line program. The jth instruction may choose between n conditions, $j+1$ then-branches, and $j+1$ else-branches. Moreover, such a BP represents at most s nonconstant functions. The proposed upper bound is better.

We consider BPs of size s as sequences $I_1, \dots, I_s$ of instructions where $I_j = (C_j, T_j, E_j)$. For C_j we have n possibilities. For T_j and E_j we allow the constants and $I_1, \dots, I_{j-1}, I_{j+1}, \dots, I_s$, hence, $s+1$ possibilities. This leads to $n^s(s+1)^{2s}$ instruction sequences, among them all syntactically correct BPs, but among them also syntactically incorrect BPs. Each BP can represent only s different functions at its inner nodes. It is not necessary to add 2 for the constant functions, since they are already counted among the other functions. Hence, we get the upper bound $sn^s(s+1)^{2s}$.

Let us now consider the BP as a graph with a numbering of the inner nodes by $1, \dots, s$. If we renumber the nodes by a permutation π on $\{1, \dots, s\}$, we may obtain a syntactically correct or incorrect BP. In the first case, it represents the same functions as before. Hence, we obtain classes of $s!$ sequences of instructions which altogether represent not more than s different functions at their inner nodes. This implies that the bound $sn^s(s+1)^{2s}$ can be divided by $s!$. □

Theorem 2.2.2. *All but an exponentially small fraction of the functions $f \in B_n$ have a BP size which is bounded below by $s(n) = 2^n n^{-1}(1 - n^{-1/2})$.*

Proof. We consider the $2^{2^n - 2^{n/2}}$ functions $f \in B_n$ with the smallest BP size and prove that at least one of them cannot be represented within BP size $s(n)$. The class of functions we consider is an exponentially small fraction of $2^{-2^{n/2}}$ of all functions in B_n.

By Lemma 2.2.1, we can represent in size s at most $sn^s(s+1)^{2s}/s!$ functions $f \in B_n$. It is sufficient to prove that this bound is smaller than $2^{2^n-2^{n/2}}$ for the chosen $s(n)$ and large n. By Stirling's formula, $s! \geq cs^{s+1/2}e^{-s}$ for some constant c and

$$\begin{aligned}\log(sn^s(s+1)^{2s}/s!) &\leq \log s + s\log n + 2s\log(s+1) - \log c \\ &\quad - \left(s + \frac{1}{2}\right)\log s + s\log e \\ &\leq s(\log s + \log n + c')\end{aligned}$$

for some constant c'. Hence, it is sufficient to prove that

$$2^n n^{-1}(1-n^{-1/2})(n - \log n + \log(1-n^{-1/2}) + \log n + c') < 2^n - 2^{n/2} = 2^n(1-2^{-n/2}).$$

This is equivalent to

$$(1-n^{-1/2})(1+n^{-1}\log(1-n^{-1/2}) + c'n^{-1}) < 1 - 2^{-n/2}.$$

The left-hand term is bounded above by

$$(1-n^{-1/2})(1+c'n^{-1}) \le 1-n^{-1/2}+c'n^{-1} < 1-2^{-n/2}$$

for large n. □

In the next section, it is shown that this lower bound is close to optimal. The lower bound tells what we have to expect with very large probability if we consider a random Boolean function, where for each input a the value $f(a)$ equals 0 or 1 with probability 1/2 independently of the other function values. If we consider some specific function, Theorem 2.2.2 tells us nothing. The strongest lower bound technique for explicitly defined functions is still due to Nechiporuk (1966). The idea is that functions with many subfunctions cannot be computed by BPs of small size. The existence of many subfunctions implies some complex structure and it is quite natural to suppose that this implies not too small complexity. But we have to be cautious, since the idea does not work for circuits. Functions with the largest possible number of subfunctions may have linear circuit size. Now we prove Nechiporuk's bound.

Definition 2.2.3. Let $S \subseteq X_n$ be a subset of the variable set. All subfunctions obtained by replacing the variables in $X_n - S$ by constants are called subfunctions on S.

Theorem 2.2.4. *Let $f \in B_n$ essentially depend on all its variables, let $S_1, \dots, S_k \subseteq X_n$ be disjoint sets of variables, and let s_i be the number of subfunctions of f on S_i. Then*

$$\mathrm{BP}(f) = \Omega\left(\sum_{1\le i\le k} (\log s_i)/\log\log s_i\right).$$

Proof. Let G be an optimal BP representing f and let t_i be the number of nodes labeled by variables from S_i. Then $\mathrm{BP}(f)$ is bounded below by the sum of all t_i and it is sufficient to prove that $t_i = \Omega((\log s_i)/\log\log s_i)$. Since f essentially depends on all its variables, $t_i \ge |S_i|$.

Each subfunction of f on S_i can be computed by a BP on $|S_i|$ variables with at most t_i inner nodes. Such a BP can be obtained from G by replacing the variables outside S_i with the appropriate constants. If x_j is replaced with $c_j \in \{0,1\}$, we can replace all edges leading to x_j-nodes v with edges leading to the c_j-successor of v and afterwards we can eliminate v. It is obvious that we obtain a BP for the corresponding subfunction of f. This BP has only inner nodes labeled by variables from S_i and at most t_i inner nodes.

The number s_i of subfunctions of f on S_i cannot be larger than the number of BPs on $|S_i|$ variables with at most t_i nodes. Using the simple upper bound from the proof of Lemma 2.2.1, we obtain

$$s_i \le t_i|S_i|^{t_i}((t_i+2)!)^2 \le t_i^{t_i+1}((t_i+2)!)^2 \le t_i^{4t_i}$$

for $t_i \geq 4$. Here we used the property $|S_i| \leq t_i$ shown above. The inequality $s_i \leq t_i^{4t_i}$ implies $t_i = \Omega((\log s_i)/\log\log s_i)$. □

We apply Nechiporuk's method to three functions.

Definition 2.2.5.

(i) The function ISA_n, where $n = 2^k$, is defined on $n+k$ variables $x_0, \ldots, x_{n-1}$, $y_0, \ldots, y_{k-1}$. ISA_n is called the *indirect storage access* function. The vector $(y_{k-1}, \ldots, y_0)$ is interpreted as the binary number $|y| \in \{0, \ldots, n-1\}$. Then we consider the k x-variables $x_{|y|}, \ldots, x_{|y|+k-1}$ (the indices are taken mod n) and interpret this vector as the binary number $\alpha(x,y) \in \{0, 1, \ldots, n-1\}$. Finally, $\text{ISA}_n(x,y) := x_{\alpha(x,y)}$.

(ii) The function DET_n is defined on n^2 variables x_{ij}, $1 \leq i,j \leq n$, and computes the *determinant* of the matrix $X = (x_{ij})$ over the field $\mathbb{Z}_2$.

(iii) The function 3-CLIQUE_n is defined on $\binom{n}{2}$ variables x_{ij}, $1 \leq i < j \leq n$ and checks whether the graph $G(x)$ whose adjacency matrix contains the entries x_{ij} contains a triangle (or equivalently a *clique* on three vertices).

Theorem 2.2.6.

(*i*) $\text{BP}(\text{ISA}_n) = \Omega(n^2/\log^2 n)$.

(*ii*) $\text{BP}(\text{DET}_n) = \Omega(n^3/\log n)$.

(*iii*) $\text{BP}(3\text{-CLIQUE}_n) = \Omega(n^3/\log n)$.

Proof.

(i) We consider $\lfloor n/k \rfloor$ variable sets $S_i = \{x_{(i-1)k}, \ldots, x_{ik-1}\}$, $1 \leq i \leq \lfloor n/k \rfloor$. To bound the number of subfunctions of ISA_n on S_i, we fix the vector y such that $|y| = (i-1)k$. Then all 2^{n-k} different replacements of the x-variables outside S_i lead to different subfunctions on S_i. Let us investigate two replacements which differ in the value of $x_j \notin S_i$. Then we may consider the input where the binary number represented by the variables in S_i equals j. This leads to different outputs for the two given replacements. Hence, $s_i \geq 2^{n-k}$ and $(\log s_i)/\log\log s_i \geq (n-k)/\log(n-k) = \Omega(n/\log n)$. Since we consider $\Omega(n/\log n)$ variable sets S_i, the lower bound follows from Theorem 2.2.4.

(ii) The proof is due to Kloss (1966) and left as an exercise.

(iii) The proof is due to Schürfeld (1983) and left as an exercise. □

It can be proved that Nechiporuk's bound cannot give better lower bounds than $\Omega(n^2/\log^2 n)$ for functions on n variables (see exercises). So Theorem 2.2.6(i) is one example of the largest lower bounds known today on the BP size of explicitly defined Boolean functions. For functions where Nechiporuk's bound leads only to trivial bounds it is difficult to obtain good lower bounds. We mention such a result of Babai, Pudlák, Rödl, and Szemerédi (1990) without proof.

Definition 2.2.7. The *majority function* $\text{MAJ}_n \in B_n$ computes 1 on input $a = (a_1, \ldots, a_n)$ iff $\|a\| = a_1 + \cdots + a_n \geq \lceil n/2 \rceil$.

Theorem 2.2.8. $\text{BP}(\text{MAJ}_n) = \Omega((n \log n)/\log\log n)$.

Altogether, the known lower bound techniques are quite weak. For random functions, we can prove exponential lower bounds. But for explicitly defined Boolean functions, the largest lower bound of size $n^2/\log^2 n$ has not been improved since 1966 and there are only a few small lower bounds which are not obtained with Nechiporuk's technique.

2.3 Upper Bound Techniques for BPs

In the last section, we proved a lower bound of size $2^n n^{-1}(1 - o(1))$ on the BP size of almost all Boolean functions $f \in B_n$. Here we look for an upper bound for all $f \in B_n$ (Shannon (1949)).

Theorem 2.3.1. *The BP size of a Boolean function $f \in B_n$ is bounded by $(3 + o(1))2^n/n$.*

Proof. For some i which is chosen later, we start with a complete binary tree with i levels, i.e., $1 + 2 + 4 + \cdots + 2^{i-1} = 2^i - 1$ nodes. The nodes on level j, $1 \leq j \leq i$, are labeled by x_j. The 2^i edges leaving the 2^{i-1} nodes on the x_i-level have to lead to nodes representing the 2^i subfunctions of f on $S = \{x_{i+1}, \ldots, x_n\}$. The number of functions on S equals $2^{2^{n-i}}$. The BP contains $2^{2^{n-i}}$ further nodes representing the different functions on S. A node for the subfunction f' is labeled by x_k, where k is the smallest index such that f' essentially depends on x_k. The edges leaving x_j-nodes for $j \geq i$ lead to the corresponding subfunctions. The BP contains $2^i - 1 + 2^{2^{n-i}}$ nodes. The first term increases exponentially while the second term decreases double exponentially. To minimize the number of nodes it would be best to choose i such that $2^i = 2^{2^{n-i}}$ or $i + \log i = n$. But we have to choose i as an integer.

Let $i{:=}n - \lfloor \log(n + 1 - \log n) \rfloor$. Then $i \in [n - \log(n + 1 - \log n), n - \log(n+1-\log n) + 1)$. Since the function $x \to 2^x + 2^{2^{n-x}}$ is first decreasing until it reaches its minimum and then increasing, it is sufficient to investigate the left and right borders of the considered interval. If $x = n - \log(n+1-\log n)$, then $2^x = 2^n/(n+1-\log n) = (1+o(1))2^n/n$ and $2^{2^{n-x}} = 2 \cdot 2^n/n$. If $x = n - \log(n+1-\log n) + 1$, then $2^x = (2+o(1))2^n/n$ and $2^{2^{n-x}} = o(2^n/n)$. This proves the theorem. □

Gröpl, Prömel, and Srivastav (1998) have investigated the behavior of $2^x + 2^{2^{n-x}}$ more exactly. The construction in the proof of Theorem 2.3.1 is even an OBDD (see Section 3). Using the full power of general BPs, Breitbart, Hunt III, and Rosenkrantz (1995) have proved an upper bound of $(1+o(1))2^n/n$ on the BP size of $f \in B_n$.

We design small-size BPs for some "important" functions in order to present fundamental methods for the construction of BPs. The hidden weighted bit function HWB_n is defined in Definition 1.1.3.

Definition 2.3.2.

(i) $\text{MUL}_{k,n} \in B_{2n}$, $0 \le k \le 2n-1$, is the function computing the bit p_k of the product $(p_{2n-1}, \ldots, p_0)$ of two numbers $(x_{n-1}, \ldots, x_0)$ and $(y_{n-1}, \ldots, y_0)$.

(ii) A function $f \in B_n$ is called *symmetric* if the output depends only on the number of ones in the input and not on the positions of the ones.

(iii) $\text{E}_{k,n} \in B_n$, $0 \le k \le n$, computes 1 iff the input contains exactly k ones (*exact counting*).

(iv) $\text{T}_{k,n} \in B_n$, $0 \le k \le n$, computes 1 iff the input contains at least k ones (*threshold function*).

(v) $\text{MOD}_{k,n} \in B_n$, $2 \le k \le n$, computes 1 iff the number of ones in the input is a multiple of k (*modular counting*).

Theorem 2.1.9 shows the existence of small-size BPs if a (Turing machine) computation does not need much storage space. This is the idea behind the BP design for HWB_n, symmetric functions, and $\text{MUL}_{k,n}$.

Theorem 2.3.3. *The hidden weighted bit function* HWB_n *can be represented by a BP of size* $\frac{1}{2}n^2 + \frac{3}{2}n + 2$.

Proof. We generalize the first design for HWB_n shown in Fig. 1.1.1. The BP consists of $n+1$ levels of inner nodes. The ith level, $1 \le i \le n+1$, consists of i nodes $v_{i,0}, \ldots, v_{i,i-1}$. If $i \le n$, $v_{i,j}$ is labeled by x_i, its 0-successor is $v_{i+1,j}$, and its 1-successor is $v_{i+1,j+1}$. Obviously, the node $v_{i,j}$, $1 \le i \le n+1$, is reached for the inputs $a = (a_1, \ldots, a_n)$ such that $a_1 + \cdots + a_{i-1} = j$. Hence, at node $v_{n+1,j}$ we have stored that $sum := a_1 + \cdots + a_n$ equals j. We label $v_{n+1,j}$ by x_j (and replace $v_{n+1,0}$ with the 0-sink). The 0-successor of $v_{n+1,j}$

	l=6	l=5	l=4	l=3	l=2	l=1	l=0
				z_{03}	z_{02}	z_{01}	z_{00}
			z_{13}	z_{12}	z_{11}	z_{10}	
		z_{23}	z_{22}	z_{21}	z_{20}		
	z_{33}	z_{32}	z_{31}	z_{30}			
p_7	p_6	p_5	p_4	p_3	p_2	p_1	p_0

(a)

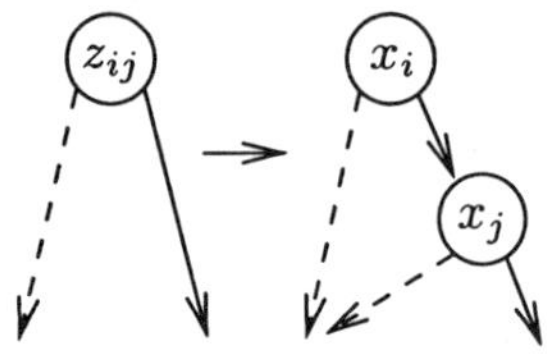

(b)

Figure 2.3.1: *The multiplication method.*

is the 0-sink and the 1-successor the 1-sink. This is a correct realization of HWB_n, since $\text{HWB}_n(a) = a_j$ if $sum = j$. The number of inner nodes equals $1 + \cdots + n + n = \frac{1}{2}n^2 + \frac{3}{2}n$ and we have to add 2 for the sinks. □

It is easy to see that we may save some nodes in the above construction (see the discussion on HWB_4 in Section 1.1) but this only leads to a minor improvement.

Theorem 2.3.4. *Each symmetric function $f \in B_n$ can be represented by a BP of size $\frac{1}{2}n^2 + \frac{1}{2}n + 2$.*

Proof. This result is implicitly contained in the proof of Theorem 2.3.3. If the path activated by input $a = (a_1, \ldots, a_n)$ reaches $v_{n+1,j}$, then $sum = a_1 + \cdots + a_n = j$. By the definition of symmetric functions, the function f takes the same output c_j for all inputs where $sum = j$ and $v_{n+1,j}$ can be replaced by a c_j-sink. The result follows, since we can merge all 0-sinks and also all 1-sinks. □

Theorem 2.3.5. *Each output bit $\text{MUL}_{k,n}$ of the multiplication of two n-bit numbers can be represented by a BP of size $4n^3$.*

Proof. We simulate the school method for multiplication. Let $z_{ij} = x_i y_j$. The product of x and y equals the sum of all $z_{ij}2^{i+j}$. Using the school method, we consider columns of z_{ij} with $l = i + j$; see Fig. 2.3.1(a) for $n = 4$.

In order to compute p_k, we sum the bits of column 0, store the carry c_0, sum c_0 and the bits of column 1, store the carry c_1, and so on until we sum c_{k-1} and the bits of column k. The output bit p_k equals the last bit of the sum of c_{k-1} and the bits of column k. We discuss our BP design with respect to the "variables" z_{ij}. At the end, each "z_{ij}-node" is replaced as described in Fig. 2.3.1(b), which increases the size by a factor of 2. In order to estimate the size of this BP we need an upper bound for c_{m-1}. The sum of the columns $l = 0, \ldots, m-1$ (a bit in the l-column represents 2^l) is bounded generously by $n(2^0 + 2^1 + \cdots + 2^{m-1}) < n2^m$, since each column contains at most n bits. Hence, the carry c_{m-1} takes a value in $\{0, \ldots, n-1\}$. To store the intermediate sums,

$2n$ nodes for the values $0, \dots, 2n-1$ are sufficient. We perform the counting in the same way as in the BP for HWB_n. If we switch from one column to the next, the sums $2j$ and $2j+1$ lead to the carry j. Altogether, we have at most n^2 z-levels of width bounded by $2n$. Since the first level only has width 1, we do not have to add 2 for the sinks. □

A more precise analysis of the BP size for multiplication is left as an exercise. Threshold functions and exact and modular counting are symmetric functions. Hence, the challenge is to beat the $O(n^2)$-bound of Theorem 2.3.4. For exact counting this is not too difficult using the *Chinese Remainder Theorem* (see Hardy and Wright (1979)).

Theorem 2.3.6. *Let $p_1, \dots, p_k$ be pairwise relatively prime, i.e.,* $\gcd(p_i, p_j) = 1$ *for $i \neq j$ (gcd $\widehat{=}$ greatest common divisor). If $a \equiv b \bmod p_i$, $1 \le i \le k$, and $0 \le a, b < p_1 * \cdots * p_k$, then $a = b$.*

The idea (Wegener (1984)) is to choose an appropriate set of prime numbers whose product is at least n and to check sequentially whether the number of ones in the input takes the right value modulo each prime. Since we do not need too many prime numbers and all prime numbers are quite small compared to n, we get a small-size BP. We need a result from number theory on the distribution of prime numbers which can be derived from Corollaries 1 and 2 in the paper of Rosser and Schoenfeld (1962).

Lemma 2.3.7. *For any constant $c > 0$, there exist constants N and $c' > 0$ such that for all $n \ge N$, there are at least $(c \log n)/\log\log n$ primes between $\log n$ and $c' \log n$.*

Lemma 2.3.8. *There is a BP with $(n+1)p$ nodes, among them p sources s_i, $0 \le i < p$, and p sinks s_i^*, $0 \le i < p$, such that for an input $a = (a_1, \dots, a_n)$ the path activated by a and starting at s_i ends in s_j^*, where $j \equiv i + a_1 + \cdots + a_n \bmod p$.*

Proof. The BP for $n = 4$ and $p = 3$ is shown in Fig. 2.3.2. In general, we have nodes $v_{i,j}$, $1 \le i \le n+1$, $0 \le j < p$; the nodes $v_{i,j}$ are labeled by x_i if $i \le n$, $s_j = v_{1,j}$, and $s_j^* = v_{n+1,j}$. The 0-edge from $v_{i,j}$ leads to $v_{i+1,j}$ and its 1-edge leads to $v_{i+1,j+1}$ if $j+1 < p$ and to $v_{i+1,0}$ if $j+1 = p$. Obviously, this BP counts mod p. □

These simple BPs will be used as building blocks.

Theorem 2.3.9. *Each exact counting function $E_{k,n}$ can be represented by a* BP *G of size $O((n \log^2 n)/(\log\log n))$.*

Proof. By Lemma 2.3.7, there exist $l := \lceil (\log n)/(\log\log n) \rceil$ primes $p_1, \dots, p_l$ such that $\log n \le p_i \le c \log n$. Then the product of all p_i is larger than

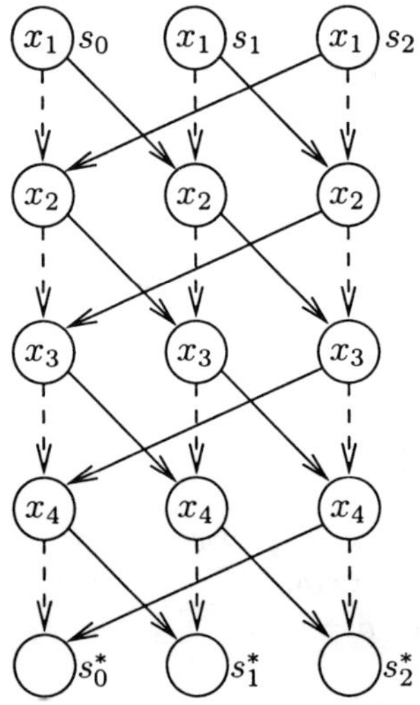

Figure 2.3.2: *A BP counting* mod 3.

$(\log n)^{(\log n)/(\log\log n)} = n$. For each p_i, we use a BP G_i constructed in Lemma 2.3.8. The source s_0 of G_1 becomes the source of G. Then the sink s_j^* of G_i such that $j \equiv k \bmod p_i$ is replaced by the source s_0 of G_{i+1} if $i < l$ and by the 1-sink if $i = l$. All other sinks of G_i are replaced by the 0-sink.

We reach the 1-sink iff $a_1 + \cdots + a_n \equiv k \bmod p_i$ for all i and this is, by the Chinese Remainder Theorem (Theorem 2.3.6), equivalent to $a_1 + \cdots + a_n = k$. The size of each G_i is bounded, by Lemma 2.3.8, by $O(n \log n)$ and, therefore, the size of G is bounded by $O((n \log^2 n)/(\log\log n))$. □

Lupanov (1965) has designed BPs of size $O(n^{3/2})$ for the threshold functions. The best lower bound for the majority and (by reduction) most of the other threshold functions is of size $\Omega((n \log n)/(\log\log n))$ (see Theorem 2.2.8). It took a long time to close this gap. The following result is due to Sinha and Thathachar (1997).

Theorem 2.3.10.

(*i*) *Each threshold function* $T_{k,n}$ *can be represented by a BP* G *of size* $O((n \log^3 n)/(\log\log n \log\log\log n))$.

(*ii*) *Each modular counting function* $\mathrm{MOD}_{k,n}$ *can be represented by a BP* G *of size* $O((n \log^4 n)/(\log\log n)^2)$.

Sketch of proof. The proofs are too complicated to be presented in full detail. We give a short description of the main ideas for the design of a BP for threshold functions. Since we are using BPs constructed in the proof of Lemma 2.3.8 as building blocks, we talk about inputs $a \in \{0, \ldots, n\}$ where a represents all inputs $b \in \{0,1\}^n$ with exactly a ones. A BP for $T_{k,n}$ has to separate inputs from $\{0, \ldots, k-1\}$ from inputs from $\{k, \ldots, n\}$. A p-box is a building block counting mod p. The situation for exact counting is much easier than for threshold functions, since all inputs which finally should lead to the 1-sink reach the same sink of a p-box. If p is not very large (otherwise the p-box would be too large), each sink is reached by inputs from $\{0, \ldots, k-1\}$ and by inputs from $\{k, \ldots, n\}$.

The idea is to construct a chain of boxes such that some sinks can be replaced by 0-sinks and others by 1-sinks, and such that only inputs from some relatively short interval reach the other sinks. Hence, the length of the interval of "undecided inputs" has decreased and we can repeat this approach until all inputs are decided. More precisely, the main tool has the following properties. Some odd primes $p_2, \ldots, p_m$ such that $p_m < p_i$ for $2 \le i \le m-1$ and some power p_1 of 2 are given.

Let $M = p_1 * \cdots * p_m$ and $M' = M/p_m$. Then there exist connections between a sequence of a p_1-box, $\ldots$, p_m-box with the following properties. If we consider inputs from an interval of length $l \le (p_m - m + 2)M'$, only inputs from a subinterval of length $l' \le (2m-2)M'$ reach sinks of the chain which cannot be replaced by 0-sinks or 1-sinks if we want to represent $T_{k,n}$.

We add some ideas of how to construct such a chain of boxes. First, we describe the connections between the boxes. Let $p_{m+1} = 1$ and $N_i = M/p_i p_{i+1}$. It is a number theoretical fact that the set of all $uN_i \bmod p_i$, $0 \le u < p_i$, is equal to $\{0, \ldots, p_i - 1\}$. The sink with number $uN_i \bmod p_i$ of the p_i-box is identified with the source with number $-uN_i \bmod p_{i+1}$ of the p_{i+1}-box.

Each input $a \in \{0, \ldots, n\}$ can be described as $a = a_1 N_1 + \cdots + a_m N_m$ with $0 \le a_j < p_j$. The input a reaches the sink $a_m M'$ of the chain constructed above. Now we consider a sink of the chain and ask which inputs reach this sink. The number of the sink can be uniquely described by some $rM' \bmod p_m$ where $0 \le r < p_m$. If an input b reaches this sink, then $b \equiv (rM' + i) \bmod M$ for some $i \in \{0, \ldots, (m-1)M'\}$. Although the intervals I_r containing all $(rM' + i) \bmod M$ for $i \in \{0, \ldots, (m-1)M'\}$ are not disjoint, it is now not too difficult to prove that the constructed chain can serve as the described main tool. □

It is somewhat surprising that the design of small BPs for simple functions such as threshold functions and modular counting is a difficult task.

2.4 Algorithms on BPs

Since BPs have a quite large representational power and since multiplication has a polynomial-size BP representation, by Theorem 1.4.3, we cannot expect that all the operations listed in Section 1.3 can be performed efficiently. It is indeed quite easy to obtain almost the same results as for circuits in Section 1.4. By $|G|$ we denote the size of a BP G.

Theorem 2.4.1. *Let G_f and G_g be BPs.*

(*i*) *Evaluation is possible in time $O(\text{depth}(G_f))$.*

(*ii*) *Synthesis is possible in time $O(|G_f| + |G_g|)$.*

(*iii*) *Replacement by constants is possible in time $O(|G_f|)$.*

(*iv*) *Replacement by functions is possible in time $O(|G_f| + |G_g|)$.*

(*v*) *Quantification is possible in time $O(|G_f| + |G_g|)$.*

Proof.

(i) This follows by definition using the first type of semantics.

(ii) This statement follows from the proof of Theorem 2.1.4. Since negation is easy, it is sufficient to consider the operations AND and EXOR. The proof of Theorem 2.1.4 contains the proof for AND and the proof for EXOR if a BP for $(g, \overline{g})$ is available. This can be obtained by copying G_g and negating the sinks of one of the copies.

(iii) If we want to replace x_i with c, we visit the nodes and edges of G_f in some order. If we consider an edge leading to a node v with label x_i, we replace this edge with an edge to the c-successor of v.

(iv) First, we compute BPs for $f_{|x_i=1}$ and for $f_{|x_i=0}$ by copying G_f and using (iii). Then we perform three binary synthesis steps to represent $ite(g, f_{|x_i=1}, f_{|x_i=0})$.

(v) This statement follows, since quantification consists of replacement by constants and binary synthesis. □

Theorem 2.4.2.

(*i*) *SAT is NP-complete for BPs.*

(*ii*) *SAT-COUNT is NP-hard and #P-complete for BPs.*

(*iii*) *The equivalence test is coNP-complete for BPs.*

(*iv*) *The redundancy test is coNP-complete for BPs.*

(*v*) *The minimization problem is NP-hard for BPs.*

Proof. We can copy (with some obvious modifications) the proof of Theorem 1.4.1. The reason is that the proof of Theorem 1.4.1 is based on the NP-completeness of SAT for CNFs. But CNFs are formulas not using EXOR or its negation. Hence, CNF formulas lead to BPs for the same function of the same size (see Theorem 2.1.3). □

We summarize the results of the last four sections. BPs are a useful computation model for Boolean functions but the BP model is too powerful to admit efficient algorithms for many important operations.

2.5 DTs

Formulas are circuits whose underlying graph is a tree if the variables may be duplicated. Since research on formulas is a well-established subject with many interesting issues, we also consider DTs, which are BPs whose underlying graph is a tree.

Definition 2.5.1. A *DT* is a BP whose underlying graph is a tree.

In Section 1.1, we discussed why *inconsistent paths* cause difficulties. For DTs, inconsistent paths can be eliminated efficiently.

Lemma 2.5.2. *From a DT T_f for $f \in B_n$, a DT T_f^* for f without inconsistent paths can be constructed in time $O(|T_f|)$. Moreover, $|T_f^*| \leq |T_f|$.*

Proof. The algorithm is based on a depth-first search (DFS) traversal of T_f starting at its root. In an array a of length n we store for each variable whether its value is 0, 1, or undefined on the path from the root to the considered node v. If v is an x_i-node and $a[i] = c_i \in \{0,1\}$, we replace the edge leading to v by an edge to the c_i-successor of v. The subtree reached via the $(1 - c_i)$-edge leaving v is no longer reachable from the source. All paths from the root to this subtree are inconsistent. □

In the following, we always assume that the given DTs do not contain inconsistent paths and implicitly we apply the algorithm of Lemma 2.5.2 to the result of all operations which change the DT.

Theorem 2.5.3. *Let T_f and T_g be DTs.*

(i) *Evaluation is possible in time $O(depth(T_f))$.*

(ii) *SAT, SAT-COUNT, and replacement by constants are possible in time $O(|T_f|)$.*

(iii) *Synthesis, equivalence test, and replacement by functions are possible in time $O(|T_f||T_g|)$.*

(iv) *Quantification and redundancy test are possible in time $O(|T_f|^2)$.*

Proof.

(i) The statement is obvious.

(ii) For SAT, it is sufficient to look in a DT without inconsistent paths for a 1-sink. For SAT-COUNT, a DFS traversal is sufficient. For each 1-sink s, we compute the length $l(s)$ of the path from the root to s. Then $2^{n-l(s)}$ inputs reach this 1-sink. We have to sum the values $2^{n-l(s)}$ for all 1-sinks s. Replacement by constants can be done as for BPs.

(iii) Since negation can be done by negating the sinks, we only have to consider the operations AND and EXOR. For AND, we replace the 1-sinks of T_f by copies of T_g. For EXOR, we replace the 1-sinks of T_f by copies of $T_{\overline{g}}$ and the 0-sinks of T_f by copies of T_g. The evaluation rule proves that this procedure is correct. Equivalence test is the same as EXOR-synthesis and the negation of SAT. Before applying the SAT algorithm, we have to eliminate the inconsistent paths. For the operation replacement by functions, the obvious synthesis procedure is not efficient enough. We have to represent $h = gf_{|x_i=1} + \overline{g}f_{|x_i=0}$. We replace the 1-sinks of T_g by copies of $T_{f_{|x_i=1}}$ and the 0-sinks of T_g by copies of $T_{f_{|x_i=0}}$. The evaluation rule proves that we obtain a DT for h.

(iv) Quantification is the synthesis of DTs for $f_{|x_i=0}$ and $f_{|x_i=1}$. The redundancy test can be performed as an equivalence test of $f_{|x_i=0}$ and $f_{|x_i=1}$. □

All operations besides minimization can be performed efficiently with simple algorithms. What about minimization? For more general DTs, Hyafil and Rivest (1976) have proved that the minimization problem is NP-complete. To the best of our knowledge, nothing is known for DTs as described in Definition 2.5.1.

While formulas, although they are restricted to trees, have some representational power, this does not hold for DTs.

Theorem 2.5.4. *If $f \in B_n$ can be represented by a DT T_f with s_0 0-leaves and s_1 1-leaves, then f can be represented by a DNF with s_1 monomials and by a CNF with s_0 clauses.*

Proof. For the construction of a DNF, we consider all paths from the root of T_f to a 1-sink v. All inputs reaching v can be described by a monomial m_v containing x_i (or $\overline{x}_i$) if the path contains an x_i-node which is left via its 1-edge (or 0-edge, respectively). The DNF is the disjunction of all monomials m_v for all 1-sinks v.

Considering the 0-sinks instead of the 1-sinks, we obtain a DNF for $\overline{f}$ with s_0 monomials. We negate this DNF to obtain a representation for f. Applying de Morgan rules twice we obtain a CNF for f with s_0 clauses (e.g., $\overline{f} = x_1\overline{x}_3 + \overline{x}_1x_2\overline{x}_4 \Rightarrow f = \neg(x_1\overline{x}_3+\overline{x}_1x_2\overline{x}_4) = (\overline{x_1\overline{x}_3})\wedge(\overline{\overline{x}_1x_2\overline{x}_4}) = (\overline{x}_1+x_3)\wedge(x_1+\overline{x}_2+x_4))$. □

DNFs are quite popular in applications. Although simple functions such as the EXOR of n variables have exponential DNF size, the class of functions with polynomial-size DNFs is "small but not very small." The class of functions with polynomial-size DNFs *and* polynomial-size CNFs is much smaller. One has to think for a while to find interesting functions in this class.

One may ask whether Theorem 2.5.4 has the counterpart that all functions with polynomial-size DNFs and CNFs can be represented by polynomial-size DTs. This has been disproved by Jukna et al. (1997). For such a result, one has to be able to prove nonpolynomial lower bounds on the DT size of functions with polynomial-size DNFs and CNFs, i.e., very simple functions. The lower bound technique is based on *spectral methods* and is interesting in itself. Results obtained by Linial, Mansour, and Nisan (1993), Brandman, Orlitsky, and Hennessy (1990), and Kushilevitz and Mansour (1991) have been combined by Jukna et al. (1997).

If one works with spectral methods, it is convenient to switch to the $(+1,-1)$-*notation*, i.e., we replace 0 with $+1$ and 1 with -1. In this notation, a monomial is a product (over the reals) of positive literals, say $x_1x_3x_7$. Considering inputs from $\{+1,-1\}$ we see that monomials represent parity functions, i.e., the EXOR of the variables in the monomial. The product equals -1 iff the number of (-1)-factors is odd and a parity function computes 1 iff the number of ones in the sum is odd.

Definition 2.5.5. Let f, g: $\{+1,-1\}^n \to \{+1,-1\}$. Then $\langle f,g\rangle := 2^{-n}\sum_{a\in\{+1,-1\}^n} f(a)g(a)$.

Lemma 2.5.6. *For $S \subseteq \{1,\ldots,n\}$ let X_S be the product of all x_i, $i \in S$. The set of all monomials X_S forms an orthonormal basis with respect to $\langle\cdot,\cdot\rangle$ for the class of all $f : \{+1,-1\}^n \to \{+1,-1\}$.*

Proof. Since $X_S(a)\cdot X_S(a) = 1$ for all $a \in \{+1,-1\}^n$, $\langle X_S, X_S\rangle = 1$. The product $X_S \cdot X_{S'}$ contains $x_i^2 = 1$ if $i \in S\cap S'$ and x_i if $i \in S \oplus S'$. If $S \neq S'$, there is at least one $i \in S\oplus S'$. Let $X_S\cdot X_{S'} = x_i\cdot$ rest. Then

$$\begin{aligned}\langle X_S, X_{S'}\rangle &= 2^{-n}\left(\sum_{a_i\in\{+1,-1\}} a_i \sum_{a_j\in\{+1,-1\};j\neq i} \text{rest}(a)\right)\\ &= 2^{-n}\left(\sum_{a_j\in\{+1,-1\};j\neq i}\text{rest}(a) - \sum_{a_j\in\{+1,-1\};j\neq i}\text{rest}(a)\right)\\ &= 0.\end{aligned}$$

Finally, each finite function and, hence, each function $f\colon \{+1,-1\}^n \to \{+1,-1\}$, can be written as a polynomial over the reals. Using the identities $x_i^2 = 1$ we obtain a representation $f = \sum_{S\subseteq\{1,\dots,n\}} a_S X_S$. □

Since the monomials are an orthonormal basis, the coefficients in the representation of f are uniquely determined. They are denoted by $\widehat{f}(S)$ and are called the *Fourier coefficients* of f.

Lemma 2.5.7. *The Fourier coefficients of f can be described as*

$$\widehat{f}(S) = \langle f, X_S\rangle = 2^{-n} \sum_{a\in\{+1,-1\}^n} f(a)X_S(a).$$

Proof. The claim follows if we can verify that

$$f = \sum_{S\subseteq\{1,\dots,n\}} \langle f, X_S\rangle X_S.$$

This follows from Lemma 2.5.6 by the following calculation:

$$\begin{aligned}\sum_{S\subseteq\{1,\dots,n\}} \langle f, X_S\rangle X_S(a) &= 2^{-n} \sum_{S\subseteq\{1,\dots,n\}} \sum_{b\in\{+1,-1\}^n} f(b)X_S(b)X_S(a)\\ &= \sum_{b\in\{+1,-1\}^n} f(b) \sum_{S\subseteq\{1,\dots,n\}} 2^{-n}X_S(b)X_S(a).\end{aligned}$$

If $b = a$, the last sum gives the value 1, since $X_S(a)^2 = 1$ for all S. If $b \neq a$, then $b_i \neq a_i$ for some index i and $a_ib_i = -1$. For each S containing i, we get $X_S(b)X_S(a) = -X_{S-\{i\}}(b)X_{S-\{i\}}(a)$ and, therefore, the sum equals 0. Hence, altogether we obtain $f(a)$. □

Let $\mathrm{DT}(f)$ denote the minimal number of leaves of a DT representing f.

Theorem 2.5.8. *For $f \in B_n$ and $S \subseteq \{1,\dots,n\}$,*

$$\mathrm{DT}(f) \geq 2^{|S|} \sum_{T\supseteq S} |\widehat{f}(T)|.$$

Proof. We consider f in the $(+1,-1)$-notation. Let T_f be a DT for f of size $\mathrm{DT}(f)$. For each leaf (or sink) l of T_f, we denote by $\mathrm{val}(l)$ the label of l, by I_l the set of indices of those variables which are tested on the path from the root to l, and by $B_l \subseteq \{+1,-1\}^n$ the set of inputs reaching l. We identify B_l with its characteristic function. The Fourier coefficients of B_l can be computed by Lemma 2.5.7 as

$$\widehat{B}_l(T) = 2^{-n} \sum_{a\in B_l} X_T(a).$$

Since each input reaches a unique leaf, $f(a)$ is the sum of all $\text{val}(l) \cdot B_l(a)$ (here we consider B_l as a function taking values in $\{0,1\}$). Since $\langle \cdot, \cdot \rangle$ is an inner product,

$$\widehat{f}(T) = \sum_l \text{val}(l) \widehat{B}_l(T).$$

If $T \not\subseteq I_l$, by the same calculation as in the proof of Lemma 2.5.6, we obtain that $X_T(a) = 1$ for exactly half the inputs $a \in B_l$ and $\widehat{B}_l(T) = 0$. If $T \subseteq I_l$, $X_T(a)$ takes for all $a \in B_l$ the same value and $|\widehat{B}_l(T)| = 2^{-n}|B_l| = 2^{-|I_l|}$. For $S \subseteq \{1, \ldots, n\}$, there are $2^{|I_l|-|S|}$ sets T such that $S \subseteq T \subseteq I_l$. Hence,

$$\sum_{T \supseteq S} |\widehat{f}(T)| \leq \sum_{T \supseteq S} \sum_l |\widehat{B}_l(T)| = \sum_l \sum_{T | I_l \supseteq T \supseteq S} 2^{-|I_l|} = \sum_l 2^{-|S|} = 2^{-|S|} \text{DT}(f). \quad \Box$$

By Theorem 2.5.8, we may obtain large lower bounds if for some large set S the corresponding Fourier coefficient is not too small or if many Fourier coefficients are not too small. Let us first consider the simple case of parity functions, i.e., monomials m_T in the $(+1,-1)$-notation. It is easy to prove directly that DTs for m_T need $2^{|T|}$ leaves. In Lemma 2.5.6, we have shown that $\widehat{m}_T(S) = 0$ if $S \neq T$ and $\widehat{m}_T(T) = 1$. Choosing $S = T$ in Theorem 2.5.8, we obtain the optimal bound $2^{|T|}$. However, parity functions do not have small DNFs or CNFs.

Definition 2.5.9. For $n = 2^h$, the function $F_h = \text{NAND}_h$ is defined by the formula whose underlying graph is a complete binary tree of depth h where the leaves are labeled by the variables $x_1, \ldots, x_n$ and all inner nodes compute binary NAND, i.e., $\overline{xy} = \overline{x} + \overline{y}$.

Theorem 2.5.10.

(*i*) *The function $NAND_h$ can be represented by a DNF with at most $2^{2^{(h+1)/2}-1} = 2^{O(n^{1/2})}$ monomials and by a CNF with at most $2^{2^{(h+2)/2}-2} = 2^{O(n^{1/2})}$ clauses.*

(*ii*) *Each DT for $NAND_h$ has $2^{\Omega(n)}$ leaves.*

(*iii*) *Let DCNF(f) be the sum of the minimal number of monomials and the minimal number of clauses to represent f. Then*

$$DT(NAND_h) = 2^{\Omega(\log^2 DCNF(NAND_h))}.$$

This theorem implies that the DT complexity is not polynomially bounded with respect to the combined DNF/CNF complexity measure DCNF. We may reformulate our result. We interpret NAND_h as a function on $N = \lfloor 2^{2^{h/2}} \rfloor$ variables where $N-n$ variables are dummy variables. Then NAND_h has polynomial-size DNFs and CNFs with respect to N but the DT complexity grows quasi polynomially like $2^{\Omega(\log^2 N)} = N^{\Omega(\log N)}$.

Proof of Theorem 2.5.10.

(i) $F_0(x_1) = x_1$ and the bounds hold for $h = 0$. Since $\mathrm{NAND}(x, y) = \overline{x} + \overline{y}$, we get $\mathrm{DNF}(F_h) \le 2 \cdot \mathrm{DNF}(\overline{F}_{h-1}) = 2 \cdot \mathrm{CNF}(F_{h-1})$. Since $\mathrm{NAND}(x, y) = \overline{xy}$, we get $\mathrm{DNF}(\overline{F}_h) \le \mathrm{DNF}(F_{h-1})^2$. Now the bounds follow by induction.

(ii) We apply Theorem 2.5.8 for $S = \emptyset$. A set T is called dense if it contains at least one variable from each subformula of depth 2, i.e., at least one variable from each set $\{x_{4i+1}, x_{4i+2}, x_{4i+3}, x_{4i+4}\}$. We calculate exactly the sum c_h of the absolute values of all Fourier coefficients for dense sets. By case inspection, we obtain that $c_2 = 27/8 = 3.375$. For larger h, we describe NAND algebraically in the $(+1, -1)$-representation. Then $\mathrm{NAND}(x, y) = \frac{1}{2}(xy - x - y - 1)$. Let $F^{(1)}_{h-1}$ and $F^{(2)}_{h-1}$ be copies of F_{h-1} on disjoint sets of variables. Then

$$F_h = \frac{1}{2}(F^{(1)}_{h-1} F^{(2)}_{h-1} - F^{(1)}_{h-1} - F^{(2)}_{h-1} - 1).$$

Since $F^{(1)}_{h-1}$ and $F^{(2)}_{h-1}$ depend only on half the variables, they do not distribute anything to the Fourier coefficients for dense sets. Hence, it is sufficient to consider $\frac{1}{2} F^{(1)}_{h-1} F^{(2)}_{h-1}$ and its Fourier coefficients for dense sets. Using Lemma 2.5.7 for a dense set $T = T_1 \uplus T_2$ (where T_1 is the set of indices of variables belonging to $F^{(1)}_{h-1}$), we can conclude that the absolute value of the Fourier coefficients of $\frac{1}{2} F^{(1)}_{h-1} F^{(2)}_{h-1}$ with respect to T is $\frac{1}{2} |F^{(1)}_{h-1}(T_1)||F^{(2)}_{h-1}(T_2)|$ and, hence, $c_h = \frac{1}{2} c^2_{h-1}$. By induction, we obtain $c_h = 2(c_2/2)^{(2^h - 2)} = 2^{\Omega(n)}$, since $c_2 > 2$.

(iii) This follows directly from (i) and (ii). □

We mention an application of Theorem 2.5.8 for $S = \{1, \ldots, n\}$, i.e., an application where $2^n |\widehat{f}(\{1, \ldots, n\})| = 2^{\Omega(n)}$. Let G_h be the function defined by a formula whose underlying graph is a complete ternary tree of depth h. The function is defined on $n = 3^h$ variables which label the leaves. Each inner node realizes the ternary majority operation, i.e., it decides whether its input contains at least two ones. Jukna et al. (1997) have shown that the DT complexity of G_h is $2^{\Omega(n)}$ and that the DCNF complexity is $2^{O(n^{1/\log 3})} = 2^{O(n^{0.631})}$. Hence, Theorem 2.5.10 gives the better separation of the two complexity measures. Can one get even stronger separation results? Ehrenfeucht and Haussler (1989) have proved that the separation result of Theorem 2.5.10 is at least almost optimal.

Theorem 2.5.11. *For all Boolean functions $f \in B_n$,*

$$DT(f) = n^{O(\log^2 DCNF(f))}.$$

Proof. Let a DNF $m_1+\cdots+m_t$ for f with the minimal number t of monomials and a DNF $m'_1+\cdots+m'_u$ for $\overline{f}$ with the minimal number u of monomials be given. By the definition of the DCNF complexity measure, $t+u=\text{DCNF}(f)$. We use a greedy approach for the construction of a DT for f.

Since f and $\overline{f}$ cannot simultaneously equal 1, each monomial m_i contains for each monomial m'_j some literal l such that $\overline{l}$ is contained in m'_j. Moreover, $m_1+\cdots+m_t+m'_1+\cdots+m'_u$ is a DNF for 1. Since all 2^n inputs are satisfying, there is at least one monomial, without loss of generality (w.l.o.g.) m_1, with at least $2^n/(t+u)$ satisfying inputs. The length of this monomial is, therefore, bounded above by $k := \log(t+u)$. By the above arguments, m_1 contains a literal l such that $\overline{l}$ is contained in at least u/k of the monomials $m'_1,\ldots,m'_u$. The variable x_i belonging to l is tested at the root of the DT which we construct for f.

We consider the subfunctions $f_{|x_i=1}$ and $f_{|x_i=0}$ which have to be represented in the subtrees of our DT. For one edge leaving the root, called the *special edge*, the DNF of f or $\overline{f}$ (in our case $\overline{f}$) has been decreased by a factor of at most $(1-\frac{1}{k}) \le (1-\frac{1}{\log(t+u)})$. Our approach is a greedy one, since we have chosen the variable at the root in such a way that we decrease one of the DNFs as much as we can guarantee. For both subtrees we follow the same approach. After at most $2\log^2(t+u)$ special edges we have decreased the size of one of the DNFs at least $\log^2(t+u)$ times by a factor of at most $(1-\frac{1}{\log(t+u)})$. Since the sizes of the DNFs for f and $\overline{f}$ are each bounded by $t+u$, we then obtain a constant subfunction of f which can be represented by a leaf. Altogether, the constructed DT contains paths whose length is bounded by n, and the number of special edges is bounded on each path by $s := 2\log^2(t+u)$. Hence, there are n choices for the length of a path, s choices for the number of special edges on a path, and at most $\binom{n}{s}$ choices where the special edges occur on the path. Here we use the assumption that $s \le n/2$. Otherwise, the bound is trivial, since $\text{DT}(f) \le 2^n$. Hence, we have constructed a DT for f with at most $ns\binom{n}{s} \le nsn^s$ leaves. This proves the theorem. □

The conclusion is that all operations besides minimization can be performed efficiently on DTs. The class of functions with polynomial DT size is a proper subclass of the class of functions which have DNFs and CNFs of polynomial size and, therefore, a very small class of simple functions.

2.6 Exercises and Open Problems

2.1.E Let BP* be the complexity measure for ite straight line programs such that the instructions $I_1,\ldots,I_{j-1}$ are also allowed as conditions of I_j. Prove that $\text{BP}^*(f) \le 2C(f)+n$.

2.2.D Prove that $\text{L}^*(f) = \text{L}(f)^{O(1)}$ for all Boolean functions f.

2.3.D Prove that Nechiporuk's lower bound for the BP size of $f \in B_n$ is bounded by $O(n^2/\log^2 n)$.

2.4.M Prove that the circuit size of the indirect storage access function ISA_n is bounded by $O(n \log n)$.

2.5.M Prove that the BP size of the indirect storage function ISA_n is bounded by $O(n^2)$.

2.6.E The direct storage access function DSA_n (or multiplexer MUX_n) is defined on $x_0, \ldots, x_{n-1}, y_0, \ldots, y_{k-1}$ where $n = 2^k$. Let $\mathrm{DSA}_n(x, y) := x_{|y|}$. Prove that $\mathrm{BP}(\mathrm{DSA}_n) = O(n)$.

2.7.D (See Kloss (1966) or Wegener (1987)). Prove that $\mathrm{BP}(\mathrm{DET}_n) = \Omega(n^3/\log n)$.

2.8.D (See Schürfeld (1983) or Wegener (1987)). Prove that $\mathrm{BP}(\text{3-CLIQUE}_n) = \Omega(n^3/\log n)$.

2.9.E Prove that $\mathrm{BP}(\text{3-CLIQUE}_n) = O(n^3)$.

2.10.E Improve the upper bound of Theorem 2.3.5 for the BP size of $\mathrm{MUL}_{k,n}$ and make the upper bound dependent on k.

2.11.E Prove that the BP size of $\mathrm{multgraph}_n$ (see Section 1.4) is bounded by $O(n^3)$.

2.12.E Prove that if f can be represented by a DT with s 1-leaves, then f has a DNF representation $m_1 + \cdots + m_s$ such that $m_i m_j = 0$ if $i \neq j$.

2.13.E Prove that if f computes 1 on s inputs, then $\mathrm{DT}(f) \leq sn$.

2.14.O Prove a lower bound of size $\Omega(n^2)$ for the BP size of explicitly defined Boolean functions $f_n \in B_n$.

2.15.O What is the BP size of the alternating tree function (see Definition 2.1.6)? (A partial answer can be obtained by applying the results of Cleve (1991).)

2.16.O What is the complexity of the minimization problem for DTs?

Chapter 3

Ordered Binary Decision Diagrams (OBDDs)

3.1 OBDDs

General BPs have a lot of representational power but many operations are hard to perform. The representational power of DTs is low and no efficient minimization algorithm is known. That all other operations can be performed efficiently is useless in this situation. We are still looking for a data structure allowing efficient operations and the compact representation of not too few functions. Bryant (1985, 1986) has introduced *OBDDs*, which have found a lot of applications.

Definition 3.1.1. A *variable ordering* π on $X_n = \{x_1, \ldots, x_n\}$ is a permutation on the index set $I = \{1, \ldots, n\}$. The position of x_i in the π-ordered list of variables is $\pi(i)$; the π-ordered list equals $x_{\pi^{-1}(1)}, \ldots, x_{\pi^{-1}(n)}$. The variable ordering π is sometimes identified with the corresponding π-ordered list of variables.

Definition 3.1.2.

(i) A π-OBDD for a variable ordering π is a BP where the sequence of tests on a path is restricted by the variable ordering π, i.e., if an edge leads from an x_i-node to an x_j-node, then $\pi(i) < \pi(j)$. The size of the smallest π-OBDD for f is denoted by π-OBDD(f).

(ii) An OBDD is a π-OBDD for some variable ordering π. The size of the smallest OBDD for f is denoted by OBDD(f).

From the definition it follows that each path in an OBDD contains at most one x_i-node. Hence, OBDDs do not contain inconsistent paths. The BPs in

Section 1.1 for the hidden weighted bit function are not OBDDs but the construction in the proof of Theorem 2.3.1 leading to an upper bound on the BP size of any $f \in B_n$ leads to an OBDD for the variable ordering $x_1, \ldots, x_n$, i.e., $\pi = id$. The same construction works for each variable ordering. Thus we obtain the following results.

Theorem 3.1.3. *The π-OBDD size of any Boolean function $f \in B_n$ is bounded for any variable ordering π by $(3 + o(1))2^n/n$.*

Breitbart, Hunt III, and Rosenkrantz (1995) have improved this result by an upper bound of size $(2 + o(1))2^n/n$.

It will turn out that all operations introduced in Section 1.3 can be performed efficiently for π-OBDDs. On the other hand, the representational power of π-OBDDs for a fixed variable ordering π is much too low. But often there exist good variable orderings. Unfortunately, it is a hard problem to find an optimal or a good variable ordering. Hence, Chapter 5 is devoted to the variable-ordering problem. In most of this chapter, we assume that a fixed variable ordering π is given. Then it is convenient to assume that $\pi = id$.

Before we present efficient algorithms on π-OBDDs, we describe structural properties of π-OBDDs in order to prove the existence of a unique π-OBDD for f with minimal size. This implies that π-OBDDs can be *reduced* (see Definition 1.3.2) and later we work only with reduced π-OBDDs.

Let $f \in B_n$ and $\pi = id$. We start with a somewhat naive approach to describing a small-size π-OBDD for f. It will turn out that the resulting π-OBDD has minimal size for f. We use the notation $f_{|x_1=a_1,\ldots,x_i=a_i}$ for the subfunction of f where x_j is replaced by $a_j \in \{0,1\}$, $1 \le j \le i$. This function is an element of B_n and depends syntactically on $x_1, \ldots, x_n$ but $x_1, \ldots, x_i$ are redundant by definition. If x_{i+1} is also redundant for $g := f_{|x_1=a_1,\ldots,x_i=a_i}$, then g, $g_{|x_{i+1}=0}$, and $g_{|x_{i+1}=1}$ all are equal. Finally, let $\alpha(f)$ denote the variable with the smallest index k such that f essentially depends on x_k, i.e., x_k is not redundant. If f is constant, then $\alpha(f) :=$ const.

The source of our π-OBDD is a node labeled by $x_k := \alpha(f)$. If $\alpha(f) =$ const, we represent f by a sink labeled by the appropriate constant. If a node is labeled by a variable x_k and has to represent a function g, the 0-successor has to represent $g_0 := g_{|x_k=0}$ and the 1-successor has to represent $g_1 := g_{|x_k=1}$. We represent them by nodes labeled by $\alpha(g_0)$ and $\alpha(g_1)$, respectively. Since by our construction x_k is not redundant for g, $g_0 \not\equiv g_1$. Obviously, we represent each function h only by one node which is labeled by $\alpha(h)$. Hence, our construction uses two simple tricks:

- Represent a function h by a node labeled by $\alpha(h)$,

- represent each function only once.

The result of our construction can be described in the following way. The π-OBDD contains a node for each subfunction $f_{|x_1=a_1,\dots,x_i=a_i}$, $0 \le i \le n$, $a_k \in \{0,1\}$ for $1 \le k \le i$, labeled by $\alpha(f_{|x_1=a_1,\dots,x_i=a_i})$ if this is a variable, or by the appropriate constant. If two such subfunctions g and h are equal (as B_n-functions), they are represented by the same node. The c-successor of a node for a nonconstant subfunction g has to represent $g_{|x_k=c}$ if $\alpha(g) = x_k$. Hence, we know how to direct the edges.

Theorem 3.1.4. *For some $f \in B_n$ let s_i be the number of different subfunctions $f_{|x_1=a_1,\dots,x_{i-1}=a_{i-1}}$ essentially depending on x_i and let s_{n+1} be the size of the image of f. Then, for $\pi = id$, a π-OBDD of minimal size for f contains s_i nodes labeled by x_i, s_{n+1} sinks, and is unique up to isomorphism.*

Proof. Our description above leads to a π-OBDD G of the appropriate size. On the other hand, let $f_{|x_1=a_1,\dots,x_{i-1}=a_{i-1}}$ be a subfunction of f essentially depending on x_i. Let us consider an input $b = (b_1,\dots,b_n)$ where $b_1 = a_1,\dots,$ $b_{i-1} = a_{i-1}$. Let us follow the path activated by the partial input $(a_1,\dots,a_{i-1})$ and let v be the final node on this path. Then v has to represent f for all inputs with the properties of b, i.e., v has to represent $f_{|x_1=a_1,\dots,x_{i-1}=a_{i-1}}$. If v is not labeled by x_i, then because of the given variable ordering, no path starting at v reaches a node labeled by x_i. Hence, v represents a function which cannot essentially depend on x_i in contradiction to the assumption that v represents $f_{|x_1=a_1,\dots,x_{i-1}=a_{i-1}}$. Altogether, we have proved that each π-OBDD for f contains at least s_i x_i-nodes representing the subfunctions $f_{|x_1=a_1,\dots,x_{i-1}=a_{i-1}}$ essentially depending on x_i. Moreover, it has to contain s_{n+1} sinks. If the π-OBDD for f contains exactly these nodes, the connections by the edges are uniquely determined. □

This structural description of a reduced π-OBDD (Sieling and Wegener (1993a)) is useful for the design of efficient algorithms on OBDDs and for the proof of bounds on the OBDD size of functions.

Up to now, we have considered single-output functions $f \in B_n$, although a BP can represent functions $g \in B_{n,m}$ by representing simultaneously the outputs $g_1,\dots,g_m$ of g. Minato, Ishiura, and Yajima (1990) have applied this idea to OBDDs. Before that, OBDDs were restricted to one source. Definition 3.1.2 of OBDDs is based on the definition of BPs which does not have this restriction. Hence, an OBDD represents $g \in B_{n,m}$ if it contains for each coordinate function g_i a node v_i representing g_i. Minato, Ishiura, and Yajima (1990) have introduced the notion *shared* BDD (SBDD), since the OBDDs for the different coordinate functions $g_1,\dots,g_m$, may *share* nodes. Since their idea is so natural, people do not distinguish between OBDDs for $f \in B_n$ and SBDDs for $g \in B_{n,m}$. It is common to use the general definition, Definition 3.1.2, for OBDDs to represent functions $g \in B_{n,m}$. For each $i \in \{1,\dots,m\}$, we use a pointer to the node representing g_i. Theorem 3.1.4 can be generalized to the case of multi-output functions $g = (g_1, \ldots, g_m) \in B_{n,m}$. Then s_i is the number of different subfunctions $g_{j|x_1=a_1,\ldots,x_{i-1}=a_{i-1}} \in B_n$ essentially depending on x_i, where $1 \le j \le m$ and $a_k \in \{0,1\}$ for $1 \le k \le i-1$.

Another idea of Minato, Ishiura, and Yajima (1990) and Akers (1978a) is the use of so-called *complemented edges.* For each edge (also for pointers to the nodes representing the output functions), we have an extra bit, the compl-bit. In order to evaluate such a generalized OBDD we follow the activated path. The label of the sink which is reached on this path has to be negated (or complemented) if the number of compl-bits which are set on the activated path is odd.

Reduced representations are unique and also called *canonical.* It seems that we are losing canonicity by introducing complemented edges. This problem can be circumvented by restricting the use of complemented edges by the following two rules:

- OBDDs contain only a 0-sink and
- only 1-edges and pointers to nodes representing outputs may carry a compl-bit.

Starting from an OBDD with arbitrary compl-bits, we easily obtain an OBDD with the above restrictions. The algorithm works bottom-up. Edges to a 1-sink are replaced by edges to the 0-sink and their compl-bits are complemented. If the compl-bit of a 0-edge (v, w) is set, then this compl-bit, the compl-bits of all edges reaching v, and the compl-bits of the 1-edge leaving v are complemented. Here it is essential that we have pointers for the output functions which have to be represented. The function represented at v may change but not the functions we point to if we take the compl-bit on this pointer into account.

Theorem 3.1.5. *The representation of Boolean functions by π-OBDDs with complemented edges in the restricted form is canonical. Let $\equiv$ be the equivalence relation on B_n where $g \equiv h$ iff $g = h$ or $g = \overline{h}$. Let $\overline{s}_i$ be the number of different equivalence classes with respect to $\equiv$ among the subfunctions $f_{|x_1=a_1,\ldots,x_{i-1}=a_{i-1}}$ of $f = (f_1, \ldots, f_m) \in B_{n,m}$ essentially depending on x_i. Then, for $\pi = id$, the minimal size of a π-OBDD with complemented edges for f contains $\overline{s}_i$ x_i-nodes and one 0-sink.*

Proof. We still have to represent the subfunctions of f. Pointing to a node representing g, we can represent g and with the compl-bit also $\overline{g}$ but nothing else. This proves the claim on the minimal size of π-OBDDs with complemented edges.

The proof of the canonicity of the representation is done by induction on the number of variables n. The case $n = 0$ is obvious. Let the claim be true for functions on $n-1$ variables and let $f \in B_n$ (w.l.o.g. f has a single output). If f

does not essentially depend on x_1, the claim follows by the induction hypothesis. Otherwise, we use the induction hypothesis for the OBDD representing $f_{|x_1=0}$ and $f_{|x_1=1}$. We represent f by a pointer to an x_1-node v whose c-successor is an edge pointing to the same node as the pointer for $f_{|x_1=c}$. This edge gets the same compl-bit as the corresponding pointer. If the 0-edge gets a compl-bit which is set, we have to eliminate it. This is only possible by complementing the pointer for f and the compl-bit of the 1-edge leaving v. □

For simplicity, we discuss only OBDDs without complemented edges, although the OBDD packages use complemented edges. All our results can easily be translated to OBDDs with complemented edges. The following result describes the relation between the size of OBDDs with and without complemented edges exactly.

Corollary 3.1.6. *For each Boolean function f it holds that π-OBDD$(f, \overline{f}) = 2 \cdot \pi$-OBDD$_{\text{ce}}(f)$ where the index* ce *means that complemented edges are allowed.*

Proof. Without loss of generality $\pi = id$. We consider the different subfunctions $g_1, \ldots, g_r$ of f where $x_1, \ldots, x_{i-1}$ have been replaced with constants and which essentially depend on x_i. Let $h_1, \ldots, h_s$ be representatives of the different equivalence classes of $g_1, \ldots, g_r$ with respect to $\equiv$. Then the reduced π-OBDD for f with complemented edges contains s x_i-nodes. The corresponding subfunctions for $(f, \overline{f})$ are $g_1, \ldots, g_r, \overline{g}_1, \ldots, \overline{g}_r$ which perhaps are not different. The different functions are $h_1, \ldots, h_s, \overline{h}_1, \ldots, \overline{h}_s$ and the reduced π-OBDD for $(f, \overline{f})$ without complemented edges contains $2s$ x_i-nodes. □

3.2 OBDDs and Deterministic Finite Automata (DFAs)

In Section 3.1, we presented the standard introduction to OBDDs. Here we discuss another approach based on the classical automata theory (see Hopcroft and Ullman (1979) for an introduction to automata theory). Only to simplify the description do we restrict ourselves to single-output functions $f \in B_n$. Such a function can be interpreted as the language $L(f)$ of all $a \in \{0,1\}^n$ such that $f(a) = 1$. Since $L(f)$ is a finite language, it is a *regular* language and can be represented by a DFA. It is well known that minimal DFAs are a canonical representation of regular languages. Nerode's Theorem describes for which strings one reaches the same node in a minimal DFA for a regular language. Two strings $v, w \in \{0,1\}^*$ are called *Nerode equivalent* with respect to L iff for each $z \in \{0,1\}^*$ either $xz \in L$ and $yz \in L$ or $xz \notin L$ and $yz \notin L$. A minimal DFA for L has a state for each equivalence class and after having read w the corresponding state is reached.

We apply this general result to the special regular languages $L(f)$ based on Boolean functions $f \in B_n$. The function f which is the constant 0 plays a special role, since $L(f)$ is empty. Hence, we assume that f is not the constant 0. Then all $a \in f^{-1}(0)$ and all a whose length $|a|$ is larger than n are (Nerode) equivalent and are represented by a nonaccepting state s_0 (the 0-sink). Edges leaving s_0 lead to s_0. Also all $a \in f^{-1}(1)$ are equivalent and are represented by the only accepting state s_1 (the 1-sink). Edges leaving s_1 lead to s_0. Let $a = (a_1, \ldots, a_i) \in \{0,1\}^i$, $0 \le i \le n-1$, and $b = (b_1, \ldots, b_j) \in \{0,1\}^j$, $0 \le j \le n-1$. If $f_{|x_1=a_1,\ldots,x_i=a_i}$ is the constant 0, then a belongs to the equivalence class represented by s_0. If a and b cannot be represented by s_0, they can only be equivalent if $i = j$. The reason is that all strings c such that ac has to be accepted have length $n-i$, while this holds for b for $n-j$ instead of $n-i$. If $i = j$, it follows directly by the definition of the equivalence relation that a and b are equivalent iff $f_{|x_1=a_1,\ldots,x_i=a_i} = f_{|x_1=b_1,\ldots,x_i=b_i}$. We obtain the following conclusion. If f is not the constant 0, the minimal DFA for $L(f)$ is very similar to an OBDD for the variable ordering $\pi = id$. Let s_i^* $(1 \le i \le n)$ be the number of different subfunctions $f_{|x_1=a_1,\ldots,x_{i-1}=a_{i-1}}$ besides the constant 0. Then the minimal DFA for $L(f)$ contains s_i^* nonaccepting states where the ith input bit is read (or, in the notation of OBDDs, tested). Moreover, the DFA contains two states, an accepting one (s_1) and a nonaccepting one (s_0). The last one is reached for all $a = (a_1, \ldots, a_i)$, $1 \le i \le n$, such that the corresponding subfunction is the constant 0 and for inputs with more than n bits. The only cycle in this DFA is the loop at s_0. If we eliminate the edges leaving s_0 and s_1 and if we label the inner nodes by the corresponding variables, we obtain an OBDD for f.

What is the difference between the reduced *id*-OBDD G for f and the OBDD G^* obtained for f from the minimal DFA for $L(f)$? The number of x_i-nodes in G equals the number of different subfunctions $f_{|x_1=a_1,\ldots,x_{i-1}=a_{i-1}}$ *essentially depending on* x_i while the number of x_i-nodes in G^* equals the number of different subfunctions $f_{|x_1=a_1,\ldots,x_{i-1}=a_{i-1}}$ which are different from the constant 0. Hence, in G we have the unrestricted possibility of omitting tests which belong to redundant variables for the corresponding subfunction of f. In G^* this is only possible if the subfunction of f is the constant 0.

DFAs and *id*-OBDDs differ only in the possibility of omitting tests. Hence, we can transfer a lot of knowledge on DFAs to *id*-OBDDs. The most important issue of OBDDs is the freedom to choose the variable ordering.

There are some situations where it is helpful to have the notion of OBDDs where omitting tests is not allowed.

Definition 3.2.1. An OBDD is called *complete* if all paths from a source to a sink have length n.

Complete OBDDs differ from DFAs only in the minor aspect that tests may not be omitted even if the corresponding subfunction is the constant 0. Hence, it is an easy exercise to derive the following result from our investigations.

Theorem 3.2.2. *The minimal-size complete π-OBDD for f contains as many x_i-nodes as there are different subfunctions $f_{|x_{\pi^{-1}(1)}=a_1,\ldots,x_{\pi^{-1}(i-1)}=a_{i-1}}$.*

This theorem implies that minimal-size complete π-OBDDs are a canonical representation of Boolean functions which are called *quasi-reduced* π-OBDDs.

Theorem 3.2.3. *The size of the quasi-reduced π-OBDD for f is at most a factor of $n+1$ larger than the size of the reduced π-OBDD for f.*

3.3 Efficient Algorithms on OBDDs

Here we present efficient algorithms for the operations listed in Section 1.3. This is very easy for all operations besides synthesis and reduction. If a circuit or a logical description is translated into an OBDD, many synthesis steps have to be performed. We have already seen that it is of particular importance to work with reduced OBDDs after each synthesis step. We discuss algorithms for synthesis and for reduction and, finally, we present the best synthesis algorithm for practical purposes where the reduction is included in the synthesis. Since we usually work with reduced π-OBDDs and since we shall present a linear-time reduction algorithm, we assume w.l.o.g. that π-OBDDs are given in reduced form.

Theorem 3.3.1. *Let f and g be given by π-OBDDs G_f and G_g on X_n. Then evaluation is possible in time $O(n)$. SAT is possible in constant time $O(1)$. SAT-COUNT, replacement by constants, and redundancy test on G_f are possible in linear time $O(|G_f|)$. The equivalence test on G_f and G_g is possible in linear time $O(|G_f|+|G_g|)$. If f and g are represented by a common π-OBDD (π-SBDD) $G_{f,g}$, the equivalence test is possible in constant time $O(1)$.*

Proof. The result on the evaluation operation is obvious. Since we work with reduced OBDDs, SAT has to be answered negatively iff the pointer for the function f leads to the 0-sink. The operation SAT-COUNT works bottom-up and only uses the fact that OBDDs do not contain inconsistent paths. For each node v, the number $c(v)$ of satisfying inputs of f_v is computed, where f_v is considered as a function on X_n. Obviously, $c(s_0)=0$ for the 0-sink s_0 and $c(s_1)=2^n$ for the 1-sink s_1. Let v be an x_i-node with the successors v_0 and v_1. An input a satisfies f_v iff $a_i=0$ and a satisfies f_{v_0} or $a_i=1$ and a satisfies f_{v_1}. Since f_{v_0} cannot essentially depend on x_i, exactly half the satisfying inputs for f_{v_0} fulfill $a_i=0$, similarly for f_{v_1}. Hence, $c(v)=(c(v_0)+c(v_1))/2$ can be computed in constant time. The result for the operation replacement by constants follows as in Theorem 2.4.1. From Theorem 3.1.4 it follows that the variable x_i is redundant for f iff G_f does not contain any x_i-node. This can be tested in linear time by DFS. Finally, f and g are the same functions iff G_f

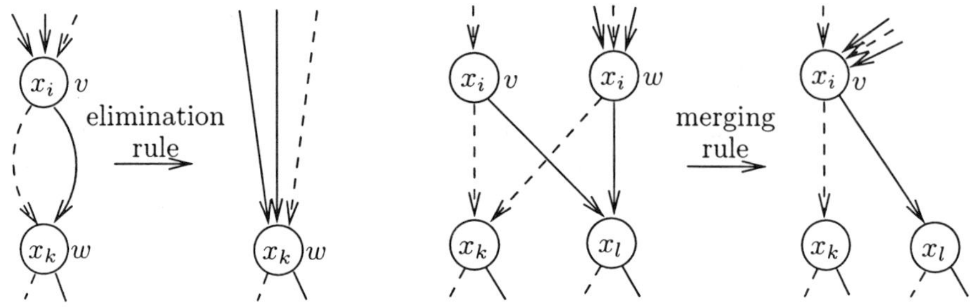

Figure 3.3.1: *The reduction rules for OBDDs.*

and G_g are isomorphic. Hence, we run simultaneously through G_f and G_g with a DFS approach (1-edges first). The OBDDs are isomorphic iff we always find nodes with the same label. If f and g are represented in $G_{f,g}$, they are equal iff the pointers for f and g point to the same node. □

Our structural description of reduced π-OBDDs in Theorem 3.1.4 does not directly lead to an efficient reduction algorithm. It is possible to adapt the reduction algorithm for DFAs. But we obtain more efficient algorithms by using the fact that OBDDs do not contain cycles. This enables us to apply a bottom-up approach. There are two simple local rules to reduce the OBDD size and these *reduction rules* are sufficient to reduce a π-OBDD (see Fig. 3.3.1).

Elimination rule: If a node w is the 0-successor and the 1-successor of a node v, all edges leading to v can be redirected to lead to w and v can be eliminated.

Merging rule: If two nodes v and w have the same label, the same 0-successor, and the same 1-successor, these nodes can be merged, i.e., all edges leading to w can be redirected to lead to v and, afterwards, w can be eliminated.

Theorem 3.3.2. *The reduction rules do not change the functions represented by a π-OBDD. A π-OBDD is reduced iff all nodes are reachable from a pointer to a represented function and no reduction rule is applicable.*

Proof. The first assertion is obvious. It is also obvious that OBDDs with nodes which are not reachable from a pointer and OBDDs where a reduction rule can be applied are not reduced. Let us now assume that the π-OBDD G representing $f \in B_{n,m}$ contains only reachable nodes and that no reduction rule is applicable to G. We claim that G is the reduced π-OBDD for f. Without loss of generality, let $\pi = id$ and let s_i be the number of different subfunctions $f_{j|x_1=a_1,\dots,x_{i-1}=a_{i-1}}$ essentially depending on x_i. Let t_i be the number of x_i-nodes of G. By Theorem 3.1.4 and its generalization to multi-output functions, we know that $t_i \geq s_i$ for all i. If G is not the reduced π-OBDD, we choose the largest i where $t_i > s_i$. Again by Theorem 3.1.4, G and the reduced π-OBDD G_f for f are isomorphic on the subgraphs consisting of the sinks and the x_j-nodes where $j > i$. Either an x_i-node v in G represents a subfunction f' not essentially depending on x_i or two x_i-nodes v and w represent the same subfunction f' essentially depending on x_i. In the first case, $f'_{|x_i=0} = f'_{|x_i=1}$. Since the part of G below the x_i-nodes is reduced, the 0-successor and the 1-successor of v are equal to each other and the elimination rule is applicable in contradiction to our assumptions. In the second case, we conclude likewise, since the lower part of G is reduced, that the merging rule is applicable to v and w which again is a contradiction of our assumptions. □

The theorem is a new structural description of reduced π-OBDDs and it is the basis of all reduction algorithms. The nodes which are not reachable from a pointer can be determined in linear time by a DFS traversal and can be eliminated. Afterwards, we can apply the reduction rules bottom-up to the set of all x_i-nodes, $i = n, \ldots, 1$, if w.l.o.g. $\pi = id$. Reducing the x_i-level does not change the levels $i+1, \ldots, n$. Hence, if the reduction rules are not applicable in the lower part, they will not be applicable later. This approach is contained in the early papers of Bryant (1985, 1986). He also describes an algorithm which reduces the x_i-level in time $O(t_i \log t_i)$ if it contains t_i nodes. Each x_i-node v is described by the pair (v_0, v_1) of its successors. With respect to this description and the lexicographical ordering on the pairs of node numbers, the x_i-nodes can be sorted in time $O(t_i \log t_i)$. It is obvious how to detect the nodes which can be eliminated. Moreover, after the sorting step the groups of nodes which can be merged can also be found easily, since they have the same description and build subblocks in the sorted sequence. The first node of each subblock can be chosen as representative.

The main idea in obtaining a linear-time reduction algorithm (Sieling and Wegener (1993b)) is to replace the usual sorting technique by some variant of bucket sort. Bucket sort is efficient if the number of possible values is not much larger than the number of elements to be sorted. In our case we denote by s the number of nodes of the given OBDD. The pair of direct successors (v_0, v_1) of v is an element of $\{1, \ldots, s\} \times \{1, \ldots, s\}$. Therefore, a two-phase bucket sort technique is applied. Nevertheless, we have the difficulty that the number t_i of x_i-nodes may be small compared to s. We have to avoid touching empty buckets. This is not possible in general. But in our situation it is not necessary to sort the pairs. It is sufficient to identify the sets of nodes v with the same pair of successors.

Algorithm 3.3.3. Reduction of the x_i-level if the x_j-levels, $i+1 \leq j \leq n$, are reduced and the OBDD has only one 0-sink and one 1-sink. The arrays *zero* and *one* of length s can contain at each place a list of nodes. These lists are initialized as empty lists. Let L_i be the list of x_i-nodes. Moreover, L is an empty list.

(1) Run through L_i. If the elimination rule is applicable to v, this information is stored at v. Otherwise, let w be the 0-successor of v. The node v is stored in the linear list (bucket) *zero*$[w]$. If v is the first entry in this bucket, w is added to a list L of nonempty buckets.

(Remark: Only nodes in the same bucket have the chance to become merged.)

(2) Run through L and consider the nonempty buckets one after the other. For the node v with 1-successor z we read *one*$[z]$. If this position is empty, v becomes the representative of its class of mergeable nodes and v is stored at *one*$[z]$. If the array position *one*$[z]$ contains v', we store at v that it will be replaced by v'. After having reached the end of one bucket, we run through this bucket for a second time and clear the *one*-array.

(3) Run through L and clear all buckets and L.

For the next phase, the reduction of the x_{i-1}-level, the arrays *zero* and *one*, and the list L are empty. If a successor is v, we use the information of whether and, possibly, how v is replaced. The analysis of the runtime of the algorithm follows from the description.

Theorem 3.3.4. *A π-OBDD G can be reduced in time and space $O(|G|)$.*

This result is mainly of theoretical interest. We can refer to it as an efficient submodule. We shall see that there is a practically efficient way to include the reduction process into synthesis algorithms. Then an AND-synthesis of f and the constant 1 efficiently produces a reduced π-OBDD for f.

Now we discuss algorithms for the binary synthesis of π-OBDDs. Let $G_f = (V_f, E_f)$ and $G_g = (V_g, E_g)$ be reduced π-OBDDs, w.l.o.g. $\pi = id$, and $\otimes \in B_2$. In order to obtain a reduced π-OBDD G_h for $h = f \otimes g$, we may apply the synthesis algorithm for DFAs with some extensions which take into account that we may omit tests. Let $G_h = G_f \times G_g = (V_h, E_h)$ where $V_h = V_f \times V_g$. A node (v, w) gets the label x_i if label$(v) = x_j$, label$(w) = x_k$, and $i = \min\{j, k\}$. If v_0, v_1, w_0, and w_1 are the successors of v and w, let $v_0^* = v_0$ if v is labeled by x_i and $v_0^* = v$ otherwise (similarly for v_1^*, w_0^*, w_1^*). The successors of (v, w) are (v_0^*, w_0^*) and (v_1^*, w_1^*). Finally, the nodes (v, w), where v and w are sinks, are sinks whose label is label$(v) \otimes$label(w). The correctness of this simple algorithm is obvious. With input a we simultaneously follow the activated paths in G_f and G_g. If in G_f an x_i-test is performed which is omitted in G_g, we "wait" in G_g (or vice versa). Finally, the operation $\otimes$ is applied to the labels of the sinks and, therefore, to $f(a)$ and $g(a)$. The construction is possible in time and space $O(|G_f||G_g|)$.

From a worst-case point of view this solution is optimal. At the end of this section, we present an example where the reduced π-OBDD G_h has size $\Omega(|G_f||G_g|)$. But in many cases, in particular in real-life applications, G_h will not be much larger than G_f and G_g. Hence, we look for an output-sensitive algorithm which is efficient with respect to $|G_f|$, $|G_g|$, and $|G_h|$. From the theory of DFAs, we know how to avoid the consideration of nodes which are not reachable. We do not go into the details of this approach, since we improve it by the inclusion of the reduction process.

We describe three modules of the algorithm before we present the algorithm itself.

(1) The module "terminal-case" checks whether the construction can be finished. A pair (v, w) of sinks is the terminal case of a sink labeled by $\text{label}(v) \otimes \text{label}(w)$. There may be more terminal cases. For example, if $\otimes =$ AND and one of the nodes is the 0-sink, the result is the 0-sink. In the case that f and g are represented in the same OBDD and h also will be represented in this OBDD, we may obtain the case that $v = w$. If $\otimes =$ AND, v is the result. If $\otimes =$ EXOR, the result is the 0-sink.

(2) The module "computed-table" is a data structure containing all pairs of nodes (v, w) such that the algorithm has been called for (v, w). In a DFS approach, we have to store the nodes already reached to avoid repeated computations. If we reach a pair again, we may return the result (the node) computed during the first visit to the pair.

(3) The module "unique-table" is a data structure containing all nodes of the OBDD as the triple (label, 1-successor, 0-successor). If a new node has w as 0-successor and 1-successor, the result of a search in the unique-table is w. If a new node is already contained in the unique-table, the result is the entry of the unique-table. Otherwise, a new node is created and is added to the unique-table and to the OBDD.

In the following algorithm, we identify nodes and the functions they represent. Hence, (f, g) denotes a pair of nodes of the given OBDDs and also a node of the new OBDD. Then $\text{label}(f, g)$ denotes the variable which is the label of this node.

Algorithm 3.3.5. Binary synthesis.

$\otimes(f, g)$;
if (terminal-case) **return** (result of terminal-case);
else if $((f, g) \in$ computed-table) **return** (computed-table(f, g));
    else $x_i := \text{label}(f, g)$;
        $t := \otimes(f_{|x_i=1}, g_{|x_i=1})$;
        $e := \otimes(f_{|x_i=0}, g_{|x_i=0})$;

$r :=$ find-or-add-unique-table (x_i, t, e);
insert-computed-table $((f, g), r)$;
return r.

We still have some freedom to implement the details. It will improve the efficiency of the algorithm if we have "many" terminal cases. Nevertheless, terminal-case and, possibly, the return of the result should take constant time. The data structures for the computed-table and unique-table have to support the operations search and insert. Using balanced search trees we only need linear space but an additional log-factor for the time. The maximal size of the computed-table is equal to the number of reachable nodes in $G_f \times G_g$ and the maximal size of the unique-table is the size of the resulting reduced OBDD. With hashing strategies it is possible to obtain linear space and expected linear time. Hence, we have obtained the following result.

Theorem 3.3.6. *The binary synthesis of G_f and G_g with respect to $\otimes \in B_2$ is possible in*

- *time and space $O(|G_f||G_g|)$,*
- *time $O(|G_h^*| \log |G_h^*|)$ and space $O(|G_h^*|)$ where G_h^* is the graph consisting of the nodes in the product graph G_h reachable from the node representing h, or*
- *expected time and space $O(|G_h^*|)$.*

Often, $|G_h^*|$ is much smaller than $|G_f||G_g|$.

The synthesis algorithm can easily be generalized to the synthesis of m functions where we have to consider nodes of the product of the m corresponding OBDDs. Replacement by functions is done by replacement by constants and a ternary synthesis for the operation ite. Quantification can be performed by replacement by constants and an OR- or AND-synthesis.

Many OBDD packages are available. They all are influenced by the implementation of Brace, Rudell, and Bryant (1990). The recent package of Somenzi (1998) has more features; in particular, it contains efficient heuristics for the variable-ordering problem (see Chapter 5).

One of the ideas implemented in the packages is to perform synthesis steps always as ternary ite synthesis. It is easy to see that each binary operation can be described as an ite operation, e.g., $x \wedge y = ite(x, y, 0)$, $x + y = ite(x, 1, y)$, and $x \oplus y = ite(x, \overline{y}, y)$. Then we do not need different terminal cases for the different operations. Some terminal cases are $ite(1, f, g) = f$, $ite(0, f, g) = g$, $ite(f, 1, 0) = f$, and $ite(f, g, g) = g$. The use of complemented edges increases the number of terminal cases. In order to improve the hit rate in the unique-table one may introduce standardized tuples, e.g., $ite(f, g, 0) = ite(g, f, 0)$. It is an exercise to prove that the simulation of a binary synthesis step by an ite synthesis does not increase the runtime.

Applications of OBDD packages are more often restricted by the available space than by the available time. We have seen that the unique-table is not larger than the resulting reduced π-OBDD for h, but the computed-table may be much larger. A solution to this dilemma is the use of a cache-based hash-table for the computed-table. For some (small) constant c, the number of entries with the same hash value h^* is restricted to c. If a further entry with hash value h^* is inserted, the oldest entry (first in first out (FIFO)) or the least recently used (LRU) entry is deleted. We cannot be sure that our computed-table is complete. This does not influence the correctness of our computations. We "only" repeat some computations, possibly, quite often. In the worst case, the synthesis algorithm needs exponential time. Hence, OBDD packages use a heuristic synthesis algorithm which is very successful in applications.

We conclude that operations on OBDDs can be performed very efficiently.

Finally, we show that the result of a synthesis of G_f and G_g can have the size $\Theta(|G_f||G_g|)$.

Theorem 3.3.7. *There are functions $f_n, g_n \in B_{n+2k}$, $n = 2^k$, such that for some variable ordering π the π-OBDD size of f_n and g_n is $\Theta(n)$ and the π-OBDD size of $f_n + g_n$ is $\Theta(n^2)$.*

Proof. Let the variable ordering π be given by $x_0, \ldots, x_{k-1}, y_0, \ldots, y_{k-1}, z_0, \ldots, z_{n-1}$. The function f_n is the multiplexer MUX on x and z, i.e., it interprets $(x_{k-1}, \ldots, x_0)$ as the binary number $|x|$ and outputs $z_{|x|}$. The function g_n is the multiplexer on y and z. The π-OBDD size of f_n (and also g_n) equals $2n+1$. We start with a complete binary tree with k levels labeled by $x_0, \ldots, x_{k-1}$. The number of nodes equals $n-1$. For each of the n edges leaving the x_{k-1}-level the value $|x|$ is known and it is sufficient to test $z_{|x|}$. Hence, additionally we have n z-nodes and 2 sinks. The resulting π-OBDD is reduced.

The upper bound on the π-OBDD size of $f_n + g_n$ follows, since $f_n + g_n$ is obtained by binary synthesis from f_n and g_n. For the lower bound, we consider the subfunctions obtained by replacing $x_0, \ldots, x_{k-1}$ and $y_0, \ldots, y_{k-1}$ with constants. We obtain the $\binom{n}{2} + n$ different functions $z_i + z_j$, $i \neq j$, and z_i, $1 \leq i \leq n$, which have to be represented at different nodes of the π-OBDD for $f_n + g_n$. □

3.4 Breadth-First Manipulation of OBDDs

The fundamental synthesis algorithm (Algorithm 3.3.5) constructs the new OBDD by a DFS approach. This leads to an efficient algorithm for several reasons. The recursion stack is bounded by the depth of the OBDD and, therefore, by the number of variables. The space for this stack is negligible. All nodes inserted in the unique-table belong to the reduced OBDD which will be computed. So there seems to be no need for an alternative approach. This is true only as long as the whole OBDD fits into the main memory and as long as we work with sequential computers.

It is well known that DFS algorithms cannot be parallelized efficiently. The following considerations are the basis of efficient parallel algorithms which will be discussed in Section 3.5. But the most severe limitation of OBDD applications is the size of the resulting OBDD. If it does not fit into the main memory, secondary memory has to be used. Then the number of page faults, i.e., the number of accesses to the secondary memory, has a major influence on the runtime of the algorithms. During a DFS traversal of an OBDD, the pattern of node accesses is quite irregular, since nodes are accessed more than once. To overcome this difficulty, breadth-first search (BFS) traversals are useful. Then all x_i-nodes can be treated as a block which, hopefully, fits into the main memory.

Evaluation and SAT are so easy that there is no need for a BFS algorithm. The algorithms for SAT-COUNT, replacement by constants, and redundancy test of Theorem 3.3.1 can easily be implemented as BFS algorithms. Since replacement by functions and quantification are mainly synthesis problems, we are left with the problem of describing a BFS synthesis algorithm.

Ochi, Ishiura, and Yajima (1991) and Ochi, Yasuoka, and Yajima (1993) present BFS synthesis algorithms for quasi-reduced π-OBDDs (see Section 3.2) with the simple extension that edges may lead directly to the sinks. If $\pi = id$, all edges from the x_i-level lead to the x_{i+1}-level or to sinks and there is no "random" access to other levels. This approach is successful if it does not increase the OBDD size too much. Ashar and Cheong (1994) have improved the approach to work on reduced π-OBDDs. We first describe the algorithm for quasi-reduced π-OBDDs and then the improvements of Ashar and Cheong (1994).

The reduction rules are only applicable to nodes whose successor nodes are already computed. Moreover, the application of these rules in the bottom part can lead to new reduction possibilities in the upper part. Hence, we have no idea how to integrate the reduction process into a BFS synthesis algorithm. This implies that the DFS approach is much more efficient if the OBDD fits into the main memory. All BFS synthesis algorithms for $\otimes(f, g)$ first compute the graph G_h^* consisting of the nodes in the product graph G_h^* reachable from the node representing h, and in a second phase G_h^* is reduced. Then we cannot gain from a cache-based computed-table. We describe the two phases called expansion phase and reduction phase.

Algorithm 3.4.1. Expansion phase of a BFS synthesis algorithm.

Input: A quasi-reduced *id*-OBDD $G_{f,g}$ for $f, g \in B_n$ and $\otimes \in B_2$.

Output: An *id*-OBDD representing f, g, and $f \otimes g$.

The levels $1, \ldots, n$ are treated sequentially. The list L_i contains the requests on level i. In particular, L_1 contains the initial request (f, g). The requests of L_i are treated sequentially. On request (h, h') on level i, the following is done:

(1) If the request (h, h') is a terminal case, the edge (or pointer) to (h, h') leads to the appropriate node.

(2) If otherwise the result of the request (h, h') can be found in the computed-table, the edge to (h, h') leads to the node found in the computed-table.

(3) Otherwise, a new x_i-node is created and the edge to (h, h') leads to this new node. The c-edge leaving this node, $c \in \{0, 1\}$, points to the new request $(h_{|x_i=c}, h'_{|x_i=c})$, which is inserted into L_{i+1}. The request (h, h') is inserted in the computed-table.

Depending on the data structure which we choose for the computed-table, we obtain the same time behavior as for the DFS approach in Theorem 3.3.6. But a cache-based table is not useful.

Algorithm 3.4.2. Reduction phase of a BFS synthesis algorithm.

Input: An *id*-OBDD G whose edges leaving x_i-nodes lead to x_{i+1}-nodes or to sinks. Without loss of generality, G contains only one 0-sink and one 1-sink.

Output: The extended quasi-reduced *id*-OBDD G' (edges may lead directly to the sinks) representing the same function as G.

The algorithm works levelwise bottom-up. For each node v on level i, the following is done.

(1) If an edge leaving v points to a "removed" node w, redirect the edge to the node which replaces w.

(2) If the edges leaving v point to the same sink w, mark v as "removed" and store the information that replaces v.

(3) If the unique-table of this level contains a node w with the same ordered pair of successors, mark v as "removed" and store the information that w replaces v.

(4) If v is not marked as "removed," enter v into the unique-table.

Algorithm 3.4.2 can be performed in the same time as Algorithm 3.4.1 (see Theorem 3.3.6 for the time bounds).

If the size of the quasi-reduced π-OBDD is much larger than the size of the corresponding reduced π-OBDD, the decrease of the runtime by avoiding page faults is smaller than its increase by considering more nodes. When trying to generalize Algorithms 3.4.1 and 3.4.2 to reduced OBDDs, one is faced with the following problems.

(1) In order to compute the label of a node for a new request (h, h'), one has to fetch the nodes for h and h' to find the corresponding labels.

(2) In order to look for requests already considered, one needs a random access to different "levels" of the unique-table.

(3) In order to reduce a level, one has to access nodes on different levels, since one has to look at how pointers to "removed" nodes have to be replaced.

Ashar and Cheong (1994) described strategies to overcome these difficulties. For the first problem, they reserved blocks for nodes of the same label. Hence, the label can easily be computed from the address without fetching the nodes. The other two problems can be solved in a similar way which we describe for the second problem. In a first pass through the request list, only those requests are processed which lead to new requests on the next level. The other requests are stored in lists with respect to the levels of the resulting new requests. Since a request creates two new requests, a request may be contained in up to two of these lists. The check whether "new" requests are contained in the unique-table then can be performed levelwise. A similar approach works for the reduction phase.

3.5 Parallel Computers and OBDDs

It is a natural question to ask whether we can reduce the runtime of our algorithms with parallel computers. This is not possible for DFS algorithms but the BFS algorithms of the preceding section can easily be parallelized. Ochi, Ishiura, and Yajima (1991) described the implementation of BFS algorithms for vector machines. Kimura and Clarke (1990) considered the DFA approach (see Section 3.2) and have designed parallel BFS algorithms with a good speedup for a small number of processors. These results are satisfying for practical purposes. From a theoretical viewpoint, the approach with BFS algorithms is limited, since it requires two phases where the OBDDs are considered levelwise. Hence, on OBDDs with n variables the runtime of the algorithms grows at least linearly with n.

In the theory of parallel algorithms, one assumes the existence of a polynomial number of processors (of a PRAM $\widehat{=}$ parallel random access machine) and looks for polylogarithmic runtimes with respect to the input size in order to obtain NC algorithms.

Sieling and Wegener (1993a) described NC-algorithms. The following bounds hold for CREW PRAMs. Only for SAT, equivalence test, redundancy test, and reduction is a CRCW COMMON PRAM used. Moreover, the algorithms which construct new OBDDs do not include the reduction.

Evaluation: time $O(\log n)$ with $|G_f|$ processors.

Binary synthesis: time $O(1)$ with $|G_f| \cdot |G_g|$ processors.

SAT: time $O(\log n)$ with $|G_f|^3$ processors.

SAT-COUNT: time $O((\log|G_f|)(\log n))$ with $\lceil |G_f|^3/\log|G_f| \rceil$ processors.

Equivalence test: time $O(\log n)$ with $|G_f|^3 \cdot |G_g|^3$ processors.

Replacement by constants: time $O(1)$ with $|G_f|$ processors.

Replacement by functions: time $O(1)$ with $|G_f|^2|G_g|$ processors.

Redundancy test: time $O(\log|G_f|)$ with $|G_f|^6$ processors.

Quantification: time $O(1)$ with $|G_f|^2$ processors.

Reduction: time $O(\log|G_f|)$ with $|G_f|^6$ processors.

Only the algorithm for SAT-COUNT needs more than logarithmic time. All algorithms are based on standard tricks from the design of efficient PRAM algorithms and, therefore, are not presented here. The reason for the huge number of processors for the operations SAT, SAT-COUNT, equivalence test, redundancy test, and minimization is the same. It is the so-called transitive closure bottleneck for PRAMs, i.e., the difficulty of deciding in logarithmic time whether a directed graph contains a path from v to w. Equivalence test, redundancy test, and minimization problems have to be solved on product graphs of size $|G_f||G_g|$ or $|G_f|^2$.

Our conclusion is that the operations on OBDDs can be performed efficiently on parallel computers from a practical point of view using BFS algorithms and by NC algorithms with a large but polynomially bounded number of processors.

3.6 Incompletely Specified Boolean Functions

A Boolean function $f \in B_n$ is completely specified, i.e., $f(a)$ is well defined for each $a \in \{0,1\}^n$. In many applications, one has to deal with *incompletely specified* Boolean functions $f\colon \{0,1\}^n \rightarrow \{0,1,*\}$ where $*$ means "don't care." Then we have the freedom to work with an arbitrary extension of f.

Definition 3.6.1. A Boolean function $g \in B_n$ is an *extension* of the incompletely specified Boolean function f if $g(a) = f(a)$ for all $a \in \{0,1\}^n$ where $f(a) \in \{0,1\}$.

Incompletely specified functions occur, e.g., in the following situations. If the inputs of a circuit are the outputs of other circuits, in most situations not all input combinations are possible. In Section 1.3, we considered the verification problem for sequential networks. If we look for the states reachable in $t+1$ but not in t steps, we may search from any set of states R such that $R' \subseteq R \subseteq R''$, where R' contains all states reachable in t but not in $t-1$ steps and R'' contains all states reachable in t steps. Mailhot and de Micheli (1993) describe how algorithms for technology mapping take advantage of don't care conditions. Finally, "Boolean division" leads to incompletely specified functions. The task is to find for given functions f and g a quotient q and a remainder r such that $f = (g \wedge q) + r$ and q and r are as simple (measured by the OBDD size) as possible; see Stanion and Sechen (1994) for an OBDD-based division algorithm.

In the following, we discuss representations of incompletely specified functions by π-OBDDs and then we consider the problem of finding an extension of minimal or almost minimal π-OBDD size.

Let f be an incompletely specified function. By $f_{\text{on}}, f_{\text{off}}$, and f_{dc} we denote the following completely specified functions: $f_{\text{on}}(a) = 1$ iff $f(a) = 1$, $f_{\text{off}}(a) = 1$ iff $f(a) = 0$, $f_{\text{dc}}(a) = 1$ iff $f(a) = *$.

(1) The most natural representation of f is the representation by a generalized OBDD with three types of sinks labeled by 0, 1, and $*$. This representation is called the $\{0, 1, *\}$-representation.

(2) Since on each input a exactly one of the three functions $f_{\text{on}}, f_{\text{off}}$, and f_{dc} computes 1, it is sufficient to represent f by a pair (g, h) of two of the three functions.

(3) A pair (g, h) can be composed by $zg + \overline{z}h$ to one function where z is a new variable. This is called the z-representation of (g, h).

(4) The (F, c)-representation is the representation by an extension F of f and the function $c := \overline{f}_{\text{dc}}$ representing the care set.

The $\{0, 1, *\}$-representation implicitly contains a representation of each of the functions $f_{\text{on}}, f_{\text{off}}$, and f_{dc}. They can be obtained by a relabeling of some sinks, more precisely: $f_{\text{on}} : * \to 0$, $f_{\text{off}} : 0 \leftrightarrow 1, * \to 0$, $f_{\text{dc}} : 1 \to 0, * \to 1$. The functions g and h are subfunctions of the z-representation of (g, h). If z is the top variable of the variable ordering, we obtain the z-representation of (g, h) from a representation of (g, h) by introducing one z-node whose outgoing edges lead to g and h, respectively. If f_{on} and f_{dc} are given, we obtain $f_{\text{off}} = \overline{(f_{\text{on}} + f_{\text{dc}})}$ by a synthesis step. Since (see the exercises) Theorem 3.3.7 can be generalized to functions f and g where $f \wedge g = 0$, the π-OBDD size of f_{off} can be as large as the product of the π-OBDD sizes of f_{on} and f_{dc}. Therefore, the $\{0, 1, *\}$-representation may have a size which is of the order of the product of the sizes of, e.g., f_{on} and f_{dc}. The size cannot be larger, since we obtain the

$\{0, 1, *\}$-representation from, e.g., the $(f_{\rm on}, f_{\rm dc})$-representation by a generalized binary synthesis step. The appropriate operator $\otimes$: $\{0,1\}^2 \to \{0, 1, *\}$ is defined by $\otimes(0,0) = 0$, $\otimes(0,1) = *$, and $\otimes(1,0) = 1$. The combination $(1,1)$ is not possible and, therefore, $\otimes(1,1)$ can be defined arbitrarily.

From an (F, c)-representation we may obtain $f_{\rm on} = F \wedge c$, $f_{\rm off} = \overline{F} \wedge c$, and $f_{dc} = \overline{c}$ by synthesis steps. An extension g may have a very small representation even if the other representations are of exponential size. If $f_{\rm dc}$ is a complicated function, $f_{\rm on} = \overline{f}_{\rm dc}$, and $f_{\rm off} = 0$, the constant 1 is a simple extension.

Hence, it is an important problem to find for a given variable ordering π and an incompletely specified function f an extension with minimal or almost minimal π-OBDD size. Let us call this the minimization problem for incompletely specified functions.

Takenaga and Yajima (1993), Hirata, Shimozono, and Shinohara (1996), and Sauerhoff and Wegener (1996) proved that the minimization problem is NP-hard. All papers reduce the graph-coloring problem to the minimization problem but the last paper leads to a better nonapproximability result. An algorithm A computes an $\alpha(s)$-approximation for our minimization problem if the ratio of the OBDD size of the extension computed by A and the OBDD size of an optimal extension is bounded by $\alpha(s)$. Here s is the size of the OBDD representation of the considered incompletely specified function. The following result holds for all considered representations and combines the reduction of Sauerhoff and Wegener (1996) and the nonapproximability result on the graph-coloring problem obtained by Feige and Kilian (1996) within the probabilistically checkable proofs (PCP) theory (see Mayr, Prömel, and Steger (1998) for an introduction). The complexity class ZPP contains all decision problems which are solvable by error-free probabilistic polynomial-time algorithms which for each input may give the answer "don't know" with probability at most 1/2.

Theorem 3.6.2. *For any $\varepsilon > 0$ it is impossible to obtain an $O(s^{1/3-\varepsilon})$-approximation for the minimization problem for incompletely specified functions in polynomial time unless ZPP = NP.*

The assumption NP $\neq$ ZPP is stronger than the usual assumption NP $\neq$ P (since, obviously, P $\subseteq$ ZPP) but it is nevertheless a well-established complexity theoretical hypothesis. Because of Theorem 3.6.2, one has to attack the minimization problem for incompletely specified functions by heuristic methods.

Since the OBDD size of the constructed extension is the minimization criterion, the heuristics try to match subfunctions represented in the given OBDD, i.e., one tries to represent different incompletely specified subfunctions by a common extension. The following features have to be taken into account. The check whether we match two subfunctions should be efficient. Moreover, the common extension should not need too many new OBDD nodes. By matching subfunctions we are forced to replace don't care values by fixed values. If we spend too many don't care values, this decreases the chance of matchings in the future.

Lemma 3.6.3. *Let (F_1, c_1) and (F_2, c_2) (or $(f^1_{\text{on}}, f^1_{\text{off}})$ and $(f^2_{\text{on}}, f^2_{\text{off}})$) be two incompletely specified functions. The functions can be matched iff $(F_1 \oplus F_2) \wedge c_1 \wedge c_2 = 0$ ($f^1_{\text{on}} \wedge f^2_{\text{off}} = 0$ and $f^1_{\text{off}} \wedge f^2_{\text{on}} = 0$). The common extension with the largest don't care set is given by $F_1c_1 + F_2c_2$ with the care set $c_1 + c_2$ $((f^1_{\text{on}} + f^2_{\text{on}}, f^1_{\text{off}} + f^2_{\text{off}}))$.*

Proof. The functions cannot be matched iff there exists an input a in the care set of both functions where the functions differ. This is expressed by the criteria given in the lemma. The minimal care set of an extension is $c_1 + c_2$, where an extension has to take the same value from $\{0, 1\}$ as one or both of the given functions. □

A match as described in Lemma 3.6.3 is called a *two-sided match* (tsm). This criterion has been applied by Chang, Cheng, and Marek-Sadowska (1994) using the z-representation of the pairs $(f_{\text{on}}, f_{\text{off}})$ and by Shiple et al. (1994) using the (F, c)-representation type. The tsm approach is a greedy one. Two functions are matched iff this is possible. With some synthesis steps it is possible to check whether the criterion is met. Considering more than two functions, we run into the problem that the tsm relation is not an equivalence relation. Consider, e.g., the following three functions on two variables: $(f^1_{\text{on}}, f^1_{\text{off}}) = (\overline{x}_1\overline{x}_2, x_1\overline{x}_2)$, $(f^2_{\text{on}}, f^2_{\text{off}}) = (\overline{x}_1 x_2, x_1 x_2)$, $(f^3_{\text{on}}, f^3_{\text{off}}) = (\overline{x}_1\overline{x}_2, \overline{x}_1 x_2)$. Then, by case inspection, (f_1 tsm f_2) and (f_1 tsm f_3) but (f_2 not tsm f_3), since $f^2_{\text{on}} \wedge f^3_{\text{off}} = \overline{x}_1 x_2$.

Considering m incompletely specified functions $f_1, \ldots, f_m$, we have to check for all pairs whether they can be matched by the tsm criterion. This leads to an undirected graph G whose edges describe the matchable pairs of functions. A set of functions has a common extension iff the corresponding vertices build a clique in G. Contradictory values are checked by the pairwise checks. In order to find the smallest number of groups of matchable sets of functions, we have to solve another NP-hard optimization problem, the problem of finding a minimal clique cover of a graph, i.e., covering the vertices of G with the minimal number of cliques. This problem is also solved by a greedy approach. Take a vertex v with maximal degree, since vertices with large degree seem to have a better chance of lying in a large clique. Then choose a vertex w with maximal degree among the neighbors of v; afterwards we consider the common neighbors of v and w, and so on. Shiple et al. (1994) refine this approach with the introduction of a heuristic measure for the "distance" of functions, and they first add the "closest" neighbors.

We still have to describe which functions we check for possible tsm. Chang et al. (1994) use the z-representation. If z is in the variable ordering between x_i and x_{i+1}, the functions represented by the z-nodes are the subfunctions obtained by replacing $x_1, \ldots, x_i$ by constants. They use an approach where z starts at the top of the variable ordering. After all possible tsm's are done for the functions represented by z-nodes with z at position i of the variable ordering, the variable z swaps one level down in the ordering and the next matching procedure is started. It is discussed in Section 5 how such changes in the variable ordering are performed efficiently. Shiple et al. (1994) work with a similar approach. In the (F, c)-representation, they traverse for increasing $i \in \{1, \ldots, n\}$ the OBDD for (F, c) with a simultaneous DFS approach. The recursion is terminated by finding pairs of nodes both labeled by variables x_j where $j \geq i$.

If the number of considered functions becomes too large to perform all pairwise matching tests, we consider only subgroups. The approach of Shiple et al. (1997) ensures some "locality" of the considered subgroups because of the DFS approach.

Another possibility is to match only special pairs of functions. We start a simultaneous DFS traversal at the nodes representing F and c (as described in the synthesis algorithm in Section 3.3). If we reach the node pair (v, w), we check whether we can match the functions represented at the 0- and 1-successors of (v, w) in the product graph. If a match is not possible, we recursively apply our approach to the 0-successor and the 1-successors. Afterwards, we create an x_i-node (where x_i is the label of (v, w) in the product graph) pointing to the results of the recursive calls.

Since the check whether the tsm criterion is fulfilled is time-consuming and since the care set of both functions is increased, Shiple et al. (1994) also considered stronger matching criteria.

If $c_1 \leq c_2$ and $(F_1 \oplus F_2) \wedge c_2 = 0$, then (F_1, c_1) and (F_2, c_2) can be matched. The new care set equals $c_1 + c_2 = c_2$ and we may use F_2 as a common extension. Such a match is called a *one-sided match* (osm). The osm relation obviously is not symmetric but it is transitive. For a set of functions $f_1, \ldots, f_m$, we obtain a directed graph where the edge (i, j) describes that $(f_i \text{ osm } f_j)$. If the functions $f_1, \ldots, f_m$ are different, the graph is acyclic. It is easy to find the sinks of this graph and each function represented by v can be matched with a function represented by a sink reachable from v. Since the result of the matching is the function (F_2, c_2) represented by the sink, we can replace $f_1, \ldots, f_m$ by the functions represented by the sinks. The special case where $c_1 = 0$ has already been considered by Coudert, Berthet, and Madre (1989a,b) under the notion *constrain*.

Two generalizations may be useful. Using complemented edges, we obviously may try to match F_1 and $\overline{F}_2$. Moreover, it may happen that $F_1 = F_2 = F$ but $c_1 \neq c_2$ if (F_1, c_1) and (F_2, c_2) are the successors of (F, c). Then the tsm criterion is fulfilled. If we work with the osm criterion, we may not match (F_1, c_1) and (F_2, c_2) and, therefore, create a new node labeled x_i, although the function F does not essentially depend on x_i. Shiple et al. (1994) add the option that in this situation a matching with result $(F, c_1 + c_2)$ is allowed. This option added to *constrain* is denoted by Coudert, Berthet, and Mache (1989a, b) as *restrict*.

All these different heuristics have their benefits in certain situations. But they also may lead to an increase in the OBDD size. Using the tsm criterion, there are examples where the increase is exponential due to synthesis steps in the computation of the extension. For the osm criterion, it can be proved that the increase is polynomially bounded (see exercises).

3.7 Exercises and Open Problems

3.1.E Prove the generalization of Theorem 3.1.4 to functions $f \in B_{n,m}$.

3.2.E Describe a linear-time algorithm to obtain from an OBDD with unrestricted complemented edges an equivalent OBDD without 1-sink and without complemented 0-edges.

3.3.E Find a Boolean function $f \in B_n$ essentially depending on all its variables and some variable ordering π such that $\pi\text{-OBDD}(f) = 2 \cdot \pi\text{-OBDD}_{\text{ce}}(f) - 1$.

3.4.M Describe a class of Boolean functions $f \in B_n$ such that $\pi\text{-OBDD}(f) = \pi\text{-OBDD}_{\text{ce}}(f) + 1$.

3.5.E Prove Theorem 3.2.2.

3.6.M Prove Theorem 3.2.3 and describe an efficient algorithm to obtain a quasi-reduced π-OBDD from a reduced π-OBDD.

3.7.E Prove by an example that the bound of Theorem 3.2.3 is optimal.

3.8.O Is it possible to design a synthesis algorithm which works in time $O(|G_f| + |G_g| + |G_h|) + o(|G_h^*|)$?

3.9.M Describe a synthesis algorithm for π-OBDDs with complemented edges.

3.10.E Generalize Theorem 3.3.7 to the disjunction of m functions.

3.11.D Generalize Theorem 3.3.7 to the case $f \wedge g = 0$.

3.12.M If a binary synthesis is performed as a ternary ite synthesis, prove that this is as efficient as the binary synthesis algorithm.

3.13.E Describe an efficient π-OBDD algorithm to check whether $f \leq g$.

3.14.E Design a parallel CREW PRAM algorithm which solves the evaluation problem for $f \in B_n$ with $O(|G_f|)$ processors in time $O(\log n)$.

3.15.M Design a parallel CRCW PRAM algorithm which solves the SAT problem for $f \in B_n$ with $O(|G_f|^3)$ processors in time $O(\log n)$.

3.16.E Design a parallel CREW PRAM algorithm which solves the synthesis problem (without reduction) with $O(|G_f||G_g|)$ processors in time $O(1)$.

3.17.D Design a parallel CRCW PRAM algorithm which solves the reduction problem with a polynomial number of processors in time $O(\log |G_f|)$.

3.18.D Prove that the size of the $\{0, 1, *\}$-representation of an incompletely specified function f is equal to the number of reachable nodes in the product graph of OBDDs for f_{on} and f_{dc}.

3.19.D Give an example where the tsm criterion applied levelwise leads to an exponential blow-up of the size.

3.20.D Prove that the blow-up caused by the levelwise application of the osm criterion is polynomially bounded.

3.21.M Design a polynomial-time algorithm to check whether the function f represented by an OBDD is monotone, i.e., $a_i \leq b_i$ for all i implies $f(a) \leq f(b)$.

Chapter 4

The OBDD Size of Selected Functions

4.1 OBDDs and Communication Complexity

In this chapter, the OBDD size of selected functions is estimated with respect to the best-known variable ordering. For some functions, we present a variable ordering π, give an upper bound for the corresponding π-OBDD size, and prove that no other variable orderings can lead to substantially better OBDDs. For some other functions, we prove that all variable orderings lead to OBDDs of exponential size. In any case, we need methods to estimate the size of the reduced π-OBDD representing f. Theorem 3.1.4 (and its generalization to multi-output functions) gives a structural description of the reduced π-OBDD and the number of its nodes. Although this result is sufficient for the purpose of this chapter, we present an alternative approach for the estimation of OBDD sizes. This approach is based on the theory of *communication complexity*.

The basic model in communication complexity is the following one. *Alice* and *Bob* want to cooperate for the evaluation of a Boolean function $f\colon \{0,1\}^m \times \{0,1\}^n \to \{0,1\}$. Alice knows the first part $a = (a_1, \dots, a_m)$ and Bob the second part $b = (b_1, \dots, b_n)$ of the input. Before getting the specific partial inputs, they reach an agreement on a *communication protocol.* Such a protocol decides who starts the communication and how the communication is organized. If Alice starts the communication, she sends the first message $m_1(a)$ depending on her input. The set of all possible messages $m_1(a)$, $a \in \{0,1\}^m$, has to be prefix-free to ensure that Bob can recognize the end of the message. Then he answers with a message $m_2(b, m_1(a))$ depending on his current knowledge, namely b and $m_1(a)$. Then Alice answers and so on until the aim of the communication is fulfilled, i.e., Alice or Bob or both know $f(a, b)$ and both know that the aim is fulfilled. The length of a protocol P for f on input (a, b) is the total length of all messages sent by Alice and Bob. Our aim is to minimize the worst-case or average-case length of a protocol for f. Many generalizations are possible: nondeterministic protocols, probabilistic protocols, and multiparty protocols with different models for the knowledge of the players. We also may restrict the number of communication rounds. The monographs of Hromkovič (1997) and Kushilevitz and Nisan (1997) present an introduction to communication complexity and the state-of-the-art knowledge in this area. The simplicity of the model and the reduction to one central aspect of computation make communication complexity a theory with many applications.

What is the communication complexity aspect of BDDs and BPs? For the general case, we get no relevant results. But in many situations a sequence s of variables is given. This can be a variable ordering π but also a sequence where certain variables appear more than once. An s-oblivious BDD G (see Chapter 7) has s levels, where the ith level is labeled with the ith variable from s. Edges only may lead to later levels. Then Alice is the owner of those levels labeled by any of the variables given to her and Bob is the owner of the other levels. We separate G into layers consisting of consecutive levels owned by one of the players Alice and Bob. If the number of layers is k, we get a protocol with k rounds and worst-case length $k\lceil \log |G| \rceil$. Without loss of generality, let Alice be the owner of the first layer. She starts to evaluate G on the given input until she reaches a node labeled by a variable whose value she does not know. Her first message is the number of this node. Afterwards, Bob continues the evaluation up to a node whose label is unknown to him. Altogether, it is sufficient to send k node numbers and, finally, Alice and Bob know the number of the sink reached by the activated path. An s-BDD G of small size with a small number of layers leads to a protocol with short messages. Therefore, lower bounds on the communication complexity lead to lower bounds for the appropriate BDD model. We refer to these general remarks in later chapters.

Here we consider the very special case of π-OBDDs and OBDDs. In the *one-way communication* model, only Alice is allowed to send a single message and then Bob has to be able to compute the value of the function. We describe $f: \{0,1\}^m \times \{0,1\}^n \to \{0,1\}$ by its *communication matrix* of size $2^m \times 2^n$. The matrix entry at position (a, b), $a \in \{0,1\}^m$, $b \in \{0,1\}^n$, is $f(a, b)$. Hence, the communication matrix is nothing else but the value table in a different form. For $m = 3$ and $n = 2$, we consider the communication matrix of the majority function (see Fig. 4.1.1).

If the communication matrix has r different rows ($r = 4$ in the example), Alice can encode the "row type" of her input with $\lceil \log r \rceil$ bits and can send this message to Bob. Then Bob decodes the message. He does not know the row number, but he knows the contents of the row vector. His input describes at which position of the row vector he finds $f(a, b)$. The number r of different row types is equal to the number of different subfunctions if we replace the first m

MAJ_5	00	01	10	11
000	0	0	0	0
001	0	0	0	1
010	0	0	0	1
011	0	1	1	1
100	0	0	0	1
101	0	1	1	1
110	0	1	1	1
111	1	1	1	1

Figure 4.1.1: *The communication matrix for the majority function* MAJ_5.

variables by constants. We summarize the relations between the size of OBDDs and the one-way communication complexity.

Theorem 4.1.1. *Let* $f\colon \{0,1\}^N \to \{0,1\}$ *and* $\pi = id$.

(*i*) *If the* π*-OBDD size of* f *equals* s *and Alice gets the first* m *input bits, the one-way communication complexity of* f *is bounded above by* $\lceil \log s \rceil$.

(*ii*) *The* π*-OBDD size of* f *is at least* 2^{r-1} *if the one-way communication complexity of* f *is bounded below by* r *in a situation where Alice gets some initial block of the input bits.*

(*iii*) *The OBDD size of* f *is at least* 2^{r-1} *if the condition of* (*ii*) *holds for each variable ordering.*

4.2 Read-Once Projections

One of the most useful concepts in complexity theory is the concept of *reductions* between problems. In our case, problems are sequences $f = (f_n)$ of Boolean functions. The following requirements should be fulfilled by a reducibility concept "$\leq$" which is appropriate for the complexity measure OBDD size:

- $f \leq f$ (reflexivity).
- $f \leq g$ and $g \leq h \Rightarrow f \leq h$ (transitivity).
- $f \leq g \Rightarrow$ the OBDD size of $f = (f_n)$ is bounded by a polynomial in the OBDD size of $g = (g_n)$.

OBDDs are very restricted BDDs and, therefore, as we shall see by the lower bounds proved in this chapter, a quite weak computation model. This implies that we have to look for an appropriately restricted reduction concept which is still strong enough to reduce, e.g., arithmetic functions to other arithmetic functions. The concept of *constant-depth reductions* has been applied successfully to arithmetic functions by Chandra, Stockmeyer, and Vishkin (1984). The even more restricted concept of *projections* has been investigated by Skyum and Valiant (1985), among others.

Definition 4.2.1. The sequence $f = (f_n)$ of Boolean functions is a (polynomial) *projection* of $g = (g_n)$, $f \leq_{\text{proj}} g$, if $f_n(x_1, \ldots, x_n) = g_{p(n)}(y_1, \ldots, y_{p(n)})$ for some polynomial p and $y_j \in \{0, 1, x_1, \overline{x}_1, \ldots, x_n, \overline{x}_n\}$. The number of j such that $y_j \in \{x_i, \overline{x}_i\}$ is called the multiplicity of x_i.

The concept of projections is defined for one-output functions. For multi-output functions, we expect to find every output bit of f_n at a specific position of $g_{p(n)}$.

In Section 4.12, we discuss functions $f = (f_n)$ and $g = (g_n)$ such that $f \leq_{\text{proj}} g$, the multiplicity of each variable is bounded by two, the OBDD size of g is linear, but the OBDD size of f grows exponentially. Only if the multiplicity is bounded by one do we get a reducibility concept (introduced and investigated by Bollig and Wegener (1996a)) with good properties.

Definition 4.2.2. A projection is called a *read-once projection*, $f \leq_{\text{rop}} g$, if the multiplicity of each variable is bounded by one.

The results in Section 4.6 are based on read-once projections. Reflexivity and transitivity are obvious for read-once projections. Moreover, if $f \leq_{\text{rop}} g$, we obtain a representation for f_n by applying the operation replacement by constants to a representation for $g_{p(n)}$ (where p is the polynomial in the definition of projections) and a renaming of the remaining variables. For all types of representations where this can be done in polynomial time (OBDDs, BPs, DTs, FBDDs, k-OBDDs, k-IBDDs, and their nondeterministic or probabilistic counterparts), read-once projections fulfill the required properties.

Lemma 4.2.3.

(*i*) *The relation* $\leq_{\text{rop}}$ *is reflexive and transitive.*

(*ii*) *If* $f \leq_{\text{rop}} g$, *the OBDD size of* $g = (g_n)$ *is bounded by the polynomial* q, *and the read-once projection uses the polynomial* p, *then the OBDD size of* $f = (f_n)$ *is bounded by the polynomial* $q \circ p$.

We add two results to emphasize that read-once projections lead to meaningful results.

Definition 4.2.4. The *circuit value problem* (CVP) works on inputs (C, a), where C is some standard encoding of a Boolean circuit on n inputs and $a \in \{0,1\}^n$. The output is $C(a)$, the result of the evaluation problem for the circuit C on the input a.

Proposition 4.2.5. *If $f = (f_n)$ has polynomial circuit size (i.e., $f \in$ P/poly), then $f \leq_{\text{rop}}$ CVP.*

Proof. The design of the read-once projection is easy. The circuit part in the input of CVP is replaced by the description of a polynomial-size circuit for $f = (f_n)$ and the input part remains variable. □

Since it is well known that CVP itself has polynomial circuit size, CVP is complete for P/poly (the class of sequences of Boolean functions with polynomial circuit size) even with respect to read-once projections. We prove an analogous result for the class P-OBDD (see Definition 1.3.1). We have to be careful with the exact definition of the complete problem. It should be representable in polynomial size for OBDDs with a fixed ordering. We run into difficulties if the jth OBDD node sometimes belongs to the x_i-level and sometimes to the x_k-level. Therefore, we consider only OBDDs on n variables with exactly n nodes on each level.

Definition 4.2.6. The OBDD *value problem* (OBDDVP) is defined on inputs (G, a) such that for some n the graph G is an *id*-OBDD on $x_1, \ldots, x_n$ with n^2 inner nodes and two sinks. The nodes $(i-1)n+1, \ldots, in$ are x_i-nodes, node n^2+1 is the 0-sink, node n^2+2 is the 1-sink, and node 1 is the source where the evaluation on the input $a \in \{0,1\}^n$ starts. An inner node of G is described by the numbers of its 0- and 1-successors. The output is the result of the evaluation problem on the OBDD G and the input a.

Theorem 4.2.7. *OBDDVP is P-OBDD-complete with respect to read-once projections.*

Proof. Let $f = (f_n) \in$ P-OBDD. Then f_n can be represented by a π_n-OBDD G'_n of polynomial size $p(n)$. Without loss of generality $p(n) \geq n$. Let G''_n be obtained from G'_n by adding dummy OBDD nodes on each level and $p(n) - n$ dummy levels at the bottom such that each of the $p(n)$ levels contains $p(n)$ nodes. Then we consider OBDDVP for the parameter $p(n)$. In order to compute $f_n(a)$, we can apply OBDDVP to the input consisting of G''_n and $a' = (a_1, \ldots, a_n, 0, \ldots, 0)$. This proves $f \leq_{\text{rop}}$ OBDDVP.

We still have to prove that OBDDVP $\in$ P-OBDD. We use the following variable ordering for the parameter n: x_1, encoding of node 1, ..., encoding of node n, x_2, encoding of node $n+1$, ..., encoding of node $2n$, x_3, and so on. Since the number of nodes equals n^2+2, the encoding of a node consists of $N := 2\lceil\log(n^2+2)\rceil$ Boolean variables which may be ordered arbitrarily. We design a gadget $G(v)$ for an x_i-node v. We start with an x_i-node, the source of $G(v)$. The 0-edge points to a complete binary tree where the $N/2$ variables encoding the 0-successor of v are tested, similarly for the 1-edge. If a leaf of this tree represents an inner node w, this leaf is replaced with the source of $G(w)$. If it represents a sink, it is replaced with the corresponding sink. These connections between the gadgets respect the given variable ordering, since label$(w) >$ label(v) in the situation described above. The resulting OBDD has size $O(n^4)$, since each gadget has size $O(n^2)$. At the source, it represents OBDDVP for the parameter n. □

The BP value problem (BPVP) can be defined in a similar way, leading to a P-BP-complete problem with respect to read-once projections (see exercises). One might think that similar results hold for all restricted types of BPs. This is not the case. Bollig and Wegener (1998b) have proved that the classes P-DT and P-FBDD do not have complete problems with respect to read-once projections. These results are interesting, since not many negative results on the existence of complete problems for natural complexity classes and reducibility concepts are known.

4.3 Storage Access

Now we investigate the OBDD size of concrete functions. We start with two functions which are abstractions of the direct and indirect access to storage.

Definition 4.3.1. The *direct storage access* function DSA_n, where $n = 2^k$, is defined on $n + k$ variables $x_0, \ldots, x_{k-1}, y_0, \ldots, y_{n-1}$. $\text{DSA}_n(x, y)$ equals $y_{|x|}$, where $|x|$ is the number whose binary representation equals x.

The function DSA often is called the *multiplexer* MUX. It has the property that control variables (the address variables x) and data variables (the variables y describing the contents of the storage) are separated, i.e., no variable is a control *and* a data variable. Then we may apply the rule "control variables first."

Theorem 4.3.2. *DSA_n (or MUX_n) has an OBDD size of* $2n+1$ *for all variable orderings where the control variables are tested first.*

Proof. The control variables are tested in a complete binary tree with $n - 1$ inner nodes. Each of the n leaves is replaced by a test of the corresponding data variable which decides about the output. We have to add 2 for the sinks. □

The ISA function (see Definition 2.2.5) has a different flavor. The y-variables are control variables, but each x-variable acts as a control variable in the computation of $\alpha(x, y)$ and as a data variable afterwards. Hence, we want to test each x-variable in the second phase (after testing y) in order to compute $\alpha(x, y)$ and again in the third phase in order to decide about the contents of the addressed storage cell. This is impossible in OBDDs. Indeed, the OBDD size of ISA grows exponentially as has been proved by Breitbart, Hunt III, and Rosenkrantz (1993, 1995).

Theorem 4.3.3. *The OBDD size of ISA_n is bounded below by* $2^{\lfloor n/\log n\rfloor - 1}$.

Proof. Let π be an arbitrary variable ordering. We use the communication complexity theoretical approach. Alice gets the (according to π) first input bits until $\lfloor n/\log n\rfloor - 1$ x-variables have been tested. We partition the set of x-variables to $\lfloor n/\log n\rfloor$ complete blocks of $k = \log n$ variables each and possibly one further block. A block is a subsequence $x_i, \ldots, x_{i+k-1}$ of the x-variables. There is at least one complete block B given completely to Bob. The communication protocol has to work in the situation where $|y|$ is the index of the first variable of B. Alice does not know anything about $\alpha(x, y)$ in this situation. Each of her data variables can be addressed by the variables in B. Hence, Alice has to communicate her whole knowledge on $\lfloor n/\log n\rfloor - 1$ x-variables. In other words, all $2^{\lfloor n/\log n\rfloor - 1}$ rows of the restricted communication matrix (after the replacement of the y-variables by constants) are different. □

4.4 Addition

The basic arithmetic functions are of particular interest.

Definition 4.4.1.

(i) The *addition* function $\mathrm{ADD}_n \in B_{2n,n+1}$ computes the sum $s = (s_n, \ldots, s_0)$ of two binary numbers $x = (x_{n-1}, \ldots, x_0)$ and $y = (y_{n-1}, \ldots, y_0)$.

(ii) The *multiple addition* function $\mathrm{MULTADD}_n \in B_{n^2, n+\lceil \log n\rceil}$ computes the sum S of n binary n-bit numbers $x^1, \ldots, x^n$.

Lemma 4.4.2. *The bit s_i of ADD_n has an OBDD size which is bounded above by $3i + 5$.*

Proof. We use the variable ordering $x_0, y_0, x_1, y_1, \ldots, x_{n-1}, y_{n-1}$ which is an *interleaved* ordering of the x- and y-variables. If y_j, $0 \le j < i$, has been tested, we get two different subfunctions of s_i essentially depending on x_{j+1}. The subfunctions distinguish whether the carry bit c_j equals 0 or 1. If x_j, $0 \le j \le i$, has been tested, we get three different subfunctions of s_i, namely we have to distinguish the three values $a \in \{0, 1, 2\}$ for $x_j + c_{j-1}$ (where $c_{-1} = 0$). The subfunction essentially depends on y_j iff $a = 1$ or $j = i$. Altogether, we have one x_0-node, two x_j-nodes, $1 \le j \le i$, one y_j-node, $0 \le j < i$, two y_i-nodes and two sinks. If $i = n$, we have no x_n- or y_n-node. □

The variable ordering $x_0, y_0, \ldots, x_{n-1}, y_{n-1}$ is very good for each output bit of ADD_n but not for the simultaneous representation of *all* bits of ADD_n. Let $j > j'$. Then each of the subfunctions of s_j after the test of x_i (or y_i) where $i < j$ essentially depends on y_j, while $s_{j'}$ (and all its subfunctions) does not essentially depend on y_j. Hence, the OBDDs for $s_{n-1}, \ldots, s_0$ according to the

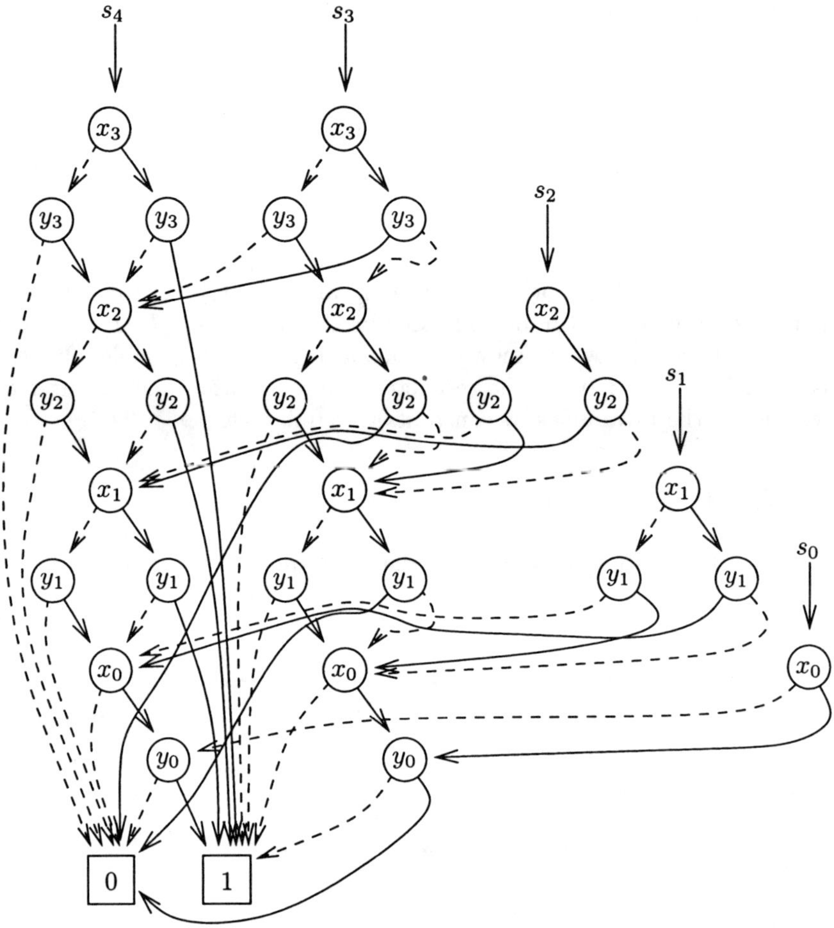

Figure 4.4.1: *An OBDD for the addition of* 4*-bit numbers.*

given variable ordering only share the sinks. This leads to an OBDD size of $\Omega(n^2)$. The variable ordering $x_0, y_0, \ldots, x_{n-1}, y_{n-1}$ is the ordering in which we treat the variables in the same ordering as the school method for addition. We can do better by a variable ordering which is more appropriate for OBDDs.

Theorem 4.4.3. *The OBDD size of ADD_n is bounded above by $9n-5$ if $n \geq 2$.*

Proof. We choose the interleaved variable ordering $x_{n-1}, y_{n-1}, \ldots, x_0, y_0$. The main effect of this variable ordering is that $s_n, \ldots, s_{i+1}$ depend on $x_i, y_i, \ldots, x_0$, y_0 only via the carry bit c_i computed at position i. The resulting OBDD for $n = 4$ is shown in Fig. 4.4.1.

The computation of s_j starts at an x_j-node (x_{n-1}-node for s_n). Hence, we have two x_{n-1}-nodes representing s_n and s_{n-1} and three x_i-nodes, $0 \le i \le n-2$, representing s_i, c_i, and $\overline{c}_i$. These are altogether $3n-1$ x-nodes. On the y_{n-1}-level, we have to represent the subfunctions of s_n and s_{n-1} after replacing x_{n-1} by a constant. These are the four different functions $c_{n-2}y_{n-1}$, $c_{n-2}+y_{n-1}$, $c_{n-2} \oplus y_{n-1}$, and $c_{n-2} \oplus \overline{y}_{n-1}$ all essentially depending on y_{n-1}. On the y_i-level, $1 \le i \le n-2$, we have to represent the subfunctions of s_i, c_i, and $\overline{c}_i$ after replacing x_i by a constant. These are the six different functions $c_{i-1} \oplus y_i$, $c_{i-1} \oplus \overline{y}_i$, $c_{i-1}y_i$, $c_{i-1}+y_i$, $\overline{c_{i-1}y_i}$, $\overline{c_{i-1}+y_i}$ all essentially depending on y_i. For the y_0-level, we obtain only the functions y_0 and $\overline{y}_0$, since $c_{-1}=0$. These are altogether $6n-6$ y-nodes if $n \ge 2$. Adding the two sinks, we obtain the result. □

Theorem 4.4.4. *The OBDD size of $MULTADD_n$ is bounded by $O(n^4)$.*

Proof. We choose the columnwise variable ordering where the least significant columns are tested first. Then, as for ADD_n, we cannot share nodes for the output bits at the positions $n-1,\ldots,0$. Each of the $n+\lceil \log n \rceil$ output bits can be represented in size $O(n^3)$. This follows along the lines of the proof of the upper bound on the BP size of multiplication (Theorem 2.3.5). □

The estimation of the OBDD size with respect to the columnwise ordering where the most significant columns are tested first is much more difficult (see exercises). If we add two numbers, the carry is a single bit c_j leading to the four functions 0, 1, c_j, and $\overline{c}_j$. Here the carry is a $\lceil \log n \rceil$-bit number and the dependence of the more significant sum bits on the carry is more complicated.

4.5 Multiplication

The school method for multiplication is nothing else but the multiple addition of the numbers $z_i = \sum_{0 \le j \le n-1} x_i y_j 2^{i+j}$. Using this idea we obtain (see Theorem 2.3.5) a leveled BP of size $O(n^4)$ for MUL_n, i.e., all output bits of the multiplication of two n-bit numbers. But each bit x_i or y_i is used in n different levels. Therefore, this approach does not lead to OBDDs of polynomial size.

Bryant (1986) has proved that for each variable ordering there is an output bit of MUL_n with exponential OBDD size. This lower bound does not imply that MUL_n needs exponential size if we use different variable orderings for different output bits. Later, Bryant (1991) also proved that the output bit $\text{MUL}_{n-1,n}$ has exponential OBDD size for all variable orderings.

Before we prove this result, we discuss it. Although multiplication "almost" is a special case of MULTADD, its OBDD size is exponentially larger. How can this happen? Multiplication is just a polynomial constant-depth reduction of MULTADD and, as the bounds on MUL and MULTADD show, this is not enough to transfer results on the OBDD size. What does this mean for the verification of combinational circuits? Using OBDDs we may verify circuits for MULTADD in polynomial time but not circuits for MUL. Since multiplication is in some sense the addition of n numbers of a special kind (the ith one is $z_i = |x|y_i 2^i$, $0 \le i \le n-1$), we may use special methods for multiplication (such as the Karatsuba–Ofman method; see Wegener (1987)). This makes the verification of multipliers difficult. If we restrict multipliers to computing the numbers $z_i = |x|y_i 2^i$ first and, in a second phase, to adding the numbers z_i, OBDDs can be used for verification (Burch (1991)) by using the polynomial-size OBDDs for MULTADD and the "new" variables $z_{ij} = x_i y_j$.

In the following, we prove an exponential lower bound of size $2^{n/8}$ on the OBDD size of the *middle bit of multiplication* $\text{MUL}_{n-1,n}$. The idea is the following. For each variable ordering, we find a subfunction of $\text{MUL}_{n-1,n}$ which essentially equals the computation of the bit at position $N-1$ of the addition of two N-bit numbers x and y where $N \ge n/8$. Moreover, the variable ordering is bad for this purpose, i.e., among the first N tested variables is exactly one of the variables x_i and y_i. First, we prove that this leads to exponential OBDD size.

Lemma 4.5.1. *Let π be a variable ordering where, for each i, exactly one of the variables x_i and y_i belongs to the first half. Then the π-OBDD size of the bit s_{n-1} of ADD_n is at least 2^n.*

Proof. Without loss of generality we assume that first all x-variables are tested. It is sufficient to prove that all 2^n subfunctions obtained by replacing the x-variables by constants are different. Let $a = (a_{n-1}, \dots, a_0)$ and $a' = (a'_{n-1}, \dots, a'_0)$ be two such different replacements of the x-variables. Let i be the maximal index such that $a_i \ne a'_i$, w.l.o.g. $a_i = 1$. Let $b = (b_{n-1}, \dots, b_0)$ be the following replacement of the y-variables: $b_j := \overline{a}_j$ for $j > i$, $b_i := 1$, and $b_j := 0$ for $j < i$. Then $s_{n-1}(a, b) = 0$ but $s_{n-1}(a', b) = 1$. □

Theorem 4.5.2. *The OBDD size of $MUL_{n-1,n}$ is at least $2^{n/8}$.*

Proof. Without loss of generality n is even. Let π be an arbitrary variable ordering. We consider the upper part $U = \{x_{n-1}, \dots, x_{n/2}\}$ and the lower part $L = \{x_{n/2-1}, \dots, x_0\}$ of the x-variables. The top part T of π contains the first $n/2$ x-variables with respect to π and the bottom part B the other $n/2$ x-variables. We want to find a distance parameter $d \in \{1, \dots, n-1\}$ such that many pairs (x_i, x_{i+d}) where x_i belongs to L and x_{i+d} belongs to U have the property that x_i is tested in the top part with respect to π iff x_{i+d} is tested in the bottom part.

Let $X_{UT} := U \cap T$, similarly X_{UB}, X_{LT}, and X_{LB}. If $k = |X_{UT}|$, then $|X_{UB}| = |X_{LT}| = n/2 - k$ and $|X_{LB}| = k$. Hence, $(X_{LB} \times X_{UT}) \cup (X_{LT} \times X_{UB})$ contains $k^2 + (n/2-k)^2 \ge n^2/8$ pairs (x_i, x_{i+d}) where $0 \le i < n/2 \le i + d \le n-1$. By the pigeonhole principle there exists some d^* such that

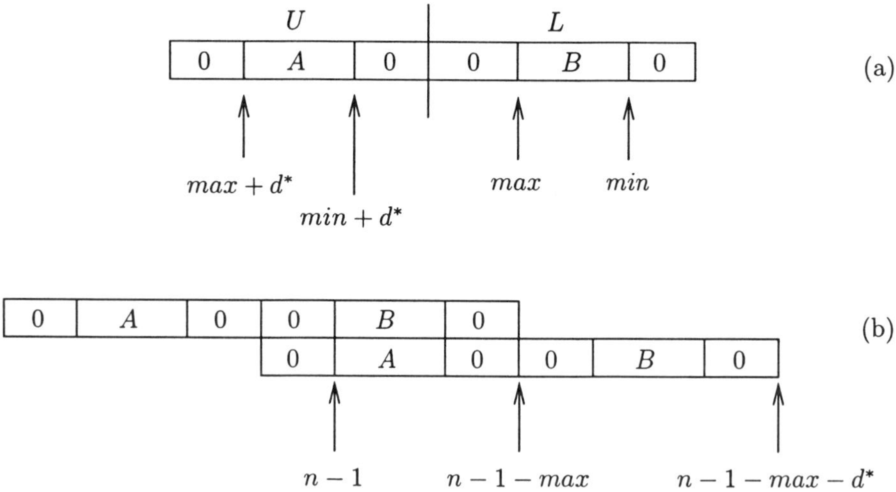

Figure 4.5.1: *The x-vector and the shifted x-vectors after the replacement of variables by constants.*

$(X_{LB} \times X_{UT}) \cup (X_{LT} \times X_{UB})$ contains at least $n/8$ pairs (x_i, x_{i+d^*}). Let I be the set of the corresponding i and let min and max be the minimal, respectively, maximal element of I. We consider the following subfunction of $\mathrm{MUL}_{n-1,n}$ on the variables x_i and x_{i+d^*}, $i \in I$:

- If $j \leq n/2 - 1$ and, additionally, $j < min$ or $j > max$, replace x_j by 0.
- If $j \leq n/2 - 1$, $j \notin I$, and $min < j < max$, replace x_j by 1.
- If $j \geq n/2$ and $j - d^* \notin I$, replace x_j by 0.
- Replace $y_{n-1-max-d^*}$ and $y_{n-1-max}$ by 1.
- Replace all other y-variables by 0.

What is the effect of these replacements? Since only two y-variables are replaced by 1, multiplication is reduced to the addition of x shifted by $n-1-max-d^*$ positions and x shifted by $n-1-max$ positions. The x-vector after the replacement is shown in Fig. 4.5.1(a). In the numbers A and B the same positions are variable. The other positions contain a 1 in B and a 0 in A. The shifted x-vectors are shown in Fig. 4.5.1(b). Hence, $\mathrm{MUL}_{n-1,n}$ equals the second most significant bit of the sum of A and B. All positions $i \notin I$ have no influence on this bit, since they contain a 1 and a 0. They propagate a carry if existent. Therefore, we have the situation of the addition of two numbers of at least $n/8$ bits. Since all (x_i, x_{i+d^*}) are in $(X_{LB} \times X_{UT}) \cup (X_{LT} \times X_{UB})$, the variable ordering has the property that in the top part exactly one bit of each position is tested. Hence, by Lemma 4.5.1 we obtain the lower bound. □

Such a lower bound is not surprising; see our discussion on Theorem 1.4.3.

4.6 Squaring and Division

Here we consider the other fundamental arithmetic functions. For subtraction, we obtain results similar to addition (see exercises).

Definition 4.6.1.

(i) The *squaring* function $\mathrm{SQU}_n \in \mathrm{B}_{n,2n}$ computes the square of an n-bit number.

(ii) The function $\mathrm{INV}_n \in \mathrm{B}_{n,n}$ computes the n most significant bits of the *multiplicative inverse* of an n-bit number x.

(iii) The *division* function $\mathrm{DIV}_n \in \mathrm{B}_{2n,n}$ computes the n most significant bits of the quotient of two n-bit numbers.

Theorem 4.6.2. *$MUL \leq_{\mathrm{rop}} SQU \leq_{\mathrm{rop}} INV \leq_{\mathrm{rop}} DIV$.*

Corollary 4.6.3. *The functions SQU_n, INV_n, and DIV_n need exponential OBDD size.*

The corollary follows directly from Theorem 4.5.2, Theorem 4.6.2, and the results of Section 4.2. Proofs without read-once projections are not known. The proof of Theorem 4.6.2 is due to Wegener (1993).

Proof of Theorem 4.6.2.

(i) In order to multiply the n-bit numbers x and y we square the $(3n+2)$-bit number $z = (x_{n-1}, \ldots, x_0, 0, \ldots, 0, y_{n-1}, \ldots, y_0)$, which is a legal input in a read-once projection. We obtain $|z|^2 = (|x|2^{2n+2} + |y|)^2 = |x|^2 2^{4n+4} + |x||y|2^{2n+3} + |y|^2$. Since $|y|^2$ and $|x||y|$ have length $2n$, the number $|x||y|$ is contained in $\mathrm{SQU}_n(z)$.

(ii) In order to square $x = (x_{n-1}, \ldots, x_0)$ we consider the number $|q| = |x|2^{-t} + 2^{-T}$ for $t = 4n$ and $T = 10n$. The input of INV_{10n} is the binary representation of $1 - |q|$. Let $(q_{-1}, \ldots, q_{-10n})$ be the binary representation of $|q|$ and $(q'_{-1}, \ldots, q'_{-10n})$ be the binary representation of $1 - |q|$. Since $q_{-10n} = 1$ by definition, we conclude that $q'_{-10n} = 1$ and $q'_{-i} = \overline{q}_{-i}$ for

$i < 10n$. Hence, the binary representation of $1 - |q|$ is a legal input in a read-once projection. Moreover,

$$\begin{aligned}
(1-|q|)^{-1} &= 1+|q|+|q|^2+|q|^3+\cdots \\
&= 1+(|x|2^{-t}+2^{-T})+(|x|2^{-t}+2^{-T})^2+(|x|2^{-t}+2^{-T})^3 \\
&\quad +\cdots \\
&= 1+|x|2^{-t}+|x|^2 2^{-2t}+2^{-T}+2|x|2^{-t-T}+2^{-2T} \\
&\quad +(|x|2^{-t}+2^{-T})^3+\cdots \\
&= 1+|x|2^{-t}+|x|^2 2^{-2t}+\text{rest}
\end{aligned}$$

and, if $n \geq 2$,

$$\text{rest} \leq 2^{-10n}+2^{-12n}+2^{-20n}+2\cdot 2^{-9n} < 2^{-8n}.$$

If $n \geq 2$, the $8n+1$ most significant bits of $(1-|q|)^{-1}$ represent $1+|x|2^{-t}+|x|^2 2^{-2t}$. Since $|x|^2 \leq 2^{2n}$, the number $|x|^2$ is contained in the binary representation of the $10n$ most significant bits of $(1-|q|)^{-1}$.

(iii) The proof is obvious by choosing 1 as numerator. □

One may conjecture that MUL, SQU, INV, and DIV are equivalent with respect to $\leq_{\text{rop}}$. In particular, one may expect that SQU $\leq_{\text{rop}}$ MUL, since squaring is a special case of multiplication. This is expressed as $\text{SQU}_n(x) = \text{MUL}_n(x,x)$, but this is only a projection with multiplicity 2. Bollig and Wegener (1996a) have proved that SQU is *not* a read-once projection of MUL. Circuits for squaring may take advantage of the fact that squaring is a very restricted form of multiplication and this may cause additional difficulties for verification.

4.7 Symmetric Functions

A symmetric function f (see Definition 2.3.2) can be described by its *value vector* $v(f) = (v_0, \ldots, v_n)$ such that $f(a) = v_i$ if $a_1 + \cdots + a_n = i$. Because of its definition, all variable orderings lead to the same OBDD size for symmetric functions.

Lemma 4.7.1. *The number of x_i-nodes in an id-OBDD for a symmetric function $f \in \text{B}_n$ is equal to the number of different nonconstant subvectors $(v_j, \ldots, v_{j+n-i+1})$, $0 \leq j \leq i-1$.*

Proof. The subfunctions of a symmetric function are also symmetric. If we have replaced j of the first $i-1$ variables by ones, the subfunction has the value vector $(v_j, \ldots, v_{j+n-i+1})$. If a symmetric function is not constant, it essentially depends on all its variables. Hence, the lemma follows from Theorem 3.1.4. □

Theorem 4.7.2.

(*i*) *If* $1 \le k \le n/2$, *the OBDD size of the threshold functions* $T_{k,n}$ *and* $T_{n+1-k,n}$ *equals* $k(n-k+1)+2$.

(*ii*) *If* $0 \le k \le n/2$, *the OBDD size of the exact counting functions* $E_{k,n}$ *and* $E_{n-k,n}$ *equals* $nk - k^2 + n + 2$.

(*iii*) *If* $2 \le k \le n/2$, *the OBDD size of the modular counting function* $MOD_{k,n}$ *equals* $nk - k^2 + 2k + 1$.

The proof of this theorem is an application of Lemma 4.7.1 and is left as an exercise.

Using de Bruijn sequences as value vectors, Wegener (1984) proved that the largest possible OBDD size of a symmetric function equals $n^2/2 - n\log n + \Theta(n)$. Also Ross, Butler, Kapur, and Mercer (1991b) and Heap (1993) investigated the OBDD size of some symmetric functions.

4.8 General Threshold Functions

Up to now, we have used the notion threshold functions only for the important but restricted situation that all weights equal 1. Threshold functions with arbitrary weights (which can w.l.o.g. be bounded exponentially) are the basic functions for discrete *neural networks*.

Definition 4.8.1. The *threshold function* $T_{w_1,\ldots,w_n,t} \in B_n$ with integer weights $w_1,\ldots,w_n$ and threshold value t computes 1 on input a iff $a_1w_1+\cdots+a_nw_n \ge t$.

These threshold functions are not necessarily symmetric. The output of a threshold function depends on $(a_1,\ldots,a_i)$ only via the partial sum $a_1w_1+\cdots+a_iw_i$. If the weights are polynomially bounded, the OBDD size for each variable ordering obviously is polynomially bounded by $O(n(|w_1|+\cdots+|w_n|))$. Hosaka et al. (1997) have proved an exponential lower bound on the OBDD size of an explicitly defined threshold function.

Definition 4.8.2. The threshold function T_n^*, where $n=k^2$ and k is even, is defined on the variables x_{ij}, $1 \le i,j \le k$. The weights are $w_{ij} = 2^{i-1}+2^{k+j-1}$ and the threshold value is $t = k(2^{2k}-1)/2$.

Theorem 4.8.3. *The OBDD size of* T_n^* *is bounded below by* $2^{\sqrt{n}/2}$.

Proof. If we consider the weights as $2k$-bit numbers, they all contain exactly a one in the first half and exactly a one in the second half of their representations. Moreover, for each position there are exactly k weights with a one at that position. The corresponding weights for a position i in the less significant half are the weights of the ith row of the weight matrix. Positions of the more significant half are represented as columns of the weight matrix. It follows that the threshold value t is half the sum of all weights.

Let a variable ordering π be fixed. Let L be the first level where variables from $k/2$ rows or $k/2$ columns have been tested. We assume that variables from $k/2$ rows have been tested on level L. The proof for the other case is similar. We consider $2^{k/2}$ assignments to the variables tested in the first L levels. Let $I(L)$ contain the $k/2$ row numbers i such that variables from the ith row are among the first L variables according to π. For each row $i \in I(L)$, we consider the first tested variable and both assignments to this variable. All other tested variables of this row are replaced by zeros. In this way, we obtain $2^{k/2}$ different partial sums, since the partial sums have $k/2$ positions (not belonging to $I(L)$) among the least k significant bits which are 0 while the other $k/2$ positions may take any combination of zeros and ones.

It is sufficient for the proposed lower bound to prove that these $2^{k/2}$ assignments lead to different subfunctions of T_n^*. Let us consider two assignments A_1 and A_2 with partial sums W_1 and $W_2 < W_1$. If there is an assignment A_1' of the remaining variables with total weight $t - W_1$, the combination of A_1 and A_1' leads to the total weight t and the output 1 while the combination of A_2 and A_1' leads to a total weight of less than t and the output 0. Hence, the subfunctions belonging to A_1 and A_2 are different. Altogether, it is sufficient to find for each of the considered $2^{k/2}$ assignments A an assignment A' to the remaining variables leading to a total weight of t.

Let L' be the first level where variables from $k/2$ columns have been tested and let $J(L')$ be the set of the corresponding column numbers. The variables x_{ij}, $i \in I(L)$, $j \in J(L')$, not yet replaced by constants, are replaced by zeros. Now exactly a quarter of the variables have been replaced by constants; these variables are arranged in a $(k/2) \times (k/2)$ submatrix. The aim is to fill the empty positions with zeros and ones in such a way that each row and each column contains exactly $k/2$ ones. By renumbering the rows and columns, we may assume that the upper left corner U of the matrix is filled. Then the matrix is filled in the following way where in $\overline{U}$ every bit of U is negated:

$$M = \begin{array}{|c|c|} \hline U & \overline{U} \\ \hline \overline{U} & U \\ \hline \end{array} \,.$$

A vector $a \in \{0,1\}^k$ whose second half is the negation of its first half obviously contains $k/2$ ones. This finishes the proof of our theorem. □

4.9 Functions with Short Disjunctive Normal Forms (DNFs)

Many functions have only exponential-size *disjunctive normal forms* (DNFs) (also called *sum of products*) but polynomial-size OBDDs. One such example is the majority function. The question is whether functions with short DNFs, i.e., DNFs of polynomial length, always have polynomial-size OBDDs. This is not the case, as has been proved by Devadas (1993). Wegener (1986) has already presented a special type of clique function which has polynomial DNF size and exponential size for FBDDs (or read-once BPs) and, therefore, also for OBDDs. Bollig and Wegener (1998a) present a much simpler function with the following properties:

- It has a DNF with $\Theta(n^{3/2})$ prime implicants,
- all prime implicants have length 2,
- all prime implicants contain only positive literals,
- the FBDD size and, therefore, the OBDD size grows exponentially.

We present this function in Section 6.2, where we discuss lower bounds for FBDDs. The conclusion is that very simple functions may need OBDDs (and FBDDs) of exponential size.

4.10 The Hidden Weighted Bit Function

The function HWB (see Definition 1.1.3) was introduced by Bryant (1991), since it is a very simple version of storage access where each variable is a control variable *and* a data variable. Therefore, it is not surprising that it has exponential OBDD size. We present here structural properties of HWB to simplify the estimation of the π-OBDD size of HWB. After the proof of the lower bound on the OBDD size of HWB, we motivate the extensive investigation of the general variable-ordering problem in Chapter 5 by showing how difficult it is to find an almost optimal variable ordering for HWB.

For the following considerations, we fix an arbitrary variable ordering π. By $N(k,s)$, $0 \le s \le k \le n$, we denote the number of different subfunctions of HWB_n obtained by replacing the first k variables according to π with constants such that s variables are replaced by 1. Each $N(k,s)$ is a lower bound on the π-OBDD size of HWB_n while the sum of all $O(n^2)$ values $N(k,s)$ gives an upper bound. If s of k tested variables have the value 1, then *sum*, the number of ones in the input, is contained in the "window" $W = \{s, \dots, s+n-k\}$. Let $w = w(k,s)$ be the number of variables x_j such that $j \in W$ and x_j belongs to the first k variables according to π. These variables are called fixed window variables. Finally, let $\binom{w}{i} = 0$ if $i < 0$ or $i > w$.

Lemma 4.10.1. $N(k,s) = \binom{w}{w-k+s} + \cdots + \binom{w}{s}$.

Proof. Let v_c be the number of fixed window variables which have the value c. By assumption, $v_1 \le s$ and $v_0 \le k-s$. Furthermore, $v_1 = w - v_0 \ge w-k+s$. Let a and a' be assignments to the first k variables assigning 1 to s variables. If all fixed window variables get the same value according to a and a', the subfunctions of HWB_n with respect to a and a' are equal. If $a_i \ne a'_i$ for some fixed window variable x_i, the extension of a and a' by $i-s$ ones and $n-i-k+s$ zeros leads to different values of HWB_n. Hence, a and a' lead to different subfunctions of HWB_n. This proves the lemma, since $\binom{w}{w-k+s} + \cdots + \binom{w}{s}$ equals the number of different assignments to the fixed window variables. □

With this lemma it is easy to obtain a slight improvement of the lower bound on the OBDD size due to Bryant (1991).

Theorem 4.10.2. *The OBDD size of HWB is* $\Omega(2^{n/5})$.

Proof. Without loss of generality $n = 10m$. Let $k = 6m$. If $s = m$, then $W = \{m, \dots, 5m\}$ and if $s = 5m$, then $W = \{5m, \dots, 9m\}$. In one of these two cases the number w of fixed window variables is at least $2m$, no matter what the variable ordering is. Without loss of generality this happens for $s = m$. Since $\binom{r+1}{l} \ge \binom{r}{l}$, the bound of Lemma 4.10.1 is at least $\binom{2m}{0} + \cdots + \binom{2m}{m} \ge 2^{2m-1} = 2^{n/5-1}$. □

Bollig, Löbbing, Sauerhoff, and Wegener (1999) have investigated different variable orderings for HWB. The variable ordering $x_1, \dots, x_n$ is called the *natural* one. All variables of HWB_n are control and data variables. More precisely, x_i is the decisive data variable for exactly $\binom{n}{i}$ inputs. The heuristic "control variables first" can be generalized to the rule to test first the variables which only in a small number of cases work as data variables. This leads to the *alternating* or *zigzag* (Jain, Abraham, Bitner, and Fussell (1992)) ordering $x_n, x_1, x_{n-1}, x_2, x_{n-2}, x_3, \dots$.

Theorem 4.10.3.

(*i*) *The OBDD size of HWB is* $\Omega(2^{0.40n})$ *and* $O(2^{0.41n})$ *for the natural variable ordering.*

(*ii*) *The OBDD size of HWB is* $\Omega(2^{0.255n})$ *and* $O(2^{0.26n})$ *for the alternating or zigzag ordering.*

Sketch of proof. We omit the tedious estimations based on Lemma 4.10.1.

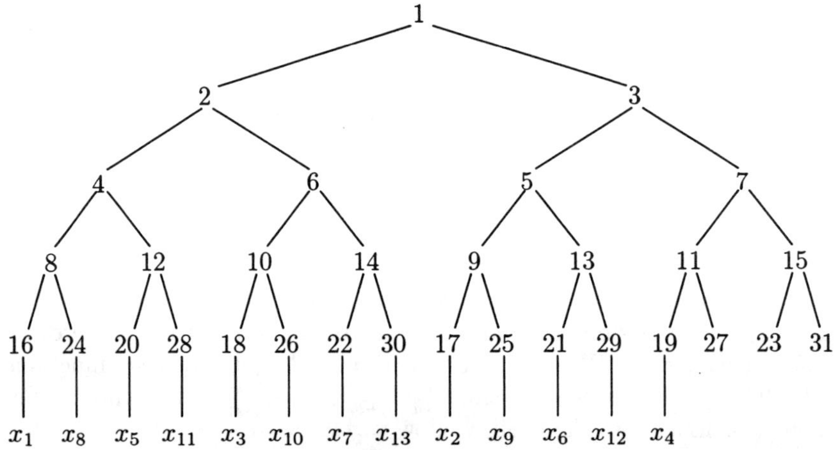

Figure 4.10.1: *The uniform variable ordering for* $n = 13$.

(i) For the lower bound, choose $k = 0.59n+1$ and $s = 0.18n$. Then $w = 0.41n$.

(ii) For the lower bound, choose $k = 0.74n+1$ and $s = 0.11n$. Then $w = 0.26n$.

□

Lemma 4.10.1 suggests minimizing the number of fixed window variables for all windows (of fixed length). A variable ordering is called *perfectly uniform* if each interval, i.e., subsequence of variables, of αn variables contains $\alpha\beta n$ variables of the first βn variables, $0 \le \alpha, \beta \le 1$. Since perfectly uniform variable orderings do not exist, we describe the so-called uniform variable ordering where intervals of length αn contain $\alpha\beta n \pm O(\log n)$ of the first βn variables.

We construct a complete binary tree with $m = \lfloor \log n \rfloor + 2$ levels $0, \ldots, m-1$. We number the nodes according to the following *scattered* numbering. The root gets the number 1. The left child of a node with number k on level l gets the number $k + 2^l$ and its right child the number $k + 2^{l+1}$. On the last level, we map the n leftmost leaves order-preserving to the indices of $x_1, \ldots, x_n$ in order to obtain the uniform ordering (see Fig. 4.10.1 for an example).

Let I be an interval $x_i, \ldots, x_j$ of αn variables. We choose at most $m-1$ complete subtrees whose leaves are mapped to the variables (in Fig. 4.10.1 the subtrees rooted at 2, 5, and 19). Let T be one of these subtrees. If it is rooted at level l, it "contains" as leaves the variables x_{k+r2^l}, $0 \le r < 2^{m-l-1}$, for some $k \le 2^l$. Hence, the variables belonging to T are equally spread and the number of variables in I belonging to T is at most 1 away from its perfect value $\alpha 2^{m-l-1}$. Since $m - 1 = \lfloor \log n \rfloor + 1$, we have proved the proposed property.

Theorem 4.10.4. *The OBDD size of HWB is* $2^{0.25n \pm O(\log n)}$ *for the uniform variable ordering.*

Proof. Since we allow an error term of $O(\log n)$ in the exponent, we can argue with the perfectly uniform variable ordering. The lower bound follows from Lemma 4.10.1 for $k = 0.5n$ and $s = 0.25n$. Then $w = 0.25n + 1$ and $N(k,s) = 2^{0.25n+1} - 2$. For the upper bound, we consider $N(k,s)$ for arbitrary $k = \alpha n$. The window has length $(1-\alpha)n + 1$ and contains $\alpha(1-\alpha)n + \alpha \leq 0.25n + 1$ variables of the first k variables according to the uniform ordering. Hence, $w \leq 0.25n + 1$ and $N(k,s) \leq 2^{0.25n+1}$. □

In order to come closer to the lower bound we define a *hybrid* variable ordering which tries to combine the good properties of the alternating and the uniform variable ordering. It starts like the alternating variable ordering until the first $0.1n$ and the last $0.1n$ variables are considered. The remaining $0.8n$ variables are ordered according to the uniform variable ordering for $0.8n$ variables.

Theorem 4.10.5. *The OBDD size of HWB is $\Omega(2^{0.2028n})$ and $O(2^{0.2029n})$ for the hybrid variable ordering.*

We omit the tedious proof of this theorem. The conclusion of our considerations is the following. Although HWB is a function with a clear and simple structure, it seems to be difficult to find an almost optimal variable ordering. The task will be even harder for functions given by circuits without known structural properties.

4.11 Read-Once Formulas

The simulation of formulas over the full binary basis by BPs (see Theorem 2.1.4) leads to a result on OBDDs if we restrict ourselves to a subclass of formulas.

Definition 4.11.1. A formula is called *read once* if the leaves of its underlying tree are labeled by different variables.

Theorem 4.11.2. *If $f \in B_n$ can be represented by a read-once formula, its OBDD size is bounded by $\alpha n^\beta + 2$, where $\alpha \leq 1.360$ and $\beta = \log_4(3+\sqrt{5}) < 1.195$.*

Proof. We use the construction of the proof of Theorem 2.1.4, which leads to a BP whose size is bounded by $\alpha n^\beta + 2$. If the read-once formula for f has the form $f = g \otimes h$, then g and h are read-once formulas on disjoint sets of variables. The construction ensures that the BP can be cut into two parts such that in the top part only the variables of g and in the bottom part only the variables of h are tested or vice versa. Since the construction works recursively on g and h, the BP indeed is an OBDD. □

In Section 5.5, we prove that this approach leads to optimal OBDDs for read-once formulas.

4.12 Selected Functions on Matrices

Functions on matrices of Boolean variables are good examples in many of the later chapters. Here we investigate the OBDD size of some of these functions. By X we denote an $n \times n$ matrix of Boolean variables and by $X_1, \ldots, X_n$ the rows of X.

Theorem 4.12.1. *Let $g, h \in B_n$ be functions with OBDD size $s(g)$ and $s(h)$, respectively. Let $f(X) = h(g(X_1), \ldots, g(X_n))$. Then the OBDD size of f is bounded above by $s(g)s(h)$.*

Proof. We start the construction with an optimal OBDD for $h(y_1, \ldots, y_n)$. Then we replace each y_i-node v by a copy of an optimal OBDD for $g(X_i)$. Each edge to the c-sink of this OBDD is replaced with an edge to the c-successor of v. Now it is obvious that we represent f with at most $s(g)s(h)$ nodes. The corresponding variable ordering is obtained by replacing y_i in an optimal variable ordering for h with an optimal variable ordering for g on X_i. □

By this simple result, we obtain a linear $O(n^2)$ bound on the OBDD size for the function ROW_n testing whether X contains a 1-row, i.e., a row consisting of ones only, and the function testing whether each row of X contains exactly one entry 1. Hence, matrix functions collecting some information about simple properties of the rows (or the columns) are easy, since we may use a rowwise (or columnwise) variable ordering. But it seems to be difficult to collect the same information about the rows *and* the columns. Let COL_n test whether X contains a 1-column.

Theorem 4.12.2. *The OBDD size of ROW_n+COL_n testing whether X contains a 1-row or a 1-column is bounded below by $2^{\sqrt{n-1}}$.*

Proof. Let us consider the situation where the first $n-1$ variables according to the chosen variable ordering have been tested. Then there are at least $m = (n-1)^{1/2}$, w.l.o.g., rows which are partially tested, i.e., at least one but not all variables of the row are tested. We distinguish 2^m different assignments where in each of m partially tested rows either all variables are replaced by 0 or all variables are replaced by 1. All other tested variables are replaced by 0. These assignments lead to different subfunctions. Let a and a' be two of these assignments which differ in the assignment to the variables of the ith row. We consider the assignment to the remaining variables where only the variables of the ith row are replaced by ones. This leads to different outputs for a and a'. □

A better lower bound is obtained in Theorem 6.2.13.

Theorem 4.12.3. *The OBDD size of the function $PERM_n$ testing whether X is a permutation matrix is $\Omega(n^{-1/2}2^n)$.*

Proof. A Boolean matrix is a permutation matrix iff each row and each column contains exactly one entry 1. (Hence, the function PERM also tests whether some connection realizes a crossbar switch.)

The following lower bound proof is due to Krause (1988). Let G_n be an OBDD for PERM_n. Each path from the source to the 1-sink has length n^2. If some variable is not tested, only one of its two possible values may extend an assignment to the other variables to a permutation matrix. Hence, shorter paths cannot lead to the 1-sink.

Let π be a permutation on $\{1, \ldots, n\}$, $M(\pi)$ the corresponding permutation matrix, and $p(\pi)$ the path activated by the input $M(\pi)$. We separate $p(\pi)$ into its initial part $p_1(\pi)$ ending in $v(\pi)$, the node reached by the $k = \lceil n/2 \rceil$th 1-edge on $p(\pi)$, and into the rest $p_2(\pi)$. Let $I(\pi)$ be the set of indices i such that $x_{i,\pi(i)}$ is tested on $p_1(\pi)$. By definition, $|I(\pi)| = k$.

Let $v(\pi) = v(\pi')$. It follows that the path consisting of $p_1(\pi)$ and $p_2(\pi')$ is a computation path leading to the 1-sink. Hence, it corresponds to a permutation π''. This is possible only if $I(\pi) = I(\pi')$ and if the sets $J(\pi) = \{\pi(i) \mid i \in I(\pi)\}$ and $J(\pi') = \{\pi'(i) \mid i \in I(\pi')\}$ are equal. There are $k!$ one-to-one mappings from $I(\pi)$ to $J(\pi)$ and $(n-k)!$ one-to-one mappings from $\{1, \ldots, n\} - I(\pi)$ to $\{1, \ldots, n\} - J(\pi)$. Hence, the number of different permutations leading to the same v-node is bounded above by $k!(n-k)!$ and the number of different v-nodes is bounded below by $n!/(k!(n-k)!)$. Since $k = \lceil n/2 \rceil$, the proposed lower bound follows by Stirling's formula. □

This lower bound technique is known as the *cut-and-paste technique.* All paths leading from the source via v to the 1-sink are cut into their initial parts from the source to v and their final parts from v to the 1-sinks. Then we may paste each initial part with each final part in order to obtain a computation path leading to the output 1.

Let $g_n(X, Y) = \text{ROW}_n(X) + \text{COL}_n(Y)$ and let $h_n(X, Y)$ be the function which tests whether each row of X contains exactly one entry 1 and each column of Y contains exactly one entry 1. The OBDD size of both functions is $O(n^2)$ using a variable ordering testing X rowwise and then Y columnwise. The upper bounds on the functions on X, respectively, Y have been presented as applications of Theorem 4.12.1 and we know that the disjunction or conjunction of functions on disjoint sets of variables leads to an addition of the OBDD sizes.

The functions $G(X) := g(X, X)$ and $H(X) := h(X, X)$ are simple projections of g, respectively, h which have multiplicity 2. They are restricted versions of g, respectively, h in the same sense as squaring is a restricted version of multiplication. Remember that $\text{SQU}(x) = \text{MUL}(x, x)$. However, G is the function considered in Theorem 4.12.2 and $H = \text{PERM}$. Hence, both functions have exponential OBDD size. This proves the claim of Section 4.2 that projections with multiplicity 2 may lead to an exponential blow-up of the OBDD size. Indeed, OBDDs are suitable for the verification of circuits for h while they (and also many of their generalizations) are not suitable for the verification of circuits for

PERM. This establishes the statement that f may be harder to verify than g although $f \leq_{\text{proj}} g$. But if $f \leq_{\text{rop}} g$, this is no longer true, showing that $\leq_{\text{rop}}$ is a suitable reduction concept for verification.

4.13 Exercises and Open Problems

4.1.M The function $\text{COM}_n \in B_{2n}$ compares two n-bit numbers x and y and outputs 1 if $|x| \leq |y|$. Prove that COM $\leq_{\text{rop}}$ ADD.

4.2.M Prove that MULTADD $\leq_{\text{rop}}$ MUL.

4.3.M The inner product function IP $\in B_{2n}$ computes $x_1y_1 \oplus \cdots \oplus x_ny_n$. Prove that IP $\leq_{\text{rop}}$ MUL.

4.4.M Define the BPVP and prove that $f \leq_{\text{rop}}$ BPVP if $f \in$ P-BP and that BPVP $\in$ P-BP.

4.5.E Discuss the use of complemented edges in OBDDs for the bit s_i of ADD_n and the variable ordering $x_0, y_0, \ldots, x_{n-1}, y_{n-1}$.

4.6.E Discuss the use of complemented edges in OBDDs for ADD_n and the variable ordering $x_{n-1}, y_{n-1}, \ldots, x_0, y_0$.

4.7.M Give bounds on the OBDD size for the addition of integers represented by their sign and absolute value.

4.8.E Estimate the OBDD size of MULTADD using a rowwise variable ordering.

4.9.D Estimate the OBDD size of MULTADD using the columnwise ordering where the most significant columns are tested first.

4.10.M Use read-once projections to obtain exponential or nonpolynomial lower bounds on the OBDD size of as many output bits of multiplication as possible.

4.11.E Describe output bits of multiplication with polynomial OBDD size.

4.12.O Is the OBDD size of the most significant output bit of multiplication exponential?

4.13.M Prove that the most significant bit of squaring has linear OBDD size.

4.14.E Give explicit exponential lower bounds on the OBDD size of SQU, INV, and DIV.

4.15.M Prove Theorem 4.7.2.

4.16.D (See Hosaka et al. (1997).) Let n be divisible by 4. Let f_n be the threshold function with the weights $w_i = 2^{i-1}$ if $i \leq n/2$, $w_i = 2^{n/2} - 2^{n-i}$ otherwise, and the threshold value $t = (w_1 + \cdots + w_n)/2$. If the variables are ordered according to descending weights, prove that the OBDD size is bounded below by $\binom{n/2}{n/4}$.

4.17.D (See Hosaka et al. (1997).) Prove that there exists a variable ordering such that the OBDD size of the function defined in the last exercise is $O(n^2)$.

4.18.M (See Devadas (1993).) Let $x = (x_0, \ldots, x_{n-1})$, $y = (y_0, \ldots, y_{n-1})$, and $z = (z_0, \ldots, z_{k-1})$, where $n = 2^k$. Let $f_n(x, y, z)$ be the disjunction of all $x_i y_{i+|z|}$ where the index is interpreted mod n. Prove that the function has polynomial-size DNFs and exponential OBDD size.

4.19.M (See Bollig, Löbbing, Sauerhoff, and Wegener (1999).) Prove that the OBDD size of HWB is bounded above by $O(n2^{n/2})$ regardless of the variable ordering.

4.20.M (See Bollig, Löbbing, Sauerhoff, and Wegener (1999).) Prove that the OBDD size of HWB is bounded below by $\Omega(n^{-1/2}2^{n/2})$ for the variable ordering $x_m, x_{m-1}, x_{m+1}, x_{m-2}, x_{m+2}, \ldots$, for $m = \lceil (n+1)/2 \rceil$.

4.21.E Estimate the OBDD size of PERM for the rowwise variable ordering.

Chapter 5

The Variable-Ordering Problem

5.1 The Variable-Ordering Problem

The importance of an optimal or at least good variable ordering is evident from our investigations in the last two chapters. The *variable-ordering problem* for a representation type R is the problem of computing an optimal (almost optimal, good) OBDD variable ordering from a representation of a Boolean function of type R. As representation types we consider circuits and OBDDs. Circuits are the usual representation type for Boolean functions and can be derived from any logical description. It will turn out that we often do not obtain a good variable ordering from a circuit representation. Then it is important to replace that variable ordering with an optimal or at least a better one. In this situation, we start with an OBDD representation.

Definition 5.1.1. The *variable-ordering spectrum* of a Boolean function f is the function $sp_f : \mathbb{N} \rightarrow \mathbb{N}$, where $sp_f(i)$ is the number of variable orderings leading to an OBDD size of i for f.

Knowing the variable-ordering spectrum, we know how many optimal, almost optimal, good, medium, bad, and very bad variable orderings f has. But usually we are unable to compute or estimate this spectrum. Therefore, we introduce other parameters.

Definition 5.1.2. Let $\max(f)$ ($\min(f)$) be the maximal (minimal) number i such that $sp_f(i) > 0$. The *sensitivity* $s(f)$ of f is defined by $s(f) = \max(f)/\min(f)$.

The sensitivity of f only tells us something about the relation between the worst and best cases. A small sensitivity implies that the variable-ordering problem is not of striking practical relevance—any variable ordering will work. We distinguish several classes of functions.

Definition 5.1.3.

(i) A function $f = (f_n)$ is called *nice* if all variable orderings lead to polynomial OBDD size.

(ii) A function $f = (f_n)$ is called *almost nice* if the function is not nice but the fraction of variable orderings leading to polynomial OBDD size converges to 1.

(iii) A function $f = (f_n)$ is called *ugly* if all variable orderings lead to nonpolynomial OBDD size and *very ugly* if all variable orderings lead to exponential OBDD size.

(iv) A function $f = (f_n)$ is called *almost ugly* if the function is not ugly but the fraction of variable orderings leading to nonpolynomial OBDD size converges to 1.

(v) A function $f = (f_n)$ is called *ambiguous* if for some $\varepsilon > 0$ there exist some polynomial p and some superpolynomially increasing function q such that for large enough n the fraction of variable orderings leading to an OBDD size for f_n bounded above by $p(n)$ is at least ε and the same holds for the fraction with an OBDD size bounded below by $q(n)$.

We investigate random functions (Section 5.2) and some important functions (Section 5.3) with respect to their sensitivity and the introduced classes of functions. Then (Section 5.4) the complexity of the variable-ordering problem is investigated. It is not surprising that it is a hard problem. Nevertheless, some approaches to computing optimal variable orderings are presented (Section 5.5). Then algorithms for coping with the variable-ordering problem are discussed. Simple heuristics are used to transform circuit representations into OBDD representations (Section 5.6). If, during the gate-by-gate transformation, the OBDD becomes too large, reordering heuristics are used to improve the variable ordering (Section 5.8). For this purpose, algorithms are necessary which construct π'-OBDDs for f from π-OBDDs for f (Section 5.7). Finally, transformed OBDDs are discussed (Section 5.9).

5.2 Random Functions

From Theorem 2.2.2 and Theorem 2.3.1 (the BP constructed in the proof is indeed an OBDD), we conclude an upper bound of $3 + o(1)$ on the sensitivity of almost all Boolean functions. This already improves a bound given by Liaw and Lin (1992). In order to get better results, we investigate *random Boolean functions*, where a random function f on X_n equals each $g \in B_n$ with probability 2^{-2^n}. Another formulation is that $f(a)$ takes the values 0 and 1 with probability 1/2 independently from the values $f(b)$, where $b \neq a$.

Theorem 5.2.1. *The expected size of the quasi-reduced* π*-OBDD for a random Boolean function* f *on* X_n *is*

$$S(n) = \sum_{0 \le i \le n} 2^{2^{n-i}}(1 - (1 - 2^{-2^{n-i}})^{2^i}).$$

Proof. The variable ordering is w.l.o.g. $\pi = id$. Let us investigate the subfunctions after the replacement of $x_1, \ldots, x_i$ by constants. Then we get $k = 2^i$ independent random functions on $n - i$ variables. Hence, the expected number $E(Y)$ of nodes in the quasi-reduced OBDD is equal to the expected number of nonempty buckets if we throw k balls independently into $m = 2^{2^{n-i}}$ buckets (the functions on $x_{i+1}, \ldots, x_n$). Let Z_j be the random variable taking the value 1 if the bucket j is nonempty and 0 otherwise. Then we are interested in $E(Y) = E(Z_1) + \cdots + E(Z_m) = mE(Z_1) = m \cdot \text{Prob}(Z_1 = 1)$, since $Z_1, \ldots, Z_m$ are identically distributed. The result follows, since $\text{Prob}(Z_1 = 0) = (1 - \frac{1}{m})^k$ and we may add the expected values for the different levels. □

Using results on random allocations by Kolchin, Sevest'yanov, and Christyakov (1978) and the second moment method, Wegener (1994b) obtained the following results which we present without proofs.

Theorem 5.2.2. *Let* $Q(f)$ *be the maximal quotient of the size of the quasi-reduced* π*-OBDD and of the reduced* π*-OBDD (maximized over all* π*). Then for all but a fraction of* $O(2^{-n/4})$ *of all Boolean functions,* $Q(f) \le 1 + O(n2^{-n/3})$.

Theorem 5.2.3. *For all but a fraction of* $O(2^{-n/4})$ *of all Boolean functions,* $s(f) \le 1 + O(n^2 2^{-n/3})$.

For most values of n, the following holds on each level. With high probability, the expected OBDD size is very close to its maximal value. This follows from the experiment described in the proof of Theorem 5.2.1. If $2^i \ll 2^{2^{n-i}}$, we throw some balls into many buckets and it is likely that almost no bucket gets two or more balls. If $2^i \gg 2^{2^{n-i}}$, we throw many balls into a small number of buckets and it is likely that almost no bucket remains empty. Since $2^{2^{n-i}}$ is decreasing double exponentially (and 2^i is increasing), there is at most one level where $2^i \approx 2^{2^{n-i}}$. If we throw k balls independently into k buckets, it is possible that no bucket is empty but the expected fraction of nonempty buckets converges to $1 - e^{-1}$. Gröpl, Prömel, and Srivastav (1998) used more involved methods to obtain even more precise estimates. Here it is sufficient to conclude that, for random functions, the quasi-reduced *id*-OBDD is almost of the same size as the optimal OBDD. Random functions are very ugly with overwhelming probability. For almost all functions there is almost nothing to gain with optimal variable orderings.

5.3 Nice, Ugly, and Ambiguous Functions

A random function is without structure and, therefore, very ugly. Here we investigate important functions and other functions with polynomial circuit size. We look for examples for the classes of functions from Definition 5.1.3.

Theorem 5.3.1. *The following functions are nice:*

- *Symmetric functions,*
- *threshold functions with polynomial weights,*
- *functions $f = (f_n)$ where $|f_n^{-1}(1)|$ grows polynomially,*
- *functions $f = (f_n)$ where, for some constant k, the function f_n equals $g(h_1(X_1), \ldots, h_k(X_k))$ for an arbitrary function $g \in B_k$ and symmetric functions $h_1, \ldots, h_k$ on subsets of the variable set,*
- *the most significant bit of the squaring function.*

Proof. The claim is obvious for symmetric functions and has been derived in Section 4.8 for threshold functions with polynomial weights. If $|f_n^{-1}(1)| \le p(n)$, an OBDD has at most $p(n)$ paths leading to the 1-sink. If a level has $p(n)+1$ or more nodes, at least one node can be replaced with the 0-sink. For the fourth class of functions, it is sufficient to store the number of ones in each of the blocks $X_1, \ldots, X_k$. There are at most $(n+1)^k$ combinations of partial sums and this is an upper bound on the number of nodes on a level. Finally, the most significant bit of SQU_n equals 1 on the input a iff $|a|^2 \ge 2^{2n-1}$ or $|a| \ge \lceil 2^{n-1/2} \rceil$. Hence, we have to compare the binary input number a with the constant number $c = (c_{n-1}, \ldots, c_0)$, where $|c| = \lceil 2^{n-1/2} \rceil$. After having tested i x-variables, it is sufficient to distinguish between $i+1$ situations. Either all tested variables agree with the corresponding c-bits or the most significant position where the current inputs a and c differ is j. Then we have to store j. Since we know c_j, we also know whether $a_j < c_j$ or $c_j < a_j$. The situation for the less significant bits does not matter. □

Even for nice functions, the OBDD size may vary considerably with the chosen variable ordering. Let, w.l.o.g., $m = n/k$ be an integer. Then consider $g(h_1(X_1), \ldots, h_k(X_k))$ for disjoint variable sets $X_1, \ldots, X_k$ of size m and

$h_1 = \cdots = h_k = \text{MAJ}_m$ and $g = \text{MAJ}_k$. Using the variable ordering where the sets $X_1, \ldots, X_k$ are tested consecutively, we need only to store the number of ones in the block X_i which is tested and the number of ones among $h_1(X_1), \ldots, h_{i-1}(X_{i-1})$. This leads to an OBDD size of $O(n^2)$ if k is a constant. But if we have tested half the variables of each block, we have to store the number of ones of each block and the OBDD size is $\Omega(n^k)$. There is a big advantage to using the better variable ordering.

Theorem 5.3.2. *The following functions are very ugly:*

- *The indirect storage access function,*
- *the middle bit of multiplication,*
- *squaring,*
- *multiplicative inverse,*
- *division,*
- *the threshold function T_n^* from Definition 4.8.2,*
- *the hidden weighted bit function,*
- *the test whether a Boolean matrix contains a* 1*-row or a* 1*-column,*
- *the permutation matrix test function.*

Proof. All these lower bounds have been proved in Chapter 4. □

The investigation of the different variable orderings for HWB in Section 4.10 has proved that for very ugly functions it also makes sense to look for (almost) optimal variable orderings.

Theorem 5.3.3. *The following functions are almost ugly:*

- *The direct storage access function or multiplexer,*
- *the most significant bit of addition,*
- *the comparison function COM deciding for n-bit numbers whether $|x| \geq |y|$,*
- *the disjoint quadratic function DQF where* $\text{DQF}_n(x_0, \ldots, x_{n-1}, y_0, \ldots, y_{n-1}) = x_0y_0 + \cdots + x_{n-1}y_{n-1}$.

Proof. All the listed functions have linear OBDD size; see Chapter 4 for DSA and ADD and use the variable ordering $x_0, y_0, \ldots, x_{n-1}, y_{n-1}$ for COM and DQF.

In the following, we investigate random variable orderings and prove that, with high probability, the OBDD size of the functions is exponential. For DSA, we get 2^i different subfunctions if we replace i data variables, i.e., y-variables, with constants. Let $i = \lfloor n^{1/2} \rfloor$. The probability that the first i variables according to a random variable ordering are data variables is

$$\prod_{0 \le j < i} \left(1 - \frac{k}{n+k-j}\right) = \prod_{0 \le j < i} \frac{n-j}{n+k-j} \ge \left(\frac{n-i}{n+k-i}\right)^i = \left(1 - \frac{k}{n+k-i}\right)^i$$
$$\ge 1 - \frac{ik}{n+k-i} = 1 - O(n^{-1/2} \log n)$$

and converges to 1 as $n \to \infty$.

The other three functions can be handled simultaneously. Let us consider the situation after the test of n variables. Among them are p pairs (x_i, y_i) and $s = n - 2p$ singletons x_i (or y_i) such that the partner y_i (or x_i) is not yet tested. For the pairs, we consider a fixed replacement, namely, $(0,1)$ for ADD and $(0,0)$ for COM and DQF. Then it is easy to see that we obtain 2^s subfunctions for the different assignments of constants to the singletons. We prove that, with high probability (exponentially close to 1), the number of singletons is at least $0.4n$ or, equivalently, the number of pairs is at most $0.3n$. We choose the n variables which are tested first in two stages. In the first stage, we only distinguish x- and y-variables. Let εn be the number of chosen x-variables. Then we choose randomly the $(1-\varepsilon)n$ y-variables. The expected number of pairs is $\varepsilon(1-\varepsilon)n \le 0.25n$. The distribution of the number of pairs is hypergeometric. If $\varepsilon \le 0.3$ or $\varepsilon \ge 0.7$, the number of pairs is certainly bounded by $0.3n$. Otherwise, deviations from the mean value are less probable than for the binomial distribution and, therefore, can be estimated by Chernoff's bound (see Motwani and Raghavan (1995)) implying that the number of pairs is bounded above by $0.3n$ with a probability of $1 - \delta_n$ where δ_n converges exponentially fast to 0. □

For almost ugly functions $f \in B_n$, we consider a search space of $n!$ variable orderings and the amount of good or even acceptable states in the search space is exponentially small.

It is not so easy to find an almost nice function. The following (until now unpublished) example is due to Pudlák.

Definition 5.3.4. The *common knowledge direct storage access* function CKDSA_n is defined on $n^2k + n$ variables, where $n = 2^k$. There are n^2 blocks of addresses of length k and n data variables $y_0, \ldots, y_{n-1}$. If all blocks contain the same address $|a|$, the output equals $y_{|a|}$. Otherwise, the output equals 0.

Theorem 5.3.5. *The common knowledge direct storage access function is almost nice.*

Proof. The OBDD size of CKDSA is $\Omega(2^n)$ if all data variables are tested first. Now let us consider a random variable ordering. We claim that the OBDD size is $O(n^3 \log n)$ if, for each address bit, the corresponding bit of at least one address block is tested before any data variable is tested. This is obvious, since we have only to store the bits of the common address and the corresponding data variable or we reach the 0-sink. For each address bit, the probability of testing one of the n^2 corresponding bits before any of the n data bits equals $n^2/(n^2+n) \geq 1 - 1/n$ and the probability that the OBDD size is bounded by $O(n^3 \log n)$ is at least $1 - (\log n)/n$. □

It is unknown whether ambiguous functions exist.

It is hard to cope with ugly or even very ugly functions. For the other functions, we state the following conjecture. Among all Boolean functions whose OBDD size is bounded by cn^d for some constants $c \geq 1$ and $d > 1$, all but a small (exponentially small) fraction of functions is almost ugly. Then the search for an optimal or only an acceptable variable ordering is the search for a rare object. But this does not imply that the search has to be difficult.

5.4 The Complexity of the Variable-Ordering Problem

The variable-ordering problem for a representation type R is defined as follows. Given a representation of a Boolean function of type R, find an optimal variable ordering. First, we consider the decision variant where we have to decide for a given size bound s whether the given Boolean function has an OBDD size bounded by s.

Theorem 5.4.1. *The decision variant of the variable-ordering problem for circuits is NP-hard.*

Proof. The result is obvious, since we may represent the inputs for SAT, i.e., a conjunction of clauses by a circuit of essentially the same size. Then we set $s = 1$. If we can solve the variable-ordering problem in polynomial time, we can decide whether the given function is constant and then it is easy to decide whether it is the constant 0. □

It is not clear whether the considered problem is contained in NP. The decision variant of the variable-ordering problem for OBDDs is in NP as can be proved using results proved in Section 5.7. The proof of the NP-hardness of this problem has to follow a different pattern than the proof of Theorem 5.4.1, since the satisfiability problem for OBDDs is easy. It took a long time until an NP-hardness result was established. Tani, Hamaguchi, and Yajima (1993) present an NP-hardness result for multi-output functions. A much more complicated proof due to Bollig and Wegener (1996b) ensures the NP-hardness also for single-output functions.

In applications, we would be satisfied with efficient approximation algorithms. Based on recent results in the PCP theory, Sieling (in the journal version of Sieling (1998a)) proved that this is not possible unless NP = P. An algorithm A produces a c-approximation for the variable-ordering problem for OBDDs if the ratio of the π-OBDD size of the considered function f with respect to the variable ordering π computed by A and the π^*-OBDD size of f for an optimal variable ordering π^* is bounded by c.

Theorem 5.4.2. *For any constant c, it is impossible to obtain in polynomial time a c-approximation for the variable-ordering problem for OBDDs unless NP = P.*

The proof of this theorem is too long to be presented in a monograph.

5.5 The Computation of Optimal Variable Orderings

Although we know from the last section that we cannot expect efficient algorithms for the computation of optimal variable orderings, we present some approaches which are much better than the explicit consideration of all $n!$ orderings.

The algorithm with the best worst-case runtime is due to Friedman and Supowit (1990). The main idea is based on the structural properties of reduced OBDDs stated in Theorem 3.1.4. Let the set of variable indices $I = \{1, \dots, n\}$ be partitioned to I_1, I_2, and I_3. The number of x_i-nodes, $i \in I_2$, is the same for all variable orderings where the variables x_i, $i \in I_1$, are tested first in an arbitrary order, then the variables x_i, $i \in I_2$, in a fixed order, and then x_i, $i \in I_3$, in an arbitrary order. Top-down approaches set $I_1 = \emptyset$ and find optimal variable orderings for the x_i, $i \in I_2$, and increasing size of I_2. Bottom-up approaches set $I_3 = \emptyset$ and optimize the variable ordering on all x_i, $i \in I_2$, and increasing size of I_2. The approach of Friedman and Supowit (1990) works bottom-up using the value table of f. Without loss of generality f is not constant.

For each $I \subseteq \{1, \dots, n\}$, the situation that all x_i, $i \notin I$, are tested before all x_i, $i \in I$, is considered. The algorithm computes an optimal variable ordering π_I for an OBDD G_I representing all subfunctions of the considered function f where the variables x_i, $i \notin I$, are replaced with arbitrary constants. Moreover, $\min(I)$ and $\text{table}(I)$ are computed. The number $\min(I)$ is the size of the π_I-OBDD G_I and $\text{table}(I)$ contains, for each assignment to the variables x_i, $i \notin I$, the information which of the min(I) nodes of G_I represents the corresponding subfunction.

Algorithm 5.5.1. The information on π_I, min(I), and table(I) is known for all I where $|I| = k$. For some I^* where $|I^*| = k+1$ the information on π_{I^*}, min(I^*), and table(I^*) has to be computed.

(1) The computation steps (2)–(5) are performed for all $i' \in I^*$.

(2) Let $I := I^* - \{i'\}$ (the best variable ordering where $x_{i'}$ is tested as the first of the I^*-variables is investigated).

(3) For all assignments to the variables x_i, $i \notin I^*$, an $x_{i'}$-node is created and its 0- and 1-successor are computed using table(I).

(4) The set of $x_{i'}$-nodes is reduced using the elimination rule and the merging rule.

(5) Let $s(i')$ be the resulting number of $x_{i'}$-nodes and $s^*(i') = s(i') + \min(I)$.

(6) Let i^* be an index with the minimal s^*-value. Then π_{I^*} is the ordering consisting of i^* followed by π_I for $I := I^* - \{i^*\}$, $\min(I^*) = s^*(i^*)$, and table(I^*) contains the information collected in the sub-OBDD resulting from the reduction of the x_{i^*}-level computed in step 4.

The correctness of this algorithm should be clear by our preliminary considerations. We compare all possible starting variables, each combined with the optimal ordering of the remaining variables. For $I = \{1, \ldots, n\}$, we obtain the desired information on the optimal variable ordering. In order to solve the variable-ordering problem for OBDDs, Algorithm 5.5.1 is performed for each nonempty set $I^* \subseteq \{1, \ldots, n\}$ in the order of increasing values of $|I^*|$.

Theorem 5.5.2. *The variable-ordering problem for OBDDs can be solved in time $O(n3^n)$ using space $O(n^{-1/2}3^n)$.*

Proof. We use the approach described above. Then we apply Algorithm 5.5.1 to $\binom{n}{k}$ inputs I^* where $|I^*| = k$. The main loop (steps 2–5) is called for k elements $i \in I^*$. If we use an array for each table and the linear-reduction algorithm (see Theorem 3.3.4), the main loop can be performed in $O(2^{n-k})$ steps. Hence, the runtime is of order

$$\sum_{1 \le k \le n} \binom{n}{k} k 2^{n-k} = n \sum_{1 \le k \le n} \binom{n-1}{k-1} 2^{n-k}$$

$$= n \sum_{0 \le k \le n-1} \binom{n-1}{k} 1^k 2^{n-1-k} = n3^{n-1}.$$

It is obvious that we may eliminate the information on sets I with $k-1$ elements when we consider sets I with $k+1$ elements. Only the information for two adjacent levels has to be stored simultaneously. We consider $\binom{n}{k}$ sets I where $|I| = k$ and the table size is bounded by 2^{n-k}. The space complexity is of the order of the largest $s(k) = \binom{n}{k}2^{n-k}$, $0 \le k \le n$. It is easy to verify that $s(k+1)/s(k) \le 1$ iff $k \le \lfloor n/3 \rfloor$. Hence, for all k, $s(k) \le s(\lfloor n/3 \rfloor) = O(n^{-1/2}3^n)$. □

A top-down approach using similar ideas is more space efficient but needs even more time (see exercises).

Until now, we have focused on the worst case. The considered approaches can be improved by adding heuristic ideas. We start with the list of the main ideas:

- Representing f by OBDDs (with changing variable orderings) instead of by its value table,
- reducing the search space by taking symmetries into account,
- reducing the search space by branch-and-bound techniques with different lower bound techniques.

OBDDs for arbitrary variable orderings are not larger than the value table but much smaller in practical situations. Even an exponential size of $2^{n/2}$ is a remarkable decrease in size with respect to the size 2^n of value tables. This idea has been used by Ishiura, Sawada, and Yajima (1991). In that part of the OBDD where the variable ordering does not matter (e.g., the top part in a bottom-up approach), one may use reordering techniques (see Section 5.8) to decrease the OBDD size. Then we need techniques to change the variable ordering. In the bottom-up approach, we often like to replace a variable x_i at level $n-k$ by a variable x_j which is in the top part. Ishiura, Sawada, and Yajima (1991) use exchange operations while Jeong, Kim, and Somenzi (1993) improve this by using jump-down operations (see Section 5.7).

Another idea is to use symmetries. Functions like addition are not symmetric functions but they may be called symmetric with respect to the variable sets $S_0 = \{x_0, y_0\}, \ldots, S_{n-1} = \{x_{n-1}, y_{n-1}\}$.

Definition 5.5.3. A Boolean function f is symmetric with respect to the set $S \subseteq X_n$ if f depends on the variables in S only via the number of ones in this part of the input. Then S is called a set of symmetric variables.

If $S = \{x_i, x_j\}$, this definition reduces to the condition $f_{|x_i=0,x_j=1} = f_{|x_i=1,x_j=0}$. We use the notation $x_i \sim x_j$ if f is symmetric with respect to $S = \{x_i, x_j\}$. Actually, this is an equivalence relation, since $x_i \sim x_j$ and

$x_j \sim x_k$ imply

$$\begin{array}{lllll} f_{|x_i=0,x_j=0,x_k=1} & = & f_{|x_i=0,x_j=1,x_k=0} & = & f_{|x_i=1,x_j=0,x_k=0} \quad \text{and} \\ f_{|x_i=0,x_j=1,x_k=1} & = & f_{|x_i=1,x_j=0,x_k=1} & = & f_{|x_i=1,x_j=1,x_k=0}. \end{array}$$

By Shannon's decomposition rule for x_j, $f_{|x_i=0,x_k=1} = f_{|x_i=1,x_k=0}$ and $x_i \sim x_k$. Hence, the equivalence classes with respect to $\sim$ are the maximal sets of *symmetric variables*. If we exchange two symmetric variables x_i and x_j in a variable ordering, it is sufficient to exchange the node labels of x_i- and x_j-nodes. Hence, we may fix a variable ordering on each maximal set of symmetric variables and have to investigate only variable orderings which are compatible with the fixed variable orderings on these maximal sets of symmetric variables. Although symmetries are unlikely for random functions (see exercises), many functions in applications have symmetries. By our considerations, it is sufficient to check for the $\binom{n}{2}$ pairs x_i and x_j whether $f_{|x_i=0,x_j=1} = f_{|x_i=1,x_j=0}$, which can be done for each pair in linear time with respect to the OBDD size of f. If x_i and x_j are adjacent in the variable ordering, the following more simple test decides whether x_i and x_j are symmetric. We assume that the OBDD is reduced and x_i is tested before x_j. Then we check whether x_j-nodes only are reached via x_i-nodes and whether for each x_i-node v we reach the same node w no matter whether we follow the partial path activated by $x_i = 0$ and $x_j = 1$ or by $x_i = 1$ and $x_j = 0$.

Möller, Mohnke, and Weber (1993) presented a list of criteria which are sufficient to prove the asymmetry of variables. These tests can be performed very efficiently and, in general, decrease the number of checks for symmetries.

Theorem 5.5.4. *A Boolean function $f \in B_n$ is asymmetric with respect to x_i and x_j if one of the following conditions is fulfilled in a reduced OBDD where x_i is tested before x_j:*

- $|f^{-1}_{|x_i=1}(1)| \neq |f^{-1}_{|x_j=1}(1)|$,
- *there is an x_i-node v such that no x_j-node can be reached from v,*
- *an x_j-node can be reached from a source without meeting an x_i-node,*
- *for an x_i-node v the following node sets V_{01} and V_{10} are different. A node is in V_{01} if it is reached from v by following the 0-edge and stopping after a 1-edge leaving an x_j-node. The node set V_{10} is defined by exchanging the roles of 0 and 1.*

The easy proof of this theorem is left as an exercise. The first criterion is very efficient. It is sufficient to run the SAT-COUNT algorithm for each variable x_i by ignoring the 0-edges leaving x_i-nodes. Then we may group the variables with respect to the different values of $c_i := |f^{-1}_{|x_i=1}(1)|$. Only variables with the same c-value can be symmetric. For groups of variables with the same c-value, we apply other criteria of Theorem 5.5.4.

Finally, we describe the main features of the branch-and-bound techniques. They are simple split-and-prune approaches. Since we work with OBDDs, we use the size of the smallest OBDD ever seen as upper bound U on the minimal OBDD size. We always consider situations where the last (or first) k variables of the ordering are fixed. If we prove a lower bound $L \geq U$ for the OBDD size with respect to all considered variable orderings, we may prune this part of the search tree. How do we efficiently obtain good lower bounds?

We assume that the function $f \in B_{n,m}$ essentially depends on all its n variables. The set of nonessential variables is easily detected in the first OBDD. First we consider bottom-up approaches for sets I with $|I| = k$. Let us consider a reduced OBDD testing at first all variables x_i, $i \notin I$, then x_j, and, finally, the remaining variables. Then the following lower bounds hold for the OBDD size for all variable orderings testing all variables x_i, $i \in I$, as last variables:

- $n - k + \min(I)$,
- $s - m + \min(I)$ where s is the number of x_j-nodes,
- $t - m + \min(I)$ where t is the number of edges between x_i-nodes, $i \notin I$, and x_k-nodes, $k \in I$.

The first lower bound is trivial, since we need for functions essentially depending on all their variables at least one x_i-node for each $i \notin I$. The second lower bound (Ishiura, Sawada, and Yajima (1991)) uses the fact that all x_j-nodes have to be reached from the m sources. The least number of nodes is used by m trees. Moreover, a binary tree with l nodes is left by $l+1$ edges. The third lower bound (Jeong, Kim, and Somenzi (1993)) improves the second bound. We need at least $t - m$ nodes in a forest of m binary trees to create t outgoing edges. The algorithm works with the maximum of the lower bounds.

Drechsler, Drechsler, and Günther (1998) use the top-down approach and the lower bound that an OBDD starting with $x_1, \ldots, x_i$ has to contain at least as many nodes labeled by x_j, $j > i$, as there are different subfunctions $f_{|x_1=a_1,\ldots,x_i=a_i}$. This easily follows from Theorem 3.1.4. Now we consider a reduced OBDD testing at first the variables x_i, $i \in I$, where $|I| = k$. Then the maximum of the following lower bounds is used:

- $n - k + \min(I)$,
- $s + \min(I)$ where s is the number of x_k-nodes, $k \notin I$, directly reached from an x_j-node, $j \in I$.

This lower bound is much stronger (for most functions) than the lower bounds used in the bottom-up approach. This makes this approach the most efficient one, although in the worst case top-down algorithms are slower than bottom-up algorithms.

One might ask whether, for some classes of Boolean functions, it is easier to obtain optimal variable orderings. The only known nontrivial class of Boolean functions of this kind is the class of functions given by read-once formulas; see Definition 4.11.1.

Definition 5.5.5. A variable ordering for a function given as a read-once formula is called *DFS ordering* if it can be obtained by a DFS traversal of the read-once formula starting at its output node.

Lemma 5.5.6. *For read-once formulas, the construction in the proof of Theorem 2.1.4 leads to an optimal variable ordering among the set of DFS orderings.*

Proof. This can be proved by induction on the number of variables. The base step $n = 1$ is obvious. For larger n, it is distinguished for which subtree the variables are tested first. Then (by the induction hypothesis) optimal orderings are used for g or $(g, \overline{g})$ and h or $(h, \overline{h})$ and the better alternative is chosen. □

Lemma 5.5.7. *For functions given as read-once formulas there is an optimal variable ordering which is a DFS ordering.*

Sketch of proof. The proof is due to Sauerhoff, Wegener, and Werchner (1996). It is done by induction on the number of variables. For the different cases of the type of the last gate and of the type of considered functions (f or $(f, \overline{f})$), for a given variable ordering π a DFS ordering π' is explicitly defined which for each variable x_i has the property that the number of x_i-nodes in the π'-OBDD is not larger than the number of x_i-nodes in the π-OBDD. The proof is done for each case and level by an application of Theorem 3.1.4. We omit this tedious case inspection. □

Combining Lemma 5.5.6, Lemma 5.5.7, and Corollary 2.1.5, we obtain the following result.

Theorem 5.5.8. *Optimal variable orderings for functions given as read-once formulas on n variables can be computed in linear time $O(n)$.*

5.6 Heuristics for the Computation of Good Variable Orderings

For not very small n and not too simple functions, we cannot hope to compute an optimal variable ordering. In the beginning of the use of OBDDs, a lot of heuristics were presented to compute a variable ordering from a circuit description of the function. These results have become less important. Today, these algorithms may be used to compute an initial variable ordering. However, during the gate-by-gate transformation of circuits into OBDDs, reordering techniques (see Section 5.8) are applied. They are more powerful, since they work on OBDDs and not on the circuit itself. In the following, we describe the DFS heuristic, the fan-in heuristic, the level heuristic, and the dynamic-weight heuristic which compute a variable ordering from a circuit description.

The *DFS heuristic* (Fujita, Fujisawa, and Kawato (1988); Fujita, Fujisawa, and Matsunaga (1993)) is based on the idea that variables should be tested first if they influence many subparts of the function and the difficult outputs of the function. Moreover, variables which are "close together" in the circuit should be close together in the variable ordering. The DFS traversal is started at an output depending on the maximal number of variables. Then the variables are ordered as they are found. For the remaining variables a DFS traversal from the other outputs is started. For each output one can use the idea to put the variables with fan-out 1 at the end of the ordering, since their "influence" seems to be local. The runtime is linear with respect to the circuit size. Sometimes, it has turned out to be useful to combine the outputs by a dummy gate and to work with this new single-output circuit.

The *level heuristic* (Malik et al. (1988)) uses the idea that the influence of variables on long computation paths is "larger" than on short computation paths. For each variable (and gate), we define its *depth* as the length of the longest path to some output (depth in circuits usually is the length of the longest path from an input to the gate). The variables with the same depth belong to the same level. The variables of the same level are grouped together and the variables with larger depth are tested earlier.

Malik et al. (1988) also introduced the *fan-in heuristic* which combines ideas of the DFS and the level heuristics. They use the DFS heuristic but they choose an ordering of the direct predecessors of a gate by descending depth. All these heuristics run in linear time.

Minato, Ishiura, and Yajima (1990) also tried to find "important" variables and to group "related" variables together. Their *dynamic weight heuristic* works on single-output circuits (perhaps with a dummy output gate). The output gate gets weight 1. A gate with weight w and p direct predecessors gives weight w/p to each predecessor. The gates sum the weights they get from the different direct successors. Then the variable with the largest weight is chosen as the first variable (see Fig. 5.6.1). It is plausible that x_4 has the largest influence on the output and that x_1, x_2, and x_3 have the same influence.

Then x_4 is chosen as the first variable and eliminated from the circuit. The recomputation of the weights increases the weight of the "adjacent" variable x_3 more than the weight of the variables x_1 and x_2, which are "far away." Hence, this method increases the chance of grouping adjacent variables together. Because of the recomputation of the weights, this method takes quadratic time.

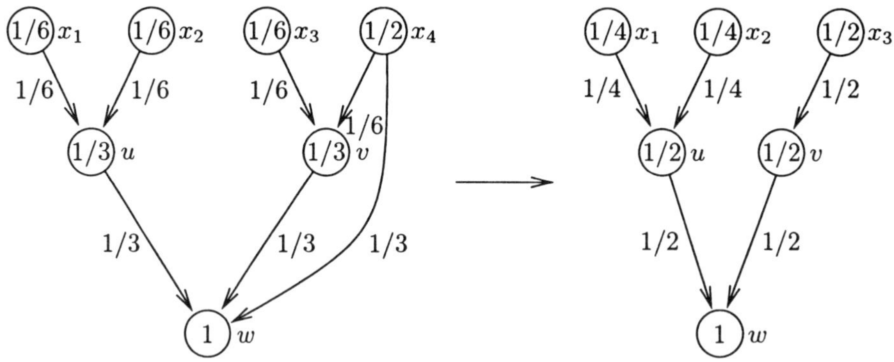

Figure 5.6.1: *Assigning weights with the dynamic weight heuristic.*

Another method to estimate the importance of variables has been suggested by Calazans et al. (1992). Butler et al. (1991) used ideas from simulation and testability theory to obtain variable orderings. The approach of Besson et al. (1993) is based on kernel analysis while Wu and Marek-Sadowska (1993) use methods from cover pattern processing.

Ross, Butler, Kapur, and Mercer (1991a) tried to compare the variable orderings computed by different heuristics before the corresponding OBDDs are constructed completely. For this reason, they choose a small size bound for OBDDs. If the size bound is reached, all edges which would lead to further nodes are gathered in a sink labeled by $*$. Hence, we get the $\{0, 1, *\}$-representation of an incompletely specified function f^* such that the given function f is an extension of f^* (see also Section 3.6). In order to treat all assignments to the first variables according to the variable ordering in a fair manner, they use a BFS approach for the construction of OBDDs (see Section 3.4). Then the number of inputs where the OBDD for f^* evaluates to 0 or 1 is a measure for the quality of the variable ordering.

One also may use special heuristics taking into account the knowledge of the special application. We refer to Fujita, Matsunaga, and Kakuda (1991); Jeong et al. (1991); and Fujii, Ootomo, and Hori (1993). For the special situation of general threshold functions (see Section 4.8), Hosaka et al. (1997) suggest the variable ordering according to descending weights. Large weights make a variable "important." This often works quite well but Exercises 4.16 and 4.17 show that this also may lead to exponential OBDD size in situations where small polynomial size is possible. This is the nature of all the heuristics.

5.7 Changing the Variable Ordering

Often, we have to improve the chosen variable ordering. One reason is that, during a gate-by-gate transformation of circuits into OBDDs, we find out that we have chosen a variable ordering leading to too large OBDDs. Another reason is that the variable orderings which lead to polynomial-size OBDDs for the outputs of a circuit may lead to exponential-size OBDDs for some functions computed at inner gates of the circuit.

We are going to describe local changes of the variable ordering which will be applied in local search algorithms such as hill climbing and also in simulated annealing algorithms (see Section 5.8). Using evolutionary algorithms we obtain variable orderings which differ a lot from the given ones. During the transformation of a specification S and a circuit C into OBDDs, these techniques may lead to a π_1-OBDD for S and a π_2-OBDD for C. In order to check them for equivalence, we may try to change the variable ordering of one OBDD to the variable ordering of the other OBDD. We list the operations on variable orderings which we investigate.

Definition 5.7.1. Let π be the given variable ordering. The variable at position i according to π is $x_{\pi^{-1}(i)}$.

(i) The operation *swap*(i), $1 \le i \le n-1$, interchanges the variables at the positions i and $i+1$.

(ii) The operation *jump-up*(i,j), $1 \le j < i \le n$, causes the variable at position i to jump to position j while the variables at the positions $j,\ldots,i-1$ are shifted one position downward.

(iii) The operation *jump-down*(i,j), $1 \le i < j \le n$, causes the variable at position i to jump to position j while the variables at the positions $i+1,\ldots,j$ are shifted one position upward.

(iv) The operation *exchange*(i,j), $1 \le i,j \le n$, interchanges the variables at the positions i and j.

(v) The operation $\pi \rightarrow \pi'$ causes a global rebuilding by replacing the variable ordering π with π'.

The operations *swap*, *jump-up*, *jump-down*, and *exchange* are called local rebuilding operators. *Swap* is a fundamental operator, since all other operators can be performed by a sequence of swaps. For the *jump* operation, $|i-j|$ swaps are sufficient while *exchange* needs $2|i-j|-1$ swaps. It is also obvious that *exchange*(i,j) can be simulated by *jump-up*(j,i) followed by *jump-down*$(i+1,j)$ if w.l.o.g. $i<j$. However, we need $|i-j|$ exchange steps to simulate a jump operation.

The effect of *swap*(i) can be illustrated easily. Without loss of generality, $\pi = id$. Because of Theorem 3.1.4, nothing has to be changed on the levels $j<i$ and $j>i+1$. At the level i, the subfunctions obtained by assigning constants to $x_1,\ldots,x_{i-1}$ and essentially depending on x_i are represented, similarly for

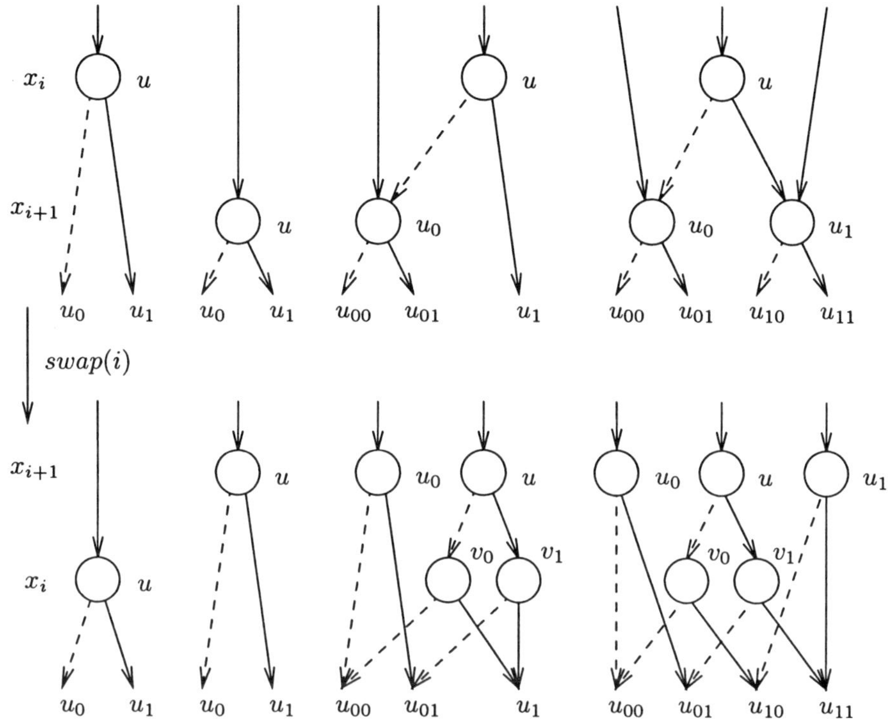

Figure 5.7.1: *The swap operator. The top row shows the levels i and $i+1$ before the operation, the bottom row shows the levels afterwards.*

level $i+1$. In Fig. 5.7.1, all possible configurations (except symmetric ones) are shown.

We see that all edges from the levels $1,\ldots,i-1$ need not be redirected if we relabel the node u in cases 3 and 4 of Fig. 5.7.1 in the right way. Moreover, we have to recompute the number of edges leading to u_0 and u_1 by subtracting the number of edges which have reached them from level i. If nodes lose all edges leading to them, they can be eliminated. The resulting OBDD has to be reduced on the levels $i+1$ and i. The whole procedure runs in linear time with respect to the size of the levels i and $i+1$. How can we bound the size of the resulting OBDD? Let s_j be the number of x_j-nodes before the swap and s_j^* afterwards. Then $s_j^* = s_j$ if $j < i$ or $j > i+1$. Each node on level i (the u-node) creates at most two new nodes (v_0 and v_1). Hence, the size is bounded by $s_1+\cdots+s_{i-1}+3s_i+s_{i+1}+\cdots+s_n+2$. Another bound is obtained as follows. A case inspection of Fig. 5.7.1 proves that $s_i^* \le 2s_i$. This also follows from the fact that each subfunction $f_{|x_1=a_1,\ldots,x_{i-1}=a_{i-1}}$ has only two subfunctions with respect to x_{i+1}. Moreover, $s_{i+1}^* \le s_1+\cdots+s_{i-1}+m$ for functions $f \in B_{n,m}$, since each x_{i+1}-node is reached by an edge from the first $i-1$ levels of the OBDD. The number of edges leaving this part is bounded by $s_1+\cdots+s_{i-1}+m$ if we have m sources, since each of the other nodes is also reached at least once.

Theorem 5.7.2. *The operation swap(i) can be performed in linear time with respect to the size of the levels i and $i+1$. The size of the resulting OBDD is bounded by $s_1+\cdots+s_{i-1}+3s_i+s_{i+1}+\cdots+s_n+2$ in any case and by $2(s_1+\cdots+s_i)+s_{i+2}+\cdots+s_n+m+2$ for functions with m outputs.*

We have seen that a jump operation can be implemented as a sequence of swap operations. But there also is a direct procedure.

Algorithm 5.7.3. Let π be the variable ordering and G the OBDD for f before the $jump(i,j)$ operation and π' the variable ordering afterwards.

(1) Construct G_0 and G_1 from G by replacing x_i by 0 and 1, respectively. Since G_0 and G_1 do not contain x_i-nodes, they are π-OBDDs as well as π'-OBDDs. The OBDD G_c represents $f_{|x_i=c}$.

(2) Apply an ite synthesis step to compute the π'-OBDD for $f=ite(x_i, f_{|x_i=1}, f_{|x_i=0})$.

Before we discuss the time complexity of this algorithm, we derive size bounds for the resulting OBDD (Bollig, Löbbing, and Wegener (1996)). It turns out that there is a remarkable difference between jump-up and jump-down operations.

Theorem 5.7.4. *Let s_k be the size of the x_k-level of an id-OBDD representing $f \in B_{n,m}$. The size of the OBDD resulting from the jump-up(i, j) operation is bounded by*

$$s_1+\cdots+s_{j-1}+3(s_j+\cdots+s_{i-1})+s_i+\cdots+s_n+2$$

and by

$$2(s_1+\cdots+s_{i-1})+s_{i+1}+\cdots+s_n+m+2.$$

Proof. We obtain the first bound by analyzing the sequence $swap(i-1),\ldots,$ $swap(j)$. If we apply $swap(k)$, the x_k-level has not been changed before and each of the s_k x_k-nodes creates at most two new nodes, leading to the first upper bound.

For the second bound, we get $s_k^*=s_k$ for $k<i$ and $k>j$ if s_k^* denotes the number of x_k-nodes after the jump-up. Moreover, $s_k^*\le 2s_k$ for $j\le k\le i-1$, since $f_{|x_1=a_1,\ldots,x_{k-1}=a_{k-1}}$ has two subfunctions with respect to x_k. Finally, $s_i^*\le s_1+\cdots+s_{j-1}+m$, since the x_i-nodes have to be reached from the top part. □

Theorem 5.7.5. *Let s_k be the size of the x_k-level of an id-OBDD. The size of the OBDD resulting from the jump-down(i, j) operation is bounded by*

$$s_1 + \cdots + s_i - 1 + (s_{i+1} + \cdots + s_n + 2)^2$$

and also by

$$s_1 + \cdots + s_n + 2 + (2^{j-i+1} - 2)s_i.$$

Proof. First, we prove the second bound resulting from an analysis of the sequence *swap*$(i), \ldots,$ *swap*$(j-1)$. A case inspection of Fig 5.7.1 shows that after *swap*(i) the x_{i+1}-level contains at most $s_i + s_{i+1}$ nodes while the x_i-level contains at most $2s_i$ nodes. Therefore, after *swap*$(i+1)$ the x_{i+2}-level contains at most $2s_i + s_{i+2}$ nodes and the x_i-level at most $4s_i$ nodes. Altogether, this leads to $s_{i+1} + \cdots + s_j + s_i(1 + 2 + \cdots + 2^{j-i-1})$ nodes on the levels $i, \ldots, j-1$ and 2^{j-i} x_i-nodes. Hence, the size increase is bounded by $s_i(2^{j-i+1} - 2)$.

For the proof of the first bound, we estimate the number of nodes on each level. Again, $s_k^* = s_k$ if $k \le i-1$ or $k \ge j+1$. For an x_k-level, $i+1 \le k \le j$, we have to consider the functions $\overline{x}_i g_0 + x_i g_1$ essentially depending on x_k, where g_c is a subfunction of $f_{|x_i=c}$ obtained by assigning constants to $x_1, \ldots, x_{i-1}, x_{i+1}, \ldots, x_{k-1}$. At least one of the functions g_0 or g_1 has to depend essentially on x_k. Hence, $s_k^* \le s_k^2 + 2s_k(s_{k+1} + \cdots + s_n + 2)$.

Finally, we have to estimate s_i^*. Again, we have to consider the functions $\overline{x}_i g_0 + x_i g_1$ essentially depending on x_i. Now $x_1, \ldots, x_{i-1}, x_{i+1}, \ldots, x_j$ are replaced with constants. The function $\overline{x}_i g_0 + x_i g_1$ essentially depends on x_i iff g_0 and g_1 are different. Hence, $s_i^* \le (s_{j+1} + \cdots + s_n + 2)(s_{j+1} + \cdots + s_n + 1)$. Putting all estimates together, we obtain the following upper bound on the size of the resulting OBDD:

$$\sum_{1 \le k \le i-1} s_k + \sum_{i+1 \le k \le j} (s_k^2 + 2s_k(s_{k+1} + \cdots + s_n + 2))$$

$$+(s_{j+1} + \cdots + s_n + 2)(s_{j+1} + \cdots + s_n + 1) + \sum_{j+1 \le k \le n} s_k + 2$$

$$= s_1 + \cdots + s_{i-1} + (s_{i+1} + \cdots + s_n + 2)^2. \qquad \square$$

Our bounds imply that jump-up may only double the OBDD size (if $m = 1$) while the size bound for jump-down grows quadratically in the size of the given OBDD. Since jump-up and jump-down are inverse operations, jump-down may only halve the size while we may hope that jump-up may decrease the OBDD size almost to the square root of the given size. Are these remarkable differences possible or due only to our estimations?

Theorem 5.7.6. *There exists a function with linear π-OBDD size and quadratic π'-OBDD size, where π' is the result of a jump-down operation on π.*

Proof. The example is based (as in the proof of Theorem 3.3.7) on the DSA function. Let

$$f(s, x, y, z) := s \wedge \mathrm{DSA}(x, z) + \overline{s} \wedge \mathrm{DSA}(y, z).$$

The OBDD size is $O(n)$ for the variable ordering s, x, y, z. Now we let s jump down to the end of the variable ordering (or only after x and y). After having tested x and y we obtain the n^2 different subfunctions $sz_i + \overline{s}z_j$, $1 \leq i, j \leq n$. □

Usually, it is very efficient to implement jump operations as sequences of swaps. However, we cannot be sure that intermediate results are not much larger than the final result. If we apply Algorithm 5.7.3, we perform one synthesis step. The time for such a synthesis is bounded by the number of reachable nodes in the product graph which can be much larger than the reduced OBDD. But in our special situation this cannot happen. It follows from the next theorem that the number of reachable nodes in the product graph is at most $2|G^*|+1$ if G^* is the final reduced result. This proves that jump operations can be performed with hashing strategies in expected linear time with respect to input and output size.

Theorem 5.7.7. *Let G be a π-OBDD representing $x_i g \in B_n$ and H be a π-OBDD representing $\overline{x}_i h \in B_n$, where g and h do not essentially depend on x_i. Let $f := x_i g + \overline{x}_i h$ and F be the π-OBDD representing f. Then the number of reachable nodes in the product graph F^* of G and H is bounded by $2|F|+1$.*

Sketch of proof. The complete proof is given by Savický and Wegener (1997). Let, w.l.o.g., $\pi = id$. It is shown that x_j-nodes in F^*, where $j < i$, are neither deleted nor merged. The intuitive reason is that x_i is not yet tested. If we assume that we can delete or merge such nodes, we can find nodes in G or H which could be deleted or merged in contradiction to our assumption that we work with reduced OBDDs. The deletion of x_i-nodes is possible but the number of x_i-nodes is only larger by 1 than the number of nodes above the x_i-level. Deletions are not possible for x_k-nodes, $k > i$. Otherwise, deletions would be possible in G or H. Mergings on x_k-levels, $k > i$, are possible only for nodes (u, v) and (u', v'), where u and v' represent the same function essentially depending on x_k and v and u' represent the constant 0 or the same with interchanged roles of u and v' on one hand and v and u' on the other hand. Hence, at most two nodes can be merged to a single one, which only can halve the size of these levels. □

The exchange operation can be implemented as a jump-down followed by a jump-up. Hence, the size can grow and shrink at most quadratically. Wang, Hwang, and Chen (1993) have shown how we can directly implement an exchange step. Let $f_{|x_i \leftarrow \overline{x}_i}$ be the function obtained from f if we replace x_i by $\overline{x}_i$. In order to apply this replacement to an OBDD it is sufficient to interchange the 0- and 1-edges leaving x_i-nodes. The algorithm for the exchange operation works as follows.

Algorithm 5.7.8. A π-OBDD G for f is transformed into a π'-OBDD G' for f where π' results from π by *exchange*(i,j).

(1) Compute a π-OBDD G^* by copying G and performing the operation $f \to f_{|x_i \leftarrow \overline{x}_i, x_j \leftarrow \overline{x}_j}$.

(2) Interchange the labels of x_i- and x_j-nodes in G and G^*. The results are π'-OBDDs H and H^* representing functions called h and h^*.

(3) Use the synthesis algorithm to compute a π'-OBDD G' for $(x_i \oplus x_j \oplus 1)h + (x_i \oplus x_j)h^*$.

Theorem 5.7.9. *The π'-OBDD G' computed in Algorithm 5.7.8 represents f.*

Proof. Let a be an input vector. It is sufficient to prove that $f(a) = h(a)$ if $a_i = a_j$ and $f(a) = h^*(a)$ if $a_i \neq a_j$. Let p and p^* be the paths in G, respectively, G^* activated by a. If $a_i = a_j$, the activated path remains unchanged if we interchange the labels of x_i- and x_j-nodes. Hence, we reach the same sink in G as in H and $f(a) = h(a)$. Now let $a_i \neq a_j$. The path p^* contains at most one x_i-node v and at most one x_j-node w. The node v is left via the a_i-edge and w via the a_j-edge, if existent. After interchanging the variable labels and the labels of the outgoing edges, the activated path reaches v (if existent) which is an x_j-node and v has to be left via its a_j-edge which is the $\overline{a}_i$-edge. Since the 0- and 1-edges have been interchanged, this is the former a_i-edge. Hence, we follow p^*. The same argument works for w (if existent). Hence, $f(a) = h^*(a)$. □

We are left with the global rebuilding problem. Let G be the given π-OBDD for f and H the resulting π'-OBDD for f. Tani and Imai (1994) have presented an algorithm which runs in time $O(|G||H^*|)$ and uses space $O(|G| \cdot W(H^*))$, where H^* is the quasi-reduced π'-OBDD for f and $W(H^*)$ its width, i.e., the maximal number of nodes with the same label. At the same time, Meinel and Slobodová (1994) have obtained a linear $O(|G|+|H|)$-space solution which runs in time $O(|G||H|^2)$. Savický and Wegener (1997) have presented a solution which is efficient with respect to space *and* time.

Theorem 5.7.10. *The π'-OBDD H for f can be computed from the π-OBDD G using space $O(|G|+|H|)$ and time $O(|G||H|\log|H|)$. Using hashing strategies the expected time is $O(|G||H|)$.*

The less space-efficient solution of Tani and Imai (1994) is more time efficient than the deterministic solution of Savický and Wegener (1997) if H^* is almost of the same size as H. But the use of hashing strategies is not unusual in OBDD algorithms.

Proof of Theorem 5.7.10. We assume that f essentially depends on n variables. By renumbering the variables we can assume that $\pi' = id$. For the construction of H we use a top-down approach, i.e., we construct H level by level as in a BFS approach. Without the possibility of applying reduction rules, we obtain a complete binary tree of exponential size. We use G to prevent the creation of nodes which would be deleted later. We also use G to perform all possible mergings at the x_i-level of H before the next level is constructed. During the construction of H, we store all edges in both directions in order to efficiently find a path from a node v to a source.

At the beginning, H consists of two sinks. Furthermore, we initialize node lists $L(1), \ldots, L(n)$. For each output of f, a candidate node is created and inserted into $L(i)$ if the output essentially depends on x_i but not on $x_1, \ldots, x_{i-1}$. The list $L(1)$ is not empty, since, by our assumptions, some output essentially depends on x_1. Then the procedure *construct*(k) is called for $k = 1, \ldots, n$. It assumes that the levels $1, \ldots, k-1$ of H are already constructed and $L(k)$ contains the candidate nodes of level k. This is fulfilled at the beginning for $k = 1$.

First, we describe *construct*(k), omitting the details.

(1) Let $v_1, \ldots, v_{r(k)}$ be the nodes in $L(k)$. It is decided which nodes have to represent the same subfunction of f. The possible mergings are performed.

(2) Let $w_1, \ldots, w_{s(k)}$ be the remaining nodes on level k. For each w_j and each $c \in \{0, 1\}$, it is decided whether the c-successor of w_j in H is the 0-sink, the 1-sink, or a node with label x_m, $k < m \le n$. Either an edge to the appropriate sink is created or an edge to a new x_m-node which is inserted into $L(m)$.

It is obvious that we construct a π'-OBDD representing f. All nodes are reachable from the sources and we prevent a reduction rule from being applicable. Hence, we construct H. For the analysis, it is important that, after calling *construct*(k), we do not eliminate edges to the x_k-level and each edge is redirected at most once.

Now we describe the details of our implementation. In step 1, we identify the node v_i with a reduced π-OBDD $G_{i,k}$. For this reason we start at v_i and run backward until we find a source. This describes a partial assignment to some of the variables $x_1, \ldots, x_{k-1}$. We obtain $G_{i,k}$ from G in time $O(|G|)$ by performing this assignment to the sub-OBDD of G with the only source we have found. The path in H has length $O(n) = O(|G|)$, since f essentially depends on n variables. We encode $G_{i,k}$ by a string of $O(|G|)$ numbers. The codeword for $G_{i,k}$ is obtained by a DFS traversal (0-edges before 1-edges) through $G_{i,k}$. Each node is described by its DFS number, its label, its 0-successor, and its 1-successor. The π-OBDDs $G_{i,k}$ and $G_{i',k}$ represent the same function iff their codewords are *identical*. Moreover, let $v_i < v_{i'}$ if the codeword for $G_{i,k}$ is smaller than the codeword for $G_{i',k}$ with respect to the lexicographical ordering.

The nodes $v_1, \ldots, v_{r(k)}$ are inserted into an empty AVL tree. The binary search tree can be replaced by hashing. We use the ordering defined above. We do not store the codewords in order to save space. They are recomputed whenever necessary. Each of the $r(k)$ insertions into the AVL tree can be done in $O(\log r(k))$ operations and each operation takes time $O(|G|)$, since we have to compute $O(1)$ codewords. Since $r(k)$ is the number of edges in H leading to x_k-nodes, the time for all operations on all levels is bounded by $O(|G||H|\log|H|)$. The extra amount of space is $O(|G|)$, since we only store two codewords at the same time. If two nodes have the same codeword, they are merged using the reversed edges. This can be implemented in such a way that each edge is redirected at most once.

Step 2 is easier. For the c-successor of w_j, we use the partial assignment found on a path from w_j to a source v^* and the additional assignment $x_k := c$. This assignment is applied to the sub-OBDD of G with the only source v^*. In the resulting OBDD $G_{j,k,c}$, we look for the minimal m such that $G_{j,k,c}$ contains an x_m-node. Then the c-successor of w_j in H is an x_m-node. If $G_{j,k,c}$ consists only of a sink, this sink is the c-successor of w_j in H. It is easy to see that step 2 can be performed for all levels in time $O(|G||H|)$. □

A runtime of $O(|G||H|)$ (using hashing strategies) looks like the runtime for a synthesis step. For synthesis steps, the corresponding bound is a worst-case bound and we hope that the algorithm usually is much faster. Here the bound $O(|G||H|)$ describes the order of the runtime in all cases, i.e., this algorithm really needs quadratic time, which is a lot for large OBDDs.

The size $|G|$ is known in advance while $|H|$ is unknown. So we may be interested in constructing H only if $|H| \le max$ for a given size bound max. Since we do not eliminate edges, we know for nonconstant functions after the creation of the $(2max - 3)$th edge that the OBDD will have more than max nodes and we can stop the algorithm. For $MAX := \min\{max, |H|\}$, the runtime of our algorithm then is bounded by $O(|G| \cdot MAX \log MAX)$ in the deterministic version and by $O(|G| \cdot MAX)$ in the randomized version.

As a corollary, we obtain the following result on the equivalence test for OBDDs with different orderings.

Corollary 5.7.11. *The equivalence of a π-OBDD G and a π'-OBDD H can be decided in time $O(|G||H|\log|H|)$ (expected time $O(|G||H|)$) using linear space $O(|G| + |H|)$.*

Proof. We apply the global rebuilding technique to construct an equivalent π'-OBDD G' from G and use the size bound $|H|$. If $|G'| = |H|$, we perform the equivalence test for OBDDs with the same variable ordering. □

5.8 Reordering Techniques

In typical situations, it is impossible to compute an optimal variable ordering (see Section 5.5) and the heuristics working on circuit descriptions (see Section 5.6) may behave very poorly. In these situations, strategies of *dynamic reordering* are applied. One starts with some variable ordering computed by some heuristic. The functions computed at the inputs and the first gates of a circuit depend only on a small number of inputs and can be represented with the given variable ordering. If the OBDD becomes too large, one tries to find a better variable ordering. Such a reordering algorithm may be started by the user or automatically, e.g., if the OBDD size has been increased by some specified factor. One can easily construct situations where such a reordering is necessary. There exist circuits where, for each gate, the function represented at that gate has small π-OBDD size for an appropriate π but there is no variable ordering π^* such that π^* is good for all gates.

In Section 5.7, we presented a pool of methods for local and global reordering techniques. We may apply all types of search algorithms such as local search, hill climbing, stochastic evolution, simulated annealing, and evolutionary algorithms.

We have to distinguish how we define the neighborhood of a variable ordering. Swap operations can be performed quite efficiently but these neighborhoods lead to a large number of local optima. With jump operations we are much more flexible than with exchange operations. With two jumps we can simulate an exchange step but we are not forced to exchange two variables. Mercer, Kapur, and Ross (1992) allow jump operations where the distance is bounded by a constant c (in their case $c = 5$). Fujita, Matsunaga, and Kakuda (1991) and Ishiura, Sawada, and Yajima (1991) present the *window permutation algorithm* with window size k. Some level $i \in \{1, \ldots, n-k+1\}$ is chosen and all $k!$ variable orderings of the variables at the positions $i, \ldots, i+k-1$ are investigated. The reorderings are realized as a sequence of swaps. It has turned out that larger jumps are necessary to escape from local optima. The *sifting algorithm* due to Rudell (1993) is currently the most popular dynamic reordering technique (some preliminary ideas are contained in Jacobi, Calazans, and Trullemans (1991)). The variables are sorted into decreasing order based on the number of OBDD nodes labeled with the same variable. Then each variable is moved through the variable ordering using swaps. It is first moved to the closer end and then backward in the other direction. By this approach, we find the optimal position under the assumption that all other variables remain in their relative order. The variable is finally moved to its optimal position. Since each variable is moved only once, a sifting procedure consists of $O(n^2)$ swaps. Rudell (1993) suggested stopping the movement of a variable in one direction if the OBDD size becomes at least doubled, i.e., is increased by the factor $c = 2$. Obviously, we get a time-space trade-off. The algorithm becomes faster if we do not allow a size increase of more than a small factor c, but we explore more possibilities for large c-values. This enables us to escape more often from local optima. It depends on the application how to choose the parameters.

We have seen in Section 5.5 that a test whether two variables x_i and x_j are symmetric is particularly simple if the variables are adjacent in the variable ordering. If x_i is moved, it becomes adjacent to each variable and it is easy to compute the set of variables which are in the same equivalence class as x_i with respect to the relation symmetry. This has been noticed by Panda, Somenzi, and Plessier (1994). They also discuss the following difficulty with the sifting approach. Let S and T be two sets of symmetric variables and U be the set of the remaining variables. Let π be a variable ordering where all S-variables are tested before all T-variables and let π' be the variable ordering where the group of T-variables jumps before the group of S-variables. If π' is better than π, sifting may not detect this. If a T-variable (or an S-variable) is moved, it may not find a better position than the current one. The reason is that it may be more important to test the T-variables in adjacent order than to test one T-variable in front of the S-variables.

Example 5.8.1. Let $f_n(x_1, \ldots, x_n, y_1, \ldots, y_n) = T_{3,n}(x) \oplus T_{4,n}(y)$. In the following, we estimate the OBDD size up to additive constants. The threshold functions are monotone. Therefore, OBDDs for $T_{k,n}$ and $\overline{T}_{k,n}$ cannot share inner nodes. Since $T_{3,n}(x)$ and $T_{4,n}(y)$ are defined on disjoint sets of variables, we may use the methods of the proof of Theorem 2.1.4 to obtain the following results. If the x-variables are tested before the y-variables, the OBDD size equals $11n \pm O(1)$. This is not optimal, since the reversed variable ordering leads to an OBDD size of $10n \pm O(1)$. Here we have used the fact (Theorem 4.7.2) that the OBDD size of $T_{k,n}$ equals $kn \pm O(1)$ for constant k. But we cannot improve the first ordering by moving a y-variable upward or an x-variable downward. For the ordering $x_1, \ldots, x_{n-1}, y_1, \ldots, y_n, x_n$, we need $15n \pm O(1)$ nodes, approximately $3n$ on the first $n-1$ levels, then $12n \pm O(1)$ y-nodes to represent $T_{4,n}(y)$, $T_{4,n}(y) \oplus x_n$, and $T_{4,n}(y) \oplus 1$. For the ordering $y_1, x_1, \ldots, x_n, y_2, \ldots, y_n$, we need $14n \pm O(1)$ nodes, approximately $6n$ x-nodes to represent $T_{3,n}(x) \oplus T_{4,n}(c, y_2, \ldots, y_n)$ for $c \in \{0, 1\}$, and then $8n \pm O(1)$ y-nodes as in the ordering x, y.

Panda, Somenzi, and Plessier (1994) "lock" variables as soon as they are identified as symmetric. Groups of symmetric variables have to stay together. They are moved only as a block. One may realize a generalized swap of two groups of symmetric variables as a sequence of simple swaps or by a special procedure (see exercises). This procedure is called *group sifting*.

We have seen in Example 5.8.1 that group sifting can be successful in situations where sifting fails. Are there also examples where no symmetric variable ordering, i.e., variable orderings where all equivalence classes of symmetric variables are tested one after the other, is optimal? The following example is due to Sieling (1998b).

Example 5.8.2. The function $f_n \in B_{n+1}$ is defined for $n = 3k+1$, $k \in \mathbb{N}$, by

$$f_n(y, x_1, \ldots, x_n) := \begin{cases} 1 & \text{if } x_1 + \cdots + x_n = k, \\ y & \text{if } x_1 + \cdots + x_n = 2k+1, \\ 0 & \text{otherwise.} \end{cases}$$

It is easy to obtain the following results on the OBDD sizes of the only two symmetric variable orderings and a better (and even optimal) nonsymmetric variable ordering (w.l.o.g., n is even):

$$x_1, \ldots, x_n, y \quad : \quad \frac{1}{3}n^2 + \frac{4}{3}n + \frac{7}{3},$$

$$y, x_1, \ldots, x_n \quad : \quad \frac{1}{3}n^2 + \frac{4}{3}n + \frac{10}{3},$$

$$x_1, \ldots, x_{n/2}, y, x_{n/2+1}, \ldots, x_n \quad : \quad \frac{1}{4}n^2 + \frac{4}{3}n + \frac{8}{3}.$$

The performance ratio $R(f)$ of f according to symmetric variable orderings is defined as the ratio of the OBDD size according to a best symmetric variable ordering and the OBDD size according to an optimal variable ordering. For our example, $R(f_n) \to \frac{4}{3}$ as $n \to \infty$. No example with a worse performance ratio is known but neither is a nontrivial upper bound on the performance ratio. The following results (Sieling (1998b) and Sieling and Wegener (1998a)) justify the approach of group sifting.

Theorem 5.8.3.

(*i*) *For a function chosen randomly among the functions with a constant number d of symmetric groups of size n, the probability that only symmetric variable orderings are optimal equals* $1 - o(1)$.

(*ii*) *For a function chosen randomly among the functions with n symmetric groups of constant size d, the probability that only symmetric variable orderings are optimal equals* $1 - e^{-\Omega(n)}$.

Theorem 5.8.3 contains the case of many symmetric groups of constant size as well as the case of a constant number of symmetric groups of equal size. The proofs are quite complicated and omitted here. This also holds for the next result about a situation where optimal variable orderings have to be nonsymmetric. But in that case the advantage of nonsymmetric variable orderings is marginal.

Theorem 5.8.4.

(*i*) *For a function chosen randomly among the functions with two symmetric groups, one of size* 1 *and one of size* $n-1$, *the probability that only nonsymmetric variable orderings are optimal equals* $1 - e^{-\Omega(n^{1/2})}$.

(*ii*) *In the same situation and for each* $\delta > 0$, *with a probability of* $1-e^{-\Omega(n^{1/2})}$, *the performance ratio is bounded by* $1+\delta$.

Such results also motivate the following approach of Panda and Somenzi (1995). Besides the known (positive) symmetries defined by $f_{|x_i=0,x_j=1} = f_{|x_i=1,x_j=0}$, they consider "negative symmetries" defined by $f_{|x_i=0,x_j=0} = f_{|x_i=1,x_j=1}$. Both types of symmetry can be checked very efficiently for adjacent variables. If x_i precedes x_j, all nodes reaching x_j-nodes have to reach x_i-nodes first and then the assignments $(0,1)$ and $(1,0)$ in the case of positive symmetry and the assignments $(0,0)$ and $(1,1)$ in the case of negative symmetry have to reach the same node. We may call the variables *almost symmetric* if the condition is fulfilled for all but an ε-fraction of all x_i-nodes. Panda and Somenzi (1995) have considered group sifting with groups of almost symmetric variables.

Incompletely specified functions may have extensions where x_i and x_j or x_i and x_k are symmetric but no extension with $\{x_i, x_j, x_k\}$ as a symmetric group. Scholl, Melchior, Hotz, and Molitor (1997) and Scholl, Möller, Molitor, and Drechsler (1999) present methods to use "possible symmetries" for the variable-ordering problem.

Panda and Somenzi (1995) also used information on the shape of the OBDD. Let $N(i)$ be the size of level i. Then a typical OBDD has the property that the N-function looks like a pear or like an onion, increasing to a single maximum and then more quickly decreasing. Let

$$S(i) := N(i+1)/N(i) - N(i)/N(i-1).$$

A negative $S(i)$ indicates that $N(i+1)$ is relatively small, i.e., many mergings take place and this may be a good reason to group the variables in level i and $i+1$. This leads to "soft groups" of variables, i.e., groups of variables which temporarily are moved together in the group-sifting process. In later phases of the process, these variables may become members of different soft groups.

The considered reordering techniques work with local changes of the variable ordering. Therefore, they are quite efficient and often lead to satisfying results. For cases where they are too time-consuming or do not lead to OBDDs of small enough size, one may try to find better starting variable orderings and more information on good variable orderings. Jain, Adams, and Fujita (1998) and Slobodová and Meinel (1998) present sample methods. For example, for some number k of variables, a sample of t random assignments is considered. This leads to subfunctions on smaller sets of variables where it is easier to find good variable orderings with the above-mentioned methods. It is also discussed how to use this information for a heuristic approach to the variable-ordering problem.

All the considered methods are hill climbers which only accept better variable orderings. In the sifting algorithm, x_i is moved through all possible positions. Hence, we may construct larger OBDDs. But, finally, x_i is moved to another position only if we obtain a smaller OBDD. We know from many optimization problems that we sometimes have to accept worse solutions to reach regions of better solutions which are far away from the present solution. Considering the variable-ordering problem, we have to cope with an additional problem. For, e.g., the traveling salesperson problem (TSP), it is easy to compute the cost of any tour, even if it is a very bad one. For the variable-ordering problem, we do not know of any other method to compute the π-OBDD size as to construct the π-OBDD.

Calazans et al. (1992) used an approach known as *stochastic evolution*. They use a hill climber based on swaps. If one gets stuck in a local optimum, one accepts random swaps which increase the OBDD size. It turns out that swaps are too local.

Ishiura, Sawada, and Yajima (1991) presented a *simulated annealing* approach based only on exchanges. We have already discussed the disadvantages of the exchange operation. A more refined simulated annealing algorithm based on jumps and exchanges (Bollig, Löbbing, and Wegener (1995)) leads to much better results. Large improvements for several benchmark functions have been obtained by this approach but the runtime for these successes is huge. It took some time until variable orderings of the same quality were obtained by a faster algorithm. We give only a brief description of simulated annealing; for more details see van Laarhoven and Aarts (1987). Starting with some variable ordering π, a local change, i.e., a jump or an exchange, is chosen randomly. If the new variable ordering π' is at least as good as π, we go on with π'. But if π' is worse than π, we also accept π' with some positive probability as the new variable ordering. This probability is given by the so-called Metropolis function $e^{-d/T}$, where d is the difference between the size of the π'-OBDD and the size of the π-OBDD and $T \geq 0$ is the current "temperature." We are more likely to accept π' if it is only a little worse than π and if the temperature is still high. One has the freedom to choose the initial temperature T_0 and the schedule when and how to decrease the temperature until the process gets "frozen" with temperature $T = 0$. For $T = 0$, we define $e^{-d/T} := 0$ and finish the process with a local search or hill climber.

The class of *evolutionary algorithms* describes a frame for search algorithms which have turned out to be successful in many areas (for introductions see Bäck (1996) and Fogel (1996)). The main idea is to work with a population of p variable orderings, where permutations are represented as lists of integers. The variable orderings π are compared by their *fitness*, which is defined as the π-OBDD size of the considered function. Essential operators are *recombination* and *mutation*. For a recombination, two parents are chosen randomly with some advantage for the fitter subjects. These parents create children. A simple crossover does not work for permutations. People who have tried to solve the TSP with evolutionary algorithms were faced with the same problem. They have suggested a lot of operators to create children. One is the operator partially matched crossover (PMX). PMX chooses two cut positions i and $j \geq i$. Let π_1 and π_2 be the parents. The first child inherits the positions $i, \ldots, j$ from π_1. For the other positions, the same order is chosen as in π_2. The second child is created in the same way with interchanged roles of π_1 and π_2.

Example 5.8.5. $\pi_1 = 14325$, $\pi_2 = 45231$, $i = 2$, $j = 3$. The first child ϱ_1 inherits the information $*43**$ from π_1. The missing indices occur in π_2 in the order 521. Hence, $\varrho_1 = 54321$ and $\varrho_2 = 15243$.

The example shows that both children typically are quite different from their parents. Hence, it is time-consuming or even impossible to evaluate the fitness of the children. The same happens for other recombination operators. Drechsler, Becker, and Göckel (1996) present a genetic algorithm for the variable-ordering problem which is based on the operator PMX. This approach can be applied only if the OBDD sizes are small. The conclusion is to do without recombinations.

Based on the above experiences, Drechsler and Göckel (1997) present an evolutionary algorithm for the variable-ordering problem. It works with small populations whose size most often is 1 and is always bounded by 5. The reason is that it is much more difficult to evaluate the fitness function than in typical applications of evolutionary algorithms. The first variable ordering (or orderings) is computed by some heuristic, since random orderings are often so bad that the corresponding OBDD cannot be constructed. The following mutation operations are applied with some positive probability. For *inversion*, two cut points i and $j > i$ are chosen randomly and the ordering of the variables at the positions $i, \ldots, j$ is inverted. The other operator is sifting, which is always applied after an inversion. Each mutation is stopped if the OBDD size is doubled. The individuals for a mutation are chosen by roulette-wheel selection. The p best of p parents and p children are selected as the population of the next generation, i.e., a $(p + p)$ evolutionary strategy is applied. This approach leads to good variable orderings (sometimes the best known) but it is quite time-consuming.

5.9 Transformations of Boolean Functions

As motivation, we consider the problem of the computation of the coefficients of a polynomial $p = p_1 \cdot p_2$ from the coefficients of the n-degree polynomials p_1 and p_2. The obvious procedure takes quadratic time $\Theta(n^2)$. But it is well known that the following procedure based on the Fourier transformation is much more efficient. Its runtime equals $\Theta(n \log n)$.

(1) Compute the Fourier transformation of p_1 and p_2 considered as polynomials of degree $2n$.

(2) Multiply the $2n$ corresponding function values computed in step 1.

(3) Apply the inverse Fourier transformation to the result of step 2.

The general message of this approach is that a problem may become easier if it is transformed into another type of representation. We have to take into account the time for the transformation and its inverse.

Definition 5.9.1. A *cube transformation* is a one-to-one mapping $\tau\colon \{0,1\}^n \to \{0,1\}^n$.

A cube transformation is nothing more than a renumbering of the vertices of the n-cube. Let $\tau = (\tau_1, \dots, \tau_n)$. Instead of asking for $x_1 = a_1, \dots, x_n = a_n$ in order to reach the sink with value $f(a)$, we now ask for $y_1 = \tau_1(a), \dots, y_n = \tau_n(a)$. The resulting OBDD has been called a *τ-transformed OBDD* (τ-TBDD) by Bern, Meinel, and Slobodová (1995).

Definition 5.9.2. Let $\tau\colon \{0,1\}^n \to \{0,1\}^n$ be a cube transformation. A τ-TBDD G is an OBDD defined on $Y_n = \{y_1, \dots, y_n\}$ with the variable ordering $y_1, \dots, y_n$. The output of the function represented at v on input a is the label of the sink reached on the path starting at v and activated by $\tau_1(a), \dots, \tau_n(a)$.

We need not care about a variable ordering on Y_n, since we may encode each variable ordering in τ by the appropriate permutation of the components of τ.

Lemma 5.9.3. *Each Boolean function $f \in B_n$ has for some τ a τ-TBDD representation of linear size.*

Proof. Let $m = |f^{-1}(1)|$. Then we choose a transformation τ which maps the elements of $f^{-1}(1)$ to the vectors $b \in \{0,1\}^n$ which are the binary representations of $0, \dots, m-1$. Hence, the τ-TBDD for f has only to check whether the value of the (transformed) input is smaller than m. □

The proof of the lemma reveals the catch. For many functions, it seems to be hard to compute or even to represent the transformation τ.

Which operations do we need to convert a circuit into a τ-TBDD for some given τ? First, we need a representation of each variable x_i, $1 \le i \le n$. If τ^{-1} is the inverse of τ, x_i is the ith component τ_i^{-1} of τ^{-1}. Hence, we only can work efficiently with τ-TBDDs if the components of τ^{-1} have a compact representation. The synthesis with integrated reduction causes no problem. Having chosen τ, we work with OBDDs based on another numbering of $\{0,1\}^n$. Hence, reduced τ-TBDDs are a canonical representation of Boolean functions. The equivalence test can be done in constant time if we work with τ-TBDDs which share nodes. For the evaluation problem, we need a compact representation of τ.

The operation replacement by constants and, therefore, also the operations replacement by functions, redundancy test, and quantification cause problems.

We present an example of a function with exponential OBDD size and linear τ-TBDD size for a transformation with a linear-size representation of τ and τ^{-1}.

Example 5.9.4. Let $f_n \in B_{n^2+1}$ be defined on the variable s and an $n \times n$ matrix X of Boolean variables. Let $\mathrm{ROW}_n(X) = 1$ iff X contains a row consisting of ones only and let COL_n be defined in a similar way for the columns of X. Then

$$f_n(s, X) := s \wedge \mathrm{ROW}_n(X) + \overline{s} \wedge \mathrm{COL}_n(X).$$

This function has exponential OBDD size, since $(\exists s) f_n = \mathrm{ROW}_n + \mathrm{COL}_n$ is the function considered in Theorem 4.12.2.

We define the transformation $\tau : (s, X) \to (t, Y)$ by

$$t := s \text{ and } y_{ij} := s x_{ij} + \overline{s} x_{ji}.$$

It is obvious that τ has a compact representation. The inverse transformation is given by

$$s = t \text{ and } x_{ij} = t y_{ij} + \overline{t} y_{ji}$$

and also has a compact representation. The chosen variable ordering is t followed by a rowwise ordering on Y. If $s = t = 1$, then $y_{ij} = x_{ij}$ and we test the variables rowwise. It is easy to decide whether X has a 1-row. If $s = t = 0$, then $y_{ij} = x_{ji}$. We test the y-variables rowwise and, therefore, the x-variables columnwise. Hence, the τ-TBDD for f_n is the OBDD testing in rowwise order whether Y contains a row consisting of ones only. The τ-TBDD contains no t-nodes and only n^2 y-nodes.

In the preceding example we could choose the "good" transformation because of our knowledge of the structure of the considered function. How can we compute good transformations automatically? One idea is to restrict the set of possible transformations.

Definition 5.9.5. A cube transformation τ is called *linear* if all components τ_i of τ are linear, i.e., EXOR-sums of variables.

Linear (cube) transformations can be described by regular matrices over the field $\mathbb{Z}_2$. The inverse of a bijective linear function is also linear. Hence, for linear

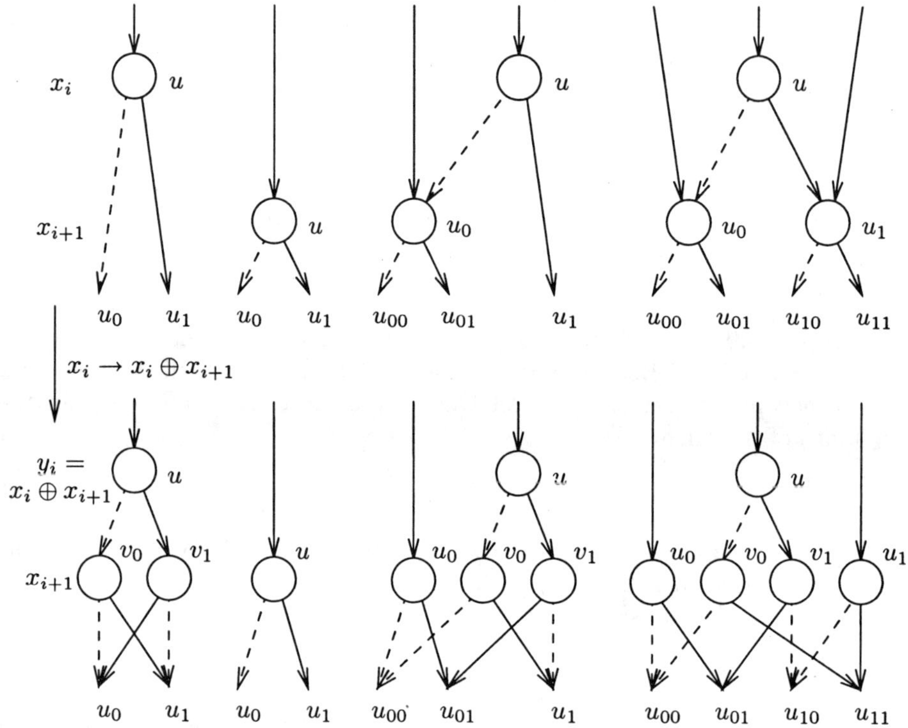

Figure 5.9.1: *An elementary linear transformation.*

transformations the requirement that τ and τ^{-1} should have compact representations is always fulfilled. Aborhey (1988) has already worked with τ-TBDDs based on linear transformations without defining the general framework. Meinel and Theobald (1996) have applied linear transformations for the encoding of the states of a DFA.

Meinel, Somenzi, and Theobald (1997) have combined the sifting algorithm with the search for linear transformations into an algorithm called *linear sifting*. It is an easy fact from linear algebra that the replacement of x_i by $y_i := x_i \oplus x_j$, $i \neq j$, is a bijective transformation and that each linear cube transformation can be obtained by such elementary transformations. During the sifting process, the variable x_i becomes adjacent to each variable x_j. The transformation $x_i \rightarrow x_i \oplus x_j$ is tried if x_i is the direct predecessor of x_j in the variable ordering. Similarly to Fig. 5.7.1, we obtain the following illustration of an elementary transformation.

An elementary transformation may decrease the size although the number of nodes in Fig. 5.9.1 is always increased. Such a size decrease is caused by the use of the reduction rules. If the size does not decrease, the elementary transformation is reversed by the application of the elementary transformation $y_i \to y_i \oplus x_j = x_i \oplus x_j \oplus x_j = x_i$.

Definition 5.9.6. A *cube embedding* is a one-to-one mapping $\tau\colon \{0,1\}^n \to \{0,1\}^m$. The function $sat(\tau)\colon \{0,1\}^m \to \{0,1\}$ outputs 1 on input b iff $\tau(a) = b$ for some $a \in \{0,1\}^n$.

Via a cube embedding τ, each function $f \in B_n$ corresponds to an incompletely specified function $f^*\colon \{0,1\}^m \to \{0,1,*\}$ in a natural way. Let $f^*(b) = f(a)$ if $\tau(a) = b$ and $f^*(b) = *$ if $sat(b) = 0$. The function f^* is well defined, since τ is one-to-one. Bern, Meinel, and Slobodová (1995) have already investigated τ-TBDDs for cube embeddings τ. This does not lead to canonical representations, since one can use any extension of f^*. Therefore, Goldberg, Kukimoto, and Brayton (1997, 1998) called the $\{0,1,*\}$-representation of f^*, together with a representation of τ, a canonical τ-TBDD. It follows from Section 3.6 that f^* has a polynomial-size OBDD representation over $\{y_1,\dots,y_m\}$ iff some extension F and $sat(\tau)$ (the description of the care set) have a polynomial-size OBDD representation (this is the so-called (F,c)-representation of f^*). We illustrate this approach by an example.

Example 5.9.7. For the hidden weighted bit function HWB_n, it is obvious to first compute the sum $s = (s_{k-1},\dots,s_0)$ of $x_1,\dots,x_n$ ($k = \lceil\log(n+1)\rceil$). Such intermediate results may be used for cube embeddings. Let $\tau\colon \{0,1\}^n \to \{0,1\}^{n+k}$ be defined by $\tau(x_1,\dots,x_n) := (s_0,\dots,s_{k-1},x_1,\dots,x_n)$. This function is one-to-one, since it contains $x_1,\dots,x_n$. Then $\mathrm{HWB}_n^*\colon \{0,1\}^{n+k} \to \{0,1,*\}$ takes the value $c \in \{0,1\}$ iff $(s_{k-1},\dots,s_0)$ is the binary representation of $sum := x_1 + \dots + x_n$ and $x_{sum} = c$, and it takes the don't care value $*$ otherwise. It is easy to see that the $\{0,1,*\}$-representation of HWB_n^* has size $O(n^3)$ for the variable ordering $s_0,\dots,s_{k-1},x_1,\dots,x_n$. We use a complete binary tree on the s-variables. If $(s_{k-1},\dots,s_0)$ represents $s \in \{0,\dots,n\}$, we check whether $x_1 + \dots + x_n = s$. In the positive case, we reach the sink with label x_s and, otherwise, the $*$-sink.

These last remarks go beyond the scope of the variable-ordering problem. We have seen that the choice of a good variable ordering is a hard problem whose solution is a key problem for the use of OBDDs. Moreover, generalizations of the basic OBDD concept offer further possibilities of obtaining a compact canonical representation of Boolean functions.

5.10 Exercises and Open Problems

5.1.E Prove that the inner product function $x_1y_1 \oplus \dots \oplus x_ny_n$ is almost ugly.

5.2.E Prove that the multiple addition function is almost ugly.

5.3.M In which of the classes nice, almost nice, ugly, very ugly, and almost ugly lies the test $\text{ROW}_n(X)$ of whether a quadratic Boolean matrix X contains a row consisting of ones only?

5.4.E Do there exist read-once formulas which are almost ugly?

5.5.O Do ambiguous functions exist?

5.6.O Which fraction of the functions $f = (f_n)$, where $\text{OBDD}(f_n) \leq n^2$, is almost ugly?

5.7.M Estimate the sensitivity of the functions considered in Theorem 5.3.1.

5.8.E Prove that the following problem is contained in NP: For an OBDD representing f and an integer s, decide whether OBDD $(f) \leq s$.

5.9.M Prove that the following problem is NP-hard. For a circuit C and a variable ordering π, one has to decide whether π is an optimal variable ordering for the function represented by C.

5.10.O Classify the complexity of the problem of Exercise 5.9 with respect to the polynomial-time hierarchy.

5.11.O Is it NP-hard for some $\varepsilon > 0$ to compute an s^{ε}-approximation for the OBDD variable-ordering problem?

5.12.M Prove that the top-down approach for the computation of an optimal variable ordering can be realized in time $O(n^2 4^n)$ and space $O(n^{1/2} 2^n)$.

5.13.E Prove Theorem 5.5.4.

5.14.M Prove that for the most significant bit s_n of ADD_n the sets $s^{-1}_{n|x_i=1}(1)$, $0 \leq i \leq n-1$, all have different sizes.

5.15.E How many functions $f \in B_n$ are symmetric with respect to x_1 and x_2?

5.16.M Determine the symmetry sets of the following functions: DSA, HWB, ADD, MUL, ROW, ROW + COL, DQF.

5.17.D Describe a function where group sifting of a set of symmetric variables increases the OBDD size significantly before it leads to smaller OBDD size.

5.18.O Do there exist functions $f_n, g_n \in B_n$ essentially depending on all their variables and variable orderings π_n (at least almost optimal for f_n and g_n) such that $\text{OBDD}(f_n + g_n) = \Omega(\pi_n\text{-OBDD}(f_n) \cdot \pi_n\text{-OBDD}(g_n))$?

5.19.D Let F_n be a formula on n variables where $O(\log n)$ variables occur on more than one leaf. Prove that $F = (F_n)$ has polynomial OBDD size.

5.20.E Realize the swap operation for OBDDs with complemented edges.

5.21.M Analyze the size increase caused by swap, jump-up, and jump-down for OBDDs with complemented edges.

5.22.E Prove that the size of the x_i-level after a *jump-up*(i, j) or a *jump-down*(i, j) operation cannot be estimated by some function depending only on the size of the x_i-level before the operation.

5.23.E Prove that there exist functions $f = (f_n)$ and $g = (g_n)$ such that $\text{OBDD}(f_n) = O(n)$, $\text{OBDD}(g_n) = O(n)$, but $\text{OBDD}(f_n + g_n)$ grows exponentially.

5.24.M Compare the OBDD size of the bit s_{n-1} of ADD_n for the variable ordering $x_0, y_0, \ldots, x_{n-1}, y_{n-1}$ with the OBDD size after a jump-up of y_{n-1} to the top.

5.25.M Design an algorithm directly swapping two sets of symmetric variables.

5.26.M Prove the OBDD size bounds stated in Example 5.8.2.

5.27.O Prove a nontrivial upper bound on the performance ratio of symmetric variable orderings.

5.28.M Design an algorithm for an elementary linear transformation for OBDDs with complemented edges.

Chapter 6

Free BDDs (FBDDs) and Read-Once BPs

6.1 Definition and Upper Bound Techniques

In Chapter 4, we proved for several important and also quite simple functions that they have exponential OBDD size. Therefore, we look for more general representations with good algorithmic behavior. Read-once BPs were introduced by Masek (1976) long before the first investigation of OBDDs. The OBDD community has independently discovered this type of representation and uses the notion FBDD.

Definition 6.1.1. An FBDD or *read-once* BP is a BP where, for each i, each path contains at most one node labeled by x_i.

Both notions capture central features of the representation type. During the evaluation, each input bit is read at most once and, at each node, the decision of which edge is chosen is "free." This means that each path from a source to a sink can be activated by some input. In other words, FBDDs are the most general BDD model without inconsistent paths.

In this section, we present some functions with polynomial FBDD size and exponential OBDD size. In Section 6.2, lower bound techniques for FBDDs are presented and applied. It turns out in Section 6.3 that many of the fundamental operations cannot be performed efficiently on FBDDs. With a restricted variant of FBDDs such as the restriction of OBDDs to π-OBDDs we obtain a new data structure for Boolean functions which is presented and investigated in Section 6.4. At the end of this chapter, we discuss OBDDs and FBDDs for search problems.

Each input a activates one computation path $p(a)$ in an FBDD and this path "defines" an ordering of the variables which are tested. This is a partial ordering $\pi(a)$ of all variables. In an OBDD, all $\pi(a)$ are consistent with some variable ordering π. In an FBDD of polynomial size, we may have an exponential number of paths such that no two of the corresponding partial orderings are consistent. How can we use this freedom? First, we consider functions where the use of different variable orderings for different inputs is apparent.

Theorem 6.1.2. *Let $f_n(s, X) = s \wedge \text{ROW}_n(X) + \bar{s} \wedge \text{COL}_n(X)$ output 1 if $s = 1$ and the $n \times n$ matrix X contains a 1-row or $s = 0$ and X contains a 1-column. Then f_n has an FBDD size of $O(n^2)$ and exponential OBDD size.*

Proof. The FBDD starts with an s-node. If $s = 1$, X is tested rowwise and if $s = 0$, X is tested columnwise.

Let $t(n) := OBDD(f_n)$. Then we obtain an OBDD of size $O(t(n)^2)$ for $(\exists s) f_n = \text{ROW}_n + \text{COL}_n$. Hence, it follows from Theorem 4.12.2 that $t(n) = 2^{\Omega(n^{1/2})}$. □

This simple FBDD construction can be used for all functions f_n of the following type. There are variables $(s_{k-1}, \ldots, s_0)$ and $N = 2^k$ functions $g_0, \ldots, g_{N-1}$ on X_n such that N is polynomially bounded in n and each g_i can be represented by a polynomial-size π_i-OBDD for some π_i. The function f_n realizes $g_{|s|}$. Such a function can be represented by a complete binary DT on the s-variables followed by π_i-OBDDs for g_i. We prove that the indirect storage access function ISA is of this type.

Theorem 6.1.3. *The FBDD size and the DT size of ISA_n are bounded by $O(n^2)$.*

Proof. The y-variables in Definition 2.2.5 of ISA_n serve as select variables $s_{k-1}, \ldots, s_0$. If $|s| = i$, we use the variable ordering π_i which starts with $x_i, \ldots, x_{i+k-1}$ (the indices are taken mod n) followed by an arbitrary ordering of the other variables. For the function g_i, i.e., ISA_n restricted to the case $|s| = i$, we take a complete binary DT for $x_i, \ldots, x_{i+k-1}$. Let v be some leaf of this tree. If we reach v, we know the value of all variables $x_i, \ldots, x_{i+k-1}$ and the value of $j = \alpha(x, y)$ whose binary representation is $(x_i, \ldots, x_{i+k-1})$. In this situation, ISA_n outputs x_j. If $j \in \{i, \ldots, i+k-1\}$, the value of x_j is known and v can be replaced by a sink. Otherwise, x_j is tested (see Fig. 6.1.1). □

The FBDDs constructed so far only use polynomially many different variable orderings. It can be proved (Sieling (1995)) that HWB still needs exponential size in such FBDDs. General FBDDs can represent HWB in polynomial size (Sieling and Wegener (1995a)).

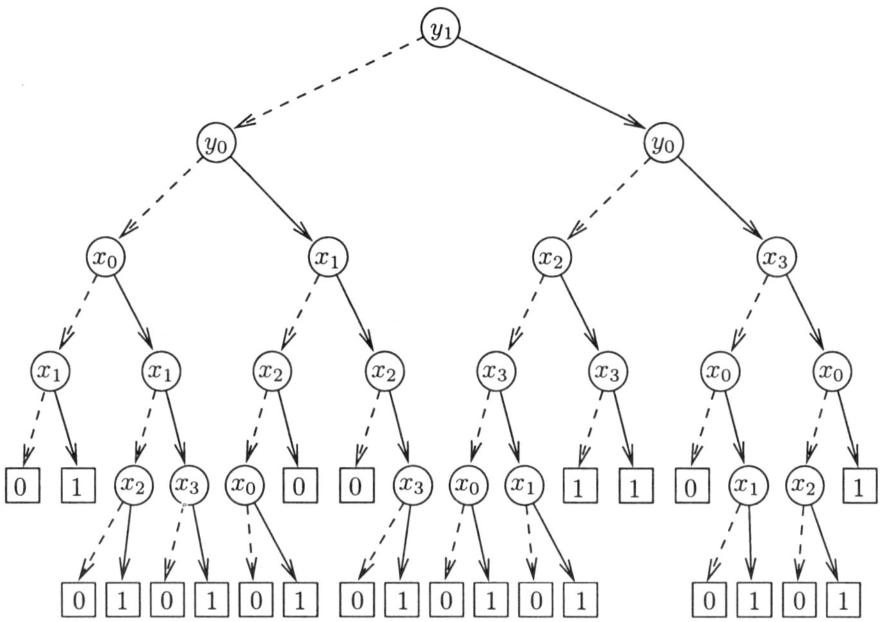

Figure 6.1.1: *An FBDD and DT for ISA_4.*

Theorem 6.1.4. *The FBDD size of HWB_n is bounded above by $n^2 - 4n + 8$ if $n \geq 3$.*

Proof. In Section 4.10, we discussed the variable-ordering problem for HWB_n. It turned out that we essentially have to store the number of ones already seen and the value of those variables which still can serve as output values. Using the notion of Section 4.10, the number w of fixed window variables should be kept small. For fixed variable orderings π, the existence of a window with at least $n/5$ fixed window variables after $3n/5$ tests leads to the lower bound of size $2^{n/5}$ for OBDDs. In FBDDs, the variable ordering for each input may react to the number of tested ones.

The inner nodes of our FBDD for HWB_n are denoted by $i[j, k]$ if the label is x_i and the window of possible sum values equals $W = \{j, \ldots, k\}$ (see Fig. 6.1.2). We use only the inner nodes $i[i, k]$, $1 \leq i < k \leq n$, $i[j, i]$, $0 \leq j < i \leq n-1$, the source $n[0, n]$, and two sinks. These are $2\binom{n}{2} + 3 = n^2 - n + 3$ nodes. Later we show how we can reduce this number. The window shrinks at its right border if a tested variable has the value 0 and it shrinks at its left border if the value is 1. We define the 0-successor of $i[i, k]$ as $(k-1)[i, k-1]$ and its 1-successor as $(i+1)[i+1, k]$. Similarly, we define the 0-successor of $i[j, i]$ as $(i-1)[j, i-1]$ and its 1-successor as $(j+1)[j+1, i]$. We prove the following properties by

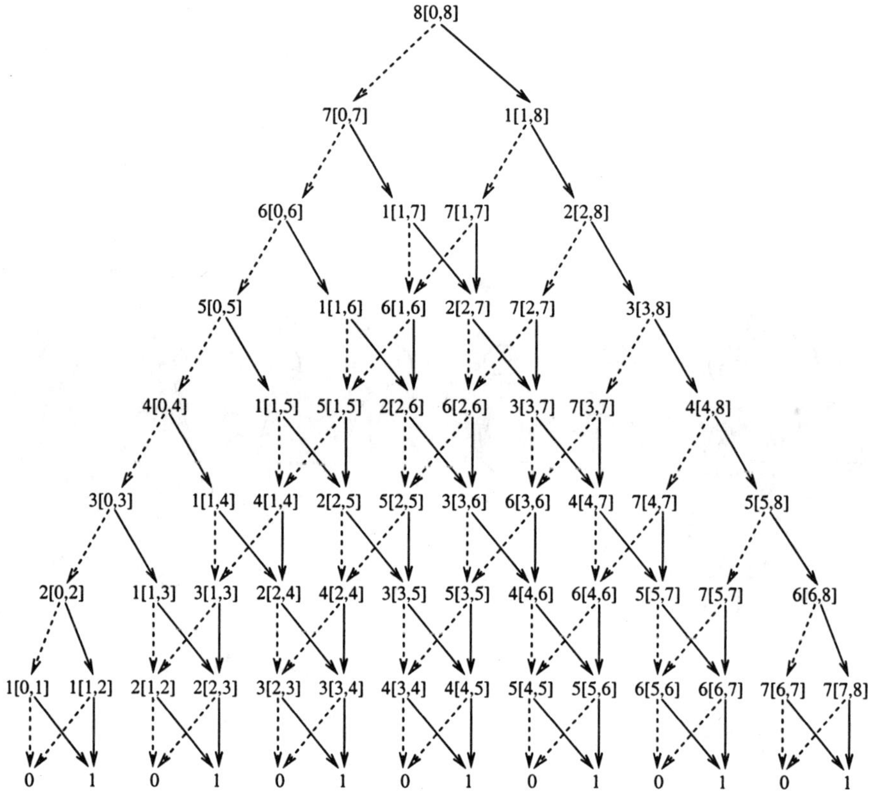

Figure 6.1.2: *An FBDD for* HWB_8.

backward induction on the length of the window:

- At nodes $i[0, i]$, no window variable has been tested.
- At nodes $i[i, k]$, the only tested window variable is x_k and its value is known to equal 1.
- At nodes $i[j, i]$ where $j \geq 1$, the only tested window variable is x_j and its value is known to equal 0.

The assertion holds for the source $n[0, n]$. Let it hold for $i[i, k]$. At this node we test x_i. If $x_i = 0$, we reach $(k-1)[i, k-1]$ and x_k is no longer a window variable. The variable x_i has become a tested window variable and its value is known to be equal to 0. If $x_i = 1$, we reach $(i+1)[i+1, k]$ and x_i is no longer a window variable. The only fixed window variable remains x_k whose value is known to be 1 by the induction hypothesis. Similar arguments work for the successors of $i[j,i]$ and $i[0,i]$. On the last level, we have the nodes $1[0,1]$, $i[i,i+1]$, and $i[i-1,i]$ whose c-successor is the sink with label c. Let us consider $i[i,i+1]$. If $x_i = 0$, HWB_n outputs $x_i = 0$ and the 0-sink is reached. If $x_i = 1$, HWB_n outputs x_{i+1} and by the above assertion we know that $x_{i+1} = 1$. Similar arguments work for $i[i-1,i]$. For the special case $1[0,1]$ and $x_1 = 0$, the input consists of zeros only and HWB_n is defined as 0. Finally, we can apply the merging rule to the nodes $i[i-1,i]$ and $i[i,i+1]$ for $i \in \{1,\ldots,n-1\}$ and then, if $n \geq 3$, eliminate the $2n-4$ nodes on the next to last level. This decreases the number of nodes by $3n-5$ to $n^2 - n + 3 - (3n-5) = n^2 - 4n + 8$. □

The FBDD for HWB_n has the shape of a triangle. OBDDs typically look like pears, since the merging rule is very powerful on the last levels and, e.g., the last level always contains at most two nodes.

6.2 Lower Bound Techniques

The first exponential lower bounds on the size of FBDDs or read-once BPs were proved independently in 1984 by Žák (1984) and Wegener (1988). Since then, many lower bounds have been presented. These results are motivated by the following problems (for an earlier survey see Razborov (1991)).

(1) Prove bounds as large as possible for functions in P (or at least in NP).

(2) Prove bounds for important functions.

(3) Prove exponential bounds for functions which are as "simple" as possible.

(4) Prove exponential bounds for functions which are simple for other generalizations of OBDDs.

First, we discuss lower bound techniques for OBDDs and then how they can be generalized to FBDDs. Lower bounds for OBDDs are obtained by proving lower bounds on π-OBDDs for arbitrary π. For each π, we may choose a level i and may count the number of different subfunctions obtained by the different replacements of the first i variables by constants, i.e., we investigate all possible 2^i paths testing the first i variables. We may generalize this idea by considering an arbitrary prefix-free set of paths $p \in P$ starting at the root, i.e., no path $p \in P$ is a lengthening (or shortening) of another path $p' \in P$. Each $p \in P$ corresponds to a subfunction f_p of f. Obviously, the number of different subfunctions with respect to P is a lower bound on the π-OBDD size of f. If we can prove that at most m of the considered subfunctions are equal, we obtain a lower bound of size $|P|/m$. This bound may be bad, since the equivalence classes corresponding to equal subfunctions may have quite different size. Then we can do better by assigning weights $w_p \geq 0$, $p \in P$, to the paths such that the sum of all weights equals 1. If the weight of each equivalence class is bounded above by ε, we need at least $\lceil \varepsilon^{-1} \rceil$ nodes to represent the equivalence classes. We need some intuition to assign weights in such a way that all equivalence classes get (almost) equal weights.

Since we may omit tests in OBDDs, there are OBDDs which are π-OBDDs as well as π'-OBDDs for some $\pi' \neq \pi$. Only in complete OBDDs (see Definition 3.2.1) is each variable tested on each path from the source to a sink. Then the chosen variable ordering can be computed by following an arbitrary computation path. The characteristic feature of an FBDD is that we may use different variable orderings for different inputs. But not every combination of variable orderings can be realized by FBDDs, e.g., only one variable can be tested at the source.

Definition 6.2.1. Let G be a complete FBDD, i.e., each variable is tested on each path from the source to a sink. Let $\pi_G(a)$ be the variable ordering on the computation path for a. An FBDD G' is called a *graph-driven* FBDD guided by G (G-FBDD) if the sequence of variables on the computation path for a in G' is consistent with $\pi_G(a)$.

Lemma 6.2.2. *For each FBDD G' on X_n there is a complete FBDD G such that $|G| \leq (n+1)|G'|$ and G' is a G-FBDD.*

Proof. According to a topological ordering of the nodes, we compute for each node v the set $V(v)$ of variables tested on some path from the source to v excluding the label of v. Using a bit vector of length n for each node, this can be done in time $O(n|G'|)$, since each edge (v, w) only has to transfer the information $(V(v), \text{label}(v))$ to $V(w)$. In a second phase, it is sufficient to add, on the edge (v, w), dummy tests of the variables in $V(w) - V(v)$ excluding the variable tested at v. □

If we investigate G-FBDDs, G determines the variable ordering $\pi_G(a)$ for each input a, but the function represented by G is insignificant. Hence, we merge the sinks to one sink with the label $*$. The resulting graph G is called a *graph ordering*. Among all graph orderings describing the variable ordering $\pi_G(a)$ for each input a, there is a unique one of minimal size (see exercises).

Now we may describe lower bound techniques for FBDDs. We have to prove lower bounds on G-FBDDs for each graph ordering G. In G we choose a prefix-free set P of paths. The lower bound methods for π-OBDDs described above only use the notion of prefix-free sets P of paths corresponding to partial inputs. Hence, these methods work as well for G-FBDDs, i.e., we get a lower bound by counting the number of different subfunctions which we obtain by considering the partial assignments corresponding to the paths in P. Again, we may use the method of assigning weights to the paths. Our description of the lower bound techniques is different from the descriptions in the literature but most ideas are borrowed from Simon and Szegedy (1993). The only differences are that we describe graph orderings explicitly and that the chosen path system P may depend on the graph ordering. One special variant of the technique, due to Jukna (1988), is useful for many functions and should be mentioned explicitly.

Definition 6.2.3. A function $f \in B_n$ is called *k-mixed* if for all $V \subseteq X_n$ such that $|V| = k$ the 2^k assignments of constants to the variables $x_i \in V$ lead to different subfunctions.

Lemma 6.2.4. *The FBDD size of k-mixed functions is bounded below by* $2^k - 1$.

Proof. Let G be a graph ordering. We consider the 2^{k-1} paths of length $k-1$ starting at the source of G. We claim that they all lead to different subfunctions. Let p and p' be two such paths. If the sets of variables tested on p and p', respectively, are the same, the claim follows from the definition of k-mixed. Otherwise, some variable x_i is tested on p' and not on p. If $f_p = f_{p'}$, the function f_p does not essentially depend on x_i. Let V be the set of variables tested on p plus the variable x_i. Then the assignments a^0 and a^1 where the variables in $V - \{x_i\}$ are fixed as on p and x_i is replaced by 0 and 1, respectively, lead to the same subfunction of f. Since $|V| = k$ and f is k-mixed, this is a contradiction. Hence, each G-FBDD starts with a complete tree of depth $k-1$ and, therefore, contains at least $2^k - 1$ nodes. □

We start with the historically first two bounds. Both are concerned with clique problems on undirected graphs $G(x)$ defined on $N = \binom{n}{2}$ Boolean variables x_{ij}, $1 \le i < j \le n$. The graph $G(x)$ contains the edge between i and j iff $x_{ij} = 1$.

Definition 6.2.5.

(i) The *clique function* $\mathrm{cl}_{k,n}$ computes 1 if the graph $G(x)$ contains a clique of size k.

(ii) The *exactly half clique function* excl_n computes 1 if the graph $G(x)$ consists of a clique of size $\lceil n/2 \rceil$ and $\lfloor n/2 \rfloor$ isolated vertices.

The clique function belongs (for certain values of $k = k(n)$) to the class of important NP-complete functions while the exactly half clique function is much simpler and can be decided in linear time. The bound for excl_n is an improvement of Savický (personal communication) of a bound due to Žák (1984) and the bound on $\mathrm{cl}_{n,k}$ is due to Wegener (1988).

Theorem 6.2.6. *The FBDD size of* excl_n *is bounded below by* $n^{-2}\binom{n}{\lceil n/2 \rceil} = 2^{n-O(\log n)} = 2^{\Omega(N^{1/2})}$.

Proof. Let G be an FBDD representing excl_n. For $a \in \text{excl}_n^{-1}(1)$, let V_a be the set of the corresponding $\lceil n/2 \rceil$ clique vertices. Let v_a be the first node on the computation path p_a for a such that, including the test at v_a, we have tested for each vertex of V_a at least one edge of the clique on V_a. This implies that inputs $b \in \text{excl}_n^{-1}(1)$ where p_b and p_a coincide up to $v_a = v_b$ have the property that V_a and V_b share those at least $\lceil n/2 \rceil - 2$ vertices j such that at least one edge adjacent to j has been tested positively before v_a. The number of those b is bounded above by $\binom{n-\lceil n/2\rceil+2}{2} \le n^2$. Let $c \in \text{excl}_n^{-1}(1)$ where $v_c = v_a$. We consider the computation path p^* starting at the source following p_c up to $v_c = v_a$ and then following p_a up to the 1-sink. This computation path describes an input $d \in \text{excl}_n^{-1}(1)$. Moreover, $d = a$, since p_d contains in its bottom part positive tests of all edges $\{i, j\}$ for some $i \in V_a$ and all other $j \in V_a$. Hence, c is one of those at most n^2 inputs from $\text{excl}_n^{-1}(1)$ which reach v_a via the path p_a. The lower bound follows, since at most n^2 of the $\binom{n}{\lceil n/2\rceil}$ inputs a from $\text{excl}_n^{-1}(1)$ share the chosen FBDD node v_a. □

Theorem 6.2.7. *Let $s = \binom{k}{2} - 1$ and $t = n - k^2 + 2$. The FBDD size of $cl_{n,k}$ is bounded below by $\binom{s+t}{s}$.*

Corollary 6.2.8. *For $k(n) = \lceil (2n/3)^{1/2} \rceil$, the FBDD size of $cl_{n,k(n)}$ is bounded below by $2^{n/3 - o(n)} = 2^{\Omega(N^{1/2})}$.*

Proof. We apply Theorem 6.2.7. Then $s = t = n/3 - o(n)$ and the result follows from Stirling's formula. □

Proof of Theorem 6.2.7. Let G be any graph ordering. We choose in G all paths starting from the root which have length $s + t$ and contain exactly s 1-edges. Let p and q be two such paths where $p \neq q$. It is sufficient to prove that f_p and f_q are different. Let G_p be the partial graph specified by p, similarly G_q for q. The graph G_p consists of existing edges (tested positively on p), forbidden edges (tested negatively on p), and undecided edges. Since p and q separate at some node, there exists an edge, w.l.o.g. $(1, 2)$, which exists in G_p and is forbidden in G_q (or vice versa).

Let A be the set of vertices $i \notin \{1, 2\}$ lying on an edge which exists in G_q. Then $|A| \le 2s = k^2 - k - 2$. Since the edge $(1, 2)$ exists in G_p, we can choose a set B of $n - k^2 + 2$ vertices $j \notin \{1, 2\}$ such that for each forbidden edge of G_p at least one vertex of this edge is contained in B. The set $\{1, \ldots, n\} - A - B$ contains at least $n - (k^2 - k - 2) - (n - k^2 + 2) = k$ vertices. Let $C \subseteq \{1, \ldots, n\} - A - B$ be a set of k vertices including 1 and 2.

We assume that $f_p = f_q$. Each assignment to the variables not tested on p or q has to produce the same constant subfunction of f_p and f_q. We consider such an assignment where exactly the edges on C are decided to exist. Since no edge on C is forbidden in G_p, the corresponding subfunction of f_p is the constant 1. Let G_q^* be the (partial) graph resulting from G_q and the chosen assignment. Only vertices in $A \cup C$ have positive degree. Since G_q contains less than $\binom{k}{2}$ edges, it does not contain any k-clique. Only edges on C are added. Hence, a k-clique D in G_q^* has to contain at least two vertices from C. Since $(1,2)$ is forbidden in G_q, D has to contain some $i \in C - \{1,2\}$. Moreover, D has to contain some node $j \notin C$, since $|C| = k$ and $(1,2)$ is forbidden. The edge (i,j) does not exist in G_q and it is not added by our assignment. Hence, G_q^* does not contain any k-clique and the corresponding subfunction of f_q is the constant 0. This contradicts the assumption $f_p = f_q$. □

The bounds of Theorem 6.2.6 and Corollary 6.2.8 are of size $2^{\Omega(N^{1/2})}$ for the number N of variables. Researchers competed for the lower bound record on the FBDD size of functions in P. The first strong exponential lower bound of size 2^{cN} for a tiny c was obtained by Babai, Hajnal, Szemerédi, and Turán (1987) for the function $\oplus\text{cl}_{n,3}$ deciding whether the number of triangles (or 3-cliques) in G is odd. The constant in the exponent was improved by Simon and Szegedy (1993) to $c = 1/2000$. Kriegel and Waack (1988) obtained the constant $c = 1/48$ for the function solving the following word problem. Let v and w be words of length n over the alphabet $\{0,1,2\}$. The question is whether the words v' and w' obtained by the elimination of all letters 2 are equal. Of course, we have to use a Boolean encoding of this function. Later on, Breitbart, Hunt III, and Rosenkrantz (1995) improved the record to $c = 1/3$ for the following function f_n on $n = 3k$ variables $x_0, \dots, x_{k-1}, y_0, \dots, y_{k-1}, z_0, \dots, z_{k-1}$. One has to compute $x_{||y||+||z||} \oplus y_{||x||+||z||} \oplus z_{||x||+||y||}$ where the indices are taken mod k. Since the FBDD size of each function $f \in B_n$ is bounded above by $O(2^n/n)$, the constant $c = 1$ is not exactly reachable. Savický and Žák (1996, 1998) have obtained a lower bound where $c = 1 - o(1)$.

Their idea is to define the function in such a way that it is k-mixed for large k. For this purpose, they use the following number theoretical result of Dias da Silva and Hamidoune (1994).

Lemma 6.2.9. *Let p be prime. For each set $A \subseteq \mathbb{Z}_p$ of size k, the set of all sums* $\pmod p$ *of exactly $h \le k$ distinct elements of A contains at least* $\min\{p, hk - h^2 + 1\}$ *elements.*

For given n, we choose p as the smallest prime larger than n, $h = \lfloor 2n^{1/2} \rfloor$, and $k = 2h$. It is well known that $p \le 2n$. Then $hk - h^2 + 1 = h^2 + 1 \ge 4n - 4n^{1/2} \ge p$ and we obtain all elements of $\mathbb{Z}_p$ as sums. We define the function *weighted sum* WS_n as follows. In $\mathbb{Z}_p$, one computes the sum $s = s(x)$ of all ix_i, $1 \le i \le n$. If $s \in \{1, \dots, n\}$, the output of WS_n equals x_s, and it is x_1 otherwise. Obviously, the function is in P. It is a kind of storage access or pointer function. The computation of the number of the cell whose contents should be read is made so complicated that we have to store the value of many variables.

Theorem 6.2.10. *The FBDD size of the weighted sum function* WS_n *is bounded below by* $2^{n-\lfloor 4n^{1/2}\rfloor-2}-1$.

Proof. We prove that WS_n is m-mixed for $m = n - \lfloor 4n^{1/2}\rfloor - 2$. Let I be some set of m variable indices and let u and v be different assignments to the corresponding variables. The goal is to prove that $f_u \neq f_v$ for the subfunctions defined by u and v. Let $J = \{1,\dots,n\} - I$ and $\Delta = \sum_{i\in I} iv_i - \sum_{i\in I} iu_i$.

We start with the simpler case where $\Delta \equiv 0 \bmod p$. Let $i^* \in I$ be chosen such that $u_{i^*} \neq v_{i^*}$. By Lemma 6.2.9 and its discussion, we can choose an input u^* with $u^*_i = u_i$ for $i \in I$ and $s(u^*) \equiv i^* \bmod p$. Let v^* be the corresponding extension of v. Since $\Delta \equiv 0 \bmod p$, we conclude that $s(v^*) \equiv i^* \bmod p$. Hence, $\mathrm{WS}_n(u^*) = u_{i^*} \neq v_{i^*} = \mathrm{WS}_n(v^*)$ and $f_u \neq f_v$.

In the following, we can assume that $\Delta \not\equiv 0 \bmod p$. We fix some $j \in J - \{1\}$. Let $l \equiv j + \Delta \bmod p$ if this implies $l \in \{1,\dots,n\}$ and $l = 1$ otherwise. Then $l \neq j$, since either $l \equiv j + \Delta \not\equiv j \bmod p$ or $l = 1 \neq j$. We define an input u^* as follows. If $l \in J$, we assign the value 0 to x_j and the value 1 to x_l. If $l \notin J$, we assign the value $1 - v_l$ to x_j. Since at least $\lfloor 4n^{1/2}\rfloor$ positions $i \notin I$ are free, we can assign values to them (by Lemma 6.2.9 and its discussion) such that we get an input u^* where $u^*_i = u_i$ for $i \in I$, u^*_j and possibly u^*_l get the values specified above, and $s(u^*) \equiv j \bmod p$. Let v^* be the corresponding extension of v. Then $s(v^*) \equiv j + \Delta \bmod p$. Hence, $\mathrm{WS}_n(u^*) = u^*_j \neq v^*_l = \mathrm{WS}_n(v^*)$ and $f_u \neq f_v$. □

Another goal is the proof of exponential lower bounds on the FBDD size of simple functions, in particular, functions with polynomial-size DNFs. Wegener (1986) has obtained the first result of this kind for a variant of the clique function (see exercises). Gál (1997) presents another function of this kind which has a clear combinatorial structure based on elementary properties of projective planes. This function has polynomial-size CNFs (the dual has polynomial-size DNFs) and the length of the clauses is approximately $n^{1/2}$. We investigate a function due to Bollig and Wegener (1998a) which also has a simple combinatorial structure, is monotone and, additionally, has prime implicants of length 2 only. Nechiporuk (1971) has already applied the solution of Kovari, Sós, and Turán (1954) to the well-known problem of Zarankiewicz to Boolean function complexity. Let $n = p^2$ for some odd prime number p. The set A_i, $0 \le i \le n-1$, where $i = a + bp$ and $a, b \in \mathbb{Z}_p$, contains all $j = c + dp$ where $c, d \in \mathbb{Z}_p$ and $c \equiv a + bd \bmod p$. It is easy to see that $|A_i| = p = n^{1/2}$. The special property is that $|A_i \cap A_j| \le 1$ if $i \neq j$. Indeed, the sets A_i are the largest possible ones with this property. Let $f^*_n(x_0,\dots,x_{n-1},y_0,\dots,y_{n-1})$ be the disjunction of all $y_i x_j$ where $j \in A_i$. The function f^*_n is monotone and has $n^{3/2}$ prime implicants of length 2. Each y_i has $n^{1/2}$ partners x_j such that $y_i x_j$ is a prime implicant of f^*_n. If $i \neq i'$, y_i and $y_{i'}$ have at most one partner in common. By the symmetry of the construction it follows that x_j has $n^{1/2}$ partners y_i and, moreover, x_j and $x_{j'}$ have at most one partner in common if $j \neq j'$.

Theorem 6.2.11. *The function f_n^* has $n^{3/2}$ monotone prime implicants of length 2 and its FBDD size is bounded below by $2^{\Omega(n^{1/2})}$.*

Proof. Let G be any graph ordering. For $N = \lfloor n^{1/2}/2 \rfloor$, we choose 2^N paths starting in G at the source and describing partial assignments. We obtain the lower bound by proving that these partial assignments lead to different subfunctions.

We have to avoid partial assignments which satisfy a prime implicant. We start at the source and, at each node, we choose both directions whenever this does not satisfy a prime implicant of f_n^*. In the other case, we choose the 0-edge. Nodes and their labels are called *free* if both directions are chosen and *bound* otherwise. Each path terminates after having passed N free nodes. Since each 1-edge creates at most $n^{1/2}$ new bound variables, we obtain 2^N different partial assignments.

Let f_u be the subfunction for one of the assignments. We claim that f_u essentially depends on all its variables not replaced by constants. Let x_i be such a variable. Since f_n^* is a monotone function with prime implicants of length 2, f_u only can be independent of x_i if for each partner y_j of x_i one of the following two conditions is fulfilled. The partner may be fixed to 0 or have become a prime implicant, since one of its partners is fixed to 1. Each of the N free nodes on the path belonging to u can do the job for at most one partner of x_i. This is obvious for variables fixed to 0. A variable x_k, $k \neq i$, fixed to 1 makes all its partners prime implicants. But x_k has at most one partner in common with x_i. We still have to consider the variables y_j fixed to 0 at bound nodes. However, then a partner of y_j has been fixed to 1 before and we have already counted the destruction of $x_i y_j$.

Now we consider two different assignments u and v. If u and v do not fix the same set of variables, then $f_u \neq f_v$, since they essentially depend on different sets of variables by the discussion above. Hence, we may assume that u and v fix the same variables. Since u and v are different, we may assume w.l.o.g. that u assigns 0 to x_i and v assigns 1 to x_i. We prove that $f_u \neq f_v$ by proving the existence of a partner y_j of x_i which is a prime implicant of f_v but not of f_u.

First, we investigate the set of variables fixed by u and v. Since $x_i = 0$ for u, the partners of x_i do not become bound for u because of x_i. If a partner of x_i is replaced by a constant, it is a free variable or bound by some x_k, $k \neq i$. Then x_k was free. The variable x_k has at most one partner in common with x_i. Hence, at most N partners of x_i are fixed by u and v and at least $n^{1/2} - N = \lceil n^{1/2}/2 \rceil$ partners of x_i are not fixed by u and v. Since $x_i = 1$ for v, all these partners are prime implicants of f_v. How many of these y-variables can be prime implicants of f_u where $x_i = 0$? Then a partner y_j is a prime implicant of f_u only if some partner x_k of y_j is set to 1. Since at most N variables are set to 1, only $N = \lfloor n^{1/2}/2 \rfloor < \lceil n^{1/2}/2 \rceil$ (since $n^{1/2} = p$ is an odd prime) y-partners of x_i are prime implicants of f_u. Hence, $f_u \neq f_v$. □

The number of further exponential lower bounds on FBDDs is immense. Dunne (1985) considered graph theoretical problems such as the existence of a Hamiltonian circuit and the computation of the permanent (mod 2). Simon and Szegedy (1993) proved an exponential lower bound for the test whether a graph on n vertices is $(n/2)$-regular, i.e., each node has degree $n/2$. For later purposes, some matrix functions are of interest.

Theorem 6.2.12. *The permutation matrix test function $PERM_n$ has an FBDD size bounded below by $\Omega(n^{-1/2}2^n)$.*

Proof. The proof of Krause (1988) presented as a proof of a lower bound on the OBDD size (see Theorem 4.12.3) indeed works for FBDDs. □

Theorem 6.2.13. *The FBDD size of the function $ROW_n + COL_n$ testing whether a Boolean matrix contains a 1-row or a 1-column is bounded below by $\Omega(n^{-7/2}2^n)$.*

Proof. The first exponential lower bound for this function is due to Bollig and Wegener (1997a). We present the improved result of Sauerhoff (personal communication) and remark that this result also improves the lower bound result on the OBDD size in Theorem 4.12.2. The key is the following observation:

$$\mathrm{PERM}_n(X) = \neg(\mathrm{ROW}_n(\overline{X}) + \mathrm{COL}_n(\overline{X})) \wedge \mathrm{E}_{n,n^2}(X).$$

If X is a permutation matrix, it contains exactly n ones and each row and column contains a 1-entry implying that $\overline{X}$ does not contain a row or column consisting of ones only. If $\mathrm{ROW}_n(\overline{X}) + \mathrm{COL}_n(\overline{X}) = 0$, we can conclude that each row and column of X contains at least a 1-entry. If, moreover, the number of ones in X equals n, the matrix X is a permutation matrix.

If the FBDD size of $\mathrm{ROW}_n(X) + \mathrm{COL}_n(X)$ equals s, the same holds for $f_n(X) = \neg(\mathrm{ROW}_n(\overline{X}) + \mathrm{COL}_n(\overline{X}))$. We easily obtain a complete FBDD of size sn^2. The FBDD size of $\mathrm{PERM}_n(X) = f_n(X) \wedge \mathrm{E}_{n,n^2}(X)$ can be bounded by $sn^2(n+1)$, since it is sufficient to distinguish whether the number of ones among the tested variables is $0, 1, \ldots, n$, or larger than n. The last case is represented by the 0-sink. Now the lower bound follows from Theorem 6.2.12. □

Now we may believe that the proof of exponential lower bounds on the FBDD size of selected functions is an easy task. Nevertheless, it took quite a long time until Ponzio (1995) was able to prove such a bound for multiplication. His best bound is of size $2^{\Omega(n^{1/2})}$. We present his simpler proof of a $2^{\Omega(n^{1/3})}$ bound.

Theorem 6.2.14. *The FBDD size of $MUL_{n-1,n}$ is bounded below by $2^{\Omega(n^{1/3})}$.*

Proof. We prove the lower bound for complete FBDDs. By Lemma 6.2.2, we obtain a lower bound for general FBDDs if we divide the obtained bound by

$2n+1$. Let S be any subset of the variables such that $|S \cap X| = m$ and $|S \cap Y| < m$ (or vice versa) for $m = \lfloor n^{1/3}/5 \rfloor$, $X = \{x_0, \ldots, x_{n-1}\}$, and $Y = \{y_0, \ldots, y_{n-1}\}$. We prove that at most $2^{|S|-m}$ assignments to the variables in S lead to the same subfunction for z_{n-1} where the bits z_i describe the output bits of multiplication.

Let G be any graph ordering. We consider all paths starting at the source until for the first time at least m x-variables or m y-variables are tested. This leads to variable sets of the type of S. Since we consider complete FBDDs, only assignments to the same set S may lead to the same FBDD node. We give the weight $2^{-|S|}$ to each considered assignment corresponding to S. Then the total weight of all considered partial assignments equals 1. If the above claim is proved, each FBDD node gathers a weight bounded by 2^{-m} and 2^m FBDD nodes are necessary to gather the total weight.

Let S be fixed. Then let s be the least index where $y_s \notin S$ and $S' = \{y_0, \ldots, y_{s-1}\} \cup (S \cap \{x_0, \ldots, x_{n-1-s}\})$. Since $|S \cap X| = m$, $|S'| \geq m$. We prove the claim by proving that assignments u and v to the variables in S lead to different subfunctions of z_{n-1} if u and v differ on S'. For a partial assignment u, the input u^* agrees with u and fixes all free variables to 0.

Claim 1. *Let u and v be assignments to the variables in S. If $z_i(u^*) = z_i(v^*)$ for all $i \in \{n-m-3, \ldots, n-1\}$ and u and v differ on S', there is a single variable outside S such that $z_i(u_1^*) \neq z_i(v_1^*)$ for some $i \in \{n-m-3, \ldots, n-1\}$ and the partial assignments u_1 and v_1 which extend u and v by fixing the chosen variable to* 1.

Hence, we get an input which leads to different outputs at position $n-1$ or "a little bit below position $n-1$." Either $z_i(u^*) \neq z_i(v^*)$ for some $i \in \{n-m-3, \ldots, n-1\}$ or the condition of Claim 1 is fulfilled. In the first case, let $u_1 = u$ and $v_1 = v$ and, in the second case, we apply Claim 1 to obtain u_1 and v_1. The following claim will ensure that we are able to force the position where u and v lead to different outputs to the desired position $n-1$.

Claim 2. *Let u_j and v_j be partial assignments to at most $3m$ x-variables and at most $3m$ y-variables. Let d be the largest index in $\{0, \ldots, n-1\}$ such that $z_d(u_j^*) \neq z_d(v_j^*)$. If $n-m-3 \leq d < n-1$ and n is large enough, there exist one x-variable and one y-variable outside S such that $z_{d+1}(u_{j+1}^*) \neq z_{d+1}(v_{j+1}^*)$ for the partial assignments u_{j+1} and v_{j+1} which extend u_j and v_j by fixing the chosen pair of variables to* 1.

It is easy to see that the claims imply the theorem. We start with at most m fixed variables of each type. Claim 1 is applied at most once and fixes one variable. Then Claim 2 is applied at most $m+2$ times. Hence, at most $m+1+m+2 \leq 3m$ (if $m \geq 3$) variables of each type are fixed and the conditions of Claim 2 are fulfilled whenever it is applied.

Proof of Claim 1. The partial assignments u and v differ somewhere on S'. We assume that they differ at least for some x-variable (the other case can be handled similarly). Then we look for a y-variable outside S which will be fixed to 1.

Since all output bits $z_n, \ldots, z_{2n-1}$ do not matter, we are counting mod 2^n, i.e., on a circle with 2^n numbers $0, \ldots, 2^n - 1$. We partition this circle to 2^{m+3} segments such that each segment contains numbers whose bits at the positions $n-m-3, \ldots, n-1$ are the same. For an input u^*, we describe by $x(u^*)$ the value of the first factor, by $y(u^*)$ the value of the second factor, and by $z(u^*)$ the value of the product. The hypothesis of the claim can be restated. The numbers $z(u^*)$ and $z(v^*)$ fall into the same segment. Changing the y_k-bit from 0 to 1 implies that the product increases by $2^k x(u^*)$. The claim is proved if, for some y_k outside S, it can be proved that $2^k(x(u^*) - x(v^*)) \bmod 2^n$ is at least 2^{n-m-1} and at most $2^n - 2^{n-m-2} - 1$. This implies that the difference is at least four segments long and also at least two segments shorter than the perimeter of the circle. Hence, $z(u^*) + 2^k x(u^*)$ and $z(v^*) + 2^k x(v^*)$ fall into different segments.

Let $x_{\text{diff}} = x(u^*) - x(v^*)$. Since u and v differ on $S' \cap X$, we conclude that there is at least one index $j \in \{0, \ldots, n-s-1\}$ where x_{diff} has a 1. Either $j = 0$ or x_{diff} has a 0 at position $j-1$. Now we choose an index $k \in \{(n-1) - j - m, \ldots, (n-1) - j\}$ such that y_k is outside S. This choice is possible. If $(n-1) - j - m \geq 0$, the set contains $m+1$ indices and S contains at most m y-variables. Otherwise, we choose $k = s$. Since $j \leq n - s - 1$, also $s \leq (n-1) - j$. Moreover, $y_s \notin S$ by the definition of s. We conclude that $2^k x_{\text{diff}}$ has a 1 at position $j+k$ and a 0 at position $j+k-1$. By our choice of k, we get $n-1-m \leq j+k \leq n-1$. The 1 at position $j+k$ implies that $2^k x_{\text{diff}} \geq 2^{n-m-1}$. The 0 at position $j+k-1 \geq n-m-2$ implies that $2^k x_{\text{diff}} \leq 2^n - 1 - 2^{n-m-2}$ (on the circle mod 2^n). □

Proof of Claim 2. Let us consider the effect of choosing the pair (x_k, y_l) where $k + l = d$. Then

$$z(u_{j+1}^*) - z(u_j^*) = 2^l x(u_j^*) + 2^k y(u_j^*) + 2^d$$

and

$$z(v_{j+1}^*) - z(v_j^*) = 2^l x(v_j^*) + 2^k y(v_j^*) + 2^d.$$

We know that $z(u_j^*)$ and $z(v_j^*)$ differ at position d. We would like to ensure that $2^l x(u_j^*) + 2^k y(u_j^*)$ and also $2^l x(v_j^*) + 2^k y(v_j^*)$ have zeros at the positions d and $d+1$ and that they do not cause a carry to position d if we add them to $z(u_j^*)$ and $z(v_j^*)$, respectively. Then we may conclude that $z(u_{j+1}^*)$ and $z(v_{j+1}^*)$ differ at position $d+1$.

We can fulfill these requirements, since we consider inputs with very few ones. Both u_j^* and v_j^* contain at most $3m$ ones in each of the factors. Then the product contains at most $9m^2$ ones. This can be proved by the following arguments using the school method for multiplication. We have to add numbers with at most $9m^2$ ones altogether. We do this columnwise from right to left. A column with r ones may produce a one in the result and the carry contains at most $r-1$ ones altogether.

It is sufficient to ensure that $2^l x(u_j^*)$, $2^k y(u_j^*)$, $2^l x(v_j^*)$, and $2^k y(v_j^*)$ have zeros at the positions $d-9m^2-2, \ldots, d+1$. The $9m^2+2$ positions $d-9m^2-2, \ldots, d-1$ then contain at most $9m^2$ ones in $z(u_j^*)$ (or $z(v_j^*)$) and the carry from earlier positions is restricted to two ones (since we add three numbers).

First, we estimate the number of bad choices of $l \in \{0, \ldots, n-1\}$. The number $x(u_j^*)$ as well as $x(v_j^*)$ has at most $3m$ ones. For each one there are at most $9m^2+4$ values l which shift this one to a forbidden position. Altogether, we get at most $6m(9m^2+4)$ bad values for l. We have to add $3m$ for the already fixed y-variables. The same estimates hold for the choice of k. Altogether, we have the choice among $d+1$ pairs (x_k, y_l) where $k+l=d$. The number of bad pairs is bounded by the sum of bad x-values and bad y-values and, hence, by $2(54m^3+27m) = 108m^3+54m$. By assumption, $d+1 \geq n-m-2$. Altogether, there exists a good pair if $n-m-2 \geq 108m^3+54m$. By the definition of m,

$$108m^3 + 55m + 2 \leq \frac{108}{125}n + 11n^{1/3} + 2 \leq n$$

for large n. □

The effect of partial assignments is harder to understand for multiplication than for all the other functions we have investigated. This makes the lower bound proof complicated.

Corollary 6.2.15. *The FBDD size of SQU_n, INV_n, and DIV_n is bounded below by $2^{\Omega(n^{1/3})}$.*

Proof. This follows from the lower bound of Theorem 6.2.14 and the read-once projections of Theorem 4.6.2, since the replacement of variables by constants and the negation of variables do not increase the size of FBDDs. □

6.3 Algorithms on FBDDs

In Chapter 3, we designed a number of efficient algorithms on π-OBDDs. All OBDD packages contain equivalence test and synthesis algorithms only for OBDDs based on the same variable ordering. The counterpart of π-OBDDs are G-FBDDs, which we investigate in detail in Section 6.4. Here we investigate the complexity of the usual operations with respect to general FBDDs.

Theorem 6.3.1. *Evaluation is possible in time $O(n)$ for FBDDs on n variables. SAT, SAT-COUNT, and replacement by constants can be performed in time $O(|G|)$ on an FBDD G.*

Proof. The result on the evaluation operation is obvious. SAT can be solved by a DFS approach checking whether the 1-sink is reachable from the source. For SAT-COUNT, we use the algorithm for OBDDs given in the proof of Theorem 3.3.1. The proof only uses the fact that all paths of an OBDD are consistent. The result for the operation replacement by constants follows as in Theorem 2.4.1. □

Theorem 6.3.2. *There are functions representable by FBDDs containing for each variable x_i only one x_i-node such that the operations synthesis and replacement by functions cause an exponential blow-up of the FBDD size. The operation quantification may cause an exponential blow-up of the FBDD size.*

Proof. The functions $f_n = \mathrm{ROW}_n$ and $g_n = \mathrm{COL}_n$ fulfill the assumptions of the theorem. An OR-synthesis leads to the function $\mathrm{ROW}_n + \mathrm{COL}_n$ whose FBDD size is bounded below by $\Omega(n^{-7/2}2^n)$ (see Theorem 6.2.13). The same holds for $s + f_n$ and the replacement of s by g_n and for $(\exists s)(sf_n + \overline{s}g_n)$. □

No deterministic polynomial-time equivalence test for FBDDs is known. The equivalence test is not coNP-complete for FBDDs if RP≠NP, since Blum, Chandra, and Wegman (1980) have designed a coRP algorithm for this problem. This result is described explicitly in the next theorem.

Theorem 6.3.3. *There is a randomized polynomial-time algorithm which for FBDDs G_f and G_g outputs "equal" if G_f and G_g are equivalent and which outputs "not equal" with a probability of at least $1/2$ if G_f and G_g are not equivalent.*

Using k independent repetitions, we may, in the second case, increase the probability of the answer "not equal" to at least $1 - (1/2)^k$. We postpone the proof of this theorem to Section 11.1, which is devoted to randomized algorithms. We conjecture that the equivalence test for FBDDs does not belong to P. This would imply also that minimization and redundancy test do not belong to P as the following theorem shows.

Theorem 6.3.4. *There are linear-time transformations from the equivalence test for FBDDs to the problems minimization and redundancy test for FBDDs and from the redundancy test to the equivalence test.*

Proof. Let G_f and G_g be FBDDs representing f and g. If s is a new variable, we obtain an FBDD for $sf + \overline{s}g$ in linear time. The source gets the label s and the edges lead to the sources of G_g and G_f, respectively. Obviously, f and g are equal iff the optimal FBDD for $sf + \overline{s}g$ does not contain any s-node or iff s is redundant for $sf + \overline{s}g$. Furthermore, x_i is redundant for f iff $f_{|x_i=0}$ and $f_{|x_i=1}$ are equal. □

Without proof we present the following result of Sieling (1999).

Theorem 6.3.5. *The minimization problem for FBDDs is NP-hard.*

We conclude that it is not possible to work efficiently with FBDDs. The same holds for OBDDs, since the functions f_n and g_n considered in the proof of Theorem 6.3.2 also have linear-size OBDDs (for different variable orderings). There is one difference. OBDDs can be checked for equivalence in polynomial time even if they use different variable orderings (see Corollary 5.7.11). This seems to be not possible for FBDDs. Fortune, Hopcroft, and Schmidt (1978) have presented a polynomial-time equivalence test between an FBDD and an OBDD. Meinel and Slobodová (1994) have obtained the following resource bounds.

Theorem 6.3.6. *The equivalence of an FBDD G and an OBDD H can be decided using space $O(|G| + |H|)$ and time $O(|G|(n + |H|))$.*

Proof. The algorithm is based on a DFS traversal of G where we store at each node via which edge it is reached for the first time. Moreover, we store the actual path. If we reach a sink with label c for the first time, we use the stored information and go back to the source. For the partial assignment defined on this path, we check whether the OBDD computes the constant c. If we reach a node v once again, we compute the partial assignments a and b for the actual path and the path which reached v for the first time. Then we test in time $O(|H|)$ whether H represents the same function if it is restricted by a and b, respectively. The time $O(n)$ may be necessary for the computation of a and b. (Usually, $n = O(|H|)$.) Hence, the time and space bounds of the algorithm are obvious. We still have to prove that G and H are equivalent iff all checks performed by the algorithm are successful.

The proof is done by induction on the number of different variables in the FBDD G. If G is a sink, we check whether H represents the same constant. Otherwise, let, w.l.o.g., x_1 be the label of the source of G. Without loss of generality, we follow the 0-edge leaving the source. By induction, we check whether G for $x_1 = 0$ represents the same function as H for $x_1 = 0$. If the DFS traverses the 1-edge leaving the source of G, we cannot apply the induction hypothesis, since we may reach nodes v which have been reached via the 0-edge leaving the source of G. Then we have checked that a partial assignment a with $a_1 = 0$ leading to v produces the subfunction g_a of the function g represented by G which is equivalent to the function h_a represented by H under the assignment a. Now we obtain a partial assignment b with $b_1 = 1$ leading to v in G and, therefore, producing the subfunction $g_b = g_a$. We check whether $h_a = h_b$. Since we know that $g_a = h_a$, we conclude in the positive case that $g_b = h_b$. By the induction hypothesis and the fact that we may stop the process at nodes seen before, we conclude that we check in the second phase whether G represents for $x_1 = 1$ the same function as H for $x_1 = 1$. Altogether, the algorithm is correct. □

Finally, we consider the transformation problem where we have to *construct* the π-OBDD H for f from an FBDD G for f. The following algorithm is due to Savický and Wegener (1997). First, all nodes not reachable from the source of the FBDD are deleted. The resulting FBDD is considered as an ite straight line program or circuit. Bottom-up, the functions f_v represented at v are transformed into π-OBDDs by ternary ite synthesis steps. Two facts make this approach much more efficient than for general circuits or BPs. The function f_v is a subfunction of the function f represented at the source. Moreover, a partial assignment leading to f_v can be found in time $O(n)$ if the reversed edges are stored. The π-OBDD for f_v is not larger than the π-OBDD for f and so we can guarantee a size bound for the intermediate results. Let v be an x_i-node and $ite(x_i, g, h)$ the corresponding synthesis step. Since G is an FBDD, g and h cannot essentially depend on x_i. Hence, Theorem 5.7.7 guarantees that the synthesis step creates a linear number of nodes with respect to input and output size and, therefore, can be performed in time $O(|H| \log |H|)$ for the π-OBDD H for f. Altogether, we obtain the time bound $O(|G||H| \log |H|)$.

Theorem 6.3.7. *The π-OBDD H for f can be computed from an FBDD G for f using space $O(|G| + n|H|)$ and time $O(|G||H| \log |H|)$. Using hashing strategies the expected time is $O(|G||H|)$.*

Proof. The time bounds follow from our observations above. The improved space bounds are left as an exercise. □

6.4 Algorithms on G-FBDDs

We are not able to work efficiently with FBDDs as we are not able to work efficiently with OBDDs. Since efficient algorithms exist for all operations on π-OBDDs and a fixed variable ordering π, we hope for efficient algorithms on G-FBDDs and a fixed graph ordering G. The following investigations are based on the independent papers of Gergov and Meinel (1994) and Sieling and Wegener (1995a).

We start with some easy conclusions from the last section. Since evaluation, SAT, and SAT-COUNT do not change the FBDD, the results from Theorem 6.3.1 for these operations hold for G-FBDDs as well as for FBDDs. We have seen in Theorem 6.3.2 that quantification may cause an exponential blow-up of the FBDD size. The given FBDD may be interpreted as a G-FBDD for some graph ordering G whose computation is described in the proof of Lemma 6.2.2. If we are forced to represent the result of quantification not only as an FBDD but even as a G-FBDD, we cannot prevent the exponential blow-up of the size.

We might expect that everything becomes easier if we restrict ourselves to G-FBDDs. This is not true, since we are forced to produce a G-FBDD as a result.

Theorem 6.4.1. *Replacement by constants may cause an exponential blow-up of the size of G-FBDDs.*

Proof. Again we consider the function $h_n = s \wedge \text{ROW}_n + \bar{s} \wedge \text{COL}_n$. The graph ordering G starts with an s-node. If $s = 1$, a rowwise ordering of the X-variables is used and if $s = 0$, a columnwise ordering. Then, obviously, h_n has linear G-FBDD size. Replacing s by 1 in h_n leads to the simple function ROW_n which is independent of s. But the graph ordering prescribes that we use a columnwise ordering of the X-variables if $s = 0$. The function ROW_n syntactically depends on s in the environment of the graph ordering G. We obtain an exponential lower bound by investigating the columnwise variable ordering for ROW_n. □

Replacement by constants is a special case of replacement by functions. Therefore, replacement by functions may also cause an exponential blow-up of the size. The redundancy test for G-FBDDs and a fixed but arbitrary graph ordering G is nothing other than the redundancy test for FBDDs, since each FBDD is a G-FBDD for some graph ordering G. Hence, G-FBDDs cannot be used efficiently if our application needs one of the operations replacement by constants, replacement by functions, quantification, or redundancy test. If we know in advance which variables are used in these operations, we may work with graph orderings with some additional properties.

Definition 6.4.2. A graph ordering G is called *x_i-oblivious* if for each x_i-node the 0-successor coincides with the 1-successor.

Theorem 6.4.3. *Replacement of oblivious variables by constants is possible in linear time for G-FBDDs.*

Proof. As in the case of general FBDDs, the algorithm replaces edges to x_i-nodes with edges to the corresponding c-successor (if x_i has to be replaced by c). We have to prove that the resulting FBDD is a G-FBDD. For inputs a where $a_i = c$, it uses the same variable ordering as before with only the exception that x_i is omitted. Hence, $\pi(a)$ is still consistent with $\pi_G(a)$. For inputs b where $b_i = \bar{c}$, the resulting FBDD uses the same variable ordering $\pi(b)$ as for b' defined by $b'_i = c$ and $b'_j = b_j$ otherwise. We have seen that $\pi(b')$ is consistent with $\pi_G(b')$. Since G is x_i-oblivious, $\pi_G(b) = \pi_G(b')$ and $\pi(b)$ is consistent with $\pi_G(b)$. □

If synthesis and equivalence test can be performed efficiently for G-FBDDs, then replacement by functions, quantification, and redundancy test can be performed efficiently for oblivious variables.

Still we are faced with the most important operations, namely synthesis, minimization, and equivalence test. We proceed in the following way. First, we prove that (up to isomorphism) there is a unique G-FBDD of minimal size for each function f, i.e., G-FBDDs are a canonical representation of Boolean functions and the operation reduction is well defined. Then we demonstrate that a G-FBDD is reduced iff neither the elimination rule nor the merging rule is applicable. This leads to a linear-time reduction algorithm and to a synthesis algorithm with integrated reduction. The representation by reduced G-FBDDs implies a constant-time equivalence test if we work with shared G-FBDDs and a linear-time equivalence test otherwise, since then the equivalence check is a simple isomorphism check.

Theorem 6.4.4. *Let $G = (V, E)$ be a graph ordering and $f \in B_{n,m}$. The minimal-size G-FBDD G^* for f (with m pointers to the nodes representing the functions $f_1, \ldots, f_m$) is unique up to isomorphism.*

Proof. For $v \in V$, we denote the graph ordering which is the subgraph of G with source v by $G(v)$. Let a be a partial assignment defined by a path from the source of G to v. Then the subfunctions of each coordinate function f_i, $1 \le i \le m$, with respect to a have to be represented by $G(v)$-FBDDs. Although $G(v)$ is defined on a subset of the variable set X_n, we assume that all considered subfunctions syntactically depend on X_n. This simplifies the notion that subfunctions are equal. We also use the notion subfunction of f for subfunctions of some f_i. Let $S(v)$ be the set of all subfunctions of f which we obtain by partial assignments leading from the source to v in G. Moreover, let A be the set of all (v, g) where $v \in V$ and $g \in S(v)$. Each G-FBDD for f contains for each $(v, g) \in A$ some node $w(v, g)$ which is the source of a $G(v)$-FBDD for g. We define a relation $\sim$ to describe which elements of A may be represented by the same G-FBDD node. Let $(v, g) \sim (v', g')$ if there exists an FBDD which simultaneously is a $G(v)$-FBDD and a $G(v')$-FBDD and represents g and g' at its source. This implies that $g = g'$.

The symmetric relation $\sim$ defines an undirected graph $A(\sim)$ with the vertex set A and edges connecting (v, g) and (v', g') if $(v, g) \sim (v', g')$. We construct a G-FBDD G^* representing f and containing exactly one node for each connected component of $A(\sim)$. By definition of $\sim$, G^* has minimal size. The uniqueness will follow from arguments given during the construction of G^*. The construction of G^* also implies that $\sim$ is an equivalence relation. Before constructing G^*, we prove a simple property of $A(\sim)$.

Claim. *Let (v, g), (v_0, g), $(v_1, g) \in A$ where v_0 and v_1 are the direct successors of v in G. If two of these elements from A are related by $\sim$, then all three elements are related by $\sim$.*

Proof of the Claim. Let x_i be the label of v. The function g cannot essentially depend on the variable x_i which is the label of v, since it can be represented by a $G(v_0)$-FBDD. If $(v_0, g) \sim (v_1, g)$, the $G(v_0)$- and $G(v_1)$-FBDD for g is also a $G(v)$-FBDD. If w.l.o.g. $(v_0, g) \sim (v, g)$, the $G(v_0)$- and $G(v)$-FBDD for g cannot contain an x_i-node, since it is a $G(v_0)$-FBDD. Since it is a $G(v)$-FBDD without an x_i-node, it is also a $G(v_1)$-FBDD. □

Now we construct G^*. We fix a topological ordering of the nodes of G. For each connected component C of $A(\sim)$, we select the pair (v, g) such that v is the last of all nodes w where (w, g) belongs to C. For each selected pair (v, g), we create a node $w(v, g)$ in G^* whose label equals the label x_i of v if v is not a sink. Let g_c be the subfunction of g for $x_i = c$ and v_c the c-successor of v in G. The pair (v_c, g_c) belongs to A. The node $w(v_c^*, g_c)$ for the selected pair (v_c^*, g_c) in the component of (v_c, g_c) is chosen as the c-successor of $w(v, g)$ in G^*. If v is the sink of G and $(v, g) \in A$, then g is a constant function and $w(v, g)$ is the corresponding sink. We prove by induction with respect to the reversed topological ordering of the nodes of G that $w(v, g)$ is the source of a $G(v')$-FBDD representing g for all (v', g) which are connected with (v, g) in $A(\sim)$.

The induction base is obvious, since we consider the sinks of G^*. For the induction step, let (v, g) be a selected pair. By our selection procedure, $(v, g) \not\sim (v_0, g_0)$ and, therefore, by the claim, $(v_0, g_0) \not\sim (v_1, g_1)$. This implies that $w(v_0^*, g_0) \neq w(v_1^*, g_1)$ and, by the induction hypothesis, these nodes are sources of a $G(v_0)$-FBDD representing g_0 and a $G(v_1)$-FBDD representing g_1, respectively. Hence, $w(v, g)$ is the source of a $G(v)$-FBDD representing g. In the next step, we consider neighbors (v', g) of (v, g) in $A(\sim)$.

Case 1. $\text{label}(v') = \text{label}(v) = x_i$.
We consider a $G(v)$- and $G(v')$-FBDD representing g. The existence of this FBDD follows from the assumption $(v, g) \sim (v', g)$ and it implies that $(v_0, g_0) \sim (v_0', g_0)$ and $(v_1, g_1) \sim (v_1', g_1)$. By the induction hypothesis, the successors of $w(v, g)$ are a $G(v_0')$-FBDD representing g_0 and a $G(v_1')$-FBDD representing g_1, respectively. Hence, $w(v, g)$ is the source of a $G(v')$-FBDD representing g.

Case 2. $\text{label}(v') \neq \text{label}(v) = x_i$ and v and v' are not connected in G.
Using the claim and the fact that (v, g) is a selected pair, we conclude that $G(v)$-FBDDs representing g start with a node labeled by x_i. Hence, by the argument of Case 1, the considered $G(v)$- and $G(v')$-FBDD is also a $G(v_0')$- and a $G(v_1')$-FBDD. We may continue these arguments until we reach successors of v' in G labeled by x_i. For each successor v'' of this type, the arguments of Case 1 are applicable. Since $w(v, g)$ has the label x_i and is the source of a $G(v'')$-FBDD for all these nodes v'', it is also a $G(v')$-FBDD representing g.

Case 3. $\text{label}(v') \neq \text{label}(v) = x_i$ and there is a path from v' to v in G.
First, we assume that v' is a direct predecessor of v in G. Without loss of generality $v = v_0'$. Since the source of a $G(v)$- and $G(v')$-FBDD is labeled by x_i, it is also a $G(v_1')$-FBDD and, therefore, $(v, g) = (v_0', g) \sim (v_1', g)$. By the above arguments, $w(v, g)$ is the source of a $G(v')$-FBDD representing g. In the general situation of Case 3, the $G(v)$- and $G(v')$-FBDD representing g proves

for all nodes v'' on the path from v' to v that $(v'', g) \sim (v, g)$. Hence, we can iterate our arguments.

Now we can use the same arguments for the neighbors of the neighbors of (v, g) in $A(\sim)$ and so on.

Altogether, $\sim$ is an equivalence relation and G^* is a minimal-size G-FBDD representing f. The nodes for the equivalence classes are necessary and there is no choice of how to direct the edges if more nodes are not used. If some edge could reach a node w_1 but also a node w_2, the corresponding node (v, g) would belong to two equivalence classes. □

For a synthesis algorithm with integrated reduction, it is essential that it is sufficient to create only nodes reachable from some pointer to a represented function and to apply the elimination rule and the merging rule.

Theorem 6.4.5. *Let H be a G-FBDD representing f and containing only nodes reachable from some pointer to a function which has to be represented. Then H is reduced iff neither the elimination rule nor the merging rule is applicable.*

Proof. It is obvious that a G-FBDD is not reduced if one of the two reduction rules is applicable. Let us now consider a G-FBDD H representing f with more nodes than the reduced G-FBDD H^* for f. We prove that one of the reduction rules is applicable. Each node w of H corresponds to a subset of some equivalence class C with respect to $\sim$ (for the definition of $\sim$ see the proof of Theorem 6.4.4). If $|H| > |H^*|$, we consider a topologically last node u of H^* such that its corresponding equivalence class $C(u)$ is represented in H by more than one node. Let W be the set of nodes of H corresponding to $C(u)$ and let $w^* \in W$ be some topologically last one of these nodes.

Case 1. There exists some $w' \in W$ such that no path from w' leads to w^*. Among the nodes w' fulfilling the assertion of this case, we choose a topologically last one. From the proof of Theorem 6.4.4 it follows that w^* and w' are labeled by the same variable. By construction, the successors of w^* and w' belong to that part of H which is isomorphic to the corresponding part in H^*. Hence, the merging rule is applicable to w^* and w'.

Case 2. For each $w' \in W$ there is a path from w' to w^*. By the arguments of the proof of Theorem 6.4.4, all nodes on the path from w' to w^* belong to W. Hence, we may assume w.l.o.g. that $w^* = w'_0$, the 0-successor of w'. By the claim in the proof of Theorem 6.4.4, the 1-successor w'_1 of w' also belongs to the same equivalence class as w'. Hence, by the assumption of this case, there is a path from w'_1 to w'_0. This only is possible if $w'_0 = w'_1$ and the elimination rule is applicable to w'. □

Based on these results, Sieling and Wegener (1995a) have designed a linear-time reduction algorithm for G-FBDDs. The difficulty is deciding how to proceed bottom-up. In the case of OBDDs, the variable ordering helps to investigate the x_i-nodes before the x_j-nodes if x_j precedes x_i in the variable ordering. Graph orderings do not lead to a unique variable ordering. Therefore, the bottom-up application of the reduction rules has to be guided quite carefully in order to guarantee the linear runtime.

We describe a synthesis algorithm with integrated reduction. Before proceeding in the same way as for π-OBDDs in Chapter 3, we give an example showing that we have to include the graph ordering in the simultaneous DFS traversal.

Example 6.4.6. Let G be a graph ordering on X_n with the following properties. Each node on level $j < n$ has two different successors and two nodes on the same level $j < n$ have different pairs as 0-successor and 1-successor. The level n contains two nodes v and w. From G we obtain the graph ordering G' on X_{n+2} where the 0-successors of v and w point to the ordering (x_{n+1}, x_{n+2}) while the 1-successors of v and w point to (x_{n+2}, x_{n+1}) (see Fig. 6.4.1).

It is obvious that the functions x_{n+1} and x_{n+2} have a G'-FBDD size of 3. The G'-FBDD for $x_{n+1}x_{n+2}$ almost looks like G'. The 0-edges leaving u_1 and u_2 lead to the 0-sink and u_3 and u_4 are replaced with nodes representing x_{n+2} and x_{n+1}, respectively. It is obvious by Theorem 6.4.5 that this is the reduced G'-FBDD for the function $x_{n+1}x_{n+2}$. Hence, a simple synthesis step of two G'-FBDDs of constant size may lead to a G'-FBDD of size $\Theta(|G'|)$.

Let $G_f = (V_f, E_f)$ and $G_g = (V_g, E_g)$ be G-FBDDs representing f and g, respectively, and let $\otimes$ be a binary Boolean operator. We want to construct a G-FBDD G_h for $h = f \otimes g$. For this reason, we start a simultaneous DFS traversal through G, G_f, and G_g. A node $(v, v_f, v_g) \in V \times V_f \times V_g$ always gets the label of v. The reason is that each path from the source to the sink in G has length n, i.e., in the graph ordering we do not omit tests. As c-successor of (v, v_f, v_g) we choose (v^c, v_f^c, v_g^c), where v^c is the c-successor of v, v_f^c is the c-successor of v_f if label(v_f) = label(v) and $v_f^c = v_f$ otherwise (we "wait" in G_f, since a test is omitted), and v_g^c is defined similarly. If v is the sink of G, v_f and v_g are sinks of G_f and G_g, respectively. Then (v, v_f, v_g) is a sink labeled by label$(v_f)\otimes$label(v_g). For each input a, we follow the corresponding computation paths in G_f and G_g and, finally, we compute the correct value of $h(a)$. The modules "terminal-case," "computed-table," and "unique-table" can be used as in Algorithm 3.3.5 for the synthesis of π-OBDDs. Similarly to Theorem 3.3.6, we obtain the following result, where G_h^* is the graph consisting of the nodes of the product graph of G, G_f, and G_g which are reachable from the pointers to h.

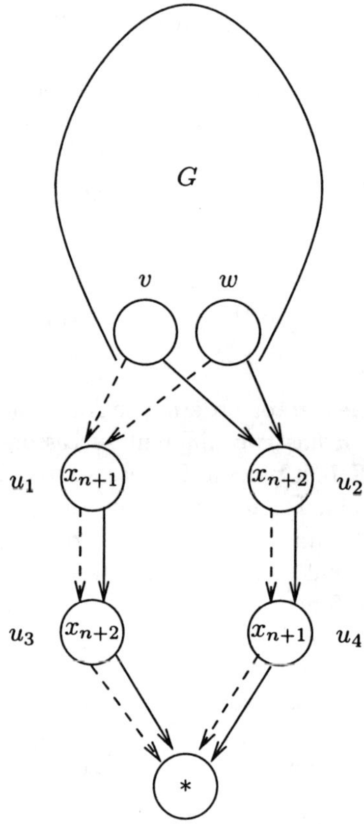

Figure 6.4.1: *The graph ordering G'.*

Theorem 6.4.7. *The binary synthesis of G-FBDDs G_f and G_g is possible in*

- *time and space $O(|G||G_f||G_g|)$,*
- *time $O(|G_h^*|\log|G_h^*|)$ and space $O(|G_h^*|)$, or*
- *expected time and space $O(|G_h^*|)$.*

Besides the considered difficulties for some operations, G-FBDDs can be handled efficiently. The additional third component G in the DFS traversal is necessary as Example 6.4.6 shows. If, for input a, each variable occurs as a label on one of the computation paths in G_f and G_g, then the consideration of G does not slow down the DFS traversal on input a.

Similarly to the variable-ordering problem for OBDDs, we are faced here with the *graph-ordering problem* for FBDDs. The problem seems to be much

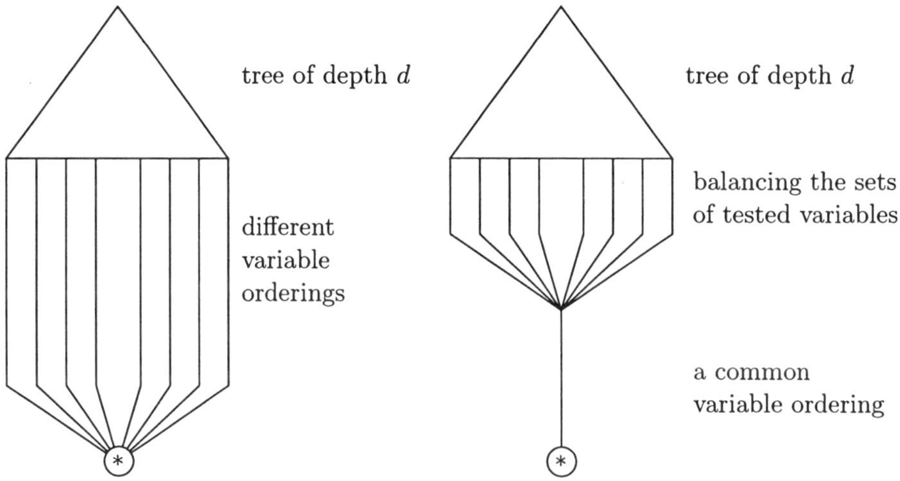

Figure 6.4.2: *Types of automatically generated graph orderings.*

harder, since we have much more freedom. Indeed, nobody is able to compute complicated graph orderings automatically like those which are necessary for a polynomial-size representation of HWB (see Section 6.1). The only graph-ordering algorithm tested in experiments is due to Bern, Meinel, and Slobodová (1996). Their approach creates graph orderings of the following kind. For a parameter d, the graph ordering starts with a complete binary tree of depth d. For each leaf, a *variable ordering* of the remaining $n - d$ variables follows.

There are several heuristics for choosing the label of the source of the graph ordering. On the given circuit C, an "important" variable x_i is computed. Then important variables for $C_{|x_i=0}$ and $C_{|x_i=1}$ are computed and used as labels of the successors of the source of the graph ordering. This approach is continued up to level d. Then, for each of the leaves, some variable-ordering algorithm is applied to the circuit restricted by the partial assignment described by the path from the source to the considered leaf.

This approach leads to graph orderings with many different variable labels on the last levels. This often implies more nodes on the last levels than for variable orderings. Therefore, Bern, Meinel, and Slobodová (1996) have refined their approach. In the tree of depth d, the use of "similar" sets of variables on the different paths is supported. Then, at the leaves, the variable orderings first test those variables not yet tested on the considered path but tested somewhere else in the tree. After some further levels, the same set of variables is tested on each path of the graph ordering and a common variable ordering can be appended. This combines the use of different variable orderings in the beginning with the use of the same variable ordering at the end (see Fig. 6.4.2).

We omit the details of these approaches, since the graph-ordering problem for FBDDs still needs further investigation.

Bern, Meinel, and Slobodová (1995) used graph orderings as cube transformations τ for τ-TBDDs. A graph ordering G defines the following function $\tau_G\colon \{0,1\}^n \to \{0,1\}^n$. For $a \in \{0,1\}^n$, the path $p(a)$ activated in G is considered. Then $\tau_G(a)$ is defined as the vector of the edge labels on $p(a)$.

Lemma 6.4.8. *For each graph ordering G, the function τ_G is a cube transformation.*

Proof. It is sufficient to prove that τ_G is one-to-one. Let $a, b \in \{0,1\}^n$ and $a \neq b$. Since G is complete and $a \neq b$, also $p(a) \neq p(b)$. Hence, there is some i such that the paths use different edges at level i. Then $\tau_G(a)$ and $\tau_G(b)$ differ at position i. □

What are the differences between G-FBDDs and τ_G-TBDDs? Let us consider a graph ordering G as a complete binary tree with 2^n leaves. We obtain the reduced G-FBDD for $f \in B_n$ by assigning the values of f to the leaves and by applying the reduction rules. The cube transformation τ_G is defined in such a way that we may use the same assignment of constants to the leaves if we relabel the inner nodes on level i, $1 \le i \le n$, with the label y_i. Afterwards, we may apply the reduction rules. In these DTs the y_i-nodes lie on the same level while the x_i-nodes lie on the same levels as in G. One might conjecture that, on the average over all functions, the merging rule is more powerful on τ_G-TBDDs than on G-FBDDs if G is not a variable ordering. But in applications, the graph ordering G is constructed knowing the function f. Both representations have the same difficulties with the operation replacement by constants and the operations based on it. The evaluation of τ_G-TBDDs is less efficient than that of G-FBDDs. The synthesis of τ_G-TBDDs is based on a simultaneous DFS traversal of G_f and G_g, while in the synthesis of G-FBDDs we also have to traverse G. The reason is the following. G-FBDDs for the variables x_i contain only one inner node and only synthesis steps (as in Example 6.4.6) introduce the structure of G into G-FBDDs. The τ_G-TBDDs for the variables already carry the structure of G. We describe the τ_G-TBDD for the variable x_i. In the graph ordering G, we replace the c-successor of x_i-nodes by c-sinks and then relabel the inner nodes such that nodes on level j, $1 \le j \le n$, get the label y_j. It is an easy exercise to prove that we obtain a τ_G-TBDD representing x_i. This τ_G-TBDD perhaps can be further reduced.

Example 6.4.9. We consider the graph ordering G' from Example 6.4.6 shown in Fig. 6.4.1. The $\tau_{G'}$-TBDD for x_{n+1} contains the whole graph G where the inner nodes are relabeled by y_j on level j. The node u_1 is replaced by a y_{n+1}-node and the node u_4 by a y_{n+2}-node whose c-edges lead to the c-sink. The nodes

u_2 and u_3 can be eliminated. By the construction of G, this is the reduced $\tau_{G'}$-TBDD for x_{n+1}. Hence, $\tau_{G'}$-TBDD$(x_{n+1}) = |G'| - 1$ but G'-FBDD$(x_{n+1}) = 3$. The same holds for the variable x_{n+2}. As the next step, we consider the $\wedge$-synthesis of x_{n+1} and x_{n+2}. As $\tau_{G'}$-TBDD, the nodes u_1 and u_2 are replaced with y_{n+1}-nodes whose 0-successors are the 0-sink and whose 1-successors are the nodes u_3 and u_4 representing y_{n+2}. We may merge u_3 and u_4 and also u_1 and u_2. Afterwards, the whole graph G can be eliminated. Hence, together with the results of Example 6.4.6,

$$G'\text{-FBDD}(x_{n+1} \wedge x_{n+2}) = |G'| - 1 \text{ but } \tau_{G'}\text{-TBDD}(x_{n+1} \wedge x_{n+2}) = 4.$$

Theorem 6.4.10. *For all functions f and graph orderings G, G-FBDD$(f) \leq |G| \cdot \tau_G$-TBDD$(f)$ and τ_G-TBDD$(f) \leq |G| \cdot G$-FBDD(f).*

The proof of this theorem is contained in Sieling and Wegener (1998b). Together with the previous examples, we have determined the maximal differences in the size of G-FBDDs and τ_G-TBDDs for the same function.

6.5 Search Problems

The difference between decision problems and search problems as well as their relations are discussed from the complexity theoretical viewpoint by Garey and Johnson (1979).

Definition 6.5.1. A *Boolean search problem* Π is given by a relation $R_\Pi \subseteq \{(x, s) \mid x \in \{0,1\}^n,\ s \in S\}$ for some finite set S of possible solutions. A function $f\colon \{0,1\}^n \to S \cup \{\varepsilon\}$ is a realization of R_Π if $f(x) = \varepsilon$ for all x such that no (x, s) is contained in R_Π and $(x, f(x)) \in R_\Pi$ for all other x.

A realization of a relation R_Π can be represented by an MTBDD (see Section 9.2), which is a BP whose sinks may be labeled by values from $S \cup \{\varepsilon\}$ (and not only from $\{0,1\}$). Incompletely specified Boolean functions (see Section 3.6) may be considered as a search problem, i.e., $(x, 0)$ and $(x, 1)$ are contained in R_Π if the value of the function at x is not specified. The main difference is that in the setting as a search problem, it is not necessary to represent the don't care set explicitly. For CAD applications, the special case of incompletely specified Boolean functions is the more natural one and has been intensively investigated. Here we motivate general search problems.

Definition 6.5.2. The *pigeonhole principle* $\text{PHP}_{m,n}$ is defined on a Boolean $m \times n$ matrix X and the corresponding relation contains (X, R_i) if the row R_i of X only contains zeros and $(X, (C_j, i_1, i_2))$ if the column C_j of X contains ones at the positions i_1 and i_2.

The well-known pigeonhole principle states that, for each X where $m > n$, the set of solutions is not empty. If the number of holes is smaller than the number of pigeons, there is either some pigeon not sitting in a hole or some hole contains at least two pigeons. This is one of the most often used arguments in combinatorics. The negation of $\text{PHP}_{m,n}$ can be described by the following set of clauses: $\overline{x}_{i,k} + \overline{x}_{j,k}$ for $i \neq j$ (a hole can contain at most one pigeon) and $x_{i,1} + \cdots + x_{i,n}$ (each pigeon sits in one hole). Obviously, this set of clauses cannot be satisfied if $m > n$. Hence, there is a resolution proof (a sequence of resolution steps refuting the set of clauses) disproving $\overline{\text{PHP}}_{m,n}$ and, therefore, proving $\text{PHP}_{m,n}$. We cite the next theorem (see Krajíček (1995)) without defining the notion of regular resolution proofs. We only want to show a result motivating the consideration of multiterminal OBDDs and FBDDs for search problems.

Theorem 6.5.3. *Let C be an unsatisfiable set of clauses and S_C the search problem which outputs the number of a clause which is not satisfied by a given input. Then the minimal length of any regular resolution proof for the unsatisfiability of C is equal to the minimal size of a multiterminal FBDD for a realization of the relation describing S_C.*

The pigeonhole principle is of particular interest, since it is the main example for proofs of lower bounds on the length of resolution proofs. If m becomes much larger than n (many more pigeons than holes), the relation describing the pigeonhole principle becomes "larger." We may forget large parts of the input and, nevertheless, we are able to find a solution. This implies that our lower bound techniques from Chapter 4 and Section 6.2 do not work.

Razborov, Wigderson, and Yao (1997) have investigated the case $m > n^2$ where no exponential lower bound is known for FBDDs nor even for OBDDs. They have proved exponential lower bounds for graph orderings where all variables of a row are tested as a block (the row model) and for graph orderings where all variables of a column are tested as a block (the column model). In order to present a lower bound technique for search problems, we describe the lower bound for the easier case of the row model.

Theorem 6.5.4. *The size of each multiterminal FBDD solving the search problem $PHP_{m,n}$ where $m > n$ and testing the variables of each row blockwise is bounded below by $2^{\Omega(n \log n)}$.*

Proof. We prove the lower bound for the restricted set of inputs where each row contains a unique 1. Since the variables of a row are tested blockwise, we describe the test of $x_{i1}, \ldots, x_{in}$ by the test of the variable x_i with values in $\{1, \ldots, n\}$ describing the position of the 1. The test of x_i may be represented by an x_i-node with n outgoing edges labeled by $1, \ldots, n$. For these inputs, we have to compute a column containing at least two ones and the positions of the two ones in this column.

Let us start with some combinatorial investigations of the considered FBDD G with row tests x_i having n outputs. For a node v of G, the set $J(v)$ is defined in the following way. It contains j if there is some unique row number $i(j)$ such that all paths from the source to v contain an edge labeled by j and leaving an $x_{i(j)}$-node. Let (u, v) be an edge of G where u is an x_i-node and the edge is labeled by j. Then $J(v) - J(u) \subseteq \{j\}$ and $|J(v)| \leq |J(u)| + 1$. An edge leaving v with label j is called *legal* if $j \notin J(v)$ and *illegal* otherwise. The essential property of FBDDs solving $\mathrm{PHP}_{m,n}$ in the row model is that no path from the source to a sink can contain legal edges only. We prove this claim by considering a path p from the source to the sink with label (C_j, i_1, i_2). Only inputs with ones at the positions i_1 and i_2 of C_j may reach this sink. Hence, the path has to contain edges labeled by j and leaving nodes with labels x_{i_1} and x_{i_2}. Let (u, v) be the last edge on p which is labeled by j and leaves an x_{i_1}- or an x_{i_2}-node and let, w.l.o.g., x_{i_2} be the label of u. If some path p' without a j-edge leaving an x_{i_1}-node reaches u, then, because of the read-once property, the inputs using p' until u and then p reach the sink (C_j, i_1, i_2). Among these inputs there are some where $x_{i_1} \neq j$ in contradiction to the assumption that G solves the pigeonhole principle. Hence, $j \in J(u)$ and the edge (u, v) is illegal.

In Section 6.2, we chose the set of prefix-free paths according to the given graph ordering G^*. This was general enough, since the reduced G^*-FBDD is uniquely determined for decision problems. The situation is different for search problems. Here we make the path system dependent on the FBDD G. Since we additionally assign weights to the paths, the description uses an underlying Markoff chain starting at the source. The terminal states of the Markoff chain are the sinks and the nodes v where $J(v) = \{1, \ldots, n\}$. At any nonterminal state u, a uniform distribution on the outgoing legal edges is used. By our considerations above, we arrive with probability 1 at a terminal v where $J(v) = \{1, \ldots, n\}$. We also have proved that $|J(v)|$ can increase along an edge at most by 1. Hence, with probability 1, there is some first node v_0 on the path where $|J(v_0)| = \lceil n/2 \rceil$. The claim is that, for each node v_0 where $|J(v_0)| = \lceil n/2 \rceil$, the probability that it is the first node of this kind reached by the Markoff chain is bounded by $2^{-\Omega(n \log n)}$. This implies the existence of $2^{\Omega(n \log n)}$ nodes in the FBDD.

To prove the claim, let $J(v_0) = \{j_1, \ldots, j_k\}$ for $k = \lceil n/2 \rceil$. Then there exist $i_1, \ldots, i_k$ such that on each path p from the source to v_0 there are j_l-edges leaving x_{i_l}-nodes. Because of the read-once property, the indices $i_1, \ldots, i_k$ are different. For the x_{i_l}-node on p, the number of outgoing legal edges is at least $\lfloor n/2 \rfloor$. This follows from the fact that $|J(v)|$ increases along an edge at most by 1 and from the definition of v_0 as the *first* node where $|J(v_0)| = \lceil n/2 \rceil$. We claim that the probability of the set of all considered paths from the source to v_0 is bounded by $\lfloor n/2 \rfloor^{-\lceil n/2 \rceil} = 2^{-\Omega(n \log n)}$. We distinguish the cases where the source is labeled by some x_{i_l}, $1 \leq l \leq k$, and by some other x-variable. In the first case, the probability of choosing the j_l-edge is bounded above by $\lfloor n/2 \rfloor^{-1}$, since the number of outgoing legal edges is at least $\lfloor n/2 \rfloor$. For the endpoint u of this edge, it is sufficient to prove that v_0 is reached with a probability of at most $\lfloor n/2 \rfloor^{-(\lceil n/2 \rceil - 1)}$. In the second case, we may choose an arbitrary edge. For the endpoint u of such an edge, it is sufficient to prove that v_0 is reached with a probability of at most $\lfloor n/2 \rfloor^{-\lceil n/2 \rceil}$. These arguments may be repeated. Since we pass through $\lceil n/2 \rceil$ nodes belonging to the first case, we obtain the proposed bound and have proved the theorem. □

6.6 Exercises and Open Problems

6.1.E How many different variable orderings are used in the FBDD for HWB_n (in Theorem 6.1.4)?

6.2.D (See Sieling (1995).) Prove that the hidden weighted bit function HWB_n has exponential FBDD size for all graph orderings G with polynomially many different variable orderings.

6.3.M (See Dunne (1985).) A function f is called k-stable if for all sets $V \subseteq X_n$ containing at most k variables and all $x_i \in V$ there exists some assignment to the variables in $X_n - V$ such that the resulting subfunction of f is x_i or $\overline{x}_i$. Prove that a function is k-mixed if it is k-stable.

6.4.D (See Dunne (1985).) Prove an exponential lower bound on the FBDD size of the determinant DET_n, i.e., the $\oplus$-sum of all $x_{1,\pi(1)} x_{2,\pi(2)} \cdots x_{n,\pi(n)}$ for all permutations π on $\{1, \ldots, n\}$.

6.5.D (See Dunne (1985).) Prove an exponential lower bound on the FBDD size of the function deciding whether an undirected graph contains a Hamiltonian circuit.

6.6.M (See Wegener (1984).) Prove that $\text{FBDD}(f) = \text{OBDD}(f)$ for symmetric functions f.

6.7.E (See Wegener (1986).) Let $f_{k(n),n}$ be the function testing whether an undirected graph on n vertices contains a $k(n)$-clique such that at least $k(n)-2$ vertices have consecutive numbers $i, i+1, \ldots, i+k(n)-3 \pmod n$. Prove that the DNF size of $f_{k(n)}$ is polynomial.

6.8.D (See Wegener (1986).) Prove for $k(n) = \lfloor n^{1/3} \rfloor$ that the FBDD size of $f_{k(n),n}$ is exponential.

6.9.E (See Wegener (1988).) Prove by reduction an exponential lower bound on the FBDD size of the function testing whether an undirected graph contains an $\lceil n/2 \rceil$-clique.

6.10.M (See Wegener (1988).) Prove that the exactly half clique function can be represented by a polynomial-size BP where each variable is tested on each path at most twice.

6.11.D Prove an exponential lower bound on the FBDD size of the general threshold function T_n^* from Definition 4.8.2.

6.12.M (See Bollig and Wegener (1997a).) Prove an exponential lower bound on the FBDD size of the function deciding for a Boolean matrix X whether X contains a 1-row and an even number of ones or a 1-column and an odd number of ones.

6.13.M Prove an exponential lower bound on the FBDD size of the function f_n^{**} obtained from f_n^* (Theorem 6.2.11) by replacing OR by EXOR.

6.14.D (See Savický (personal communication).) Define $\text{HWB}_n^*(x)$ as x_w where

$$w \equiv 2(x_0 + \cdots + x_{\lceil (2n)/3 \rceil - 1}) + x_{\lceil 2n/3 \rceil} + \cdots + x_{n-1} \mod n.$$

Prove a $2^{n/3-O(1)}$ lower bound on the FBDD size of HWB_n^*.

6.15.O Decide whether there exist Boolean functions $f = (f_n)$ with polynomial-size DNFs and CNFs and nonpolynomial FBDD size.

6.16.M (See Breitbart, Hunt III, and Rosenkrantz (1993).) Prove that the function testing whether $n \times n$ Boolean matrices contain two identical rows is $(n-1)$-mixed.

6.17.D Prove that the function *equal adjacent rows* EAR_n testing whether $n \times n$ Boolean matrices have two identical *adjacent* rows has polynomial OBDD size.

6.18.M (See Savický and Wegener (1997).) Prove the space bounds stated in Theorem 6.3.7 for the FBDD $\rightarrow$ OBDD transformation problem.

6.19.O Decide whether the equivalence test of FBDDs is contained in P.

6.20.M (See Fortune, Hopcroft, and Schmidt (1978).) Prove that the test $f \leq g$ for functions given by FBDDs (or even OBDDs and different variable orderings) is coNP-complete.

6.21.E Design an algorithm which checks in time $O(|G||H|)$ whether H is a G-FBDD.

6.22.M (See Sieling and Wegener (1995a).) Prove that graph orderings have a canonical representation which can be obtained by eliminating nodes not reachable from the source and by the application of the merging rule.

6.23.M Let G be a graph ordering with polynomially many different variable orderings. Prove that the equivalence of a G-FBDD H and an FBDD I can be checked in polynomial time.

6.24.M How large is the τ_G-TBDD for HWB_n and the graph ordering G used for the proof of Theorem 6.1.4?

6.25.M How large is the τ_G-TBDD for ISA_n and one of the graph orderings G used for the proof of Theorem 6.1.3?

6.26.E Design and analyze an algorithm for the evaluation problem on τ_G-TBDDs.

6.27.E (See Bern, Meinel, and Slobodová (1995).) Design and analyze an algorithm for the construction of a τ_G-TBDD for x_i.

6.28.M Prove that the relation describing the pigeonhole principle can be realized by an FBDD (even in the row model) of size $2^{O(n \log n)}$.

6.29.O (See Razborov, Wigderson, and Yao (1997).) Prove an exponential lower bound on the size of OBDDs and FBDDs representing a realization of the pigeonhole principle, in particular if $n > m^2$.

Chapter 7

BDDs with Repeated Tests

7.1 The Landscape between OBDDs and BPs

From a complexity theoretical point of view, we are still far from good lower bound techniques for unrestricted BPs, but we know a number of methods for proving exponential lower bounds on the size of OBDDs and read-once BPs (or FBDDs). In such a situation, the usual procedure in complexity theory is to develop lower bound techniques for other models between OBDDs and BPs. The investigation of such models is also motivated by applications. Here our motivation is that BPs do not allow the efficient realization of important operations while OBDDs and FBDDs need exponential size for too many functions.

The most important restrictions on the possibility of repeating tests are the following:

- There is a bound k (possibly depending on the number of variables) such that the length of each path is bounded above by kn,
- there is a bound k (possibly depending on the number of variables) such that each path contains at most k x_i-nodes for each x_i,
- there is a bound k (possibly depending on the number of variables) such that each path may contain more than one x_i-node for at most k variables x_i,
- there is some ordering restriction as in OBDDs.

Moreover, we can distinguish whether the restrictions are *syntactic*, i.e., they have to hold on each path of the BP, or *semantic*, i.e., they have to hold only on paths activated by some input. The last distinction becomes essential, since we investigate BPs with repeated tests and, therefore, with inconsistent paths.

Definition 7.1.1. Let $s = (s_1, \ldots, s_l)$ be a sequence of variables from $X_n = \{x_1, \ldots, x_n\}$. An *s-oblivious BDD* is a BP G such that, for each path, the sequence of node labels is a subsequence $s_{i_1}, \ldots, s_{i_r}$, $1 \leq i_1 < \cdots < i_r \leq l$, of s. The *length* of an s-oblivious BDD is the length l of the corresponding sequence s of variables.

Definition 7.1.2.

(i) A k-OBDD with respect to a variable ordering π is an s-oblivious BDD of length kn where s is the concatenation of k copies of π.

(ii) A k-IBDD with respect to k variable orderings $\pi_1, \ldots, \pi_k$ is an s-oblivious BDD of length kn where s is the concatenation of the variable orderings given by $\pi_1, \ldots, \pi_k$.

Definition 7.1.3. A *read-k-times* BP (k-BP) is a BP where each path contains for each variable x_i at most k nodes labeled by x_i.

Definition 7.1.4. A $(1, +k)$-BP is a BP where for each path p there is some set $V(p)$ of k variables such that p contains for all variables $x_i \in X_n - V(p)$ at most one node labeled by x_i.

The notion of oblivious BDDs adopts the notion of oblivious from many computation models in complexity theory. Exponential lower bounds on the size of oblivious BDDs of linear length can be proved by methods from the theory of communication complexity. The more restricted models of k-OBDDs and k-IBDDs are interesting also from the practical point of view. The sequence $s = (s_1, \ldots, s_l)$ can be understood as a cube embedding (see Definition 5.9.6) and, therefore, oblivious BDDs are τ-TBDDs for special cube embeddings.

The models of k-BPs and $(1, +k)$-BPs are motivated by the aim to develop more general lower bound techniques. Both models are bridges between read-once BPs, i.e., 1-BPs, $(1, +0)$-BPs, or FBDDs, and general BPs, i.e., ∞-BPs or $(1, +n)$-BPs. Our definitions of k-BPs and $(1, +k)$-BPs use *syntactic restrictions*, since the restrictions have to hold on all paths.

Definition 7.1.5. A *semantic restriction* on BPs is a restriction which only has to hold on computation paths, i.e., on paths activated by some input.

Using the notion read-k-times BP for k-BP one can interpret the restriction in the way that a variable can be read or tested at most k times. If only inconsistent paths contain more than k nodes labeled by x_i, then x_i is not read more than k times, since we never follow such paths. Whenever we discuss semantic restrictions, we mention this explicitly.

In Section 7.2, we present several upper bound techniques in order to show which type of functions can be represented efficiently by which type of BDD with repeated tests. Then we investigate in Section 7.3 which operations can be performed efficiently. We are faced with a new problem, the *consistency problem*. Given some BP H, it has to be decided whether H is consistent with the chosen restrictions. This problem has not been considered before, since it is easy to check whether H is a π-OBDD or a G-FBDD (Exercise 6.19) for given π or G, resp., and also easy to check whether H is a π-OBDD or a G-FBDD for some π or G and to construct in the positive case some appropriate π or G (see exercises). Afterwards, lower bound techniques are presented for $(1, +k)$-BPs (Section 7.4), k-OBDDs, k-IBDDs, and oblivious BDDs (Section 7.5), and for k-BPs (Section 7.6). New results on the size of depth-bounded BPs are presented without proofs in Section 7.7.

7.2 Upper Bound Techniques

How can we make use of the freedom to repeat the test of some variables or even all variables a limited number of times? We partially answer this question by the investigation of several typical examples.

Definition 7.2.1. A function $f \in B_n$ is called a *k-pointer function* if $f(x_1, \dots, x_n) = g(x_{p(1)}, \dots, x_{p(k)})$ such that the function $p = (p(1), \dots, p(k)) : \{0,1\}^n \to \{1, \dots, n\}^k$ can be represented by a polynomial-size OBDD with sinks labeled with elements from $\{1, \dots, n\}^k$.

Generalized OBDDs where sinks are labeled with elements from a set A are called MTBDDs and are investigated in Section 9.2. Pointer functions can be hard for FBDDs even for $k = 1$ and simple functions $p(1)$. The reason is that we have to forget the value of many variables during the computation of $p(1)$ and at the end we know $p(1)$ but not $x_{p(1)}$. Then it should be sufficient to repeat the test of $x_{p(1)}$. The functions HWB_n, ISA_n, and WS_n (see Theorem 6.2.10) are 1-pointer functions.

Theorem 7.2.2. *HWB_n and WS_n can be represented in size $O(n^2)$ by BPs which simultaneously are 2-OBDDs and $(1, +1)$-BPs. The function ISA_n can be represented in size $O(n^2)$ by a BP which simultaneously is a 2-OBDD and an FBDD.*

Proof. The constructions for HWB_n and WS_n work in the same way. We use an arbitrary variable ordering and store, for HWB_n, the number of tested ones and, for WS_n, the sum mod p (for the chosen $p \le 2n$) of those ix_i where x_i has been tested. Hence, the width is bounded by $n + 1$ and $2n$, respectively. This multiterminal OBDD (see also Section 9.2) computes $p(1)$. At the sink where $p(1) = i$, the variable x_i is tested once more.

For ISA_n, we choose the variable ordering $y_{k-1}, \ldots, y_0, x_0, \ldots, x_{n-1}$. We start with a DT computing $|y|$. For $|y| = i$, we continue with a DT on the variables $x_i, \ldots, x_{i+k-1}$ (the indices are taken mod n) with respect to the variable ordering $x_0, \ldots, x_{n-1}$ and compute $\alpha(x, y)$. Finally, if $x_{\alpha(x,y)}$ is not known, it is sufficient to test this variable. □

This representation of HWB_n is much more natural than the representation by the polynomial-size FBDD presented in the proof of Theorem 6.1.4. For pointer functions, it is a good idea to separate the computation of the pointers from the test of the chosen variables. The following example is due to Sieling and Wegener (1995b) and is used in Section 7.4 for a hierarchy result.

Definition 7.2.3. The k-pointer function $p_{k,n} \in B_n$ computes $x_{p(1)} \oplus \cdots \oplus x_{p(k)}$ for the following functions $p(1), \ldots, p(k)$. Let m be the largest number such that $mk\lceil \log n \rceil \le n$ and let the variables $x_0, \ldots, x_{n-1}$ be partitioned into k groups each consisting of m numbers of bit length $\lceil \log n \rceil$. Then $p(j)$ is the sum (mod n) of the numbers of the jth group.

Theorem 7.2.4. *The function $p_{k,n}$ can be represented by a 2-OBDD of size $O(n^{k+1})$ and by a $(1, +k)$-BP of size $O(n^2)$.*

Proof. For the 2-OBDD, we may choose an arbitrary variable ordering. For each group, we store the partial sums (mod n) of the already tested variables. A variable x_i at position r in its number contributes $x_i 2^r \bmod n$ to the partial sum of its group. The width is bounded by n^k, since we have to distinguish at most n values for each of the k groups. At the end, $(p(1), \ldots, p(k))$ is known and it is easy to compute $x_{p(1)} \oplus \cdots \oplus x_{p(k)}$ in the second layer.

For the $(1, +k)$-BP, we compute $p(1)$ by testing the variables of the first group. Width n is sufficient for this purpose. Then we compute and store $x_{p(1)}$. Afterwards, $p(2)$ is computed, $x_{p(2)}$ is tested, and $x_{p(1)} \oplus x_{p(2)}$ is stored, and so on. Altogether, we obtain $n + k$ levels of width $2n$ each. □

Here we have seen the advantage of nonoblivious BDDs. We may perform the additional tests whenever it seems to be useful. Semantic $(1, +k)$-BPs (or other BP variants) are less restricted than their syntactic counterparts. Does this increase the representational power? It seems to be hard to imagine that one may use the distinction between computation paths and inconsistent paths during the design of a $(1, +k)$-BP. The following rather artificial example is due to Sieling (1998c). There are no natural functions known where we may gain something by deterministic semantic variants (for the nondeterministic case see Chapter 10).

Example 7.2.5. The function $f_n \in B_n$ is defined on $n = (7m^2 - 5m)/2$ variables a_i, $1 \le i \le m$; b_{ij}; c_{ij}; d_{ij}, $1 \le i, j \le m$ and $i \ne j$; and e_{ij}, $1 \le i < j \le m$.

The variable a_i describes the color of the vertex i of a complete graph on m vertices. The auxiliary variables b_{ij}, c_{ij}, and d_{ij} support the computation and e_{ij} describes the color of the edge $\{i, j\}$. The function f_n computes 1 iff the auxiliary variables carry the information of vertex i, i.e., $b_{ij} = c_{ij} = d_{ij} = a_i$ for all j, and e_{ij} distinguishes whether the vertices i and j have the same color or not, i.e., $e_{ij} = a_i \oplus a_j$.

The intuition is that all variables describing the color of vertex i have to be tested (almost) as a block in order to check whether they have the same value. But the information on vertex i has to be used in connection with the information on each vertex j in order to check the correct value of e_{ij}. Sieling (1998c) has indeed proved that $(1, +k)$-BPs representing f_n need exponential size if $k \leq n^{1/2}/(6 \log n)$. Nevertheless, it is possible to represent f_n by a semantic $(1, +1)$-BP G_n of linear size $O(n)$.

We construct G_n as a conjunction of the vertex components V_i, $1 \leq i \leq m$, and the edge components E_{ij}, $1 \leq i < j \leq m$ (see Fig. 7.2.1). The conjunction is done as for general BPs: the 1-sink of one component is replaced by the source of the next component and only the 1-sink of the last component remains as a 1-sink. Hence, the size of G_n is the sum of the sizes of the components. The size is linear, since each component E_{ij} has constant size and V_i has size $O(m)$.

First, we prove that G_n represents f_n. Afterwards, it is shown that G_n is a semantic $(1, +1)$-BP.

The BP G_n computes 1 iff the following conditions are fulfilled:

$a_i = b_{i1} = \cdots = b_{im} = 0$ or $a_i = c_{i1} = \cdots = c_{im} = 1$ for $1 \leq i \leq m$ and

$(e_{ij} = c_{ij} = c_{ji} = d_{ij} = d_{ji} = 0)$ or $(e_{ij} = 0$ and $b_{ij} = b_{ji} = d_{ij} = d_{ji} = 1)$ or $(e_{ij} = b_{ji} = d_{ji} = 1$ and $c_{ij} = d_{ij} = 0)$ or $(e_{ij} = b_{ij} = d_{ij} = 1$ and $c_{ji} = d_{ji} = 0)$ for $1 \leq i < j \leq n$.

It is obvious that G_n computes 1 if $f_n(a, b, c, d, e) = 1$. Now we assume that G_n computes 1. First, we claim that $b_{ij} = c_{ij} = d_{ij} = a_i$ for all j. Let $a_i = 0$ (the other case can be handled similarly). Then $b_{i1} = \cdots = b_{im} = 0$, since the V_i-component accepts the input. The component E_{ij} for $i < j$ accepts an input with $b_{ij} = 0$ only if $c_{ij} = d_{ij} = 0$. For $i > j$, we consider E_{ji} which accepts inputs with $b_{ij} = 0$ only if $c_{ij} = d_{ij} = 0$. Hence, $c_{i1} = d_{i1} = \cdots = c_{im} = d_{im} = 0$. Considering inputs where $a_i = b_{ij} = c_{ij} = d_{ij}$ for all j, it is easy to see that E_{ij} accepts the input only if $e_{ij} = a_i \oplus a_j$.

By the definition of G_n and its components, only the variables b_{ij} and c_{ij} occur twice on some graph theoretical path. If $i < j$, b_{ij} occurs in the V_i-component and in the E_{ij}-component (if $i > j$, the E_{ji}-component), similarly for c_{ij}. If a variable is tested twice, we have reached the 0-sink, e.g., for c_{ij} and $i < j$, we reach for $c_{ij} = 0$ the 0-sink of E_{ij} and for $c_{ij} = 1$ the 0-sink of V_i. Hence, there is no possibility to test on a path activated by some input a second variable twice. □

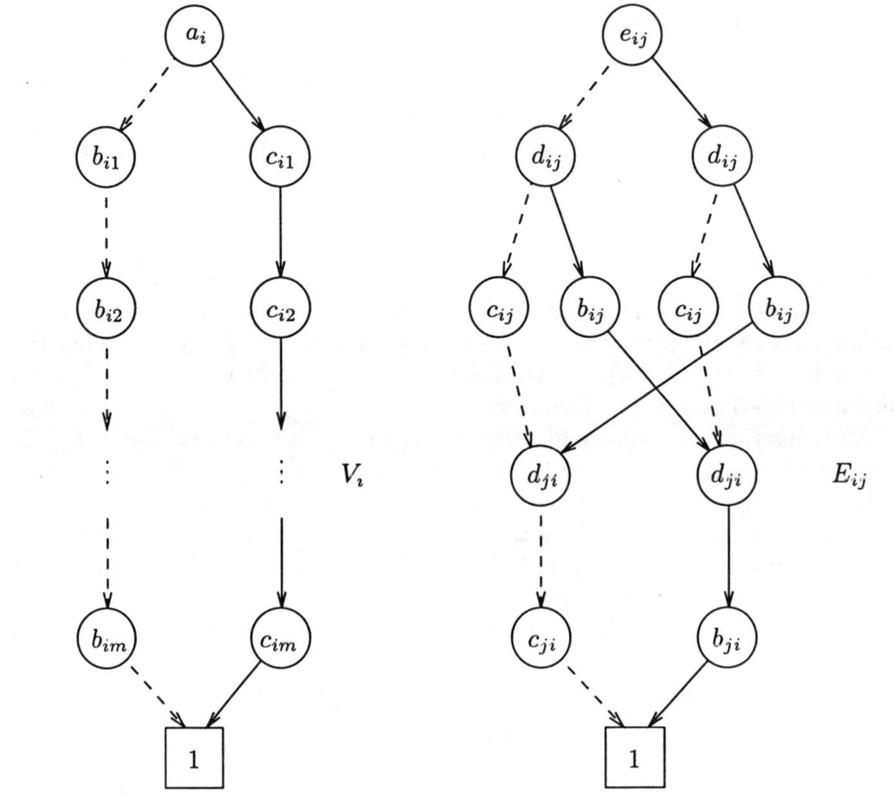

Figure 7.2.1: *The components of* G_n. *All missing edges lead to the* 0-*sink.*

Functions f on graphs which can be expressed as simple functions h on the degree of the vertices are efficiently representable by oblivious 2-BPs. The idea is to start with a DD (where the inputs may take more than two values) which realizes the simple function h and then to replace each x_i-node by a (perhaps multiterminal) OBDD computing some function g_i about the degree of vertex i. We construct polynomial-size oblivious 2-BPs for two functions whose FBDD size is known to grow exponentially. To test whether a graph is $\lceil n/2 \rceil$-regular, or more generally k-regular, is equivalent to testing whether each vertex has degree k. Hence, we may choose g_i as the symmetric Boolean function on the variables describing edges $\{i, j\}$ which checks whether the number of ones equals k and h as AND of n variables. Hence, we obtain an oblivious 2-BP of size $O(n^3)$. The function excl_n (exactly half clique function) tests whether $\lfloor n/2 \rfloor$ vertices of the graph are isolated, i.e., have degree 0, and the other $\lceil n/2 \rceil$ vertices are a clique, i.e., they necessarily have degree $\lceil n/2 \rceil - 1$. Here we may choose g_i as the symmetric function on the variables describing edges $\{i, j\}$ which outputs

0 if vertex i is isolated, 1 if the degree of i is $\lceil n/2 \rceil - 1$, and 2 otherwise. The function h checks whether the input does not contain a 2 and contains exactly $\lceil n/2 \rceil$ ones. Also h is symmetric and can be represented in size $O(n^2)$. Hence, excl_n can be represented in size $O(n^4)$. We have proved the following results.

Theorem 7.2.6. *The test whether a graph is k-regular can be represented by an oblivious 2-BP of size $O(n^3)$. The function $excl_n$ can be represented by an oblivious 2-BP of size $O(n^4)$.*

These functions have generalizations to k-uniform hypergraphs where each hyperedge $\{i_1, \dots, i_k\}$ combines k (instead of two) different vertices. The test whether such a hypergraph is $\lceil \binom{n-1}{k-1}/2 \rceil$-regular or consists of a hyperclique of size $\lceil n/2 \rceil$ and $\lfloor n/2 \rfloor$ isolated vertices can be represented efficiently by oblivious k-BPs (see exercises) and one may conjecture that they cannot be represented efficiently by $(k-1)$-BPs. This conjecture is still open. These generalizations of excl_n were the first functions suggested as examples to separate the class of functions representable by polynomial-size k-BPs from the corresponding class for polynomial-size $(k-1)$-BPs (Wegener (1988)).

Each variable x_{ij}, $1 \le i < j \le n$, in a graph description belongs to the vertex i and to the vertex j. This is similar to variables x_{ij}, $1 \le i, j \le n$, in a matrix that belong to row i and to column j. The following result is, therefore, very simple and can be generalized to simple functions on simple properties on the rows and columns of a matrix.

Theorem 7.2.7. *The test $ROW_n + COL_n$ of whether a matrix contains a 1-row or a 1-column and the test $PERM_n$ of whether a matrix is a permutation matrix can be represented by 2-IBDDs of size $O(n^2)$.*

Proof. We use the variable orderings π_1 and π_2 testing X rowwise and columnwise, respectively. For $\text{ROW}_n + \text{COL}_n$, we test whether a row or column contains ones only and combine the results by an OR. For PERM_n, we test whether each row or column contains exactly one entry 1 and combine the results by an AND. □

The examples on matrices are a little bit more structured than the examples on graphs. The reason is that each pair of different rows (or columns) has no variable in common. This leads to 2-IBDDs and not only oblivious 2-BPs. We have generalized graphs to k-uniform hypergraphs. In the same way, we can generalize (ordinary two-dimensional) matrices to k-dimensional matrices $X = (x_{i_1,\dots,i_k})_{1 \le i_1,\dots,i_k \le n}$ with $N = n^k$ entries. Instead of rows and columns we investigate hyperplanes. The jth hyperplane X_j^d in direction d consists of all variables where $i_d = j$. The number of hyperplanes equals nk. It is easy to see that the test of whether there exists a hyperplane consisting of ones only and the test of whether each hyperplane contains exactly one 1 entry can be represented by k-IBDDs of size $O(kn)$. The functions are also candidates to separate the class of functions representable by k-BPs (k-IBDDs) from the corresponding class for polynomial-size $(k-1)$-BPs ($(k-1)$-IBDDs). Such a separation has been proved by Thathachar (1998a) using the following functions.

Definition 7.2.8. Let $H_d(X)$ be the polynomial over the field $\mathbb{Z}_q$ (for an odd prime number q) which is defined on a k-dimensional matrix X in the $(+1,-1)$-notation and a dimension $d \in \{1,\dots,k\}$. The polynomial $H_d(X)$ computes the sum of the n monomials consisting of the variables of the jth hyperplane X_j^d in direction d, i.e.,

$$H_d(X) = \sum_{1\le j\le n} \prod_{i_d=j} x_{i_1,\dots,i_k}.$$

(Remember that a monomial in $(+1,-1)$-notation corresponds to the parity of the considered variables.) Then the *hyperplanar sum-of-products* predicate HSP_q^k outputs 1 iff $H_1(X)+\cdots+H_k(X)\equiv 0 \bmod q$ and the *conjunction of hyperplanar sum-of-products* predicate CHSP_q^k outputs 1 iff $H_d(X)\equiv 0 \bmod q$ for all $d\in\{1,\dots,k\}$.

Proposition 7.2.9. *The functions HSP_q^k and $CHSP_q^k$ are defined on $N=n^k$ variables and can be represented by k-IBDDs of size $O(kN)$.*

Proof. In the dth variable ordering, the variables are ordered according to the hyperplanes in direction d. Then it takes linear size to compute $H_d(X) \bmod q$. The combination of the different layers is performed in the natural way. □

We have discussed typical examples which can be represented efficiently under some restriction on how tests may be repeated. At the end of this section, we mention some other functions considered in the other chapters. For NP-complete functions such as the clique function, we cannot expect BPs of polynomial size. Functions such as multiplication, the threshold function T_n^* from Definition 4.8.2, and the function f_n^* considered in Theorem 6.2.11 have polynomial-size BPs which are $k(n)$-BPs for $k(n)=\lceil n/\log n\rceil$, $\lceil n/\log n\rceil$, and $\lceil n^{1/2}/\log n\rceil$, respectively (see exercises), and one may conjecture that they cannot be represented by polynomial-size $k'(n)$-BPs and $k'(n)\ll k(n)$. All these conjectures are open.

7.3 Efficient Algorithms and NP-Hardness Results

In this section, we discuss which of the considered BP variants may be used as data structure, i.e., which variants allow efficient algorithms for the important operations. The satisfiability test is NP-complete for most types of BDDs with repeated tests.

Theorem 7.3.1. *The satisfiability test is NP-complete for 2-IBDDs and* $(1, +n^{\varepsilon})$*-BPs for* $\varepsilon > 0$.

Proof. It is easy to guess an input and to verify that it is satisfying. Moreover, there is a standard polynomial transformation from 3-SAT to SAT for 2-IBDDs. First, the conjunction of the clauses is transformed in the usual way into a BP. Then the jth node labeled by x_i is replaced with an $x_{i,j}$-node. This leads to an OBDD for some ordering of the $x_{i,j}$-variables, since each variable is the label of only one node. Afterwards, the 1-sink is replaced with a linear-size OBDD testing for each i whether all variables $x_{i,\cdot}$ take the same value. Altogether, we obtain a 2-IBDD. The whole transformation can be performed in polynomial time. The resulting BP G is satisfiable iff all variables $x_{i,\cdot}$ take the same value $a_i \in \{0, 1\}$ and the vector $(a_1, \dots, a_n)$ satisfies all given clauses. Hence, the given 3-SAT instance is satisfiable iff G is satisfiable.

The result on $(1, +n^{\varepsilon})$-BPs follows easily by a standard padding argument, since a 2-IBDD is also a $(1, +n)$-BP. □

Savický (1998a) has proved that the satisfiability problem for $(1, +k)$-BPs can be solved in time $|G|^{O(k)}$ and, therefore, in polynomial time for constant k. The situation becomes worse for the semantic variants as shown by Sieling (1998c).

Theorem 7.3.2. *The satisfiability test is NP-complete for semantic* $(1, +1)$*-BPs.*

Proof. Let $c_1, \dots, c_m$ be 3-SAT clauses over the variables $x_1, \dots, x_n$. This instance is transformed into a $(1, +1)$-BP G which is the conjunction of the variable components $V_i, 1 \le i \le n$, and the clause components $C_j, 1 \le j \le m$. The construction resembles that of Example 7.2.5. So the conjunction is performed by replacing the 1-sink of one component by the source of the next one. Let $p(i)$ be the number of clauses containing x_i as positive literal and $q(i)$ the corresponding number for the negative literal $\overline{x}_i$. Then the BP G works on the variables $a_1, \dots, a_n, d_1, \dots, d_m, e_1, \dots, e_m, b_{i,1}, \dots, b_{i,p(i)}, c_{i,1}, \dots, c_{i,q(i)}$, $1 \le i \le n$. Each literal of each clause corresponds to one specific b- or c-variable. The component V_i is the same as in Fig. 7.2.1 (replacing b_{im} by $b_{i,p(i)}$ and c_{im} by $c_{i,q(i)}$). The component C_j works on d_j, e_j, and the b- and c-variables of the clause c_j. If, e.g., $c_j = x_{i_1} + \overline{x}_{i_2} + x_{i_3}$ and this is the l_1th occurrence of x_{i_1}, the l_2th of $\overline{x}_{i_2}$, and the l_3th of x_{i_3}, then C_j accepts an input iff

- $d_j = 0$ and $b_{i_1,l_1} = 1$ or
- $d_j = 1$, $e_j = 0$, and $c_{i_2,l_2} = 0$ or
- $d_j = e_j = 1$ and $b_{i_3,l_3} = 1$.

In general, d_j and e_j are control variables leading to the variables representing the literals of c_j. Then b-variables have to equal 1 and c-variables to equal 0.

If $(a_1, \ldots, a_n)$ is a satisfying input, we follow the corresponding edge in V_i. If $a_i = 1$, we set $c_{i,1} = \cdots = c_{i,q(i)} = 1$ and all $b_{i,\cdot}$-variables may be set to 1 in order to satisfy the C_j-components for clauses containing x_i. Hence, we can satisfy G. If G is satisfied, we claim that the assignment to $a_1, \ldots, a_n$ (interpreted as an assignment to $x_1, \ldots, x_n$) satisfies all clauses. If a C_j-component is satisfied, since some $b_{i,\cdot}$-variable equals 1, we have to use in the V_i-component the path starting with $a_i = 1$. Therefore, $a = (a_1, \ldots, a_n)$ satisfies all clauses. The BP G is a semantic $(1, +1)$-BP. If a variable is tested twice, this is a b- or c-variable in a V_i- and a C_j-component. If a b-variable is 0, the 0-sink is reached in the C_j-component. The same happens in the V_i-component if the value of the b-variable is 1. Hence, we reach the 0-sink immediately after the first variable is tested for the second time (or even earlier). □

Based on the construction of this proof, it is not too difficult to show that the consistency test for semantic $(1, +1)$-BPs, i.e., the test whether a given BP is a semantic $(1, +1)$-BP, is coNP-complete. The same holds for the consistency test for $(1, +k)$-BPs where k belongs to the input (see exercises). The consistency test is simple for the other syntactic models such as s-oblivious BDDs and, therefore, k-OBDDs with respect to π or k-IBDDs with respect to $\pi_1, \ldots, \pi_k$ and, moreover, also for k-BPs.

Although k-OBDDs may contain exponentially many inconsistent paths, the satisfiability test for k-OBDDs and constant k can be performed in polynomial time (Bollig, Sauerhoff, Sieling, and Wegener (1998)).

Theorem 7.3.3. *The satisfiability test for k-OBDDs G with respect to π is possible in time $O(|G|^{2k-1})$ and space $O(|G|^k)$.*

Proof. The first aim is to partition G into k layers $G_1, \ldots, G_k$ such that in each layer the variable ordering π is respected and edges leaving G_i lead to one of the layers $G_{i+1}, \ldots, G_k$. We denote an edge as π-legal if the edge is allowed in a π-OBDD. We start with the iterative construction of $G_k = (V_k, E_k)$ and initialize V_k with the two sinks and E_k as an empty set. If a node has two successors in V_k which are reached via π-legal edges, it is included in V_k and the edges in E_k. For the construction of $G_{k-1} = (V_{k-1}, E_{k-1})$, we initialize V_{k-1} with all nodes whose successors are in V_k and E_{k-1} with the edges leaving the nodes in V_{k-1}. The construction of $G_{k-2}, \ldots, G_1$ is done in an analogous way.

For $v \in V_k$, let $G(v)$ denote the π-OBDD which is obtained from G by choosing v as the source and eliminating all nodes not reachable from v. For $v \in V_i$ and $w \in V_j$, where $i < j$, let $G(v, w)$ denote the π-OBDD which is obtained from G by choosing v as the source, replacing each node except w

from the layers V_l, $l > i$, with the 0-sink and w with the 1-sink, and eliminating all nodes not reachable from v.

If v_1 is the source of G and a is a satisfying input, the path activated by a starts at v_1 in $G_{l(1)}$ and leads through some layers $l(1) < \cdots < l(r) = k$ such that $G_{l(i)}$ is reached for the first time at some node v_i and, finally, the 1-sink is reached. This is equivalent to the condition that a satisfies $G(v_i, v_{i+1})$, $1 \le i < r$, and $G(v_r)$. Hence, there exists a satisfying input iff there exists some sequence v_1 (the source), $v_2, \ldots, v_r$ of vertices in some layers $l(1) < \cdots < l(r) = k$ such that $G(v_i, v_{i+1})$ and $G(v_r)$ are satisfied by the same input.

It is easy to see that the number of possibilities for r and $v_2, \ldots, v_r$ is bounded by $|G|^{k-1}$. For each choice, we obtain up to k π-OBDDs $G(v_i, v_{i+1})$ and $G(v_r)$. We may combine them by AND-synthesis to a π-OBDD $G(v_1, \ldots, v_r)$ whose size is bounded by $O(|G|^k)$ and then we can apply the satisfiability test for π-OBDDs. The time bound follows directly and the space bound follows from the fact that it is sufficient to store G and one $G(v_1, \ldots, v_r)$. □

This proof shows that we may describe the function $f \in B_n$ represented by a k-OBDD G with respect to π as a disjunction of at most $|G|^{k-1}$ π-OBDDs $G(v_1, \ldots, v_r)$ of size $O(|G|^k)$ each. Even for $k = 2$ it is not always possible to obtain a polynomial-size π-OBDD for f, since the disjunction of many π-OBDDs may lead to an exponential blow-up of the size. Moreover, each input a can satisfy at most one of the π-OBDDs $G(v_1, \ldots, v_r)$. Hence, we can solve SAT-COUNT in the same resource bounds as SAT by adding the results of SAT-COUNT for each $G(v_1, \ldots, v_r)$.

In the following, we briefly discuss the other important operations. Evaluation can be performed efficiently even for general BPs. The second fundamental operation in addition to the satisfiability test is the synthesis problem.

Theorem 7.3.4. *The synthesis of k-BPs and $(1, +k)$-BPs can cause an exponential blow-up of the size as long as $k = o(\log^{1/2} n)$ and $k = o(n^{1/3}/\log^{2/3} n)$, respectively.*

Sketch of Proof. Thathachar (1998a) proved that the conjunction of hyperplanar sum-of-products predicate $\mathrm{CHSP}_q^{k+1}(X)$ needs exponential-size k-BPs if $k = o(\log^{1/2} n)$; see also Section 7.6. The function $g(X)$ testing on the $(k+1)$-dimensional matrix X whether $H_d(X) \equiv 0 \bmod q$ for all $d \le k$ and the function $h(X)$ testing whether $H_d(X) \equiv 0 \bmod q$ for $d = k+1$ both have polynomial-size k-BPs (using the method of the proof of Proposition 7.2.9). The result on k-BPs follows, since $\mathrm{CHSP}_q^{k+1}(X) = g(X) \wedge h(X)$.

Sieling (1996) proved that $p_{k+1,n}$ (see Definition 7.2.3) needs exponential-size $(1, +k)$-BPs if $k = o(n^{1/3}/\log^{2/3} n)$. Using the ideas of the proof of Theorem 7.2.4 it follows that the sum $g(x) = x_{p(1)} \oplus \cdots \oplus x_{p(k)}$ of the first k of the $k+1$ terms of $p_{k+1,n}$ as well as the last term $h(x) = x_{p(k+1)}$ can be represented in $(1,+k)$-BPs of size $O(n^2)$. The result on $(1,+k)$-BPs follows, since $p_{k+1,n}(x) = g(x) \oplus h(x)$. □

Theorem 7.3.5. *The synthesis problem for s-oblivious BDDs (k-OBDDs and k-IBDDs) G and H can be performed in time and space $O(|G||H|)$.*

Proof. Let $s = (s_1, \ldots, s_l)$. As in the proof of Theorem 7.3.3, we can partition G and H into $l+1$ layers such that the ith layers G_i and H_i, $1 \le i \le l$, only contain nodes labeled by s_i and the outgoing edges lead to layers with a larger index or to the sinks which form the layer $l+1$. We consider G and H as π-OBDDs G^* and H^* on the new variables $y_1, \ldots, y_l$ and the variable ordering $\pi = id$. Then we apply the OBDD synthesis algorithm and, afterwards, we replace y_i by $s_i \in \{x_1, \ldots, x_n\}$.

The correctness of this procedure can be proved easily. The π-OBDD G^* represents a Boolean function $g^* \in B_l$ while G represents $g \in B_n$. By construction, $g(a) = g^*(a^*)$ where $a_j^* := a_i$ if $s_j = x_i$. The result of the $\otimes$-synthesis represents $g^* \otimes h^*$ where $g^*(a^*) \otimes h^*(a^*) = g(a) \otimes h(a)$ for the inputs $a^* \in \{0,1\}^l$ corresponding to some $a \in \{0,1\}^n$. After the relabeling of the nodes, we follow for the input a the same path as for a^* and, therefore, $g \otimes h$ is represented. The results on k-OBDDs and k-IBDDs follow, since they are s-oblivious BDDs. □

The equivalence test is difficult if the satisfiability test is difficult. For k-OBDDs, an equivalence test is possible in time $O(|G_f|^{2k-1}|G_g|^{2k-1})$ and space $O(|G_f|^k|G_g|^k)$ as $\oplus$-synthesis followed by a satisfiability test. For $(1,+k)$-BPs and constant k, Savický (1998a) presents a polynomial-time probabilistic non-equivalence test with one-sided error. It is obvious that replacement by constants can be performed in linear time for all considered models. Replacement by functions and quantification basically are done by replacement by constants followed by synthesis. The corresponding results can be transferred.

Only k-OBDDs for constant k admit polynomial-time algorithms for the important operations. There are two obstacles for the practical use of k-OBDDs. The algorithms for satisfiability and equivalence test are efficient only for small k. Moreover, k-OBDDs of minimal size are not canonical. A minimal-size 2-OBDD for $x_1 + \cdots + x_n$ and the variable ordering $\pi = id$ contains exactly one x_i-node for each x_i. In the following way, we obtain 2^{n-1} different 2-OBDDs of minimal size. We test some subset of $\{x_1, \ldots, x_{n-1}\}$ in the natural ordering, then x_n and, finally, the remaining variables in the natural ordering. This is the main reason that no efficient minimization algorithm exists.

Hence, only heuristic ideas are used to transform a circuit into a k-OBDD and, moreover, for the decision about the appropriate k. One idea is to create new layers if the synthesis leads to a large size increase. The synthesis of two k-OBDDs (see the proof of Theorem 7.3.5) can be interpreted as the synthesis of the corresponding layers of the given k-OBDDs G_g and G_h. If this causes a large increase of the size of, e.g., the jth layer, we may interpret G_g and G_h as

$(k+1)$-OBDDs where in G_g layer $j+1$ is empty and in G_h layer j is empty. The result of the synthesis of these $(k+1)$-OBDDs is a $(k+1)$-OBDD G_f which may be much smaller than the k-OBDD resulting from the synthesis of the given k-OBDDs.

Another idea is to represent the possible values of some "important" gates $c_1, \ldots, c_r$ at the edges leaving the first layer. Important gates may be computed by techniques available in CAD tools or by the following heuristic. A gate c is called expensive if, during the gate-by-gate transformation of a circuit into an OBDD, the OBDD size increases significantly during the synthesis step corresponding to c. Then the direct predecessors of expensive gates may be called important, since their replacement by constants makes an expensive gate inexpensive. The first important gates have, by definition, small OBDDs. We now apply a synthesis operator which concatenates the inputs, i.e., $(a, b) \to ab$, to obtain a multiterminal OBDD where each sink represents for some string $a_1 \cdots a_r$ from $\{0,1\}^r$ all inputs where the value at c_i equals a_i, $1 \le i \le r$. In this way, we obtain the first layer of a k-OBDD. Then, at the sink with label $a_1 \cdots a_r$, we have to represent the function represented by the circuit where c_i is replaced by the constant a_i. This makes the first expensive gates, by definition, inexpensive. We may proceed in the same way. In this approach, it may be natural to use different variable orderings in the different layers and to create k-IBDDs. Such heuristics are presented by Jain, Bitner, Abadir, Abraham, and Fussell (1997) together with a heuristic satisfiability test for k-IBDDs and even general BPs. We describe the main ideas of their satisfiability test.

The algorithm is based on a labeling of the edges. Each edge gets a set of labels from $\{0, 1, *\}$ for each variable x_i. The label 0 indicates the existence of a path from the source to this edge with at least one x_i-node and the property that all x_i-nodes are left via the 0-edge, similarly for the label 1. The label $*$ indicates the existence of a path from the source to this edge without any x_i-node. This labeling can be computed efficiently by a top-down approach. If an edge gets, for some variable, no label, it does not lie on any computation path and can be replaced with an edge to the 0-sink. If an edge to an x_i-node v has only the label 0 for x_i, it can be redirected to the 0-successor of v (similarly for the label 1). Edges to the sinks and edges to x_i-nodes with only the label $*$ for x_i are called stationary. A path from the source to the 1-sink consisting only of stationary edges implies the existence of a satisfying input. The redirection of edges leads to a recomputation of the labels of certain edges and to the elimination of nodes and edges which are no longer reachable. This local simplification procedure is successful if it proves the existence of a satisfying input or it eliminates the 1-sink as not reachable. Otherwise, the algorithm chooses one of three global rebuilding techniques. The techniques are only applied to variables x_j such that G is not read once with respect to x_j.

The procedure free(x_i) produces a BP which is read once with respect to x_i. By a top-down approach, edges are copied if they have more than one label with respect to x_i. Let s_i^{free} be the size of the resulting BP after the application of the reduction rules. The computation of s_i^{free} is stopped if some intermediate result becomes too large. Let s_i^{jump} be the size of the BP resulting from a modified jump$(i, 1)$-operation (see Section 5.7). The source gets the label x_i and the outgoing edges point to $G_{|x_i=0}$ and $G_{|x_i=1}$ and then reduction rules are applied. If the smallest of the s_i^{free}- and s_i^{jump}-values is less than $(1+\alpha)|G|$ for some given parameter α, we continue with the resulting BP. Otherwise, let $s_{i,0}$ and $s_{i,1}$ be the size of $G_{|x_i=0}$ and $G_{|x_i=1}$, respectively. The satisfiability test is applied to $G_{|x_i=0}$ and $G_{|x_i=1}$ for that i such that $s_{i,0}^2 + s_{i,1}^2$ is minimal. This ensures that the resulting BPs are small and not of too different size. We start with the smaller one of $G_{|x_i=0}$ and $G_{|x_i=1}$. Whenever we find a satisfying input, we stop the whole procedure. This satisfiability test has led to quite good results in some applications, although its worst-case runtime obviously is exponential.

Altogether, the use of BDDs with repeated tests in applications is limited. They are more important for the development of lower bound techniques, which are discussed in the following sections.

7.4 Lower Bound Techniques for $(1, +k)$-BPs

Before presenting a lower bound technique for semantic $(1, +k)$-BPs, we give a short overview on the known results. One may conjecture that polynomial-size $(1, +k)$-BPs are more powerful than polynomial-size $(1, +(k-1))$-BPs. This has been proved for a wide range of k in the syntactic and semantic models.

Theorem 7.4.1.

(*i*) *There exist Boolean functions $f_n^k \in B_n$ representable by polynomial-size $(1, +k)$-BPs but not by polynomial-size $(1, +(k - 1))$-BPs as long as $k \le n^{1/2}/(2 \log n)$.*

(*ii*) *There exist Boolean functions $f_n^k \in B_n$ representable by polynomial-size $(1, +k)$-BPs but not by polynomial-size semantic $(1, +(k-1))$-BPs as long as $k \le n^{1/6}/(2 \log^{1/3} n)$.*

The results are due to Savický and Žák (1998) (some weaker results have been obtained by Sieling (1996) and Savický and Žák (1997b), respectively). Their function is also a k-pointer function where the different pointers are computed as in the definition of $p_{k,n}$ on disjoint sets of variables. The variables are partitioned into k groups of almost the same size and the pointer $p(j)$ is the weighted sum function WS (see Theorem 6.2.10) on its group of variables. If the pointers are $p(1), \dots, p(k)$, the function computes $x_{p(1)} \oplus \cdots \oplus x_{p(k)}$. It is not surprising that k-pointer functions are well suited to hierarchy results.

The function of Example 7.2.5 has linear size for semantic $(1, +1)$-BPs but is not representable by polynomial-size $(1, +k)$-BPs as long as $k \le n^{1/2}/(6 \log n)$.

The first exponential lower bounds on the size of semantic $(1,+k)$-BPs for explicitly defined functions were presented by Žák (1995) and Savický and Žák (1997a). The following improved methods and results are due to Jukna and Razborov (1998).

Definition 7.4.2. A Boolean function f is called *d-rare* if different inputs $a, b \in f^{-1}(1)$ differ at least at d positions. The function is called *m-dense* if we have to replace at least m variables by constants in order to obtain a subfunction which is the constant 0.

For a d-rare function, 1-inputs have a large Hamming distance. If $d \geq 2$ (otherwise, the notion is meaningless), all prime implicants of the function have length n and all variables have to be tested before the 1-sink may be reached. The notion m-dense is equivalent to the notion that all prime clauses have a length of at least m. This implies that at least m variables have to be tested before the 0-sink may be reached. Hence, computation paths for d-rare and m-dense functions contain tests of at least m different variables and, for 1-inputs, even tests of all variables. If all computation paths are long and the BP is not too large, a lot of computation paths split and join again. At that node where they join again, some information is lost on the inputs leading to this node. If too much information is lost and not too many variables may be tested for a second time, it is not possible to compute the correct value of the function. These vague ideas are now made precise. In particular, the notion of losing information is formalized. For partial inputs $a \in \{0,1,*\}^n$, we denote the support of a by $S(a) = \{i \mid a_i \neq *\}$.

Definition 7.4.3. Let v be a node of a BP G. The pair (a, b) of different partial inputs with the same support belongs to $L(v)$ (lost at v) if the computation paths for a and b pass through v and, on both computation paths from the source to v, all bits where $a_i \neq b_i$ have been read. The set L is the union of all $L(v)$.

If $(a, b) \in L(v)$, the partial inputs are separated during the computation and the information that they are different is lost and can be re-established only by reading variables again. The following lemma is the main technical part of the lower bound technique (Savický and Žák (1997a)).

Lemma 7.4.4. *Let G be a BP where, on each computation path, at least m different variables are tested. Let $s \leq \lfloor m/(2\log|G|+1)\rfloor$. Then we obtain pairwise disjoint sets $I_j \subseteq \{1,\ldots,n\}$, $1 \leq j \leq s$, and pairs (a_j, b_j) of different partial inputs whose support is I_j such that $|I_j| \leq 2\log|G|+1$ and $(a_j^*, b_j^*) \in L$, where a_j^* is the partial input with support $I_1 \cup \cdots \cup I_j$ which is the common extension of $a_1, \ldots, a_j$ and b_j^* is defined similarly for $a_1, \ldots, a_{j-1}, b_j$.*

Proof. We follow all possible computation paths until $r = \lfloor \log|G| \rfloor + 1$ different variables have been tested. We obtain $2^r > |G|$ partial computation paths.

Hence, at least two of them, for the partial inputs a'_1 and b'_1, lead to the same node. We extend a'_1 and b'_1 to partial inputs a_1 and b_1 with support $S(a'_1) \cup S(b'_1)$. Since $S(a'_1) \cap S(b'_1)$ contains at least the index of the variable tested at the source, $|S(a'_1) \cup S(b'_1)| \leq 2r - 1$. The new assignments in a_1 and b_1 are chosen in such a way that a_1 and b_1 differ at most on variables in $S(a'_1) \cap S(b'_1)$. By this definition, $(a_1, b_1) \in L$.

For the next step, we restrict G with respect to a_1 and construct (a_2, b_2) in the same way on the restricted BP $G_{|a_1}$. This procedure can be continued as long as we can be sure that all computation paths have a length of at least r, i.e., for j times, if $j(2r-1) \leq m$. By the choice of s, we can repeat the procedure s times. □

Theorem 7.4.5. *Each semantic* $(1, +k)$*-BP* G *for a function* $f \in B_n$ *which is* d*-rare and* m*-dense has a size bounded below by*

$$M(d, m, k) = \min\{2^{(d-1)/2}, 2^{(m/(k+1)-1)/2}\}.$$

Proof. The result is obvious for $d < 2$. Hence, let $d \geq 2$. We assume that G is a semantic $(1, +k)$-BP representing f with less than $M(d, m, k)$ nodes. By our previous discussion, we know that on each computation path at least m variables are tested. Since $|G| < 2^{(m/(k+1)-1)/2}$, we can apply Lemma 7.4.4 for $s = k + 1$. The partial input a^*_{k+1} has a support whose size is less than m and, since f is m-dense, a^*_{k+1} has an extension a^* such that $f(a^*) = 1$.

By the pigeonhole principle and the assumption that G is a $(1, +k)$-BP, there is one set I_j, $1 \leq j \leq k + 1$, such that all variables x_i, $i \in I_j$, are tested only once on the computation path for a^*. Now let v be the node such that $(a^*_j, b^*_j) \in L(v)$. All variables where a_j and b_j differ are tested before v and tests of these variables are not repeated after v if we follow the computation path for a^*. Hence, the BP G computes, on the input b^* obtained from a^* by replacing a_j with b_j, the same value as on a^*, namely $f(b^*) = 1$. The inputs a^* and b^* differ at most at $|I_j| \leq 2 \log |G| + 1$ positions. Now we apply the second upper bound on $|G|$, namely $|G| < 2^{(d-1)/2}$, leading to $|I_j| < d$. This contradicts the assumption that f is d-rare. □

Based on this lower bound technique, Jukna and Razborov (1998) have obtained superpolynomial lower bounds for $k = o(n/\log n)$. They have considered the characteristic functions of *linear codes*, which are linear subspaces of the vector space $\mathbb{Z}_2^n$. In the theory of error-correcting codes, one tries to obtain codes C with many codewords which pairwise have a large Hamming distance. Let $f_C(a) = 1$ iff $a \in C$. If the minimal distance of C is d, then f_C is d-rare. It is easy to see that f_C is m-dense if the minimal distance of the dual space or dual code $C^\perp$ is m. Hence, we can directly apply the deep theory on error-correcting codes (see MacWilliams and Sloane (1977)) to obtain functions which have large semantic $(1, +k)$-BP size by Theorem 7.4.5, even for large k. The proposed bounds are obtained for the characteristic functions of some Reed-Muller codes and some Bose–Chaudhuri–Hocquenghem codes.

In order to come closer to lower bounds for semantic 2-BPs, some further BP variants have been investigated. For so-called gentle BPs, we refer to Žák (1997) and Jukna and Žák (1998) and, for BPs based on so-called corrupting Turing machines, we refer to Jukna and Razborov (1998).

7.5 Lower Bound Techniques for Oblivious BDDs

Lower bound techniques for k-OBDDs, k-IBDDs, or oblivious BDDs of bounded length $l = kn$ (where k is a constant or can depend on n) have been developed by Jukna (1987), Alon and Maass (1988), Krause (1991), Krause and Waack (1991), and Babai, Nisan, and Szegedy (1992). Although not always stated explicitly, all lower bound techniques can be expressed most naturally in the language of communication complexity (see also Section 4.1).

Let G be an s-oblivious BDD of length l representing $f \in B_n$. Let X_n be partitioned into $A(n)$ and $B(n)$. Let Alice know the input bits for the variables in $A(n)$ and let Bob know the bits corresponding to $B(n)$. As in the proof of Theorem 7.3.5, we partition G into l levels such that the nodes of the ith level are labeled by s_i. Alice is the owner of the levels labeled by variables from $A(n)$ and Bob the owner of the other levels. A layer is a maximal block of consecutive levels owned by the same player. We denote by $ld(G)$ the number of layers of G (layer depth) with respect to the given bipartition of X_n. Alice and Bob can agree upon the following communication protocol. The owner of the first layer starts the communication and follows the computation path up to the first node v labeled by a variable of the other player. The player communicates v. Then the other player goes on in the same way until a player reaches a sink and communicates its number. Then both players know the value of the function on the considered input. The communication takes at most $ld(G)$ rounds and the length of the communication is bounded by $ld(G)\lceil \log |G| \rceil$. (We may save one round if it is sufficient that one player knows the value of the function.)

If the communication complexity of f is denoted by $\mathrm{C}(f)$, then

$$\mathrm{C}(f) \le ld(G)\lceil \log |G| \rceil \qquad \text{or}$$

$$|G| > 2^{\mathrm{C}(f)/ld(G)-1}.$$

The theory of communication complexity (see the monographs of Hromkovič (1997) and Kushilevitz and Nisan (1997)) provides us with lower bound techniques for the communication complexity of f. Since $\mathrm{C}(f) \le n$ in all relevant models, we have to ensure that $ld(G)$ is not too large. Moreover, if $ld(G)$ is known to be bounded by r, we know that the number of communication rounds is bounded by r and may apply lower bounds for communication protocols which are restricted to r rounds.

For k-OBDDs and a fixed variable ordering π, we look for lower bounds on $2k$-round protocols where $A(n)$ contains for some i the first i variables according to π. If π is not fixed, we may choose some i and have to look for a lower bound which holds for all bipartitions of X_n where $|A(n)| = i$.

The situation for k-IBDDs is more difficult. If $A(n)$ or $B(n)$ is small, we cannot expect large lower bounds on the communication complexity. If one player communicates all his or her knowledge, the other one can compute the value of f. Hence, the communication complexity is bounded above by $\min\{|A(n)|, |B(n)|\} + 1$. If $A(n)$ and $B(n)$ are too large, $ld(G)$ cannot be bounded by a small upper bound. The solution is to find subsets of not too few variables such that the number of layers with respect to these variables is small.

In the following, we consider the set $I = \{1, \ldots, n\}$ of indices of the variables. Let (A_0, B_0) be a partition of $I_0 = I$ into sets whose size is at least $n_0 = \lfloor n/2 \rfloor$. Let $s = (i_1, \ldots, i_{kn})$ be the sequence of variable indices of the levels of a k-IBDD for f. We look for "large" sets $A_k \subseteq A_0$ and $B_k \subseteq B_0$ such that the number of layers with respect to (A_k, B_k) is bounded by $2k$. Then we may apply lower bounds from communication complexity for the bipartition (A_k, B_k) of all variables of a subfunction f^* of f which is obtained by assigning well-chosen constants to all variables x_i, $i \notin A_k \cup B_k$. The sets A_k and B_k can be constructed by the following simple combinatorial approach. Let A_i and B_i be given such that $|A_i|$, $|B_i| \geq n_i$. Then we look at the sequence $(j_1, \ldots, j_n)$ belonging to the variable ordering π_{i+1}. Let r be chosen in such a way that $(j_1, \ldots, j_r)$ contains n_i of the indices in $A_i \cup B_i$. If $(j_1, \ldots, j_r)$ contains at least $\lfloor n_i/2 \rfloor$ elements of A_i, we define $A_{i+1} = A_i \cap \{j_1, \ldots, j_r\}$ and $B_{i+1} = B_i \cap \{j_{r+1}, \ldots, j_n\}$. Otherwise, $B_{i+1} = B_i \cap \{j_1, \ldots, j_r\}$ and $A_{i+1} = A_i \cap \{j_{r+1}, \ldots, j_n\}$. In both cases $|A_{i+1}|$, $|B_{i+1}| \geq \lfloor n_i/2 \rfloor$. Altogether, $|A_k|$, $|B_k| \geq \lfloor n/2^{k+1} \rfloor$. By construction, it is obvious that the number of layers with respect to A_k and B_k is bounded by $2k$. There are at most two layers in each block for a variable ordering π_i. It may even happen that adjacent layers belong to the same player and can be merged.

For s-oblivious BDDs with at most k levels labeled by the same variable, the situation becomes even more difficult. We cannot argue about the blocks which are given for k-IBDDs by the division into k variable orderings. Nevertheless, a similar result to that shown above can be obtained by the following fundamental lemma due to Alon and Maass (1988) and proved by arguments borrowed from Ramsey theory.

Lemma 7.5.1. *Let $s = (s_1, \ldots, s_l)$ be a sequence of variables from X_n such that no variable appears more than k times. For each bipartition $X_n = A \cup B$ there* *exist sets $A' \subseteq A$ and $B' \subseteq B$ such that $|A'| \geq |A|/2^{2k-1}$, $|B'| \geq |B|/2^{2k-1}$, and the number of layers in s with respect to A' and B' is bounded by $2k+1$.*

For s-oblivious BDDs of length kn it is not guaranteed that variables occur at most k times in s. But a simple counting argument proves that at least $\lfloor n/2 \rfloor$ variables occur at most $2k$ times in s. Hence, we can apply Lemma 7.5.1 for the parameter $2k$ and a subset of at least $\lfloor n/2 \rfloor$ variables.

The result of these investigations is that we obtain lower bounds on the size of k-OBDDs, k-IBDDs, and s-oblivious BDDs of length kn for those functions which have large communication complexity even for subfunctions with a support of approximately $n/2^{2k}$ variables. We only have limited control on the support of the subfunctions but we are free to choose the assignment to the other variables.

Obviously, we cannot develop here the theory of communication complexity. In order to give the reader some intuitive feeling, we present two basic lower bound techniques. Communication protocols can be represented as binary trees. The communication starts at the root, which contains the information who communicates the first bit (here we cut messages into single bits). Then we have two outgoing edges, one for the message 0 and the other for 1. The protocol determines for each bit who has to send the second one, and so on. The leaves are labeled by Boolean constants. All inputs a leading to a c-leaf v fulfill the property that $f(a) = c$. Let $L(v)$ be the set of inputs leading to the leaf v. The fundamental property is that $L(v)$ is a *rectangle* with respect to the communication matrix (see Section 4.1), i.e., if $f\colon \{0,1\}^n \times \{0,1\}^m \to \{0,1\}$ and Alice holds the first n input bits, then $L(v) = A \times B$ for some $A \subseteq \{0,1\}^n$ and $B \subseteq \{0,1\}^m$. It is obvious that a set $R \subseteq \{0,1\}^n \times \{0,1\}^m$ is a rectangle iff $(a_1, b_1) \in R$ and $(a_2, b_2) \in R$ imply $(a_1, b_2) \in R$. From this characterization it is easy to prove that $L(v)$ is a rectangle. If (a_1, b_1), $(a_2, b_2) \in L(v)$, then, on input (a_1, b_2), Alice starts the communication as on (a_1, b_1) or Bob starts the communication as on (a_2, b_2). By induction, Alice gets for (a_1, b_2) the same messages from Bob as for (a_2, b_2) and, therefore, acts on (a_1, b_2) in the same way as on (a_1, b_1) which, by assumption, is the same behavior as on (a_2, b_2). Bob gets on (a_1, b_2) and (a_1, b_1) the same messages and acts on (a_1, b_2) and (a_2, b_2) in the same way. Hence, the communication follows the same path on (a_1, b_2) as on (a_1, b_1) and (a_2, b_2). We call a rectangle R *monochromatic* (with respect to f) if f is constant on R. Summarizing our investigations and taking into account that the communication protocol is represented by a binary tree, we have proved the following result.

Theorem 7.5.2. *The sets $L(v)$ for the leaves of the binary tree representing a communication protocol for $f\colon \{0,1\}^n \times \{0,1\}^m \to \{0,1\}$ are a partition of the input set into monochromatic rectangles. If a partition of the input set into monochromatic rectangles requires t rectangles, the communication complexity* *of f (with respect to a given partition of the variable set) is bounded below by* $\lceil \log t \rceil$.

This general bound leads directly to the *fooling-set* technique.

Definition 7.5.3. Let $f\colon \{0,1\}^n \times \{0,1\}^m \to \{0,1\}$. A set $S \subseteq \{0,1\}^n \times \{0,1\}^m$ is called a fooling set for f if $f(a,b) = c$ for all $(a,b) \in S$ and some $c \in \{0,1\}$ and if for different pairs (a_1,b_1), $(a_2,b_2) \in S$ at least one of $f(a_1,b_2)$ and $f(a_2,b_1)$ is not equal to c.

Theorem 7.5.4. *If $f\colon \{0,1\}^n \times \{0,1\}^m \to \{0,1\}$ has a fooling set of size t, the communication complexity of f is bounded below by* $\lceil \log t \rceil$.

Proof. If a rectangle R contains the different pairs (a_1,b_1), $(a_2,b_2) \in S$, it contains also (a_1,b_2) and (a_2,b_1). By the definition of fooling sets, R is not monochromatic. Hence, each partition of the input set into monochromatic rectangles has to contain at least t rectangles and we can apply Theorem 7.5.2. □

The second lower bound technique is based on the rank of the communication matrix.

Definition 7.5.5. The *rank* of $f\colon \{0,1\}^n \times \{0,1\}^m \to \{0,1\}$ is denoted by $\text{rank}(f)$ and defined as the rank of the communication matrix over the field $\mathbb{R}$.

Theorem 7.5.6. *The communication complexity of $f\colon \{0,1\}^n \times \{0,1\}^m \to \{0,1\}$ is bounded below by* $\lceil \log \text{rank}(f) \rceil$.

Proof. Let L be the set of leaves of the binary tree representing a communication protocol for f. For $v \in L$, let $M(v)$ be the communication matrix of the characteristic function of the rectangle $R(v)$ representing the inputs leading to v. The rank of $M(v)$ is 1 (if $R(v) \neq \emptyset$). Moreover, the communication matrix of f is the sum of all $M(v)$ for the 1-leaves $v \in L$. By the subadditivity of the rank function, we obtain

$$\text{rank}(f) \leq \sum_{v \in L,\ v \text{ 1-leaf}} \text{rank}(M(v)) = \#\{v \in L \mid v \text{ is 1-leaf}\}.$$

Hence, the partition of the input set obtained by the communication protocol contains at least $\text{rank}(f)$ monochromatic rectangles, all colored by 1. □

For later purposes, we stress the fact that this lower bound actually holds for the number of rectangles covering all ones.

It is easy to obtain large lower bounds on the communication complexity if the input set is partitioned in the right way. For example, let $\text{EQ}\colon \{0,1\}^n \times \{0,1\}^n \to \{0,1\}$ be defined by $\text{EQ}(a,b) = 1$ iff the vectors a and b are equal.

The communication matrix is the identity matrix, its rank is 2^n, and we need 2^n rectangles of size 1×1 to cover the 2^n ones by monochromatic rectangles. This leads to large lower bounds for k-OBDDs for badly chosen variable orderings. Let EQ$'$ be the same function as EQ, but Alice gets the first halves a' and b' of a and b and Bob the second halves a'' and b'' of a and b. The communication matrix of EQ$'$ contains a one at positions where rows with $a' = b'$ meet columns with $a'' = b''$. Hence, all ones are a rectangle and rank(EQ$'$) = 1. It is indeed easy to obtain a linear-size OBDD for EQ. Hence, the crucial thing is to obtain lower bounds on the communication complexity for different partitions of the variable set and for subfunctions of the given function. The following result based on the fooling-set technique was obtained by Krause (1991).

Theorem 7.5.7. *The k-OBDD size of the permutation matrix test function $PERM_n$ is $2^{\Omega(n/k)}$.*

Proof. Without loss of generality, n is divisible by 16. For a given variable ordering π, the set A of the first $n^2/2$ variables is given to Alice and the set B of the other variables is given to Bob. At least $n/4$ rows contain at least $n/4$ A-variables. Otherwise, the number of variables given to Alice can be bounded by $\frac{n}{4}n + \frac{3}{4}n\frac{1}{4}n < \frac{n^2}{2}$. The same holds for Bob. Hence, we find $n/8$ rows with at least $n/4$ A-variables and $n/8$ *other* rows with at least $n/4$ B-variables. Then we choose a permutation σ on $\{1, \ldots, n\}$ such that $n/8$ variables $x_{i,\sigma(i)}$ are given to Alice and $n/8$ variables $x_{j,\sigma(j)}$ are given to Bob. Because of the symmetry of PERM$_n$ with respect to permutations of rows and columns, we can assume that σ is the identity; $x_{i,i}$, $1 \le i \le n/8$, are A-variables; and $x_{i,i}$, $n/8 < i \le n/4$, B-variables. Now we consider the variables $x_{i,i+n/8}$, $1 \le i \le n/8$. At least half of them belong to the same player, w.l.o.g., $x_{i,i+n/8}$, $1 \le i \le n/16$, belong to Alice. Now we are able to describe a fooling set S of size $2^{n/16}$. For each $d \in \{0,1\}^{n/16}$, the permutation matrix $X(d)$ based on the following permutation σ_d belongs to S. If $d_i = 0$, then $\sigma_d(i) = i$ and $\sigma_d(i + n/8) = i + n/8$. If $d_i = 1$, then $\sigma_d(i) = i+n/8$ and $\sigma_d(i+n/8) = i$. For all $j \notin \{1, \ldots, n/16, n/8+1, \ldots, 3n/16\}$, $\sigma_d(j) = j$.

Obviously, S contains $2^{n/16}$ different permutation matrices. Now let $d \ne d'$, in particular, w.l.o.g. $d_1 = 0$ and $d'_1 = 1$, and let us consider the input where the A-variables are fixed according to $X(d)$ and the B-variables according to $X(d')$. Then $x_{1,n/8+1}$ gets the value 0, since it is an A-variable which uses σ_d where $\sigma_d(1) = 1$. Moreover, $x_{n/8+1,n/8+1}$ gets the value 0, since it is a B-variable which uses $\sigma_{d'}$ where $\sigma_{d'}(n/8+1) = 1$. Hence, the column $n/8+1$ is a 0-column leading to the output 0. □

Theorem 7.5.8. *The k-OBDD size of the function ROW_n+ COL_n testing whether X contains a 1-row or a 1-column is bounded below by $n^{-3}2^{\Omega(n/k)}$.*

Proof. We use the relation

$$\mathrm{PERM}_n(X) = \neg(\mathrm{ROW}_n(\overline{X}) + \mathrm{COL}_n(\overline{X})) \wedge E_{n,n^2}(X)$$

derived in the proof of Theorem 6.2.13. From a k-OBDD for $\mathrm{ROW}_n(X)+\mathrm{COL}_n(X)$ of size s, we obtain a k-OBDD for $\neg(\mathrm{ROW}_n(\overline{X}) + \mathrm{COL}_n(\overline{X}))$ of the same size. The OBDD size of $E_{n,n^2}(X)$ is $O(n^3)$ for each variable ordering. By Theorem 7.3.5, we thus obtain a k-OBDD for $\mathrm{PERM}_n(X)$ of size $O(sn^3)$. Finally, we apply Theorem 7.5.7. □

Corollary 7.5.9. *The function ROW_n+ COL_n has polynomial-size 2-IBDDs but no polynomial-size k-OBDDs if $k = o(n/\log n)$.*

Corollary 7.5.10. *The function $sROW_n$+ $\overline{s}COL_n$ has polynomial-size FBDDs but no polynomial-size k-OBDDs if $k = o(n/\log n)$.*

Proof. The upper bound is proved in Theorem 6.1.2. Let t be the k-OBDD size of $s\mathrm{ROW}_n + \overline{s}\mathrm{COL}_n$. Then ROW_n and COL_n have k-OBDDs for the same variable ordering whose size is bounded by t. This implies by Theorem 7.3.5 that the k-OBDD size of ROW_n+ COL_n is bounded by $O(t^2)$ and we can apply Theorem 7.5.8. □

Up to now, we have obtained lower bounds for the k-OBDD size of functions with polynomial-size 2-IBDDs. For multiplication, we obtain exponential lower bounds for k-IBDDs and oblivious BDDs if k (or the length) is not too large. This proof due to Gergov (1994) combines the OBDD lower bound technique for multiplication and Lemma 7.5.1.

Theorem 7.5.11. *The size of oblivious BDDs of length $2kn$ representing $MUL_{n-1,n}$ is bounded below by $2^{\Omega(n/k^3 2^{4k})}$ and not polynomial for $k = o(\log n/\log\log n)$.*

Proof. We consider s-oblivious BDDs with $s = (s_1, \dots, s_{2kn})$. Without loss of generality, we assume that the number of x-levels is bounded by kn. Since each variable has to be the label of one level in order to represent $MUL_{n-1,n}$, there exists a set X' of at least $n/(k+1)$ x-variables which each appears at most k times as the label of some level. We partition X' into sets A and B of at least $n/2(k+1)$ variables such that $x_i \in A$ and $x_j \in B$ implies $i < j$. Now we apply Lemma 7.5.1 to obtain sets $A' \subseteq A$ and $B' \subseteq B$ such that $|A'| \geq n/(k+1)2^{2k}$, $|B'| \geq n/(k+1)2^{2k}$, and the number of layers with respect to A' and B' is bounded by $2k+1$.

The set of pairs $P = \{(x_i, x_j) \mid x_i \in A', x_j \in B'\}$ has at least $n^2/(k+1)^2 2^{4k}$ elements. By a counting argument, we find some set $I \subseteq \{0, \dots, n-1\}$ and some distance parameter d such that $P' = \{(x_i, x_{i+d}) \mid i \in I\} \subseteq P$, $|P'| = |I| \geq n/(k+1)^2 2^{4k}$, and $\max(I) < \min(I) + d$. Now we are in a situation similar to the proof of Theorem 4.5.2. Therefore, we replace all variables except x_i and x_{i+d}, $i \in I$, in the following way:

	0	1	2	3	4	5	6	7
0	0	0	0	0	0	0	0	0
1	0	0	0	0	0	0	0	1
2	0	0	0	0	0	0	1	1
3	0	0	0	0	0	1	1	1
4	0	0	0	0	1	1	1	1
5	0	0	0	1	1	1	1	1
6	0	0	1	1	1	1	1	1
7	0	1	1	1	1	1	1	1

Figure 7.5.1: *The communication matrix of the carry problem.*

- x_j is replaced by 1 iff $j \notin I$ and $\min(I) < j < \max(I)$,
- all other x_j except x_i, x_{i+d}, $i \in I$, are replaced by 0,
- y_j is replaced by 1 for $j = n - \max(I) - 1$ and $j = n - \max(I) - d - 1$,
- all other y_j are replaced by 0.

By the same arguments as in the proof of Theorem 4.5.2, we are left with the problem of computing the second-most significant bit of the sum of two numbers whose length is $l \geq n/(k+1)^2 2^{4k}$. The bits of one number are given to Alice and the bits of the other number are given to Bob. For both numbers, we set the most significant bit to 0. Then we get the problem of computing the carry of the sum of two numbers of length $l-1$. This makes it easier to determine the rank of the communication matrix. We identify the inputs with their binary value. For numbers of length 3 we obtain the matrix in Fig. 7.5.1. Eliminating the first row and the first column we obtain a matrix of full rank. Hence, the rank is $2^{l-1} - 1$ for $l \geq n/(k+1)^2 2^{4k}$ and the communication complexity at least $l-1$. Now the general lower bound on the size of oblivious BDDs can be applied and yields the desired bound, since the layer depth is bounded by $2k+1$. □

Again, we obtain by read-once projections similar lower bounds for squaring, multiplicative inverse, and division.

Our lower bound techniques based on communication complexity are powerful. Up to now, we have proved bounds on the k-OBDD size and arbitrary constant k. In the following, we would like to prove large lower bounds on the $(k-1)$-OBDD size of functions which have efficient k-OBDD representations. In Chapter 4, we used lower bounds on one-round communication games (often without explicitly mentioning it). This has led to large lower bounds on the OBDD size of functions which have 2-OBDD representations of small size. In order to distinguish the class of functions with polynomial-size k-OBDDs

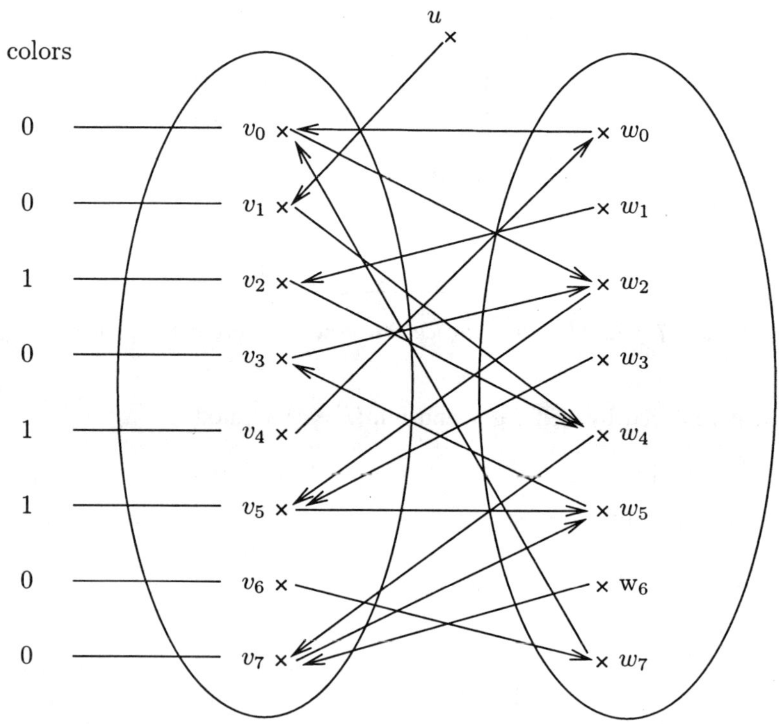

Figure 7.5.2: $PJ_{3,8}(z, x_0, \ldots, x_7, y_0, \ldots, y_7) = 1$, *since* $p = (u, v_1, w_4, v_7, w_5, v_3, w_2, v_5)$ *and* v_5 *is colored by* 1.

(or k-IBDDs) from the class of functions with polynomial-size $(k-1)$-OBDDs (or $(k-1)$-IBDDs), we need results from communication complexity which are very sensitive to the number of allowed communication rounds. Such a bound due to Nisan and Wigderson (1993) has been applied by Bollig, Sauerhoff, Sieling, and Wegener (1998) to obtain hierarchy results. The definition of the pointer-jumping function which is used for the hierarchy results is a little bit complicated. For an illustration see Fig. 7.5.2.

Definition 7.5.12. The *pointer-jumping function* $\mathrm{PJ}_{k,n}$ is defined for $n = 2^l$ on $(2n+1)l+n$ Boolean variables $x_{i,j}$, $y_{i,j}$, z_j, c_i, $0 \le i \le n-1$ and $0 \le j \le l-1$. The x-, y-, and z-variables describe a directed graph on the vertex set $U \cup V \cup W$, where $U = \{u\}$, $V = \{v_0, \ldots, v_{n-1}\}$, and $W = \{w_0, \ldots, w_{n-1}\}$, in the following way. Each vertex has outdegree 1. Pointers from vertices in V reach vertices in W and $x_i = (x_{i,l-1}, \ldots, x_{i,0})$ is the binary representation of the index of the W-node reached from v_i. Pointers from vertices in W and U reach vertices in V where $y_i = (y_{i,l-1}, \ldots, y_{i,0})$ describes the index of the V-node which is reached from w_i, and $z = (z_{l-1}, \ldots, z_0)$ does the same for u. The variable c_i describes the color of v_i. For the evaluation of $PJ_{k,n}$, we follow the unique path p of length $2k+1$ starting at u and the output is the color of the last node on this path.

Proposition 7.5.13. *The k-OBDD size of $PJ_{k,n}$ is bounded by $O(kn^2)$.*

Proof. This upper bound follows in the obvious way. We use the variable ordering $z, x_0, \ldots, x_{n-1}, y_0, \ldots, y_{n-1}, c_0, \ldots, c_{n-1}$ with arbitrary orderings of the variables in the vectors describing the pointers. We follow the path. In the first layer, we start in u and can follow three steps, in all following layers we follow two further steps. We always store the current vertex, which increases the width by a factor of n. The test of a pointer can be done by a complete binary DT of depth l and size n. At the end, it is sufficient to test one color variable. Since we follow $2k+1$ pointers, the upper bound follows. □

In order to represent $\mathrm{PJ}_{k,n}$ in $(k-1)$-OBDDs or $(k-1)$-IBDDs, we have to gather more information on the path p in some layers than in the k-OBDD described in the proof of Proposition 7.5.13. This seems to be impossible in polynomial size. The lower bounds are based on an obvious generalization of a fundamental result in communication complexity due to Nisan and Wigderson (1993). We only describe this result here. We consider a scenario similar to that of Fig. 7.5.2 and call it the *pointer-jumping scenario.* The differences are as follows. The vertex u does not exist and p starts at v_0. The vertex sets V and W have $N \geq n$ vertices but the pointer from a vertex $v \in V$ may only reach a vertex in a set $W(v) \subseteq W$ with n vertices, and similarly the pointer from a vertex $w \in W$ may only reach a vertex in a set $V(w) \subseteq V$ with n vertices. Moreover, the coloring of the vertices in V is fixed such that for each w exactly half the vertices in $V(w)$ are colored by 0. Alice gets all pointers starting in V and Bob all pointers starting in W. They have to compute the color of the vertex reached by the path p of length $2k$ starting at v_0. Obviously, a protocol of length $2k\lceil \log n \rceil$ and $2k$ rounds of communication are sufficient to solve the problem. Alice sends the first message in this "natural" protocol. Nisan and Wigderson (1993) have proved that they cannot do the same job if Bob starts the communication.

Theorem 7.5.14. *If the pointer-jumping scenario has to be solved within $2k$ rounds of communication and Bob has to start the communication, then, for $\varepsilon = 1/20\,000$, a protocol length of $\varepsilon n - 2k \log n$ is not sufficient.*

Now we are able to prove lower bounds on the size of $(k-1)$-OBDDs and $(k-1)$-IBDDs by reducing these problems to the pointer-jumping scenario.

Theorem 7.5.15. *The size of $(k-1)$-OBDDs for $PJ_{k,n}$ is bounded below by $2^{\Omega(n^{1/2}/k)}$ and not polynomial if $k = o(n^{1/2}/\log n)$.*

Proof. Let G be a $(k-1)$-OBDD of size s representing $\mathrm{PJ}_{k,n}$. We assume that the z-variables are tested only at the top of the first layer. We can ensure this by increasing the size at most by a factor of n (remember that the number of z-variables is $l = \log n$). Let L be the list representing the ordering of the x- and y-variables used by G. For each i, we mark the $l/2$th x_i-variable in L and do the same for y_i. Now we break L after the nth marked variable into L_1 and L_2. If L_1 contains at least $n/2$ marked x-variables, Alice obtains the x_i-variables belonging to L_1 such that the marked x_i-variable also belongs to L_1 and Bob obtains those y_j-variables of L_2 such that the marked y_j-variable also belongs to L_2. In the other case, L_1 contains at least $n/2$ marked y-variables and Bob gets the corresponding y-variables and Alice x-variables from L_2 in an analogous way. Let $V' \subseteq V$ and $W' \subseteq W$ be the sets of vertices such that some of the corresponding variables are given to some player. By definition, $|V'| \geq n/2$ and $|W'| \geq n/2$.

In the following, we assign constants to certain variables. By the formulation "w_j is reachable from v_i," we indicate that the remaining variables can be fixed such that the pointer from v_i leads to w_j. We assign constants to all variables not given to some player. The z-variables are fixed such that the pointer from u reaches some vertex in V'. Variables belonging to vertices in $V - V'$ or $W - W'$ are set to 0. Let $v_i \in V'$ (W'-nodes are handled in an analogous way). There are $r \leq n^{1/2}$ different ways to fix the x_i-variables not given to some player. This gives a partition of W into r subsets of equal size $n/r \geq n^{1/2}$. Since $|W'| \geq n/2$, we can choose an assignment to the x_i-variables not given to some player such that at least $n^{1/2}/2$ vertices in W' are reachable from v_i. Finally, we investigate a random coloring of the vertices in V'. By Chernoff's bound, we conclude that there is a coloring such that for each $w_j \in W'$ there exists a set $V'_j \subseteq V'$ of $n^{1/2}/3$ reachable vertices such that exactly half the vertices in V'_j are colored by 0.

Now we consider the communication problem where Alice and Bob have to evaluate the obtained subfunction of $\mathrm{PJ}_{k,n}$ in $2k-2$ rounds of communication. By the choice of the variables given to Alice and Bob, the given $(k-1)$-OBDDs can be divided into at most $2k-2$ layers, each belonging completely to one player. This leads to an upper bound of $(2k-2)\lceil \log s \rceil$ on the protocol length. The lower bound of Theorem 7.5.14 holds for $2k-1$ rounds of communication even if Alice may start the conversation. The parameter n has to be replaced by $n^{1/2}/3$, since we restrict ourselves to inputs where pointers from w_j lead to vertices in V'_j. Hence,

$$\varepsilon n^{1/2}/3 - 2k \log n \leq (2k-2)\lceil \log s \rceil$$

and, since $s \geq n$, $s = 2^{\Omega(n^{1/2}/k)}$. □

Taking into account that $\mathrm{PJ}_{k,n}$ is defined on $\Theta(n \log n)$ variables, we obtain the following corollary.

Corollary 7.5.16. *If* $k = o(n^{1/2}/\log^{3/2} n)$*, there are functions on* n *variables representable by polynomial-size* k*-OBDDs but not by polynomial-size* $(k-1)$*-OBDDs.*

Theorem 7.5.17. *If* $k \le (1-\delta)\log\log n$ *for some* $\delta > 0$*, the* $(k-1)$*-IBDD size of* $PJ_{k,n}$ *is not polynomially bounded and there are functions representable by polynomial-size* k*-OBDDs but not by polynomial-size* $(k-1)$*-IBDDs.*

Proof. Let G be a $(k-1)$-IBDD representing $\mathrm{PJ}_{k,n}$ in size s. We follow the same approach as in the proof of Theorem 7.5.15 but we have to work harder until we can partition G into $2k-2$ layers with each completely belonging to one of the players. We start with all x- and all y-variables. If we consider the variable ordering π_i, we still consider at least $(\log n)/2^{i-1}$ Boolean variables for each of at least $n/2^{i-1}$ vertices in V and for each of at least $n/2^{i-1}$ vertices in W. Then we perform the same procedure as in the proof of Theorem 7.5.15 with respect to π_i and the still considered variables. The number of considered vertices in V and also in W is halved as the number of considered bits of each pointer. At the end, for each of the at least $n/2^{k-1}$ vertices in $V' \subseteq V$, Alice knows at least $(\log n)/2^{k-1}$ variables of its pointer and Bob has the same information for the vertices in $W' \subseteq W$ where $|W'| \ge n/2^{k-1}$.

The assignment to the variables not given to Alice or Bob is also done in a way similar to the proof of Theorem 7.5.15. By the pigeonhole principle, we fix the x_i-variables not given to Alice for some $v_i \in V'$ such that at least

$$N(k) := \frac{\frac{n}{2^{k-1}}}{n^{1-2^{-(k-1)}}} = n^{2^{-(k-1)}}2^{-(k-1)}$$

vertices in W' are reachable from v_i, similarly for nodes $w_j \in W'$. Since $k \le (1-\delta)\log\log n$, $N(k) = 2^{\Omega(\log^\delta n)}$ and $N(k)$ grows faster than any polylogarithmic function. By Chernoff's bound, we obtain, for some $\alpha > 0$, an upper bound of $(n/2^{k-1})2^{-\alpha N(k)}$ on the probability that, for a random coloring of the vertices of V', for each $w_j \in V'$ at least a third of the vertices reachable in V' have the color 0 and also at least a third have the color 1. Since $n2^{-\alpha N(k)} < 1$ for large n, we can choose a coloring with the described properties.

Again, we get an upper bound of $(2k-2)\lceil \log s \rceil$ on the length of the protocol resulting from G for the evaluation of the obtained subfunction of the pointer-jumping function. The lower bound on the protocol length from Theorem 7.5.14 is in this situation $\Omega(2^{\Omega(\log^\delta n)} - 2k\log n)$. Altogether, $\log s = 2^{\Omega(\log^\delta n)}$ and s is not polynomially bounded. □

Communication complexity has turned out to be the strongest tool for proving lower bounds on the size of oblivious BDDs.

7.6 Lower Bound Techniques for k-BPs

The problem of obtaining exponential lower bounds on the size of 2-BPs for explicitly defined functions was attacked for a long time until it was solved, even for slowly increasing values of k, independently by Borodin, Razborov, and Smolensky (1993) and Okol'nishnikova (1993). A first step to hierarchy results was done by Okol'nishnikova (1997a, 1997b), but then Thathachar (1998a) obtained the result that the classes of functions representable by polynomial-size k-BPs form a proper hierarchy. Here we start with the lower bound technique of Borodin, Razborov, and Smolensky (1993), which has influenced all later developments.

Definition 7.6.1. A Boolean function $g \in B_n$ is called a (k, a)-*rectangle* if it can be represented as a conjunction of functions g_i, $1 \le i \le ka$, such that g_i essentially depends only on variables from $X(i)$ where $|X(i)| \le \lceil n/a \rceil$ and each variable x_j belongs to at most k of the $X(i)$-sets.

The rectangles considered in the previous section are $(1, 2)$-rectangles in this generalized notion. The communication complexity is small if $f^{-1}(1)$ and $f^{-1}(0)$ can be partitioned into a small number of rectangles which are based on the same disjoint sets $X(1)$ and $X(2)$. Since k-BPs are not oblivious, a small-size k-BP representing f only leads to a covering of $f^{-1}(1)$ (and also $f^{-1}(0)$) by a "small" number of generalized rectangles based on *different* variable sets $X(i)$. Although the theory of communication complexity cannot be applied to k-BPs, the following method can be seen as an appropriate generalization of communication complexity.

Theorem 7.6.2. *Let G be a k-BP representing f with size s. For $r = (2s)^{ka}$, the function f can be represented as a disjunction of at most r (k, a)-rectangles.*

Proof. In this proof it is convenient to regard edges leaving x_i-nodes as x_i-edges. For two nodes v and v' of G, we denote by $X(v, v')$ the set of all variables which appear as an edge label on a path starting at v and ending at v'. The function $f_{v,v'}$ is defined on $X(v, v')$ and computes 1 on input a if the path activated by a and starting at v reaches v'.

Let p be a path from the source v_0 to the 1-sink. On p, we look for w_1, the last node where $|X(v_0, w_1)| < \lceil n/a \rceil$. We continue with v_1, the direct successor of w_1 on p, and look for the last node w_2 where $|X(v_1, w_2)| < \lceil n/a \rceil$. This procedure leads to a sequence of nodes $v_0, w_1, v_1, \dots, w_l, v_l$ and the 1-sink w_{l+1} such that $|X(v_i, w_{i+1})| < \lceil n/a \rceil$ for $0 \le i \le l$ but $|X(v_i, v_{i+1})| \ge \lceil n/a \rceil$ for $0 \le i \le l-1$. The sequence $t = (e_1, \dots, e_l)$, where $e_i = (w_i, v_i)$, is called the *trace* of p.

For such a trace t, let $f_{t,i}$, $1 \le i \le l$, be the conjunction of f_{v_{i-1},w_i} and the literal corresponding to the edge e_i, i.e., x_j for a 1-edge leaving an x_j-node and $\overline{x}_j$

for a 0-edge leaving an x_j-node, and let $f_{t,l+1} = f_{v_l,w_{l+1}}$. Then the conjunction f_t of all $f_{t,i}$, $1 \le i \le l+1$, computes 1 for all inputs whose computation paths lead from the source to the 1-sink and which have the trace t. Therefore, f is the disjunction of all f_t. We claim that the number of traces is bounded by r and that each f_t is a (k,a)-rectangle.

Each function $f_{t,i}$ essentially depends only on less than $\lceil n/a \rceil$ variables from $X(v_{i+1}, w_i)$ and additionally, on the label of e_i, altogether at most $\lceil n/a \rceil$ variables. If more than k functions $f_{t,i}$, $1 \le i \le l+1$, essentially depend on some variable x_j, then we can glue together paths from v_i to v_{i+1}, $1 \le i \le l$, and from v_{l+1} to w_{l+1} such that we obtain a path from the source to a 1-sink containing more than k x_j-edges in contradiction to the assumption that G is a k-BP. (This path is not necessarily a computation path. Hence, there would be no contradiction for semantic k-BPs.) By the construction of a trace, $|X(v_i, v_{i+1})| \ge \lceil n/a \rceil$ for $0 \le i \le l-1$. Hence, the sum of all $|X(v_i, v_{i+1})|$ and $|X(v_l, w_{l+1})|$ can be bounded below by $ln/a + |X(v_l, w_{l+1})|$. It also can be bounded above by kn, since each variable is contained in at most k of these sets. Hence, $ln/a + |X(v_l, w_{l+1})| \le kn$. If $|X(v_l, w_{l+1})| > 0$, then $ln/a < kn$ and $l < ka$, which implies $l+1 \le ka$. Hence, f_t is the conjunction of $l+1 \le ka$ functions $f_{t,i}$. If $|X(v_l, w_{l+1})| = 0$, we only conclude that $l \le ka$. But in this case $f_{t,l+1}$ does not essentially depend on any variable. Therefore, it is a constant. If it is 0, f_t equals 0 and can be dropped. If it is 1, we obtain f_t as the conjunction of $l \le ka$ functions $f_{t,i}$. The number of traces can easily be bounded by $(2s)^{ka}$, since the BP G has less than $2s$ edges and traces are sequences of edges whose length is bounded by ka and can be made equal to ka by repeating the last edge often enough. Altogether, we have proved the theorem. □

In order to apply Theorem 7.6.2 it is sufficient to prove that each (k,a)-rectangle covers at most a fraction ε of $f^{-1}(1)$. Then we need at least ε^{-1} (k,a)-rectangles. Hence, $\varepsilon^{-1} \le (2s)^{ka}$ and $s \ge 2^{(\log \varepsilon^{-1})/ka - 1}$. Jukna (1995) has followed this approach.

Lemma 7.6.3. *Let* $\alpha = \binom{ak}{k}^{-1}$ *and* $\beta = 1 - k/a - k/n$. *Each* (k,a)*-rectangle* $g \in B_n$ *can be represented as* $g(X) = g_0(X_0) \wedge g_1(X_1)$ *such that* $X = \{x_1, \dots, x_n\}$, $|X_0 - X_1| \ge \alpha n$, *and* $|X_1 - X_0| \ge \beta n$.

Proof. Since g is a (k,a)-rectangle, $g = g_1 \wedge \dots \wedge g_{ka}$ such that g_i essentially depends on $X(i)$, where $|X(i)| \le \lceil n/a \rceil$ and each variable x_j is contained in at most k $X(i)$-sets. We prove the lemma by a probabilistic argument. Let $I \subseteq \{1, \dots, ka\}$ be chosen randomly among all sets of size k. Let X_0 be the union of all $X(i)$, $i \in I$, and X_1 the union of all $X(i)$, $i \notin I$. Let $J(j) = \{i \mid x_j \in X(i)\}$. Then, by the above properties, $|J(j)| \le k$ and

$$\text{Prob}(x_j \in X_0 - X_1) = \text{Prob}(J(j) \subseteq I) \ge \alpha,$$

since at least one of the $\binom{ak}{k}$ choices of I is good. Hence, the average size of $X_0 - X_1$ is at least αn and we fix a set I such that $|X_0 - X_1| \geq \alpha n$. Moreover, X_0 is the union of k $X(i)$-sets all of size at most $\lceil n/a \rceil$. This implies that the size of X_0 is bounded above by $k\lceil n/a \rceil$. Each variable not in X_0 is contained in $X_1 - X_0$ which, therefore, has a size of at least $n - k\lceil n/a \rceil \geq \beta n$. □

Lemma 7.6.4. *Let $f \in B_n$ be $(2d+1)$-rare (see Definition 7.4.2). If $a \geq k+1$ and $g \in B_n$ is a (k,a)-rectangle such that $g \leq f$ (g covers only inputs from $f^{-1}(1)$), then*

$$|g^{-1}(1)| \leq 2^n/\Delta_{a,k}(f)$$

where

$$\Delta_{a,k}(f) = \left(\frac{n^2}{d^2 a^k e^{k+1} k}\right)^d .$$

Proof. Let $D_r(f)$ be the maximum of all $|f'^{-1}(1)|$ for subfunctions f' of f obtained by assigning constants to r variables. Such an assignment determines a subcube of dimension $n-r$ and f' is regarded as a function on $n-r$ variables. Inputs from $f^{-1}(1)$ have a Hamming distance of at least $2d+1$. This also holds for inputs from $f'^{-1}(1)$. Hence, the Hamming balls with radius d around inputs from $f'^{-1}(1)$ are disjoint. Since each ball contains more than $\binom{n-r}{d}$ elements (this is the number of elements on the sphere with distance d) and f' is defined on 2^{n-r} inputs, we conclude that

$$D_r(f) \leq 2^{n-r}\binom{n-r}{d}^{-1}.$$

Now we turn our attention to g and its representation proved in Lemma 7.6.3. We choose $Y_0 \subseteq X_0 - X_1$ and $Y_1 \subseteq X_1 - X_0$, where $|Y_0| = \alpha n$ and $|Y_1| = \beta n$. Let $Z = X - (Y_0 \cup Y_1)$. Subfunctions g' of g where the variables of Z are replaced by constants can be represented as $g' = h_0(Y_0) \wedge h_1(Y_1)$. The crucial property is that Y_0 and Y_1 are, by construction, disjoint. Hence, we may consider the communication matrix of g' if Alice gets the inputs from Y_0 and Bob gets the inputs from Y_1. It follows that $|g'^{-1}(1)| = |h_0^{-1}(1)||h_1^{-1}(1)|$, where we consider h_0 as a function on Y_0 and h_1 as a function on Y_1. We obtain h_0 from g' by assigning constants to the variables in Y_1 such that $h_1(Y_1) = 1$. Hence, h_0 is a subfunction of g where $(1-\alpha)n$ variables are replaced by constants. This implies that $|h_0^{-1}(1)| \leq D_{(1-\alpha)n}(g) \leq D_{(1-\alpha)n}(f)$. The last inequality follows from the assumption $g \leq f$. Similarly, we obtain $|h_1^{-1}(1)| \leq D_{(1-\beta)n}(f)$. Since we may consider $2^{(1-\alpha-\beta)n}$ subfunctions g' of g, we obtain

$$\begin{aligned} |g^{-1}(1)| &\leq 2^{(1-\alpha-\beta)n} D_{(1-\alpha)n}(f) D_{(1-\beta)n}(f) \\ &\leq 2^n \binom{\alpha n}{d}^{-1} \binom{\beta n}{d}^{-1}. \end{aligned}$$

Hence, it is sufficient to prove $\binom{\alpha n}{d}\binom{\beta n}{d} \geq \Delta_{a,k}(f)$. Since

$$\binom{\alpha n}{d}\binom{\beta n}{d} \geq \left(\frac{\alpha n}{d}\right)^d \left(\frac{\beta n}{d}\right)^d,$$

it is sufficient to prove that

$$\alpha^d \beta^d \geq \left(a^{-k} e^{-(k+1)} k^{-1}\right)^d \qquad \text{or}$$

$$\alpha\beta a^k e^{k+1} k \geq 1,$$

which can be proved by Stirling's formula and $a \geq k+1$. □

In order to apply this lemma, we need functions f where $|f^{-1}(1)|$ is large, although all inputs in $f^{-1}(1)$ have a large Hamming distance. As already mentioned in Section 7.4, the characteristic functions of Bose–Chaudhuri–Hocquenghem codes have these properties. Their parameters can be chosen such that we obtain functions $f_{n,d}$ which are $(2d+1)$-rare and compute the value 1 for at least $2^n/(n+1)^d$ inputs. Choosing $a = k+1$, Lemma 7.6.4 implies that a $(k, k+1)$-rectangle $g \leq f_{n,d}$ can cover at most $2^n/\Delta_{k+1,k}(f_{n,d})$ inputs from $f_{n,d}^{-1}(1)$. This is a fraction ε of at most $(n+1)^d/\Delta_{k+1,k}(f_{n,d})$ of the 1-inputs. Then

$$\log \varepsilon^{-1} \geq d\,(2\log n - \log(n+1) - 2\log d - (k+1)\log(k+1) - (k+1)\log e)\,.$$

For $d = \lceil [(n-1)/(2(k+1)^{k+1}e^{k+1})]^{1/2} \rceil$, $\log \varepsilon^{-1} = \Omega(d)$ and the lower bound on the k-BP size is $2^{\Omega(d/k^2)}$ or $2^{\Omega(n^{1/2}/k^k)}$ according to Theorem 7.6.2.

Theorem 7.6.5. *There is an explicitly defined linear code such that the k-BP size of its characteristic function is bounded below by $2^{\Omega(n^{1/2}/k^k)}$.*

Borodin, Razborov, and Smolensky (1993) have applied their method to another class of functions.

Definition 7.6.6. Let A be an $n \times n$ matrix with entries from the field $\mathbb{Z}_q$ (q prime). Then the *A-bilinear function* $f_A\colon \mathbb{Z}_q^{2n} \to \{0,1\}$ computes 1 for the input $(x,y) \in \mathbb{Z}_q^n \times \mathbb{Z}_q^n$ if $xAy^\top \equiv 0 \bmod q$.

Definition 7.6.7. Let $n = 2^d$. Then the $n \times n$ *Sylvester matrix* S has entries from $\mathbb{Z}_3$. For the row with number $a \in \{0,1\}^d$ and the column with number $b \in \{0,1\}^d$, the entry S_{ab} equals $+1$ if the sum of all $a_i b_i$ equals 0 mod 2 and -1 otherwise. The S-bilinear function f_S is called the *bilinear Sylvester function* SYL_n.

The crucial property of Sylvester matrices is that submatrices with t rows and u columns have a large rank; more exactly (see Borodin, Razborov, and Smolensky (1993)), their rank is at least $ut/(2n\ln(2n/u))$. Borodin, Razborov, and Smolensky also have shown how such a property can be combined with Theorem 7.6.2 to obtain large lower bounds on the k-BP size of the bilinear Sylvester function. To be more precise, they have considered MDDs with three-valued decision nodes (see Section 9.1). We only cite their results.

Theorem 7.6.8. *The size of k-BPs with three-valued decision nodes representing the bilinear Sylvester function is bounded below by $2^{\Omega(n/4^k k^3)}$.*

Thathachar (1998a) used the same technique in a much more sophisticated and technically involved way and obtained the following result.

Theorem 7.6.9. *Each $(k-1)$-BP representing the hyperplanar sum-of-products predicate HSP_q^k (which is defined on N variables) has a size bounded below by $2^{\Omega(N^{1/k}2^{-2k}k^{-4})}$. For the conjunction of hyperplanar sum-of-products predicate $CHSP_q^k$, the size of $(k-1)$-BPs is bounded below by $2^{\Omega(N^{1/k}2^{-2k}k^{-3})}$.*

Corollary 7.6.10. *For $k = o(\log^{1/2} n)$, there are functions representable by polynomial-size k-IBDDs but not by polynomial-size $(k-1)$-BPs.*

Proof. This result follows from Proposition 7.2.9 and Theorem 7.6.9. □

The discussed lower bound technique will be generalized to nondeterministic and randomized BPs (see Chapters 10 and 11).

7.7 Lower Bounds for Depth-Restricted BPs

Many BP models have the property that the depth is restricted. For example, the depth of k-OBDDs, k-IBDDs, and k-BPs is restricted to kn. These BDD variants have the additional restriction that each path contains at most k x_i-nodes, $1 \le i \le n$. Here, we consider BPs where only the depth is restricted.

Definition 7.7.1. A *depth-(k,n) BP* is a BP where the length of each computation path is bounded by kn.

Even for functions essentially depending on all n variables, it is not obvious how to prove exponential lower bounds for BPs whose depth is bounded by n. Such bounds can be obtained in the following way. If all prime implicants of a function f have length n (which is equivalent to the property that f is 2-rare), each depth-$(1,n)$-BP can be replaced by a 1-BP or FBDD for the same function which is not larger than the given depth-$(1,n)$-BP. The reason is that all paths leading to the 1-sink have to be read once, since we have to test all variables in order to know that the function computes 1. Hence, if a variable is tested for a second time at node v, the node v can be replaced with the 0-sink. This implies that all exponential lower bounds for the FBDD size of 2-rare functions also hold for depth-1-BPs. The *permutation matrix test* function PERM is 2-rare and has exponential FBDD size (Theorem 6.2.12).

Beame, Saks, and Thathachar (1998) were the first to obtain exponential lower bounds for depth-$(1+\varepsilon, n)$-BPs and positive ε; more precisely, $\varepsilon = 0.0178$. Quite recently, Ajtai (1999) proved an exponential lower bound for the depth-(k, n)-BP size (k any constant) and the following function PSS_n (pairwise sums of a subset). The input $a \in \{0,1\}^n$ is interpreted as the set $A \subseteq \{1, \dots, n\}$ of all i where $a_i = 1$. Then $N(A)$ is the number of pairs (p, q) such that $p, q \in A$, $p < q$, and $p+q \in A$. Finally, $\mathrm{PSS}_n(a) = 1$ iff $N(A)$ is odd. It is easy to see that this function has polynomial size for general BPs. For the proof of the following lower bound, we refer to Ajtai (1999).

Theorem 7.7.2. *For each constant k there exists an $\varepsilon(k) > 0$ such that, for large n, the depth-(k, n)-BP size of PSS_n is bounded below by $2^{\varepsilon(k)n}$.*

7.8 Exercises and Open Problems

7.1.E Prove that for a BP G, it can be checked in time $O(n|G|)$ whether it is an FBDD.

7.2.M Prove that for a BP G, it can be checked in time $O(|G|)$ whether it is an OBDD.

7.3.E If $f \in B_n$ can be represented by a polynomial-size BP, then prove that f can be represented by a polynomial-size $(1, +(n - O(\log n)))$-BP.

7.4.E Prove that the test of whether an undirected graph on n vertices is k-regular for some k can be represented by oblivious 2-BPs of size $O(n^4)$.

7.5.M Let k be a constant. Prove that the test of whether a k-uniform hypergraph is $\lceil \binom{n-1}{k-1}/2 \rceil$-regular can be represented by oblivious k-BPs of polynomial size.

7.6.M Let k be a constant. Prove that the test of whether a k-uniform hypergraph consists of a hyperclique on $\lceil n/2 \rceil$ vertices and $\lfloor n/2 \rfloor$ isolated vertices can be represented by oblivious k-BPs of polynomial size.

7.7.O Decide whether the functions considered in Exercises 7.5 and 7.6 can be represented by polynomial-size $(k-1)$-BPs.

7.8.M (See Thathachar (1998a).) Prove that the predicates HSP_q^k and CHSP_q^k can be represented by $(k-1)$-BPs of size $2^{O(N^{1/k})}$.

7.9.M Prove that multiplication can be represented by polynomial-size $\lceil n/\log n \rceil$-BPs.

7.10.O Determine for which k (or $k(n)$) multiplication can be represented by polynomial-size k-BPs.

7.11.M Prove the following: Each threshold function with weights bounded by 2^n can be represented by polynomial-size $\lceil n/\log n \rceil$-BPs. Squaring can be represented by polynomial-size $\lceil n/\log n \rceil$-BPs.

7.12.M Let G be a graph ordering and $G(a)$ the variable ordering for input a. A BP is a k-G-FBDD if, for each input a, the sequence of tested variables is a subsequence of the sequence which repeats $G(a)$ k times. Design an efficient synthesis algorithm for k-G-FBDDs and a polynomial-time satisfiability test.

7.13.M Prove that the consistency test for semantic $(1,+1)$-BPs is coNP-complete. Hint: Use the construction from the proof of Theorem 7.3.2.

7.14.D (See Sieling (personal communication).) Prove that the consistency test for $(1,+k)$-BPs is coNP-complete if k is a part of the input.

7.15.E Design efficient consistency tests for s-oblivious BDDs with given s, k-OBDDs with given π, k-IBDDs with given $\pi_1, \ldots, \pi_k$, and k-BPs.

7.16.M (See Breitbart, Hunt III, and Rosenkrantz (1995).) Let f_n be defined on $n = 3k$ variables $x_0, \ldots, x_{k-1}, y_0, \ldots, y_{k-1}, z_0, \ldots, z_{k-1}$ by $f_n(x,y,z) = x_{||y||+||z||} \oplus y_{||x||+||z||} \oplus z_{||x||+||y||}$, where the indices are taken mod k. Prove that this function can be represented by a polynomial-size BP which is simultaneously a 2-OBDD and a $(1,+3)$-BP.

7.17.D Prove exponential lower bounds on the k-OBDD size (k is a constant) for the function from exercise 4.18.

7.18.D (See Bollig, Sauerhoff, Sieling, and Wegener (1998).) Let $k = o(n/\log n)$. Prove a nonpolynomial lower bound on the $(k-1)$-OBDD size of $\mathrm{PJ}_{k,n}$ if the variables describing a pointer have to be tested blockwise.

7.19.D (See Bollig, Sauerhoff, Sieling, and Wegener (1998).) Let $k \le (1-\varepsilon)\log n$ for some $\varepsilon > 0$. Prove a nonpolynomial lower bound on the $(k-1)$-IBDD size of $\mathrm{PJ}_{k,n}$ if the variables describing a pointer have to be tested blockwise.

Chapter 8

Decision Diagrams (DDs) Based on Other Decomposition Rules

BDD nodes are evaluated as ite instructions. In this chapter, decision diagrams (DDs) are investigated which have the same syntax as BDDs but different semantics.

8.1 Zero-Suppressed Binary Decision Diagrams (ZBDDs)

The invention of ZBDDs was motivated by Minato (1993, 1994) by applications where sets have to be manipulated. If the universe $U = \{1, \ldots, n\}$ is fixed, a subset $S \subseteq U$ can be described by its characteristic vector $s = (s_1, \ldots, s_n) \in \{0,1\}^n$, where $s_i = 1$ iff $i \in S$. In this notation, a set corresponds to a minterm. A collection C of subsets is a member of the power set $\mathcal{P}(U)$ of U and can be described by the Boolean function $f\colon \{0,1\}^n \to \{0,1\}$, where $f(a) = 1$ iff the subset A described by a belongs to C. The usual set theoretic operations correspond to Boolean operations, e.g., union $\leftrightarrow$ OR, intersection $\leftrightarrow$ AND, symmetric difference $\leftrightarrow$ EXOR, complement $\leftrightarrow$ NOT. Hence, BDDs and, in particular, OBDDs can be used to work with elements of the power set of some universe. In several applications, the universe U is not fixed and can be extended to U'. If G is an OBDD for $C \in \mathcal{P}(U)$, we have to add nodes to obtain an OBDD for the same collection C as an element of $\mathcal{P}(U')$. If $i \in U' - U$, it is necessary to add x_i-nodes and the 1-edges leaving these nodes point to the 0-sink. ZBDDs are defined in such a way that these nodes are not necessary.

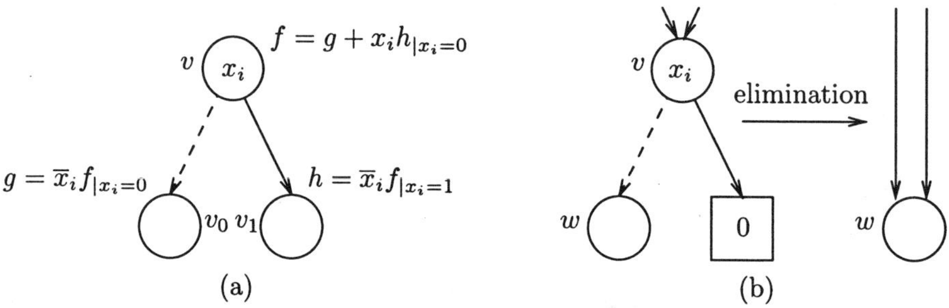

Figure 8.1.1: (a) *The decomposition rule at ZBDD nodes.* (b) *The elimination rule for ZBDDs.*

Definition 8.1.1. A ZBDD shares its syntax with OBDDs. To evaluate the function f_v represented at a node v of a ZBDD on input a and $X_n = \{x_1, \ldots, x_n\}$, follow the same computation path as in the OBDD. Then $f_v(a) = 1$ iff the computation path reaches the 1-sink *and* $a_i = 0$ for all i such that the computation path does not contain an x_i-node.

This definition has the property that a ZBDD on $X_n = \{x_1, \ldots, x_n\}$ without any x_i-node represents a function f such that $f_{|x_i=1} \equiv 0$. In the set theoretical setting it represents a collection of sets not containing i. This implies that nothing has to be done if we extend the universe. We use the notion π-ZBDD if the variable ordering π is fixed. The following results are due to Schröer and Wegener (1998). We start with some simple properties illustrated in Fig. 8.1.1(a).

Proposition 8.1.2. *Let f be represented at an x_i-node v of a π-ZBDD for $\pi = id$ and let g and h be represented at its 0-successor v_0 and 1-successor v_1, respectively. Then $g = \overline{x}_i f_{|x_i=0}$, $h = \overline{x}_i f_{|x_i=1}$, and $f = g + x_i h_{|x_i=0} = \overline{x}_i g + x_i h_{|x_i=0}$.*

Proof. The following property is equivalent to $g(a) = 1$ as well as to $(\overline{x}_i f_{|x_i=0})(a) = 1$. There exists a path from v_0 to the 1-sink which contains a 1-edge leaving an x_j-node, $i+1 \le j \le n$, iff $a_j = 1$ and, furthermore, $a_1 = \cdots = a_i = 0$. The statement $h = \overline{x}_i f_{|x_i=1}$ follows in the same way. We may prove the last statement by the following calculation: $\overline{x}_i g = g$ and $g + x_i h_{|x_i=0} = \overline{x}_i f_{|x_i=0} + x_i(\overline{x}_i f_{|x_i=1})_{|x_i=0} = \overline{x}_i f_{|x_i=0} + x_i f_{|x_i=1} = f$. □

Obviously, the OBDD merging rule can be applied to ZBDDs, since it does not change the tests on computation paths. The ZBDD *elimination rule* is described in Fig. 8.1.1(b). The correctness of this elimination rule follows from Definition 8.1.1 or from Proposition 8.1.2, since in this case $h \equiv 0$ and this is equivalent to $f_{|x_i=1} \equiv 0$.

Definition 8.1.3. A Boolean function f is called 1-*simple* with respect to x_i if $f_{|x_i=1}$ is equal to the constant 0.

Since an x_i-node v can be eliminated iff the function represented at v is 1-simple with respect to x_i, the OBDD elimination rule is no longer applicable. Using Proposition 8.1.2, the following results can be obtained in the same way as the corresponding results for OBDDs in Chapter 3. A complete ZBDD is a ZBDD where each computation path starting at the source contains tests of all variables.

Theorem 8.1.4.

(*i*) *There is (up to isomorphism) a unique complete π-ZBDD of minimal size representing $f \in B_n$. This is called the quasi-reduced π-ZBDD for f. It can be obtained in linear time from each complete π-ZBDD for f (without nodes not reachable from the source) by the application of the merging rule. It contains, if $\pi = id$, x_i-nodes representing the different functions $\overline{x}_1 \cdots \overline{x}_{i-1} f_{|x_1=a_1,\ldots,x_{i-1}=a_{i-1}}$, where $a_j \in \{0,1\}$.*

(*ii*) *There is (up to isomorphism) a unique π-ZBDD of minimal size representing $f \in B_n$. This is called the reduced π-ZBDD for f. It can be obtained in linear time from each π-ZBDD for f (without nodes not reachable from the source) by the application of the merging rule and the ZBDD elimination rule. It contains, if $\pi = id$, x_i-nodes representing the different functions $\overline{x}_1 \cdots \overline{x}_{i-1} f_{|x_1=a_1,\ldots,x_{i-1}=a_{i-1}}$ which are not 1-simple with respect to x_i.*

Corollary 8.1.5. *The quasi-reduced π-ZBDD for f is isomorphic to the quasi-reduced π-OBDD for f.*

Proof. The number of different functions $\overline{x}_1 \cdots \overline{x}_{i-1} f_{|x_1=a_1,\ldots,x_{i-1}=a_{i-1}}$ is equal to the number of different functions $f_{|x_1=a_1,\ldots,x_{i-1}=a_{i-1}}$. □

Using these structural results, it is not too difficult to compare the sizes of reduced π-OBDDs and reduced π-ZBDDs. Let π-ZBDD(f) denote the size of the reduced π-ZBDD representing f.

Theorem 8.1.6.

(*i*) π-*OBDD*(f) $\leq (n+1) \cdot \pi$-*ZBDD*(f) $+ 1$ *for all* $f \in B_n$.

(*ii*) π-*ZBDD*(f) $\leq (n+1) \cdot \pi$-*OBDD*(f) *for all* $f \in B_n$.

Proof. The proof of both statements follows along the same lines. Starting from the reduced π-ZBDD (or π-OBDD), we construct the quasi-reduced π-ZBDD (π-OBDD) which, by Corollary 8.1.5, may be interpreted as a quasi-reduced π-OBDD (π-ZBDD). The reduced π-OBDD (π-ZBDD) can only be smaller.

Let G be the reduced π-ZBDD. If not existent, we add a 0-sink. Without loss of generality, let $\pi = id$. We consider sinks as "x_{n+1}-nodes." For each x_i-node v, we add dummy nodes $v_1, \ldots, v_{i-1}$, where v_j is labeled by x_j. The 0-edge leaving the dummy node v_j leads to v_{j+1} if $j < i-1$ and to v if $j = i-1$. The 1-edge leaving v_j leads to the dummy node labeled by x_{j+1} and created for the 0-sink. Finally, an edge from the x_j-node w to the x_i-node v is replaced with an edge to v_{j+1} if $i > j+1$. Here the edge to the source is considered as an edge from an "x_0-node." Altogether, we obtain a complete π-ZBDD or π-OBDD for f. The size is bounded by $(n+1)(\pi\text{-ZBDD}(f)+1)$, since we have perhaps created a 0-sink and then at most n dummy nodes per node. The upper bound of (i) follows, since the n dummy nodes for the 0-sink can be eliminated by the OBDD elimination rule. The second statement follows in a similar way but both edges leaving a dummy node v_j lead to v_{j+1} or v. Hence, it is not necessary to create a new sink. □

These bounds are tight. The π-ZBDD (w.l.o.g. $\pi = id$) consisting only of the 1-sink represents $\overline{x}_1\overline{x}_2 \cdots \overline{x}_n$. The π-OBDD size of this function obviously is $n+2$. The π-OBDD consisting only of the 1-sink represents the constant 1. The π-ZBDD size of this function equals $n+1$. We need an x_i-node whose outgoing edges lead to an x_{i+1}-node if $i < n$ and to the 1-sink if $i = n$. The sizes of π-OBDDs and π-ZBDDs are polynomially related. Hence, we do not need new arguments for exponential lower bounds. Nevertheless, a size decrease by a linear factor $\Theta(n)$ may be remarkable for applications. We show that such a size decrease is possible also for more complicated functions than in the example above.

Example 8.1.7. Fig. 8.1.2 shows an ordered DD on $x_0, \ldots, x_{k-1}, y_0, \ldots, y_{n-1}$ for $n = 2^k$ and $n = 4$. It is a π-OBDD for the multiplexer or direct storage access function. Its size for arbitrary n equals $2n+1$ (see Theorem 4.3.2). We obtain the quasi-reduced π-OBDD for the multiplexer by inserting i dummy nodes for y_i, $0 \le i \le n-1$, and $n-1$ dummy nodes for each of the sinks, altogether $2(n-1)+1+\cdots+(n-1) = n^2/2+3n/2-2$ dummy nodes. Only the dummy nodes for the 0-sink can be eliminated by the ZBDD elimination rule. Therefore, $\pi\text{-OBDD}(\text{MUX}_n) = 2n+1$ and $\pi\text{-ZBDD}(\text{MUX}_n) = \frac{1}{2}n^2+\frac{5}{2}n$. We may interpret the DD of Fig. 8.1.2 as ZBDD. If the address is $a = |x|$, we obtain the output 1 iff $y_a = 1$ and $y_j = 0$ for all other j. Let us call this function ZMUX_n (zero-suppressed multiplexer). In order to obtain a quasi-reduced π-ZBDD, we have to add the same number of dummy nodes as above. Again only the dummy nodes for the 0-sink can be eliminated, now by the OBDD elimination rule. Therefore, $\pi\text{-ZBDD}(\text{ZMUX}_n) = 2n+1$ and $\pi\text{-OBDD}(\text{ZMUX}_n) = \frac{1}{2}n^2 + \frac{5}{2}n$.

Finally, we discuss operations on π-ZBDDs. Evaluation and satisfiability test are performed in the obvious way. The same holds for the equivalence test, since reduced π-ZBDDs are a canonical form. SAT-COUNT is even easier than for OBDDs. If an x_i-node is missing, we obtain satisfying inputs only for

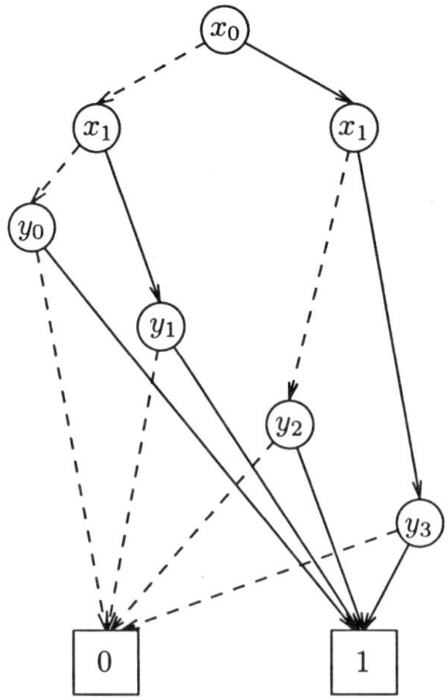

Figure 8.1.2: *An ordered DD.*

$x_i = 0$. Hence, the number of satisfying inputs equals the number of paths from the source to the 1-sink and can be computed in linear time.

The synthesis algorithm for π-OBDDs cannot be used directly for π-ZBDDs. The π-ZBDD consisting only of the 1-sink represents $g = \overline{x}_1 \cdots \overline{x}_n$. Then $\overline{g} = x_1 + \cdots + x_n$ and π-ZBDD$(\overline{g}) = 2n + 1$ (see Fig. 8.1.3 for an example).

It has turned out that the following property decides the difficulty of the synthesis operation.

Definition 8.1.8. A Boolean operator $\otimes\colon \{0,1\}^m \to \{0,1\}$ is called 0-*preserving* if $\otimes(0, \ldots, 0) = 0$.

First, we describe a synthesis algorithm for binary 0-preserving operators $\otimes$ working on π-ZBDDs G_f and G_g. By G_f^0 and G_g^0 we denote the resulting π-ZBDD after adding 0-sinks if not already existing. The result G of the $\otimes$-synthesis is defined on the vertex set $V = V_f^0 \times V_g^0$. The construction is identical to the first synthesis algorithm for π-OBDDs (described after Theorem 3.3.4) with one exception. Let (v_1^*, w_1^*) be the 1-successor of the node (v, w) with label x_i and v_1 and w_1 the 1-successor of v and w in G_f and G_g, respectively. Then $v_1^* = v_1$ as in the OBDD case if label$(v) = x_i$, but v_1^* is the 0-sink of G_f otherwise (similarly for w_1^*). This is caused by the elimination rule for ZBDDs. A missing x_i-test indicates that we should reach the 0-sink if $a_i = 1$.

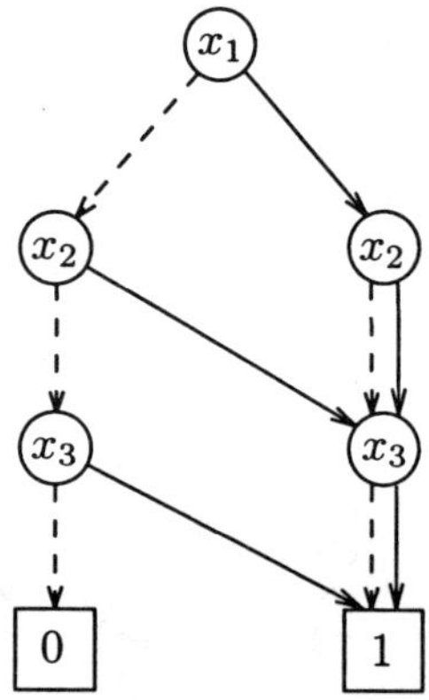

Figure 8.1.3: *A reduced π-ZBDD for $x_1 + x_2 + x_3$.*

We prove the correctness of this π-ZBDD synthesis algorithm. If $f(a) = 1$, the computation path for a in G_f reaches the 1-sink and no x_j-test where $a_j = 1$ is omitted. Let s_f^c be the c-sink of G_f. Then we reach a sink $(s_f^1, \cdot)$ in G, similarly if $g(a) = 1$. If $f(a) = 0$, the computation path for a in G_f reaches the 0-sink or the 1-sink. In the first case, we reach in G a sink $(s_f^0, \cdot)$. In the second case, we may reach a sink $(s_f^0, \cdot)$ or a sink $(s_f^1, \cdot)$ in G. The second subcase occurs only if we have omitted an x_j-test in G_f where $a_j = 1$ and the same test is omitted on the computation path for a in G_g. Summarizing, there are two possibilities. Either we reach the sink $(s_f^{f(a)}, s_g^{g(a)})$ in G which has by definition the right label $f(a) \otimes g(a)$, or $f(a) = g(a) = 0$ and we omit the x_j-test in G_f and G_g for some variable x_j, where $a_j = 1$. Then we omit the x_j-test in G and that implies by the semantics of ZBDDs that G represents a function h^*, where $h^*(a) = 0$. This is equal to $f(a) \otimes g(a) = 0 \otimes 0$ only if $0 \otimes 0 = 0$, i.e., if $\otimes$ is 0-preserving. This implies the following result.

Theorem 8.1.9. *If $h = f \otimes g$ for a 0-preserving operator $\otimes$, then π-ZBDD$(h) \leq$ π-ZBDD$^0(f) \cdot \pi$-ZBDD$^0(g)$, where the superscript 0 denotes that we consider π-ZBDDs with a 0-sink.*

We may use the same hashing techniques as for π-OBDDs to integrate the reduction into the synthesis process. This leads to the same time and space bounds for the synthesis of π-ZBDDs as for the synthesis of π-OBDDs (see Section 3.3). It is only necessary to start with ZBDDs with a 0-sink. In a straightforward way, we can generalize the synthesis algorithm to m-ary 0-preserving operators. This is also the key to the synthesis for non-0-preserving operators $\otimes\colon \{0,1\}^m \to \{0,1\}$. We define $\otimes^*\colon \{0,1\}^{m+1} \to \{0,1\}$ by $\otimes^*(a_1, \dots, a_m, 0) = 0$ and $\otimes^*(a_1, \dots, a_m, 1) = \otimes(a_1, \dots, a_m)$. By definition,

$\otimes^*$ is 0-preserving. Instead of a $\otimes$-synthesis of $G_1, \ldots, G_m$, we can perform a $\otimes^*$-synthesis of $G_1, \ldots, G_m$, and the π-ZBDD G^* representing the constant 1 with $n+1$ nodes. Although G^* does not contain a 0-sink, it is not necessary to add a 0-sink. The reason is that in G^* no test is omitted.

Theorem 8.1.10. *If $h = f \otimes g$ for a non-0-preserving operator $\otimes$, π-ZBDD$(h) \leq \pi$-ZBDD$^0(f) \cdot \pi$-ZBDD$^0(g) \cdot (n+1)$. Moreover, π-ZBDD$(\overline{f}) \leq \pi$-ZBDD$^0(f) \cdot (n+1)$.*

These bounds are tight. The function $f = \overline{x}_1 \cdots \overline{x}_n$ is represented by the 1-sink. Hence, π-ZBDD$(f) = 1$ and π-ZBDD$^0(f) = 2$. We have already seen that π-ZBDD$(\overline{f}) = 2n+1$ for $\overline{f} = x_1 + \cdots + x_n$.

To obtain algorithms for the remaining operations, it is sufficient to consider the operation replacement by constants.

Theorem 8.1.11. *The operation replacement by constants can be performed in linear time $O(|G|)$. Moreover, if $f \in B_n$, π-ZBDD$(f_{|x_i=1}) \leq \pi$-ZBDD$^0(f)$ and π-ZBDD$(f_{|x_i=0}) \leq \frac{3}{2}\pi$-ZBDD$^0(f) + \frac{1}{2}$.*

Proof. We ensure that each path from a source to a sink contains an x_i-node. Without loss of generality $\pi = id$. It is sufficient to add for each sink v (including the added 0-sink) and each inner node v with label x_j, $j > i$, a dummy x_i-node v' whose 0-successor is v and whose 1-successor is the 0-sink. Edges crossing the x_i-level and leading to v are replaced by edges to v'. This does not change the function which is represented. To replace x_i by c, it is now sufficient to replace $\overline{c}$-edges leaving x_i-nodes v' with edges to the c-successor of v'. The correctness is as obvious as the time estimate.

If $c = 1$, each new x_i-node can be eliminated, since both edges lead to the 0-sink. The 0-sink possibly added cannot be eliminated. If $c = 0$, perhaps no elimination is possible. The number of new x_i-nodes is bounded by s, the number of nodes below the x_i-level. Since we have to add x_i-nodes v' only for nodes v directly reached by an edge from a node above the x_i-level, the number of new x_i-nodes is also bounded by t, the number of edges crossing the x_i-level. Since we consider functions with one output and, therefore, ZBDDs with one source, $t-1$ is bounded by the number of nodes above the x_i-level. Hence, $s + t - 1 \leq \pi$-ZBDD$^0(f)$ and $\min\{s,t\} \leq \frac{1}{2}(\pi$-ZBDD$^0(f) + 1)$. □

It is possible that the replacement of a variable x_i by 0 increases the ZBDD size by a factor of almost 3/2 (see exercises), but a sequence of such replacements cannot lead to a larger size than $(n+1) \cdot \pi$-ZBDD$^0(f)$, since this is an upper bound for the size of the quasi-reduced π-ZBDD for f and also for its subfunctions. We emphasize that already a small change in the semantics may cause a lot of changes in the behavior of the representation type.

8.2 Ordered Functional Decision Diagrams (OFDDs)

The evaluation of BDDs (except for ZBDDs) is based on Shannon's decomposition rule $f = \overline{x}_i f_{|x_i=0} + x_i f_{|x_i=1}$. Its advantage is that each input activates exactly one path called its computation path. In some applications, one works with the representation of Boolean functions by $\mathbb{Z}_2$-polynomials, i.e., $\oplus$-sums of monomials of positive literals. This representation is related to *Reed-Muller's decomposition rule* $f = f_{|x_i=0} \oplus x_i(f_{|x_i=0} \oplus f_{|x_i=1})$, whose correctness follows by the consideration of the cases $x_i = 0$ and $x_i = 1$. The decomposition is unique. If $f = g \oplus x_i h$ for functions g and h not essentially depending on x_i, then $f_{|x_i=0} = g$, $f_{|x_i=1} = g \oplus h$, and, therefore, $h = f_{|x_i=0} \oplus f_{|x_i=1}$. This was the motivation for Kebschull, Schubert, and Rosenstiel (1992) to introduce OFDDs. A little bit later, Kebschull and Rosenstiel (1993) proposed OFDDs as a general synthesis tool and as an alternative to OBDDs (see also Tsai and Marek-Sadowska (1996) for the use of OFDDs for the detection of symmetric variables).

Definition 8.2.1. An OFDD shares its syntax with OBDDs. The c-sink represents the constant c. If f_0 and f_1 are represented at the 0-, respectively, 1-successor of the x_i-node v, then v represents $f = f_0 \oplus x_i f_1$.

We use the notion π-OFDD if the variable ordering is fixed, and in general discussions we assume, w.l.o.g., that $\pi = id$. If each path from the source to a sink contains an x_i-node for each x_i, the OFDD is called complete. An input a can activate more than one path of an OFDD. At x_i-nodes where $a_i = 0$, it is sufficient to consider the 0-successor and only the 0-edge is activated. But at x_i-nodes where $a_i = 1$, we have to consider both successors and to take the $\oplus$-sum of the results. Then both outgoing edges are activated. An input a with j ones activates in complete OFDDs exactly 2^j paths and, if the OFDD represents f, then $f(a)$ is the $\oplus$-sum of the 2^j labels of the sinks reached by the activated paths. We use the notation $b \le a$ iff $b_i \le a_i$ for all i (where $0 \le 1$). Then the input a activates in the complete OFDD G all paths which are activated by the inputs $b \le a$ if G is interpreted as a complete OBDD. This characterization will be formalized.

Definition 8.2.2. The *τ-operator* $\tau\colon B_n \to B_n$ is defined by

$$(\tau f)(a) = \bigoplus_{b \le a} f(b).$$

Becker, Drechsler, and Werchner (1995) were the first to observe the following easy but essential relationship between complete OFDDs and complete OBDDs (which are equivalent to complete ZBDDs).

Proposition 8.2.3. *Let G be a complete ordered DD representing f as an OBDD (or ZBDD). Then G represents τf as an OFDD.*

We need the following simple properties of the τ-operator.

Lemma 8.2.4.

(i) $\tau(f \oplus g) = \tau f \oplus \tau g$ *for all* $f, g \in B_n$.

(ii) $\tau\tau f = f$ *for all* $f \in B_n$.

(iii) τ *is one-to-one on* B_n.

Proof. The first property follows by definition, since $\oplus$ is commutative. This implies by definition that

$$(\tau\tau f)(a) = \tau\left(\bigoplus_{b\leq a} f(b)\right) = \bigoplus_{b\leq a}(\tau f)(b) = \bigoplus_{b\leq a}\bigoplus_{c\leq b} f(c) = \bigoplus_{c\leq b\leq a} f(c).$$

If c differs from a at i positions, the number of inputs b such that $c \leq b \leq a$ equals 2^i and we obtain the term $f(c)$ 2^i times. If $i \geq 1$, then 2^i is even and the terms cancel each other. Only if $c = a$, we get $i = 0$ and the term $f(a)$ once. Finally, τ is one-to-one on the finite set B_n, since it has an inverse, namely τ itself. □

By Lemma 8.2.4, a complete OFDD representing f represents τf as an OBDD or ZBDD. Now it is quite easy to obtain the following result.

Theorem 8.2.5. *There is (up to isomorphism) a unique complete π-OFDD of minimal size representing f. This is called the quasi-reduced π-OFDD for f. It can be obtained in linear time from each complete π-OFDD (without nodes not reachable from the source) by the application of the merging rule. It contains, if $\pi = id$, x_i-nodes representing for $S \subseteq \{1,\ldots,i-1\}$ the different functions $f_{i,S}$, where $f_{i,S}$ is the $\oplus$-sum of all $f_{|x_1=a_1,\ldots,x_{i-1}=a_{i-1}}$ such that $a_j = 0$ if $j \notin S$ and $a_j \in \{0,1\}$ if $j \in S$.*

Proof. We only have to prove the claim on the functions represented at x_i-nodes. By our considerations on Reed-Muller's decomposition rule, we have to represent $f_{i,S}$ at the node reached by 0-edges leaving x_j-nodes if $j \in \{1,\ldots,i-1\} - S$ and 1-edges leaving x_j-nodes if $j \in S$. The merging rule only merges nodes representing the same function. □

The proof also shows that a π-OFDD for f has to contain nodes representing $f_{i,S}$, $1 \leq i \leq n+1$ and $S \subseteq \{1,\ldots,i-1\}$. Since each node represents only one function, the π-OFDD can be reduced further only if $f_{i,S} = f_{j,S'}$ for some $i \neq j$, w.l.o.g., $i < j$. Since $f_{j,S'}$ can essentially depend only on $x_j, \ldots, x_n$, this may happen only if $f_{i,S}$ does not essentially depend on x_i. Then $f_{i,S|x_i=0} = f_{i,S|x_i=1}$ and $f_{i+1,S\cup\{i\}} \equiv 0$. Semantically, an x_i-node v can be eliminated iff the function represented at v does not essentially depend on x_i. This is the same as in OBDDs and it is equivalent to the statement that the function reached by the 1-edge leaving v represents the constant 0. Hence, syntactically, the ZBDD elimination rule (see Fig. 8.1.1(b)) can be used as the OFDD *elimination rule.* Now let us assume that an x_i-node v of a π-OFDD represents a function f' not essentially depending on x_i and let g' and h' be the functions represented at the successors. Then $f' = g' \oplus x_i h'$ and it follows that $h' \equiv 0$, i.e., h' does not essentially depend on $x_{i+1}, \ldots, x_n$. Repeating the same argument, the path starting at v and using only 1-edges leads to the 0-sink. All nodes on this path including v but excluding the 0-sink can be eliminated by the OFDD elimination rule. Hence, we obtain a minimal-size π-OFDD for f by the application of the merging rule and the OFDD elimination rule. These rules are syntactically identical to the reduction rules of ZBDDs. This leads to the following result.

Theorem 8.2.6. *There is (up to isomorphism) a unique π-OFDD of minimal size representing f. This is called the reduced π-OFDD for f. It can be obtained in linear time from each π-OFDD (without nodes not reachable from the source) by the application of the merging rule and the OFDD elimination rule. It contains, if $\pi = id$, x_i-nodes representing for $S \subseteq \{1, \ldots, i-1\}$ the different functions $f_{i,S}$ essentially depending on x_i.*

Corollary 8.2.7.

(*i*) π-*ZBDD*$(f) = \pi$-*OFDD*(τf) *for all* $f \in B_n$.

(*ii*) π-*OBDD*$(f) \le (n+1) \cdot \pi$-*OFDD*$(\tau f) + 1$ *for all* $f \in B_n$.

(*iii*) π-*OFDD*$(f) \le (n+1) \cdot \pi$-*OBDD*(τf) *for all* $f \in B_n$.

Proof. The first result is based on Proposition 8.2.3. Complete π-ZBDDs for f and complete π-OFDDs for τf coincide. Reduced π-ZBDDs for f and reduced π-OFDDs for f are obtained by the syntactically same reduction rules. Then the other two results follow from Theorem 8.1.6. □

Similar results can be obtained for free FDDs (FFDDs) and FBDDs with and without graph ordering (see exercises). We do not need new lower bound techniques for OFDDs, since we can apply the OBDD techniques to τf instead of f.

Theorem 8.2.8. *If f is symmetric, then τf is symmetric and its π-OFDD size is bounded above by $O(n^2)$.*

Proof. We know that $(\tau f)(a)$ is the $\oplus$-sum of all $f(b)$ where $b \leq a$. If a and a' have j ones, the number of $b \leq a$ with i ones is equal to the number of $b' \leq a'$ with i ones. Since f is symmetric, $f(b) = f(b')$ if b and b' have i ones. This implies that $(\tau f)(a) = (\tau f)(a')$ for all a and a' with the same number of ones. The size of quasi-reduced π-OBDDs for symmetric functions is $O(n^2)$. □

The OBDD size of read-once formulas has been investigated in Section 4.11. Drechsler, Becker, and Jahnke (1998) mention the following result for OFDDs.

Theorem 8.2.9. *The OFDD size of functions representable by read-once formulas on n variables is bounded above by $n \log n + n + 2$.*

Proof. The upper bound is proved by induction on n for OFDDs with the special property that the path consisting only of 0-edges is complete and all nodes (except perhaps the sink) on this path have indegree 1. We use the notion that such OFDDs have an isolated 0-path. The claim holds for $n = 1$. Negations are easy, since it is sufficient to change the sink reached by the isolated 0-path. The size does not change. Therefore, it is sufficient to consider the cases $f = g \wedge h$ and $f = g \oplus h$, where g and h are read-once formulas on disjoint sets of variables. In the case $f = g \wedge h$ we assume, w.l.o.g., that g depends on at least as many variables as h. Then (see Fig. 8.2.1) we start with an OFDD with an isolated 0-path for g. The 1-sink is replaced with the source of an OFDD with an isolated 0-path for h. The number of paths from the source to the 1-sink activated by some input a is the product of the corresponding numbers for g and for h and, therefore, odd iff the corresponding numbers for g and h are odd. This implies $f = g \wedge h$. In order to obtain an isolated 0-path, we have to copy the isolated 0-path of h and to attach the copy to the isolated 0-path (without sink) of g. Altogether, we obtain two sinks, have to sum up the sizes of the OFDDs for g and h, and obtain by the choice of the variable ordering at most $n/2$ additional nodes. The case $f = g \oplus h$ is easier if the isolated 0-path for g ends at the 0-sink. If this assumption is not fulfilled, we compute f by $\overline{g} \oplus \overline{h}$ and the desired property is fulfilled for the OFDD representing $\overline{g}$. Then we start with the OFDD for g (or $\overline{g}$) and attach the OFDD for h (or $\overline{h}$) to the isolated 0-path of the OFDD for g (or $\overline{g}$). The number of paths from the source to the 1-sink activated by some input a is equal to the sum of the corresponding paths for g and h (or $\overline{g}$ and $\overline{h}$). This implies $f = g \oplus h$. Let size(f) be the number of inner nodes of the constructed OFDD representing f. In both cases, we obtain

$$\text{size}(x_i) = 1 \text{ and } \text{size}(f) \leq \text{size}(g) + \text{size}(h) + n/2.$$

A variable provides the contribution 1 in the beginning and later the contribution 1 if it belongs to the OFDD with the smaller number of variables. This can happen for each variable at most $\log n$ times. Hence, the total contribution of each variable is bounded above by $\log n + 1$. □

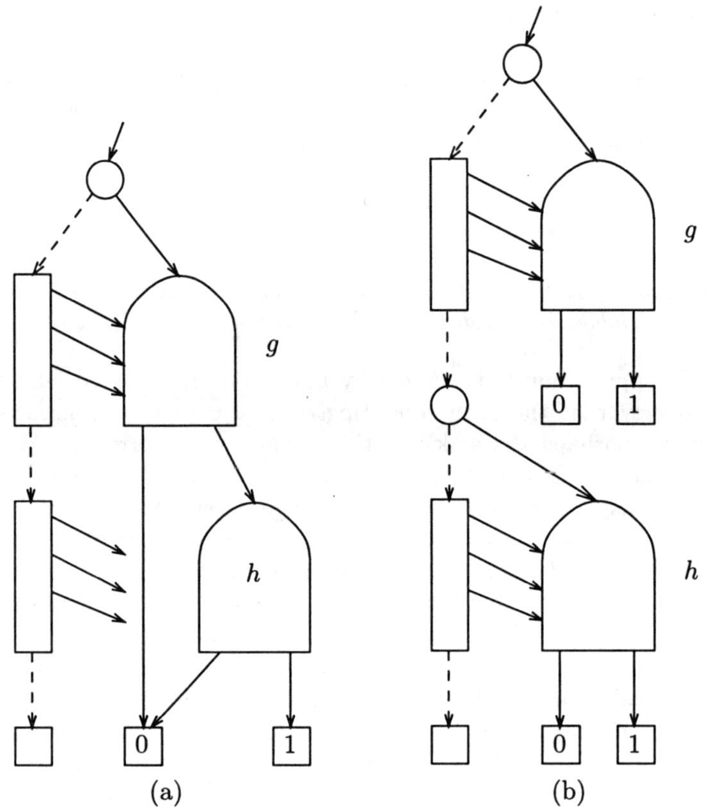

Figure 8.2.1: *OFDDs for* (a) $f = g \wedge h$ *and* (b) $f = g \oplus h$ *if* $g(0, \ldots, 0) = 0$.

As the next example, we investigate the OFDD size of the hidden weighted bit function HWB_n. For this purpose, we consider OBDDs for τHWB_n. We know that we obtain lower bounds by counting the different subfunctions obtained from τHWB_n by assigning constants to k of the n variables. The main observation is the following. If a and a' contain i ones and i is the minimal index with $a_i \neq a'_i$, then $\text{HWB}_n(a) \neq \text{HWB}_n(a')$ and also $\tau\text{HWB}_n(a) \neq \tau\text{HWB}_n(a')$. The last statement follows since $a_j = a'_j$ for $j < i$ implies that the number of $b \leq a$ with $j < i$ ones and $b_j = 1$ is the same as the number of $b' \leq a'$ with $j < i$ ones and $b'_j = 1$.

Theorem 8.2.10. *The OFDD size of HWB_n is bounded below by $\Omega(2^{n/5}n^{-3/2})$.*

Proof. Without loss of generality $n = 10m$. Let π be some variable ordering and S the set of indices i such that x_i belongs to the first $6m$ variables according to π. Let $A := S \cap \{m, \ldots, 5m\}$ and $B := S \cap \{5m, \ldots, 9m\}$. Then (compare the proof of Theorem 4.10.2) $|A| \geq 2m$ or $|B| \geq 2m$.

If $|A| \geq 2m$, choose $A' \subseteq A$ with $|A'| = 2m$. We consider all assignments to the variables x_i, $i \in S$, such that m variables $x_i, i \in A'$, get the value 1 and all other variables $x_i, i \in S$, get the value 0. These are $\binom{2m}{m} = \Omega(2^{n/5} n^{-1/2})$ assignments and the lower bound follows from Corollary 8.2.7 if we can prove that these assignments lead to different subfunctions. Let a and a' be two different of these considered assignments and let i be the smallest index such that a and a' assign different values to x_i. By definition, $i \in A'$ and $m \leq i \leq 5m$. The remaining $4m$ variables are replaced by constants, among them $i - m \in \{0, \ldots, 4m\}$ ones. Altogether, we obtain two inputs b and b' with i ones, $b_i \neq b'_i$, and i is the smallest index where b and b' differ. By the consideration above, $\tau\mathrm{HWB}_n(b) \neq \tau\mathrm{HWB}_n(b')$.

If $|B| \geq 2m$, choose $B' \subseteq B$ with $|B'| = 2m$. Now we consider the $\binom{2m}{m}$ assignments to the variables $x_i, i \in S$, such that exactly m variables $x_i, i \in B'$, get the value 0 and all other variables $x_i, i \in S$, get the value 1. Now we proceed in an analogous way. In this case $5m \leq i \leq 9m$ for the smallest index i where two assignments differ. Since we already have $5m$ ones, we may choose an extension with an additional $i - 5m \in \{0, \ldots, 4m\}$ ones. □

Becker, Drechsler, and Werchner (1995) also proved that the OFDD size of multiplication is exponential. We obtain this result as a corollary of a much more general lower bound presented in Chapter 10. Here we consider an example due to Becker, Drechsler, and Werchner (1995) showing that OBDDs and OFDDs may differ exponentially in their size.

Definition 8.2.11.

(i) The function $\oplus cl_{n,3}$ decides whether the number of triangles in an undirected graph is odd.

(ii) The function $1cl_{n,3}$ decides whether an undirected graph consists of $n-3$ isolated vertices and one triangle.

We have already mentioned in Section 6.2 that the FBDD size of $\oplus cl_{n,3}$ is $2^{\Omega(N)}$ for the number $N = \binom{n}{2}$ of variables. It is easy to show that the OBDD size of $1cl_{n,3}$ is $O(N^2) = O(n^4)$. This holds for each variable ordering and quasi-reduced OBDDs. We show that the width (maximal number of nodes with the same label) of the OBDD is bounded by $2N+3$. We need one node for the case that we have not found any edge. There are at most N different situations when we have seen one edge. If we have seen two edges, then we either know that the output is 0 (if the edges do not share one node) or we have to distinguish at most N different situations (which is the missing edge to form a triangle). If we have seen three edges, then we either know that the output is 0 (if the edges do not form a triangle) or we have to check that no further edge exists. If we have seen four edges, we know that the output is 0.

Theorem 8.2.12.

(*i*) $\tau(1cl_{n,3}) = \oplus cl_{n,3}$.

(*ii*) *The OBDD size of* $1cl_{n,3}$ *is* $O(N^2)$ *for each variable ordering while the OFDD and even the FFDD size of* $1cl_{n,3}$ *is bounded below by* $2^{\Omega(N)}$.

(*iii*) *The OFDD size of* $\oplus cl_{n,3}$ *is* $O(N^2)$ *for each variable ordering while the OBDD and even the FBDD size of* $\oplus cl_{n,3}$ *is bounded below by* $2^{\Omega(N)}$.

Proof. It is sufficient to prove the first property. The OBDD upper bound has been proved above and the FBDD lower bound has been cited in Section 6.2. This implies the bounds for OFDDs and FFDDs by the first property and the fact that the upper bound for OBDDs holds for complete OBDDs (the upper bound for OFDDs on $\oplus cl_{n,3}$ can be improved to $O(N^{3/2})$ (see exercises)). Let $G(x)$ be a graph described by the input x. To evaluate $\tau(1cl_{n,3})$ we have to consider all graphs G' obtained from $G(x)$ by eliminating edges and have to decide whether the number of such graphs consisting of $n-3$ isolated vertices and a triangle is odd. For each triangle in $G(x)$ we can eliminate all other edges to obtain such a graph and this is the only way to obtain "isolated triangles." Hence, $\tau(1cl_{n,3})$ decides whether the number of triangles in $G(x)$ is odd. □

We know that π-OFDDs are canonical and have seen how we can bound the π-OFDD size of selected functions. We still have to investigate which of the important operations can be performed efficiently.

The evaluation is no longer possible in time $O(n)$, since the input $a = (1, \ldots, 1)$ activates the whole OFDD. But a simple DFS traversal and, therefore, time $O(|G|)$ is sufficient. If the value computed at both successors of v is known, the value at v can be computed in constant time. We have seen that the reduction of π-OFDDs is possible in linear time and we may consider reduced π-OFDDs. Then the satisfiability test can be performed in constant time, since the reduced π-OFDD for the constant 0 contains only the 0-sink. The equivalence test can be performed as a simple isomorphism check in linear time and even in constant time if the π-OFDDs share nodes. The satisfiability problem has some interesting features. Let $\pi = id$. In a reduced π-OFDD, we find the lexicographically smallest satisfying input if we follow the path which chooses the 0-edge whenever it does not lead to the 0-sink and the 1-edge otherwise. This partial assignment is completed by zeros for the variables not tested. The resulting input a has the property that the "a-path" leads to the 1-sink while all "b-paths" for $b \leq a$ and $b \neq a$ lead to the 0-sink. This implies that the π-OFDD outputs 1 on a. Werchner et al. (1995) have shown that SAT-COUNT is #P-complete for OFDDs.

The next step is to investigate the fundamental synthesis problem.

Theorem 8.2.13. *Let $f, g \in B_n$ be represented by π-OFDDs G_f and G_g, respectively. Then $\overline{f}$ can be represented by a π-OFDD of size $|G_f^0| + n$ (the superscript denotes that we add a 0-sink if not existent), which can be computed in time $O(n)$, and $f \oplus g$ can be represented by a π-OFDD of size $|G_f^0||G_g^0|$, which can be computed in time $O(|G_f^0||G_g^0|)$.*

Proof. For the negation, we use an approach already considered in the proof of Theorem 8.2.9. With at most n nodes it is possible to create an isolated 0-path. Then it is sufficient to negate the sink reached by this path. The number of inner nodes increases by at most n and, if there was no 0-sink, we create one.

For the $\oplus$-synthesis (which includes the negation as the special case $g \equiv 1$), we consider the π-OFDDs G_f^0 and G_g^0 with a 0-sink. The result is defined on the vertex set $V = V_f^0 \times V_g^0$. Without loss of generality $\pi = id$. Let $v \in V_f^0$ be an x_i-node representing f^v and $w \in V_g^0$ an x_j-node representing g^w. First, we assume that $i = j$. Then we consider the direct successors v_0,v_1,w_0, and w_1 representing $f_0^v = f^v_{|x_i=0}$, $f_1^v = f^v_{|x_i=0} \oplus f^v_{|x_i=1}$, g_0^w, and g_1^w, respectively. At (v, w) we want to represent

$$f^v \oplus g^w = f_0^v \oplus g_0^w \oplus x_i(f_1^v \oplus g_1^w).$$

The node gets the label x_i, (v_0, w_0) as 0-successor, and (v_1, w_1) as 1-successor. If the $\oplus$-synthesis is done correctly for these successors, it is done correctly for (v, w). Without loss of generality $i \leq j$. If $i < j$, w represents a function g^w not essentially depending on x_i, i.e., $g^w = g^w_{|x_i=0} \oplus (x_i \wedge 0)$. Again, we label (v, w) by x_i and choose (v_0, w) as 0-successor and (v_1, s_g^0) as 1-successor, where s_g^0 is the 0-sink of G_g^0. The same is done if w is a sink. The last case is that v and w are sinks. Then (v, w) is a sink with label $label(v) \oplus label(w)$. Since in this final case $f^v \oplus g^w$ is represented, this holds by bottom-up induction for all nodes (v, w). □

The first remark is that we may use the well-known tricks for the $\oplus$-synthesis if we pay attention to the fact that 1-successors of omitted tests implicitly are 0-sinks. There are examples where the blow-up of the size cannot be prevented. The function $\overline{x}_1\overline{x}_2 \cdots \overline{x}_n$ has the π-OFDD size $n+1$. If $\pi = id$, it consists of an x_i-node whose edges lead to the x_{i+1}-node if $i \leq n$ and to the 1-sink otherwise. Its negation is $x_1 + \cdots + x_n$, whose π-OFDD size equals $2n + 1$. For a bad example for the $\oplus$-synthesis see the exercises. Becker, Drechsler, and Werchner (1995) have shown that the $\wedge$-synthesis may lead to an exponential blow-up of the π-OFDD size.

Theorem 8.2.14. *There are functions $f_N, g_N \in B_N$ with polynomial π-OFDD size such that $f_N \wedge g_N$ has exponential size even for FFDDs.*

Proof. Let $N = \binom{n}{2}$. Then we choose $\oplus cl_{n,3}$ as f_N and the negative threshold function $\overline{T}_{4,N}$ as g_N. The functions f_N (see Theorem 8.2.12) and g_N (see

Theorem 8.2.8) have polynomial π-OFDD size for each variable ordering but $\oplus cl_{n,3} \wedge \overline{T}_{4,N} = 1cl_{n,3}$. If a graph has at most three edges, it can contain a triangle only if the other $n-3$ vertices are isolated. We know from Theorem 8.2.12 that the FFDD size of $1cl_{n,3}$ is exponential. □

Nevertheless, π-OFDDs are used in applications. Kebschull and Rosenstiel (1993) present an algorithm for the $\wedge$-synthesis where in G_f and in G_g all paths to the 1-sink are considered. It is easy to see that a path represents a monomial of positive literals. The conjunction of two such monomials is again a monomial. Then the resulting monomials are combined by a $\oplus$-sum. Let us investigate the situation that f and g are represented by π-OFDDs whose sources are x_1-nodes. Then we have representations of $f_0 := f_{|x_1=0}$, $f_1 := f_{|x_1=0} \oplus f_{|x_1=1}$, g_0, and g_1. Moreover,

$$\begin{aligned} h &= f \wedge g = (f_0 \oplus x_1 f_1) \wedge (g_0 \oplus x_1 g_1) \\ &= f_0 g_0 \oplus x_1 (f_0 g_1 \oplus f_1 g_0 \oplus f_1 g_1). \end{aligned}$$

For the 0-successor it is sufficient to work with a recursive call leading to a representation of $f_0 g_0$, but the situation for the 1-successor is more difficult. We have to perform three recursive calls whose results have to be combined by $\oplus$-synthesis.

The operation replacement by constants also leads to difficulties. If we want to replace x_i by 0, we may redirect all edges to x_i-nodes v representing f^v by edges to the 0-successor of v where, by definition, the correct subfunction $f^v_{|x_i=0}$ is represented. In order to obtain $f^v_{|x_i=1}$ we may apply the $\oplus$-synthesis algorithm to $f^v_{|x_i=0}$ and $f^v_{|x_i=0} \oplus f^v_{|x_i=1}$, the functions represented at the direct successors of v. This may lead to a π-OFDD of size $\Theta(|G_f|^2)$. Bollig, Löbbing, Sauerhoff, and Wegener (1996) have shown that such a size increase can happen in each of a number of replacement steps.

Theorem 8.2.15. *There is a function f_n on $O(n^2)$ variables representable by an OFDD of polynomial size but the subfunction obtained by the replacement of $O(\log n)$ variables by the constant 1 has exponential FFDD size.*

Proof. We define the function $f_n \in B_N$, where $N = \binom{n}{2} + \lceil \log \binom{n}{3} \rceil$, by the description of an OFDD representing f_n. At the top, we have a complete binary tree of depth $\lceil \log \binom{n}{3} \rceil$. For each triple (i,j,k), $1 \le i < j < k \le n$, we choose one leaf where we represent the minterm Δ_{ijk} describing the graph on n vertices with the triangle $\{i,j,k\}$ and $n-3$ isolated vertices. Minterms have OFDDs of linear size for each variable ordering. The remaining leaves represent the constant 0. Now we replace all $\lceil \log \binom{n}{3} \rceil$ variables in the complete binary tree with the constant 1. Then we obtain the $\oplus$-sum of the functions represented at the leaves of the tree and this is $1cl_{n,3}$, whose FFDD size is $2^{\Omega(n^2)}$. □

All problems considered for OBDDs can be investigated also for OFDDs, although this is sometimes technically involved. Bollig, Löbbing, Sauerhoff, and Wegener (1996) have shown that the OFDD variable-ordering problem is NP-complete and that it is NP-hard to obtain a good approximation for the π-OFDD minimization problem for incompletely specified functions (see Theorem 3.6.2 for the corresponding OBDD problem). Moreover, they have shown that the maximal size increase caused by the swap, jump, and exchange operations is the same as for the OBDD case. Hence, the sifting algorithm can be used for OFDDs.

It is easier to work with OBDDs than with OFDDs but, if the OFDD size of a function is much smaller than its OBDD size, it makes sense to work with OFDDs.

8.3 Ordered Kronecker Functional Decision Diagrams (OKFDDs)

One may ask whether the list of possible decomposition types is arbitrarily long. Becker, Drechsler, and Theobald (1997) have shown that the list of binary decomposition types is very short if we demand that the functions represented at the successors of an x_i-node v do not essentially depend on x_i and that the function represented at v can be computed by some operation $op \in B_3$ from x_i and the functions represented at the direct successors of v. If, moreover, we do not distinguish operations leading to the same graph structure of reduced π-DDs, we obtain only three decomposition types:

$$
\begin{aligned}
f &= \overline{x}_i f_{|x_i=0} + x_i f_{|x_i=1} && \text{(Shannon type)},\\
f &= f_{|x_i=0} \oplus x_i (f_{|x_i=0} \oplus f_{|x_i=1}) && \text{((positive) Reed-Muller type)},\\
f &= f_{|x_i=1} \oplus \overline{x}_i (f_{|x_i=0} \oplus f_{|x_i=1}) && \text{(negative Reed-Muller type)}.
\end{aligned}
$$

The theory of DDs based on the negative Reed-Muller decomposition can be developed in the same way as the theory of OFDDs. The new possibility is to choose different decomposition types for different variables.

Definition 8.3.1. An OKFDD shares its syntax with OBDDs and has additionally a *decomposition-type list* $dt \in \{S, pRM, nRM\}^n$. Then the c-sink represents the constant c and at x_i-nodes v the evaluation rule $f^v = \overline{x}_i f_0^v + x_i f_1^v$ (if $dt_i = S$), $f = f_0 \oplus x_i f_1$ (if $dt_i = pRM$), or $f = f_0 \oplus \overline{x}_i f_1$ (if $dt_i = nRM$) has to be applied where f_0 and f_1 are the functions represented at the corresponding direct successors.

In Section 9.4, we get to know relations between DDs and the Kronecker product of matrices. This will explain the choice of the notion OKFDD. Drechsler et al. (1994) presented OKFDDs as a data structure and Becker,

Drechsler, and Theobald (1997) generalized the τ-operator to a τ_{dt}-operator to describe the transformation of the function described by a complete ordered DD as OBDD to the function represented by the same diagram now interpreted as OKFDD with decomposition-type list *dt*. This leads to upper and lower bounds of the size of *dt*-OKFDDs representing selected Boolean functions. Drechsler, Becker, and Jahnke (1998) describe a simple heuristic algorithm for the choice of the decomposition-type list. All results on OKFDDs can be obtained easily with the methods we have presented for OBDDs and OFDDs.

8.4 Exercises and Open Problems

8.1.E How many functions $f \in B_n$ are 1-simple with respect to x_1?

8.2.M Prove Theorem 8.1.4.

8.3.M Let $f_n(x_1, \ldots, x_n, y_1, \ldots, y_n) = 1$ iff $y_i = x_i = 1$ and $x_1 = \cdots = x_{i-1} = 0$ for some i. Compare the OBDD size and the ZBDD size of f_n for the variable ordering $x_1, \ldots, x_n, y_1, \ldots, y_n$.

8.4.M Prove that the difference of the OBDD size and the ZBDD size of a symmetric function $f \in B_n$ is bounded by $\pm O(n)$.

8.5.O Investigate the ZBDD size of the multiplexer and arbitrary variable orderings. Is it possible to obtain subquadratic size?

8.6.E Design an algorithm for the swap operation on ZBDDs.

8.7.M Estimate the maximal size increase caused by the jump operations on ZBDDs.

8.8.E Design a linear-time redundancy test on ZBDDs.

8.9.D Prove that the replacement of a variable by the constant 0 can cause an increase of the ZBDD size by a factor of almost 3/2.

8.10.E Define the semantics of ZBDDs and OFDDs in three different ways as in Definition 1.1.1 for BDDs or OBDDs.

8.11.M Define FFDDs and generalize Proposition 8.2.3 to (complete) FBDDs and FFDDs of minimal size.

8.12.D Determine the π-OFDD size for addition and the variable orderings $x_{n-1}, y_{n-1}, \ldots, x_0, y_0$ and $x_0, y_0, \ldots, x_{n-1}, y_{n-1}$.

8.13.M Prove that the OFDD size for $\oplus cl_{n,3}$ and each variable ordering is bounded by $O(N^{3/2}) = O(n^3)$.

8.14.O Determine the OFDD size of the alternating tree function (see Definition 2.1.6).

8.15.D Design a polynomial-time algorithm to compute the lexicographically largest input satisfying a π-OFDD for $\pi = id$. (Hint: Use EXOR-OBDDs; see Chapter 10.)

8.16.M Present an example of π-OFDDs G_f and G_g such that the π-OFDD size of $f \oplus g$ equals $\Theta(|G_f| \cdot |G_g|)$. (Hint: Use the construction in the proof of Theorem 3.3.7 as a syntactic model.)

8.17.M (See Becker, Drechsler, and Werchner (1995).) Give examples of polynomial-size OFDDs such that replacement by functions and quantification lead to π-OFDDs of exponential size.

8.18.D Is there an algorithm for the $\wedge$-synthesis of π-OFDDs which needs polynomial time with respect to the input size and the size of the reduced π-OFDD representing the result? (Hint: Use EXOR-OBDDs; see Chapter 10.)

8.19.E Describe an algorithm to change a dt-OKFDD for f into a dt'-OKFDD for f, where $dt'_j = dt_j$ for all j except $j = i$.

Chapter 9

Integer-Valued DDs

BDDs work with Boolean variables, Boolean edge labels, and Boolean sink labels. In this chapter, we investigate what can be gained by different generalizations to integer-valued labels and variables. We distinguish *bit-level* DDs representing functions with Boolean outputs and *word-level* DDs representing functions whose outputs can be arbitrary integers.

9.1 Multivalued Decision Diagrams (MDDs)

Variables which can take a finite number r of different values are called *multivalued variables*. The case $r = 2$ is the special case of a Boolean variable.

Definition 9.1.1. A *multivalued decision node* for a multivalued variable x_i, taking values from the finite set A_i, is a node with label x_i and $r_i = |A_i|$ outgoing edges labeled by the different values from A_i. An MDD on $X_n = \{x_1, \ldots, x_n\}$, where x_i may take values from A_i, is a DD whose inner nodes are multivalued decision nodes. An input $a = (a_1, \ldots, a_n) \in A_1 \times \cdots \times A_n$ activates the path starting at the source and choosing the a_i-edges leaving x_i-nodes.

For a function $f\colon A_1 \times \cdots \times A_n \to \{0,1\}$, we may investigate different types (ordered, free, etc.) of MDDs. But it is also possible to replace x_i with $\lceil \log r_i \rceil$ Boolean variables and to consider BDDs for the resulting Boolean function $f \in B_m$, where $m = \lceil \log r_1 \rceil + \cdots + \lceil \log r_n \rceil$. If $\log r_i$ is not an integer, the function f is incompletely specified (see Section 3.6). Is there an advantage to using MDDs instead of BDDs? In order to compare MDDs and BDDs, we have to count the number of edges, since the outdegree of inner nodes is no longer restricted to 2. In the extreme case, each $f \in B_n$ is a function $f\colon A \to \{0,1\}$ for $A = \{0,1\}^n$ and can be represented by one node with outdegree 2^n.

If we replace a multivalued variable with an encoding using Boolean variables, there is more freedom; in particular, there are more variable orderings. A multivalued decision node with outdegree r can always be simulated by a binary DT of depth $d = \lceil \log r \rceil$ with $2^d - 1$ Boolean decision nodes and 2^d leaves. We may lose a lot by using MDDs; in particular, if a variable combines information which should be scattered in a DD of small size. The considerations above show that we do not lose too much by choosing BDDs. As examples we investigate Boolean functions on $X_n = \{x_1, \dots, x_n\}$ which are symmetric (see Definition 5.5.3) with respect to the sets $S_1, \dots, S_k$. Let $s_i = |S_i|$. Then we may replace the variables from S_i by a variable y_i taking values in $\{0, \dots, s_i\}$ with the interpretation that y_i represents the sum of the variables in S_i. Using symmetric variable orderings (see Section 5.8), it is more efficient to work with MDDs on $y_1, \dots, y_k$ than with the Boolean variables $x_1, \dots, x_n$. We have seen that it is not always optimal to work with symmetric variable orderings. What happens if we replace the multivalued variable y_i with Boolean variables $z_{i,0}, \dots, z_{i,t_i-1}$ for $t_i = \lceil \log(s_i + 1) \rceil$? The problem is that the original input is some $a \in \{0,1\}^n$ with the meaning that $x_j = a_j$. Hence, it is easy to evaluate an x_j-node. In order to evaluate a y_i-node, we have to sum up the variables from S_i, and we also process the full information about these variables. In order to evaluate a $z_{i,l}$-node, we also have to compute the value of y_i, but we are using only one bit of this information.

Hence, it depends on the problem whether it is better to work with MDDs. MDDs are useful if the considered function has a natural description with multivalued variables. Since it causes no problems to generalize results on BDDs to MDDs, we investigate in the following only the case of Boolean variables.

9.2 Multiterminal BDDs (MTBDDs)

Definition 9.2.1. An MTBDD is a generalized OBDD where the sinks may be labeled by integers (or even reals). It represents the function $f\colon \{0,1\}^n \to \mathbb{Z}$, where $f(a)$ is the label of the sink reached by the path activated by a.

The idea of generalized sink labels can be applied also to general BDDs, FBDDs, and all other considered BDD variants. In several considerations of the previous chapters, we have already implicitly used MTBDDs as submodules of BDDs.

Another name for an MTBDD is ADD which reflects the fact that MTBDDs or ADDs are applied for computations in algebras, while the notion MTBDD reflects the structure of the model.

Many results from the OBDD theory can easily be generalized to MTBDDs, among them Theorem 3.1.4 and Theorem 3.2.2 on the canonicity of reduced and quasi-reduced OBDDs with a fixed variable ordering and the description of the functions represented in these OBDDs (see exercises). Reduction can be performed with the same reduction rules, and also the algorithms for the variable-ordering problem can be used. Lower bound arguments for OBDDs also work for MTBDDs.

Since the image of the functions represented by MTBDDs is contained in $\mathbb{Z}$ (or $\mathbb{R}$), Boolean operations are no longer applicable. Instead of Shannon's decomposition rule, we use *Boole's decomposition rule* $f = (1-x_i)f_{|x_i=0}+x_i f_{|x_i=1}$, where $+$ stands for addition instead of OR and multiplication replaces AND. The synthesis problem is defined for operations on $\mathbb{Z}$ like addition, subtraction, multiplication, min (minimum), and max (maximum). For these operations, we apply the frame of the π-OBDD synthesis algorithm with the new interpretation that a node (v, w), where v and w are sinks, is a sink with label $\mathit{label}(v) \otimes \mathit{label}(w)$ for the considered operator $\otimes$. Moreover, we have to choose appropriate terminal cases.

Altogether, we do not need a new theory for MTBDDs. We may even ask whether MTBDDs are useful at all. Since $\{0,1\}^n$ is finite, there is a parameter m for each $f\colon \{0,1\}^n \rightarrow \mathbb{Z}$ such that the image of f is contained in $\mathbb{Z}(m) := \{0, \pm 1, \ldots, \pm(2^m - 1)\}$. Such functions can be described as functions $f^* \in B_{n,m+1}$, where the last output bit describes the sign of the output (for the sake of uniqueness zero has a positive sign) and the other outputs describe the binary representation of the absolute value of the output. We call f^* the *bit variant* of f (one may also use other representations of integers such as two's-complement). The bit variant of f can be represented by π-OBDDs (with shared nodes).

Proposition 9.2.2. *If $f\colon \{0,1\}^n \rightarrow \mathbb{Z}(m)$ can be represented by a π-MTBDD of size s, its bit variant can be represented by a π-OBDD of size $s(m+1)$.*

Proof. Let G be the π-MTBDD of size s. We obtain a π-OBDD for the ith output f_i^* of the bit variant f^* by identifying all sinks whose labels have the same sign (for the last output) or the same bit at position i of the binary representations of their absolute values. □

The bit representation is at most a little bit larger than the given MTBDD but can be much smaller. Let *binrep*: $\{0,1\}^n \rightarrow \{0,\ldots,2^n-1\}$ compute $a_0 2^0 + a_1 2^1 + \cdots + a_{n-1}2^{n-1}$ on input $a = (a_0, \ldots, a_{n-1})$. The bit representation only needs n inner nodes representing $x_0, \ldots, x_{n-1}$ while an MTBDD needs 2^n different sinks and, therefore, $2^n - 1$ inner nodes. We conclude that MTBDDs are of small size only if the number of different sink labels or, equivalently, the size of the image of f is small. Then MTBDDs can be a little bit smaller than π-OBDDs for the bit representation. The main advantage of MTBDDs is that operations like addition or min can be performed with one binary synthesis operation. It is more complicated if f and g are represented by π-OBDDs for their bit representations and if we want to represent the sum $f+g$ by its bit representation. The carry bit essentially depends on all bits of

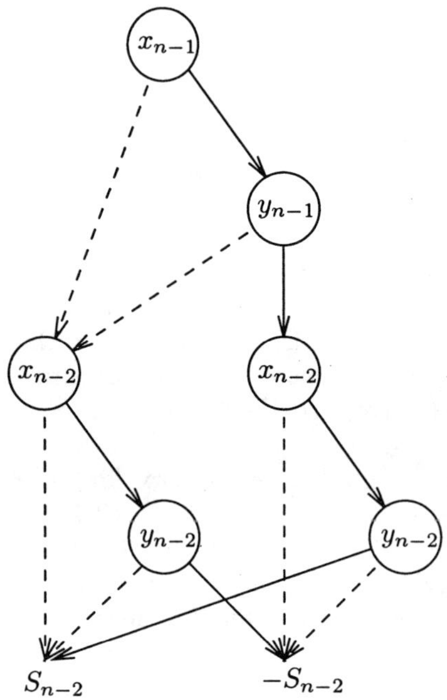

Figure 9.2.1: *An MTBDD for the Sylvester matrix* S_n.

f and g. Either we perform a synthesis step with many inputs or we simulate a circuit for addition. Hence, MTBDDs are a more natural type of representation if we want to perform operations on integers.

MTBDDs can represent $2^n \times 2^m$ matrices M with $n + m$ variables where $x_0, \ldots, x_{n-1}$ describe the row number and $y_0, \ldots, y_{m-1}$ describe the column number, i.e., $M(x, y)$ is the matrix entry at position (x, y). If the number of rows or columns is not a power of 2, we may add dummy rows or columns whose entries have to be chosen appropriately. Matrices describe linear transformations, systems of linear equations, and graphs with edge weights.

Some well-known matrices with many applications have very small MTBDD size. The Sylvester matrix S (see Definition 7.6.7), also known as the Walsh matrix, is defined recursively by

$$S_0 = [1] \quad \text{and} \quad S_n = \begin{bmatrix} S_{n-1} & S_{n-1} \\ S_{n-1} & -S_{n-1} \end{bmatrix}.$$

For the representation of S_n, we use the interleaved variable ordering $x_{n-1}, y_{n-1}, \ldots, x_0, y_0$ and obtain the MTBDD of Fig. 9.2.1 whose size is $4n$.

The addition of matrices can be performed as binary synthesis with addition as an operator on the MTBDDs representing the matrices. The same holds if we want to compute $D = (d_{ij})$, where $d_{ij} = a_{ij}b_{ij}$, from $A = (a_{ij})$ and $B = (b_{ij})$. But the matrix product $C = (c_{ij})$ of A and B is defined in a more complicated way, namely $c_{ij} = \sum_k a_{ik}b_{kj}$. Matrix multiplication is an essential operation and there are different methods of decomposing this "complicated" operation.

If we want to multiply a $2^n \times 2^m$ matrix A and a $2^m \times 2^k$ matrix B, the matrices may be described on disjoint sets of variables. Because of the definition of matrix multiplication, we identify the variables describing the columns of A and the variables describing the rows of B, denoted by $A_{x,z}$ and $B_{z,y}$. The product is defined on x and y, i.e., $C = C_{x,y}$.

Clarke et al. (1997) propose the following approach for matrix multiplication. In a first step, A and B are considered as MTBDDs on (x, y, z) and with the operator multiplication an MTBDD for $f(x, y, z) = A_{x,z} * B_{z,y}$ is computed. Afterwards, we consider the subfunctions for $z_{m-1} = 0$ and $z_{m-1} = 1$ and compute the sum of the resulting functions, which are considered as functions which do not depend (also not syntactically) on z_{m-1}. This process is repeated for $z_{m-2}, \ldots, z_0$.

Fujita, McGeer, and Yang (1997) follow the school method for matrix multiplication. They assume that $k = m = n$ and use the interleaved variable ordering $x_{n-1}, y_{n-1}, z_{n-1}, \ldots, x_0, y_0, z_0$. The eight assignments to x_{n-1}, y_{n-1}, and z_{n-1} split A, B, and C into four $2^{n-1} \times 2^{n-1}$ matrices (C depends on (x, y), A on (x, z), B on (z, y)):

$$\begin{bmatrix} C_{00} & C_{01} \\ C_{10} & C_{11} \end{bmatrix} = \begin{bmatrix} A_{00} & A_{01} \\ A_{10} & A_{11} \end{bmatrix} * \begin{bmatrix} B_{00} & B_{01} \\ B_{10} & B_{11} \end{bmatrix}.$$

C_{00} is computed by recursive calls to compute $A_{00} * B_{00}$ and $A_{01} * B_{10}$ and a call to compute the sum of the results, similarly for C_{01}, C_{10}, and C_{11}. Then the resulting MTBDD starts with a tree of depth 2 testing x_{n-1} and y_{n-1} and the four leaves point to $C_{00}, \ldots, C_{11}$. Fujita, McGeer, and Yang (1997) work with quasi-reduced MTBDDs and not with reduced ones. The reason is the following. If A and B are represented by a 1-sink, they are matrices whose entries are all 1. Then $A + B$ is a matrix whose entries are all 2. Also $A * B$ is a matrix whose entries all have the same value but it depends on the dimension of the matrices which is the right value. In the general case the value is 2^m. Hence, we also have to take into account the omitted variables.

Bahar, Frohm, Gaona, Hachtel, Macii, Pardo, and Somenzi (1997) follow another approach. They can work with arbitrary variable orderings and use two recursive calls with respect to the two values of the considered variable. At the end of the recursive calls for an x_i-variable, an x_i-node is created to point to the results of the recursive calls (if the node cannot be eliminated). The same is done for y_i-variables. For z_i-variables, we call the algorithm for the addition of the two matrices computed during the recursive calls. Since we work with reduced MTBDDs, we have to store the number of z-variables which lie in the variable ordering between the z-variable used in the recursive call and the first variable considered in the recursive call. If this number is p, we have 2^p equal blocks and have to scale up the result of the recursive call by the factor 2^p. This is done by a simple algorithm to multiply a matrix by a scalar. The following terminal cases are used. If one matrix is the constant 0, the result is the constant 0. If both matrices are constant matrices, the result is a constant matrix whose entries have the value which is the product of the A-entry, the B-entry, and the scaling factor.

Based on these modules, algorithms for the LUP factorization of matrices (Fujita et al. (1997)) and the solution of systems of linear equations by Gaussian elimination (Bahar et al. (1997)) have been developed and applied. For matrices with a few hundred rows and columns, conventional algorithms, e.g., for multiplication of sparse matrices, are faster than MTBDD-based algorithms. The MTBDD approach becomes superior for very large matrices with a simple structure.

We have described matrix multiplication over $\mathbb{Z}$ or $\mathbb{R}$, which are rings with respect to addition and multiplication. But the algorithms work for all semirings. Let us consider the computation of shortest paths in weighted graphs. If the matrix $A[k] = (a_{ij}^k)$ contains the lengths of shortest paths using at most 2^k edges, the entry a_{ij}^{k+1} of $A[k+1]$ is the minimum of all $a_{im}^k + a_{mj}^k$. For graphs on 2^n vertices it is sufficient to compute $A[n]$ (this approach is known as the Bellman–Ford algorithm (see Cormen, Leiserson, and Rivest (1990))). The matrix $A[k+1]$ is the square of $A[k]$, where the matrix multiplication uses the addition of numbers as "multiplication" and the operation min as "addition." Moreover, min is associative and commutative and the distributivity law $a + \min\{b, c\} = \min\{a+b, a+c\}$ holds. Hence, we may use the considered algorithms for matrix multiplication. A scaling is not necessary, since scaling replaces the "addition" of identical matrices. Here, min is used as addition and $\min\{a, a\} = a$.

Minato (1997) has described a package of algorithms to work with MTBDDs and OBDDs for the corresponding bit variants. His applications cover some combinatorial problems, the timing analysis of logic circuits, and scheduling problems in data path synthesis.

9.3 Binary Moment Diagrams (BMDs)

Boole's decomposition rule is the arithmetization of Shannon's decomposition rule. The result of an arithmetization of Reed-Muller's decomposition rule is $f = f_{|x_i=0} + x_i(f_{|x_i=1} - f_{|x_i=0})$. Its correctness and uniqueness follow by the consideration of the cases $x_i = 0$ and $x_i = 1$. We obtain the decomposition of f into its *constant moment* $f_{|x_i=0}$ and its *linear moment* $f_{|x_i=1} - f_{|x_i=0}$. This terminology is based on the consideration of f as a linear function with respect to x_i. Using the field $\mathbb{Z}_2$, addition and also subtraction equals EXOR

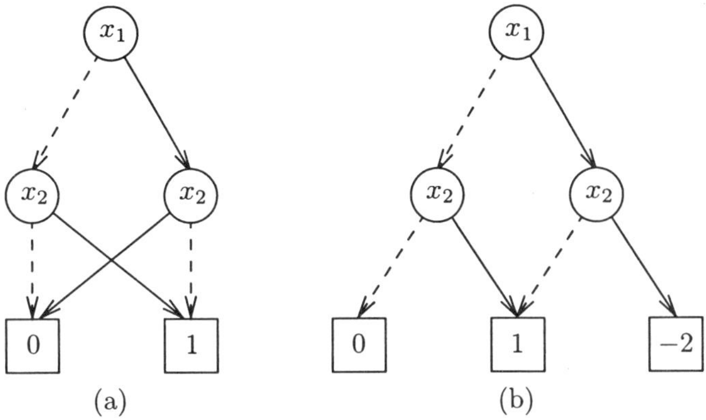

Figure 9.3.1: *BMDs for* $x_1 \oplus x_2$ *over* (a) $\mathbb{Z}_2$ *and* (b) $\mathbb{Z}$.

and we obtain Reed-Muller's decomposition rule. Bryant and Chen (1995) have introduced BMDs as generalizations of OFDDs in the same way as MTBDDs are generalizations of OBDDs.

Definition 9.3.1. A BMD over the ring R shares its syntax with MTBDDs where the sinks may be labeled by elements from R. A sink with label c represents the constant c. If f_0 and f_1 are represented at the 0-, respectively, 1-successor of the x_i-node v, the node v represents the function $f\colon \{0,1\}^n \to R$, where $f = f_0 + x_i f_1$ and the computations are performed in R.

We restrict our attention to the rings $\mathbb{Z}$, $\mathbb{Z}_m$, $\mathbb{Q}$, and $\mathbb{R}$. The choice of the ring is important. As an example we investigate BMDs for $x_1 \oplus x_2$ over $\mathbb{Z}_2$ and $\mathbb{Z}$ (see Fig. 9.3.1).

Both BMDs are of minimal size. For Boolean functions, all computations can be done in $\mathbb{Z}_2$ instead of $\mathbb{Z}$. Then $-2 \equiv 0 \mod 2$ and the 0-sink and the (-2)-sink can be merged and we obtain the BMD over $\mathbb{Z}_2$. For Boolean functions we cannot gain anything by choosing larger rings than $\mathbb{Z}_2$ and then BMDs are OFDDs (remember that MTBDDs for Boolean functions are OBDDs).

Many results on OFDDs can be generalized to BMDs without additional effort. For a fixed ring and a fixed variable ordering π, the representation by π-BMDs is canonical. The reduction of π-BMDs is possible by the application of the merging rule and the OFDD elimination rule. In order to compute a representation for $f + c$ from a representation for f, it is sufficient to construct a BMD with isolated 0-path (see the proof of Theorem 8.2.9) and to replace the label d of the sink reached by this path by the label $c + d$. The synthesis of two BMDs is well defined for the addition $+$ and the multiplication $*$ defined for the ring. Let f and g be represented by π-BMDs whose sources are x_1-nodes. Then the successors represent $f_0 = f_{|x_1=0}$, $f_1 = f_{|x_1=1} - f_{|x_1=0}$, g_0, and g_1. Since $x_1 \in \{0,1\}$, $x_1^2 = x_1$ in all rings,

$$\begin{aligned} f + g &= (f_0 + x_1 f_1) + (g_0 + x_1 g_1) \\ &= (f_0 + g_0) + x_1(f_1 + g_1) \text{ and} \\ f * g &= (f_0 + x_1 f_1) * (g_0 + x_1 g_1) \\ &= f_0 * g_0 + x_1(f_0 * g_1 + f_1 * g_0 + f_1 * g_1). \end{aligned}$$

This implies that the synthesis for addition can be performed with the usual synthesis algorithm while the synthesis for multiplication is more difficult. For multiplication, four recursive calls are necessary and, for the 1-successor, three of the results have to be combined by addition. We have seen for π-OFDDs, i.e., BMDs over $\mathbb{Z}_2$, that one synthesis step may cause an exponential size blow-up.

A function $f\colon \{0,1\}^n \to \mathbb{Z}$ (or $f\colon \{0,1\}^n \to \mathbb{Z}_m$) can be represented as a π-BMD over the corresponding ring, and its bit variant (see Section 9.2) can be represented as a π-OFDD (with many sources). In Proposition 9.2.2, we proved the fact that small π-MTBDD size leads to small π-OBDD size for the bit variant. The proof is very simple, since each input activates one computation path. This proof cannot be generalized to the π-BMD/π-OFDD situation, since in this case computations are performed on the labels of the sinks reached by the paths activated by an input. On the contrary, there are examples where the π-OFDD size of the bit variant is exponentially larger than the π-BMD size of the function. Hence, the gain by the use of word-level DDs instead of bit-level DDs is limited for Shannon's decomposition rule while it may cause an exponential size decrease for Reed-Muller's decomposition rule.

Theorem 9.3.2. *The function word-level multiplication $MUL_n^w\colon \{0,1\}^{2n} \to \mathbb{Z}$ has BMDs of size $O(n^2)$ while its bit variant has exponential OFDD size.*

Proof. The result on the OFDD size of the bit variant of MUL_n^w, which is the ordinary multiplication MUL_n, has already been stated in Section 8.2 and will be proved in Chapter 10. For BMDs, we choose the variable ordering $x_{n-1}, \ldots, x_0, y_{n-1}, \ldots, y_0$ (see Fig. 9.3.2 for the case $n = 3$).

In the general case, we have n y_{n-1}-nodes which are sources of π-BMDs representing $2^i|y|$, $0 \le i \le n-1$. For this purpose, we need n^2 y-nodes and $2n$ sinks, one labeled by 0 and the others by 2^j, $0 \le j \le 2n-2$. We can describe the multiplication by

$$\begin{aligned} \mathrm{MUL}_n^w(x,y) &= \sum_{0 \le i \le n-1} x_i * (2^i|y|) \\ &= \sum_{0 \le i \le n-2} x_i * (2^i|y|) + x_{n-1} * (2^{n-1}|y|). \end{aligned}$$

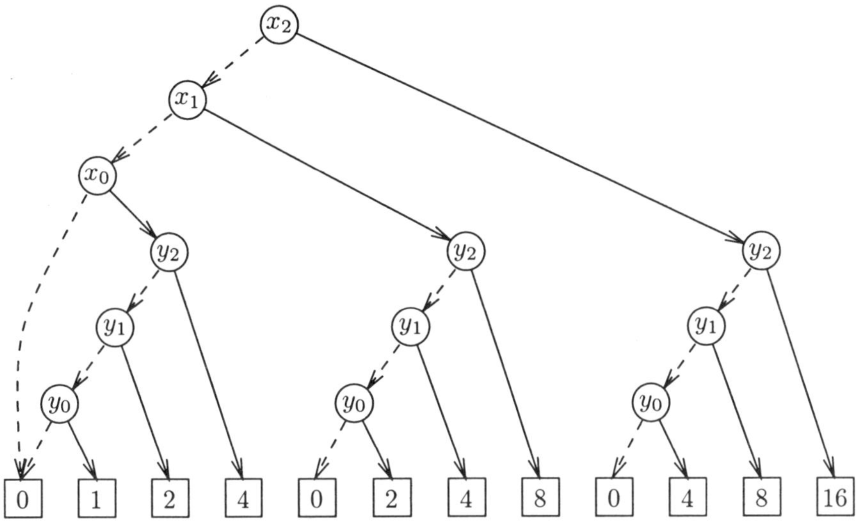

Figure 9.3.2: *A BMD for MUL_3^w. For a better overall view, sinks with the same label are not merged.*

This decomposition with respect to x_{n-1} leads to the π-BMD representation with an x_{n-1}-source whose 0-successor represents the multiplication of $(x_{n-2}, \ldots, x_0)$ with $y = (y_{n-1}, \ldots, y_0)$ and whose 1-successor represents $2^{n-1}|y|$. Hence, n further x-nodes are sufficient. Altogether, the π-BMD has size n^2+3n. □

We will discuss in Chapter 13 how this result can be used in circuit verification. A circuit is a representation on the bit level. During a gate-by-gate transformation, we obtain a representation of the bit variant of multiplication which needs exponential size and then we can obtain the small π-BMD by combining the bits $z_{2n-1}, \ldots, z_0$ of the result by $z_0 + 2z_1 + \cdots + 2^{2n-1}z_{2n-1}$. A small-size word-level DD does not lead directly to an efficient verification algorithm.

In the following, we derive relations between BMDs and MTBDDs which are similar to the relations between OFDDs and OBDDs based on the τ-operator and presented in Section 8.2. Let G be a complete π-DD, where $\pi = id$, representing g as a π-MTBDD. The sink labels are from some ring R. By f we denote the function which is represented by G as π-BMD. For $n = 1$ and $n = 2$, we obtain the following relations:

$$\begin{bmatrix} f(0) \\ f(1) \end{bmatrix} = \begin{bmatrix} 1 & 0 \\ -1 & 1 \end{bmatrix} \begin{bmatrix} g(0) \\ g(1) \end{bmatrix},$$

$$\begin{bmatrix} f(0,0) \\ f(0,1) \\ f(1,0) \\ f(1,1) \end{bmatrix} = \begin{bmatrix} 1 & 0 & 0 & 0 \\ -1 & 1 & 0 & 0 \\ -1 & 0 & 1 & 0 \\ 1 & -1 & -1 & 1 \end{bmatrix} \begin{bmatrix} g(0,0) \\ g(0,1) \\ g(1,0) \\ g(1,1) \end{bmatrix}.$$

Clarke, Fujita, and Zhao (1995b) have observed how the transformation matrices T_n (for functions $f \in B_n$) can be described concisely.

Definition 9.3.3. The *Kronecker product* of two matrices A and B is defined by

$$A \otimes B = \begin{bmatrix} a_{11} & \dots & a_{1n} \\ \vdots & & \vdots \\ a_{m1} & \dots & a_{mn} \end{bmatrix} \otimes B = \begin{bmatrix} a_{11}B & \dots & a_{1n}B \\ \vdots & & \vdots \\ a_{m1}B & \dots & a_{mn}B \end{bmatrix}.$$

(If A is an $m \times n$ matrix and B an $m' \times n'$ matrix, $A \otimes B$ is an $mm' \times nn'$ matrix.)

The above example shows that $T_2 = T_1 \otimes T_1$.

Lemma 9.3.4. *$T_n = T_1^{\otimes n} := T_1 \otimes \cdots \otimes T_1$ (n factors).*

Proof. The proof is done by induction on n, and the case $n = 1$ is obvious. The induction hypothesis can be applied to the complete π-DDs G_0 and G_1 whose sources are the 0-successor and the 1-successor of the source of G. Then $f_0 = T_1^{\otimes(n-1)} g_0$ and $f_1 = T_1^{\otimes(n-1)} g_1$. We get $f_{|x_1=0} = 1 * f_0 + 0 * f_1$ and $f_{|x_1=1} = (-1) * f_0 + 1 * f_1$. The list of all $f(a)$, $a \in \{0,1\}^n$, is the concatenation of the list of all $f_{|x_1=0}(a)$, $a \in \{0,1\}^n$ and $a_1 = 0$, and the list of all $f_{|x_1=1}(a)$, $a \in \{0,1\}^n$ and $a_1 = 1$; the same holds for g. Putting all these relations together we conclude that $T_n = T_1 \otimes T_1^{\otimes(n-1)} = T_1^{\otimes n}$. □

In order to obtain $f(a)$ from all values of g, it is sufficient to compute the inner product of the row of T_n corresponding to a and the value table of g. In the case of the field $\mathbb{Z}_2$, the matrix T_1 is equal to $T_1' = \left[\begin{smallmatrix} 1 & 0 \\ 1 & 1 \end{smallmatrix}\right]$. The τ-operator (see Definition 8.2.2) describes nothing more than the inner product of the a-row of $T_n' = (T_1')^{\otimes n}$ and the value table of g. The fact $\tau\tau g = g$ follows from the easy fact that the inverse of $T_1^{\otimes n}$ equals $\left(T_1^{-1}\right)^{\otimes n}$. In the general case, $T_1 = \left[\begin{smallmatrix} 1 & 0 \\ -1 & 1 \end{smallmatrix}\right]$ and $T_1^{-1} = \left[\begin{smallmatrix} 1 & 0 \\ 1 & 1 \end{smallmatrix}\right]$ and for the special case of $\mathbb{Z}_2$ we get $T_1 = T_1^{-1}$.

We conclude that the quasi-reduced π-MTBDD for g is isomorphic to the quasi-reduced π-BMD for $T_n g$, where g is given as a column vector representing its value table. The reduced π-BMD for g is isomorphic to the quasi-reduced π-MTBDD for $T_n^{-1} g$. Hence, lower and upper bounds for the π-BMD size of g can be obtained by proving bounds for the π-MTBDD size of $T_n^{-1} g$.

9.4 Hybrid Decision Diagrams (HDDs)

In Section 8.3, we discussed a common generalization of π-OBDDs and π-OFDDs, namely π-OKFDDs, where for each variable one of three possible decomposition types is chosen. Clarke, Fujita, and Zhao (1995a) followed the same approach for word-level DDs. If $A_1 = A$ is a regular 2×2 matrix over the ring R, we may use A_1 as the base of a decomposition. An x_1-node pointing with its 0-successor to a c_0-sink and with its 1-successor to a c_1-sink represents the function f defined by $f(0) = a_{11}c_0 + a_{12}c_1$ and $f(1) = a_{21}c_0 + a_{22}c_1$. Since A is regular, this decomposition is unique and each function $f \in B_1$ can be represented. Then we take $A_n = A_1^{\otimes n}$ as a transformation matrix, e.g., a complete π-DD with sink labels from the ring R, and representing g as π-MTBDD represents $f = A_n g$ as a π-DD based on A. We obtain an HDD if we allow different transformation matrices for the different variables. If we choose $\pi = id$ and the matrix S_i as the transformation matrix for x_i, then, for a complete π-DD, the Kronecker product $S_1 \otimes \cdots \otimes S_n$ describes the transformation of the interpretation as π-MTBDD to the interpretation as $(S_1, \ldots, S_n)$-π-HDD. Again, we have the difficulty of choosing suitable transformation matrices for the variables.

The number of different regular 2×2 matrices over $\mathbb{Z}$ is infinite. If we restrict ourselves to matrix entries from $\{-1, 0, 1\}$, we obtain the following six essentially different transformation matrices: $\left[\begin{smallmatrix}1&0\\0&1\end{smallmatrix}\right]$, $\left[\begin{smallmatrix}1&0\\-1&1\end{smallmatrix}\right]$, $\left[\begin{smallmatrix}1&0\\1&1\end{smallmatrix}\right]$, $\left[\begin{smallmatrix}0&1\\-1&1\end{smallmatrix}\right]$, $\left[\begin{smallmatrix}0&1\\1&1\end{smallmatrix}\right]$, $\left[\begin{smallmatrix}1&1\\-1&1\end{smallmatrix}\right]$.

OKFDDs are the special case of HDDs for the field $\mathbb{Z}_2$. Shannon's decomposition rule can be described with $\oplus$ instead of $+$ (see Definition 1.1.1). Then all three decomposition types are described over $\mathbb{Z}_2$. If we consider the six matrices above as $\mathbb{Z}_2$-matrices, the second matrix is equal to the third one, the fourth matrix is equal to the fifth one, and the last matrix is not regular. Hence, we obtain three different transformation matrices corresponding to the three decomposition types of Section 8.3. The first matrix corresponds to the Shannon type, the third one to the positive Reed-Muller type, and the fifth one to the negative Reed-Muller type. Hence, the value table of the function represented by a complete π-DD interpreted as a π-OKFDD with decision list dt can be obtained as the product of the transformation matrix, which is the Kronecker product of the involved transformation matrices, and the value table of the function represented by the π-DD as a π-MTBDD. This explains the choice of the notion OKFDD in Section 8.3.

9.5 Edge-Valued Binary Decision Diagrams (EVBDDs)

In Section 9.2, we have seen that the very simple function *binrep*: $\{0,1\}^n \rightarrow \{0, \ldots, 2^n - 1\}$ computing $a_0 2^0 + a_1 2^1 + \cdots + a_{n-1} 2^{n-1}$ on $a = (a_0, \ldots, a_{n-1})$ needs exponential MTBDD size. Using the variable ordering $x_{n-1}, \ldots, x_0$, we get, on the x_{i-1}-level, the subfunctions $x_0 2^0 + \cdots + x_{i-1}2^{i-1} + c2^i$, where $0 \le c \le 2^{n-i} - 1$. These are 2^{n-i} different subfunctions which only differ by a constant additive term. We can represent these subfunctions at the same node if we allow that the edges leading to a node may carry an additive weight. EVBDDs introduced by Lai and Sastry (1992) and Lai, Pedram, and Vrudhula (1994) are based on this idea. In order to obtain a canonical representation, they allow only one sink which is labeled by 0, additive weights only on edges labeled by 1, and additive edges to nodes representing the considered function. (For the simple generalization with additional multiplicative weights, called factored EVBDD, see Tafertshofer and Pedram (1997).)

Definition 9.5.1. An EVBDD G shares its syntax with OBDDs but it has only one sink whose label is 0. The 1-edges have an additional integer label called weight. The 0-edges implicitly have the weight 0. Additional weighted edges leading from nowhere to a node are possible. The sink represents the constant 0. An edge with weight w to a node representing g represents $w + g$. An x_i-node v, whose 0-successor represents f_0, whose 1-successor represents f_1, and whose outgoing 1-edge carries the weight w, represents the function $f = (1 - x_i)f_0 + x_i(f_1 + w)$.

To evaluate the function represented at a node v or at an edge e, we follow the path activated by the considered input a (if the function is represented at e, the edge e is activated) and compute the sum of the weights on this path. This follows directly from Definition 9.5.1.

Lemma 9.5.2. *For each function $f\colon \{0,1\}^n \to \mathbb{Z}$ and each variable ordering π, there is a unique complete π-DT with an additional edge to the source and a unique choice of the weights such that f is represented at the additional edge if the DT is interpreted as a π-EVBDD.*

Proof. Without loss of generality, $\pi = id$. The proof is done by induction on n. For $n = 0$, the input set contains only the empty word ε. The function $f(\varepsilon) = c$ is represented by an edge which leads to the sink and carries the weight c. For the induction step, we consider the path activated by the all-zero input. This input activates a path where all edges except the edge to the source carry, by definition, the weight 0. Hence, it is necessary that the edge to the source carries the weight $c = f(0,\ldots,0)$. Now we have to represent the function $g(x) := f(x) - c$ at the source. At the outgoing 0-edge we have to represent $g_{|x_1=0}$. This is possible in a unique way by the induction hypothesis even under the restriction that the first edge carries the weight 0. Here we need the fact that $g_{|x_1=0}(0,\ldots,0) = 0$. At the outgoing 1-edge, we have to represent $g_{|x_1=1}$, which is uniquely possible by the induction hypothesis. □

Which function is represented at the node v reached by the partial input $(a_1,\ldots,a_i)$, $0 \le i \le n$? By the proof of Lemma 9.5.2, this is the function $f_{|x_1=a_1,\ldots,x_i=a_i} - c(a_1,\ldots,a_i)$. The constant $c(a_1,\ldots,a_i)$ is the sum of the weights on the path to v. We can describe this constant more precisely. At nodes, we can only represent functions computing 0 on the all-zero input. Therefore, $c(a_1, \ldots, a_i) = f(a_1, \ldots, a_i, 0, \ldots, 0)$.

In a π-EVBDD, it is sufficient to represent each function only once. All the functions described above have to be represented. The EVBDD merging rule allows us to merge nodes with the same label, the same 0-successor, the same 1-successor, and the same weight on the 1-edge. Applying this rule to the complete π-DT, we obtain a complete π-EVBDD where all x_i-nodes represent different functions. By our usual arguments, it follows that this is the (up to isomorphism) unique quasi-reduced π-EVBDD for f. If an x_i-node v and an x_j-node w, where $j > i$, represent the same function $g(x_i, \ldots, x_n)$, this function does not essentially depend on x_i. Then $g_{|x_i=0} = g_{|x_i=1}$. In particular, $g(1, 0, \ldots, 0) = g(0, 0, \ldots, 0) = 0$ and the 1-edge leaving v carries the weight 0. Moreover, the node reached by the 0-edge is, in the quasi-reduced π-EVBDD, the same node as the node reached by the 1-edge. It is easy to see that this is the situation where an elimination is possible. The EVBDD elimination rule allows the elimination of nodes where both outgoing edges point to the same node and where the outgoing 1-edge carries the weight 0. Our considerations lead to the following result.

Theorem 9.5.3. *There is (up to isomorphism) a unique π-EVBDD of minimal size representing $f\colon \{0,1\}^n \to \mathbb{Z}$. This is called the reduced π-EVBDD for f. It can be obtained in linear time from each π-EVBDD (without nodes and edges not reachable from the edge representing f) by the application of the EVBDD merging rule and the EVBDD elimination rule. If $\pi = id$, it contains x_i-nodes representing the different functions $f_{|x_1=a_1,\ldots,x_{i-1}=a_{i-1}} - f(a_1, \ldots, a_{i-1}, 0, \ldots, 0)$ essentially depending on x_i.*

It is possible to replace $\mathbb{Z}$ by an arbitrary ring. For $\mathbb{Z}_2$, all weights are 0 or 1. The property $f(a) = 1$ is equivalent to the property that the number of edges with weight 1 on the path activated by a is odd. Comparing this remark with the discussion in Section 3.1, we obtain the following result.

Proposition 9.5.4. *The representation of a Boolean function $f \in B_n$ by its reduced π-EVBDD over $\mathbb{Z}_2$ is isomorphic to its representation by the reduced π-OBDD with complemented edges.*

We have seen that the only difference between MTBDDs and EVBDDs is that subfunctions differing only by a constant additive term can be represented in EVBDDs at the same node, while this is not possible in MTBDDs. The function *binrep* shows that this may cause an exponential size decrease.

Theorem 9.5.5. *The function binrep: $\{0,1\}^n \to \mathbb{Z}$, where binrep$(a_0, \ldots, a_{n-1}) = a_0 2^0 + a_1 2^1 + \cdots + a_{n-1} 2^{n-1}$, has exponential MTBDD size. Its π-EVBDD size is $n+1$. More generally, each affine function essentially depending on n variables has π-EVBDD size $n+1$.*

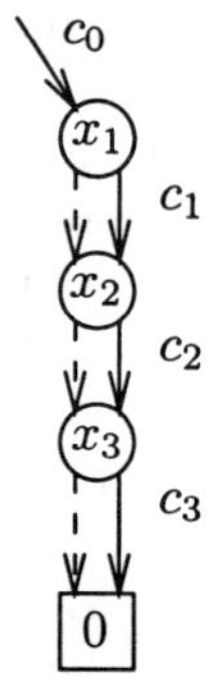

Figure 9.5.1: *An EVBDD for* $f(x_1, x_2, x_3) = c_0 + c_1x_1 + c_2x_2 + c_3x_3$.

Proof. The bound on the MTBDD size was proved in Section 9.2. The function *binrep* is linear. We consider an arbitrary affine function $c_nx_n + \cdots + c_1x_1 + c_0$ and $\pi = id$. This function is represented by a starting edge with weight c_0. The 1-edge leaving the only x_i-node has weight c_i and both edges leaving the x_i-node lead to the x_{i+1}-node if $i < n$ and to the sink otherwise (see Fig. 9.5.1). □

Theorem 9.5.5 implies that it is not possible to apply lower bound techniques based on one-way communication complexity. The reason is that, in the usual protocol, Alice has to send the first node of the activated path lying in Bob's part of the π-EVBDD and, additionally, the sum of the weights on the part of the activated path belonging to her part of the EVBDD. Such an approach can lead to good lower bounds only for EVBDDs with limited weights. In the general case, we may apply Theorem 9.5.3.

Theorem 9.5.6. *The EVBDD size of the function word-level multiplication* MUL_n^w *is bounded below by* 2^n.

Proof. Let π be an arbitrary variable ordering and w.l.o.g. let y_j be the last variable with respect to π. We consider the subfunction where $y_k = 0$ for all $k \neq j$. Then the subfunctions with respect to the 2^n assignments to x are ay_j2^j, $0 \leq a \leq 2^n - 1$. The difference of two of these subfunctions is of the type $(a - b)y_j2^j$ and, if $a \neq b$, not a constant. Hence, the lower bound follows from Theorem 9.5.3. □

In the following, we discuss the algorithmic properties of π-EVBDDs. Since reduced π-EVBDDs are a canonical representation, the satisfiability test (is there an input a with $f(a) \neq 0$?) and the equivalence test can be performed efficiently. In the following, we investigate the synthesis problem for binary operators, in particular, addition and multiplication. Later, in Section 15.1 we present an integer-programming solver using EVBDDs as representation type. This requires further operations on EVBDDs. Lai, Pedram, and Vrudhula (1996) describe a general synthesis algorithm. This algorithm is based on a simultaneous DFS traversal of the given π-EVBDDs G_f and G_g. For π-OBDDs, it is sufficient to remember the pair of nodes (v, w) reached in both graphs. Here we have to take into account the weights "already seen." A *situation* is described by $((c, v), (d, w))$ with the interpretation that we have reached v in G_f and have seen the weight c, similarly for (d, w) and G_g. The initial situation consists of the weights on the starting edges and the sources. Depending on the operator, we have terminal cases, among them always $((c, s_f), (d, s_g))$ for the sinks s_f and s_g. Then the result is $(c \otimes d, (s_f, s_g))$, where $\otimes$ is the considered operation and (s_f, s_g) is the sink of the resulting π-EVBDD G_h. The computed-table and the unique-table are managed with respect to situations instead of node pairs as in the OBDD case. Let us consider the case that $((c, v), (d, w))$ is not contained in the computed-table. If *label*(v) is in the variable ordering behind *label*(w), we "wait" in G_f, i.e., both the 0-successor and 1-successor are (c, v). Otherwise, we compute the successor (c_0, v_0) and (c_1, v_1) where $c_0 = c$, v_0 is the 0-successor of v, c_1 is the sum of c and the weight on the 1-edge leaving v, and v_1 is the 1-successor of v. The pair (d, w) is treated in a similar way. The difference with the π-OBDD case is that we add the weight on the 1-edge to c. As one may see for the example of *binrep*, there may be exponentially many pairs $(\cdot, v)$ which we have to consider. If we reach the sink, we consider the pair (c, s_f), where c equals the label of the sink which is reached by the same input for the corresponding π-MTBDD. Implicitly, we run through an "unfolding of the π-EVBDD" which is the corresponding π-MTBDD. Since we always consider one path, we never store the π-MTBDD explicitly. The synthesis algorithm is applied recursively to $((c_0, v_0), (d_0, w_0))$ and $((c_1, v_1), (d_1, w_1))$. Let the results be $(b_0, (v_0, w_0))$ and $(b_1, (v_1, w_1))$. If the results are equal, we apply the elimination rule and return $(b_0, (v_0, w_0))$. Otherwise, we create a new node whose label is the smaller of the labels of v and w with respect to π. The 0-successor is $(0, (v_0, w_0))$, since 0-edges have to carry the weight 0. The 1-successor is defined as $(b_1 - b_0, (v_1, w_1))$ in order to have the same "missing" weight b_0. This missing weight is transferred to the incoming edge. The result is $(b_0, (v, w))$. This synthesis algorithm returns the reduced π-EVBDD for $h = f \otimes g$.

Theorem 9.5.7. *The synthesis of two π-EVBDDs with respect to multiplication may lead to an exponential size blow-up.*

Proof. The π-EVBDD size of *binrep*(x) and *binrep*(y) is linear. Applying multiplication to these π-EVBDDs, the result is the word-level multiplication which needs, by Theorem 9.5.6, exponential EVBDD size. □

We discuss multiplication on π-EVBDDs in more detail. The functions represented at x_1-nodes of G_f and G_g can be described and multiplied in the following way:

$$[(1-x_1)f_0 + x_1(f_1+c)] * [(1-x_1)g_0 + x_1(g_1+d)]$$
$$= (1-x_1)f_0g_0 + x_1(f_1g_1 + cg_1 + df_1 + cd).$$

The situation for the 0-successor is easy. For the 1-successor, we implicitly need three recursive calls to compute f_1g_1, cg_1, and df_1. Although the last two calls are special cases where one term is constant, we have to compute the sum of three results. Hence, with regard to similar results for π-OFDDs and π-BMDs, it is not surprising that this leads to an exponential size blow-up. The situation for addition seems to be easier:

$$[(1-x_1)f_0 + x_1(f_1+c)] + [(1-x_1)g_0 + x_1(g_1+d)]$$
$$= (1-x_1)(f_0+g_0) + x_1(f_1+g_1+c+d).$$

Theorem 9.5.8. *The synthesis of two π-EVBDDs G_f and G_g with respect to addition leads to a π-EVBDD G_h whose size is bounded by $|G_f||G_g|$ and which can be computed in time $O(|G_f||G_g|)$.*

Proof. An x_i-node v of G_f is reached for partial inputs $(a_1, \ldots, a_{i-1})$ such that the subfunctions $f_{|x_1=a_1,\ldots,x_{i-1}=a_{i-1}}$ differ only by a constant additive term. A similar remark holds for an x_j-node w of G_g. If we reach, for some partial inputs, v in G_f and w in G_g, the corresponding subfunctions of $f+g$ also differ only by a constant additive term. All these subfunctions are represented in the π-EVBDD for $h = f+g$ at the same node; see Theorem 9.5.3. We obtain the result for the runtime by a simple modification of the synthesis algorithm. The recursive calls for $((c,v), (d,w))$ are replaced by calls for (v,w). The computed-table and the unique-table are managed with respect to node pairs instead of situations. At the end of the recursive call, we may add $c+d$ to the resulting weight. □

9.6 Edge-Valued Binary Moment Diagrams (*BMDs)

Bryant and Chen (1995) have also noticed the advantage of edge weights and have presented *BMDs as BMDs with multiplicative edge weights.

Definition 9.6.1. A *BMD (or *multiplicative* BMD) over the ring R is a BMD whose edges may carry weights from R. A sink with label $c \in R$ represents the constant c. If f_0 and f_1 are represented at the 0-, respectively, 1-successor of the x_i-node v and if w_0 and w_1 are the corresponding edge weights, the node v represents the function $f\colon \{0,1\}^n \to R$ where $f = w_0f_0 + x_iw_1f_1$.

For Boolean functions, we may perform the computations in $\mathbb{Z}_2$. Odd weights can be replaced with 1 and even weights with 0, i.e., the corresponding edge may point to the 0-sink. Then we obtain BMDs and even OFDDs. Hence, Boolean functions with exponential OFDD size also have exponential *BMD size. Bryant and Chen (1995) have shown that *BMDs can represent word-level functions like multiplication, squaring, and exponentiation in small size.

Theorem 9.6.2.

(*i*) *The function word-level multiplication has *BMDs of linear size for the variable orderings* (*a*) $x_{n-1}, \ldots, x_0, y_{n-1}, \ldots, y_0$, (*b*) $x_{n-1}, y_{n-1}, \ldots, x_0, y_0$, *and* (*c*) $x_0, y_0, \ldots, x_{n-1}, y_{n-1}$.

(*ii*) *The function word-level squaring has *BMDs of size* $O(n^2)$.

(*iii*) *The function word-level exponentiation* $x \to 2^{|x|}$ *has linear-size *BMDs.*

Proof. The last result is the simplest. We choose the variable ordering $x_{n-1}, \ldots, x_0$. Let $|X_i| = x_i 2^i + \cdots + x_0 2^0$. Then $2^{|x|} = 2^{|X_{n-1}|}$. For $n = 1$, we get $2^{|x|} = 1 + x_0$, which can be realized with two nodes. For the induction step, we obtain

$$2^{|X_{n-1}|} = 2^{|X_{n-2}| + x_{n-1}2^{n-1}} = 2^{|X_{n-2}|} + x_{n-1} 2^{|X_{n-2}|}\left(2^{2^{n-1}} - 1\right).$$

This is a representation suitable for *BMDs. We create an x_{n-1}-node. Both outgoing edges point to the node representing $2^{|X_{n-2}|}$, the 0-edge gets the weight 1, and the 1-edge gets the weight $2^{2^{n-1}} - 1$. The total *BMD size is $n + 1$.

For multiplication and the variable ordering $x_{n-1}, \ldots, x_0, y_{n-1}, \ldots, y_0$, we start with the BMD of Fig. 9.3.2. The subgraphs whose sources are the n y_{n-1}-nodes are isomorphic with the exception that the sink labels are doubled from one subgraph to the next. In *BMDs, the leftmost y-subgraph is sufficient. The edge to the y-subgraph whose sink labels are $0, 2^i, \ldots$ is replaced with an edge to the remaining y-subgraph with weight 2^i. Then the size is $3n + 1$, since we have n x-nodes, n y-nodes, and $n+1$ sinks whose labels are $0, 1, 2^1, \ldots, 2^{n-1}$.

See Fig. 9.6.1 for an example of the variable ordering $x_{n-1}, y_{n-1}, \ldots, x_0, y_0$. The left column contains $2n - 1$ inner nodes.

Moreover, we have n y-nodes, $n - 1$ x-nodes, and $n + 1$ sinks, altogether $5n - 1$ nodes. In general, we use the following decomposition. If $x_{n-1} = 0$, we have to multiply $(x_{n-2}, \ldots, x_0)$ by $(y_{n-1}, \ldots, y_0)$. The same result has to be considered for $x_{n-1} = 1$, but then we have to add $2^{n-1}|y|$. Hence, the 1-edge gets the weight 2^{n-1} and points to a node representing $|y|$. Then we go on in the same way. The construction of the *BMD for the third variable ordering is left as an exercise.

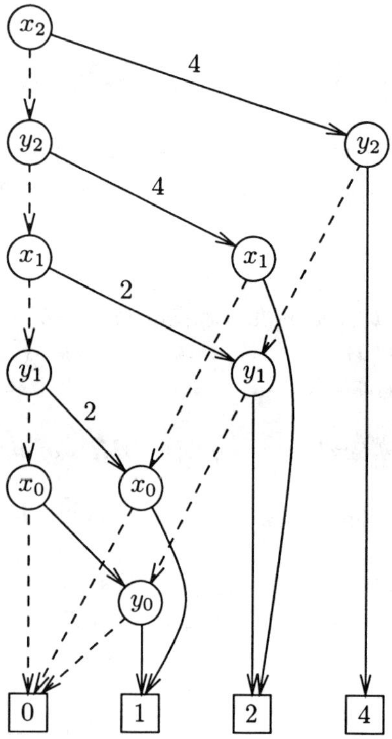

Figure 9.6.1: *A $*$BMD for MUL_3^w. All weights not explicitly described are* 1.

Finally, we represent the function word-level squaring with respect to the variable ordering $x_{n-1}, \ldots, x_0$. With the same notation as above,

$$\begin{aligned} |X_{n-1}|^2 &= \left(2^{n-1}x_{n-1} + |X_{n-2}|\right)^2 \\ &= 2^{2n-2}x_{n-1}^2 + 2^n x_{n-1}|X_{n-2}| + |X_{n-2}|^2 \\ &= |X_{n-2}|^2 + x_{n-1}2^n\left(2^{n-2} + |X_{n-2}|\right). \end{aligned}$$

This decomposition can be used in $*$BMDs. The 0-edge leaving the x_{n-1}-source points to a sub-$*$BMD representing $|X_{n-2}|^2$ and has weight 1. The 1-edge gets the weight 2^n and points to a sub-$*$BMD for the affine function $2^{n-2} + |X_{n-2}|$, which can be represented with $n-1$ inner nodes. The number of inner nodes altogether is $n + (n-1) + \cdots + 1 = O(n^2)$. □

Our results suggest a new approach for the factorization problem. For a constant c, it is easy to obtain a $*$BMD of linear size for the word-level function $xy - c$. If $2^{n-1} \leq c < 2^n$, we consider $(n-1)$-bit numbers x and y. If we can decide in polynomial time whether a *BMD computes 0 for some input and if, in the positive case, we can construct such an input, we have solved the factorization problem. Hence, the following result is not surprising.

Theorem 9.6.3. *The decision whether an EVBDD, a BMD, or a *BMD evaluates to* 0 *on some input is NP-complete.*

Proof. We may guess an input a and can evaluate the diagram for a in polynomial time. The problem PARTITION is known to be NP-complete (see Garey and Johnson (1979)). For positive integers $s_1, \ldots, s_n$, it has to be decided whether some subset of the integers has half the weight of the whole set, i.e., whether its sum is $S/2$ for $S = s_1 + \cdots + s_n$. In linear time, we can construct an EVBDD, BMD, or *BMD for the function $-S/2 + x_1 s_1 + \cdots + x_n s_n$. The partition problem has a solution iff the constructed diagram evaluates to 0 on some input. □

In order to obtain a canonical representation, we have to restrict the freedom of assigning weights to the edges (as for π-EVBDDs and π-OBDDs with complemented edges).

Definition 9.6.1* (refined definition of *BMDs). Weighted edges leading from nowhere to a node are allowed. An edge with weight w pointing to a node representing f represents wf. Only one sink with label 1 is allowed. Edges with weight 0 point to the sink. If one edge leaving a node has the weight 0, the other one has the weight 1. For other cases, the following restrictions hold:

(a) If we work with the ring $\mathbb{Z}$, the gcd of the weights on the edges leaving a node is 1. The weight on the 0-edge is positive.

(b) If we work with the ring $\mathbb{Q}$ or $\mathbb{R}$, 0-edges have the weight 1.

We have used the more general Definition 9.6.1 to simplify the description of the examples. By a bottom-up approach, we can replace a general *BMD with a *BMD of the same size which also fulfills Definition 9.6.1*. The idea is to divide the weights on the edges leaving some node v by the same factor w^* and to multiply the weights of the edges leading to v by this factor w^*. In fields such as $\mathbb{Q}$ and $\mathbb{R}$, we may divide by the weight of the 0-edge if it is not 0, and in $\mathbb{Z}$ we choose the positive or negative gcd. This procedure can be applied also to nodes representing an output, since we allow additional edges to these nodes.

The BMD *merging rule* allows us to merge nodes with the same label, the same 0-successor, the same 1-successor, the same weight on the 0-edge, and the same weight on the 1-edge. The BMD *elimination rule* allows the elimination of a node whose 1-edge carries the weight 0. The correctness of these rules is obvious. We may also use the well-known arguments to prove that the refined π-*BMDs are canonical. For $n = 0$, we have constant functions.

The minimal-size representation of the constant c is an edge with weight c leading to the sink. Let $f_1, \ldots, f_m \colon \{0,1\}^n \to \mathbb{R}$. We represent these functions uniquely as polynomials which are linear with respect to each single variable. For $g \colon \{0,1\}^n \to \mathbb{R}$ and $\pi = id$, the decomposition $g = g_{|x_1=0} + x_1(g_{|x_1=1} - g_{|x_1=0})$ is unique. We apply this decomposition to $f_1, \ldots, f_m$. By the induction hypothesis there is a unique minimal-size $*$BMD representing $f_{j|x_1=0}$ and $f_{j|x_1=1} - f_{j|x_1=0}$, $1 \le j \le m$. If $f_{j|x_1=1} - f_{j|x_1=0} \equiv 0$, the edge representing f_j is equal to the edge representing $f_{j|x_1=0}$. Otherwise, we create an x_1-node whose outgoing 0-edge gets the same weight as the edge representing $f_{j|x_1=0}$ and leads to the same node; the 1-edge is defined similarly. The refined definition of $*$BMDs prescribes how to recalculate the weights on these edges and how to define the weight on the edge leading to this node and representing f_j. Nodes representing the same function can be merged because of the uniqueness of the construction.

Now it is easy to multiply a function f by a constant c. If $c = 0$, we return an edge with weight 0 leading to the sink. Otherwise, the weight on the edge representing f is multiplied by c. Already the addition of two functions f and g represented by π-$*$BMDs causes difficulties. Similarly to the general synthesis algorithm for π-EVBDDs, we have the difficulty that we implicitly run through an "unfolding of the π-$*$BMD" which is the corresponding π-BMD. We never store the π-BMD explicitly but we store the product of the weights already seen. This may lead to exponentially many different weights such that we have to consider one node v for all these weights. For π-EVBDDs we have used the trick of replacing the addition of $(w' + f) + (w'' + g)$ with $(w' + w'') + (f + g)$, leading to a polynomial-time synthesis algorithm. The weights of $*$BMDs work multiplicatively and $w'f + w''g \not\equiv w'w''(f+g)$. Hence, the addition of π-$*$BMDs may lead to an exponential size blow-up (see exercises). For multiplication, the weights can be treated similarly to the addition of π-EVDDs, since $(w'f)(w''g) = (w'w'')(fg)$. But as we know from the discussion of π-OFDDs and π-BMDs, we have to perform four recursive calls for multiplications. For the function represented at the 1-successor, we have to add three of the results and this may cause an exponential size blow-up. The operations on π-$*$BMDs cannot be performed in polynomial time but applications show that the manipulation is often efficient enough, since we gain much by the small size of the representations.

Drechsler, Becker, and Ruppertz (1996) have introduced *Kronecker* $*$BMDs as a hybrid representation type where some nodes work in the EVBDD style and others in the $*$BMD style. Then we may have additive and multiplicative weights. Enders (1995) described in very general form how one obtains transformations between the different representation types and Becker, Drechsler, and Enders (1997) have compared all these models. We have already discussed how to realize such an approach.

Finally, one may ask whether division has a polynomial-size representation in one of these models. Scholl, Becker, and Weis (1998) and Thathachar (1998b) have answered this question negatively. Their results are presented in Chapter 10, since they are based on methods developed in that chapter.

9.7 Exercises

9.1.E Generalize Theorem 3.1.4 and Theorem 3.2.2 to π-MTBDDs.

9.2.E Describe terminal cases for the π-MTBDD synthesis and the operations addition, subtraction, multiplication, min, and max.

9.3.D Prove that matrix multiplication may cause an exponential size blow-up for the representation by π-MTBDDs.

9.4.M Design an algorithm to obtain the π-MTBDD representation of $f\colon \{0,1\}^n \to \{0,\ldots,2^m-1\}$ from the π-OBDD representation of its bit variant.

9.5.E Design an algorithm to compute the π-BMD representation of $-f$ from the π-BMD representation of f.

9.6.M Prove that word-level squaring, i.e., $x \to x^2$, can be represented by BMDs of quadratic size.

9.7.D Prove that the BMD size of exponentiation $x \to 2^x$ is exponential.

9.8.M Prove that there are only six essentially different 2×2 transformation matrices with entries from $\{-1,0,1\}$.

9.9.M Design an algorithm for the change of the transformation type of a variable x_i in π-HDD representations.

9.10.M Design an algorithm for the operation replacement by constants for π-EVBDDs.

9.11.E Represent the functions $\mathrm{ADD}(x,y,z) = x+y+z$, $c(x,y,z) = T_{2,3}(x,y,z)$ (threshold function), and $s(x,y,z) = x \oplus y \oplus z$ by π-EVBDDs. Use the synthesis algorithm to compute a representation of $2c+s$ and perform the equivalence test with ADD.

9.12.M Prove that the EVBDD size of word-level squaring is exponential.

9.13.D Prove that the EVBDD size of exponentiation is exponential.

9.14.D Prove that the $*$BMD size for word-level multiplication and the variable ordering $x_0, y_0, \ldots, x_{n-1}, y_{n-1}$ is linear.

9.15.E Prove that π-$*$BMDs in their refined form are canonical.

9.16.D Prove that the operation addition on π-$*$BMDs may lead to an exponential size blow-up.

9.17.M (See Bryant and Chen (1995).) Design an algorithm for $*$BMDs which replaces the variable x_i by $wx_i + c$ (called affine replacement).

9.18.D Analyze the size blow-up of $*$BMDs caused by the affine replacement of one variable.

Chapter 10

Nondeterministic DDs

Nondeterminism is one of the most powerful concepts in computer science. In this chapter, we investigate nondeterministic DDs. In the introductory section, we present models of nondeterministic BPs and distinguish between the usual existential nondeterminism and other modes of nondeterminism. Functions with small-size nondeterministic OBDDs and FBDDs are presented in Section 10.2 and, in Section 10.3, we generalize several lower bound techniques from deterministic to nondeterministic models. It is not surprising that nondeterministic models lead to interesting complexity theoretical classifications but there are two nondeterministic π-OBDD models which even allow polynomial-time algorithms for important operations. These models are presented in Sections 10.4 and 10.5.

10.1 Different Modes and Models of Nondeterminism

A lot of models of nondeterministic BPs have been presented and all these models are polynomially related. We have chosen an approach (Meinel (1990)) which is very general and quite natural.

Definition 10.1.1. Let Ω be a set of Boolean functions. An *Ω-branching program* is a directed acyclic graph with decision nodes for Boolean variables and *nondeterministic nodes*. Each nondeterministic node is labeled by some function $g \in \Omega$. If $g \in B_r$, the node has r outgoing edges labeled by $1, \dots, r$. A c-sink represents the constant c. Shannon's decomposition rule is applied at decision nodes. If f_i, $1 \le i \le r$, is represented at the successor reached via the edge with label i leaving a nondeterministic node v labeled by $g \in \Omega \cap B_r$, the function $g(f_1(x), \dots, f_r(x))$ is represented at v.

The following nondeterministic nodes are of particular interest.

- $\Omega = \{\text{OR}\}$ leading to the usual *existential nondeterminism* or OR-nondeterminism.
- $\Omega = \{\text{AND}\}$ leading to *universal nondeterminism* or AND-nondeterminism.
- $\Omega = \{\text{EXOR}\}$ leading to *parity nondeterminism* or EXOR-nondeterminism.
- $\Omega = \{\text{AND}, \text{OR}\}$ leading to *alternating nondeterminism.*

Either we consider only binary nondeterministic nodes or nondeterministic nodes of unbounded fan-out. In the second case, a nondeterministic node of fan-out r contributes $r-1$ to the BP size in order to obtain comparable results. Moreover, we may consider majority nondeterminism based on the Boolean majority function and MOD_m-nondeterminism based on the modular counting function. In these last cases, nondeterministic nodes of unbounded fan-out are allowed. Since all considered functions for nondeterministic nodes are symmetric, we can do without the labeling of the outgoing edges.

Meinel (1990) has shown by case inspection that each $\Omega \subseteq B_2$ is equivalent to one of the five cases $\Omega = \emptyset$ (deterministic BPs), $\Omega = \{\text{OR}\}$, $\Omega = \{\text{AND}\}$, $\Omega = \{\text{EXOR}\}$, and $\Omega = \{\text{AND}, \text{OR}\}$. Since this holds for all considered BP variants, we restrict ourselves to these cases. The case $\Omega = \{\text{AND}, \text{OR}\}$ is not interesting, since we obtain the representational power of circuits.

Proposition 10.1.2. *Boolean circuits are polynomially related to {AND, OR}-OBDDs and {AND, OR}-BPs.*

Proof. Since decision nodes are ite instructions (see Definition 1.1.5), it is obvious that a circuit of size $3|G|$ can simulate an alternating BP G. It is also well known (see Wegener (1987)) that it is sufficient to consider Boolean circuits whose inputs are $x_1, \overline{x}_1, \ldots, x_n, \overline{x}_n$ and which work with AND- and OR-gates. Then we obtain an alternating OBDD by replacing the inputs by $2n$ decision nodes representing the literals and reversing the edge directions. The output gate of the circuit corresponds to the source of the BP. □

In spite of this result, Plessier, Hachtel, and Somenzi (1994) have proposed $\{\text{AND}, \text{OR}\}$-OBDDs under the notion *extended* BDD (XBDD) as a representation of Boolean functions. The idea is to use as "little nondeterminism" as possible. The synthesis is easy, since we have Boolean gates available and the satisfiability or equivalence test is done heuristically. We do not follow this approach here.

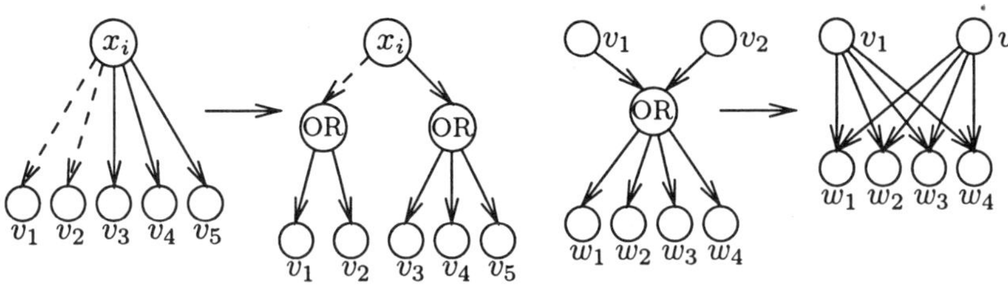

Figure 10.1.1: *Simulations between different models of nondeterministic nodes.*

We describe one alternative model for nondeterministic nodes for the case $\Omega = \{\text{OR}\}$, $\Omega = \{\text{AND}\}$, or $\Omega = \{\text{EXOR}\}$. All inner nodes are labeled by Boolean variables and may have an unbounded number of outgoing 0-edges and 1-edges. An input a activates all a_i-edges leaving x_i-nodes. The function f_v represented at v computes 1 on input a iff there is at least one activated path from v to the 1-sink (if $\Omega = \{\text{OR}\}$), all activated paths starting at v reach the 1-sink (if $\Omega = \{\text{AND}\}$), or the number of activated paths leading from v to the 1-sink is odd (if $\Omega = \{\text{EXOR}\}$). In Fig. 10.1.1, we have shown how such a "nondeterministic decision node" can be simulated by a decision node followed by nondeterministic nodes. The size measured as the number of edges increases at most by a constant factor. The reverse simulation is also possible (see Fig. 10.1.1) and does not increase the number of nodes. If the BP starts with a nondeterministic node v, we may use the same trick by inserting a dummy decision node whose edges lead to the successors of the nondeterministic node. This procedure may cause difficulties if we have restrictions on repeating the tested variables. In the OBDD case, we may choose the first variable of the ordering. Then we obtain two levels with the same variable and it is easy to simulate these levels by one level. In each of the three cases OR, AND, and EXOR, it is not useful to connect two nodes with more than one edge with the same label. For OR and AND, the number of "identical edges" can be reduced to one without changing the represented functions, and for EXOR two "identical edges" cancel each other. Hence, in all cases the number of edges grows at most as the square of the number of nodes. It will always be clear from the context which of the two models we consider.

In Theorem 2.1.9, we have shown that polynomial-size BPs correspond to nonuniform Turing machines with logarithmic space. Babai, Hajnal, Szemerédi, and Turán (1987) have proved that polynomial-size FBDDs correspond to nonuniform so-called eraser Turing machines with logarithmic space. Similar results hold for nondeterministic BPs and their variants and corresponding Turing machine models (Meinel (1990), Krause, Meinel, and Waack (1991)). Hence, the famous result of Immerman (1988) and Szelepcsényi (1988) has the following consequence for nondeterministic BPs.

Theorem 10.1.3. *The class of functions representable by polynomial-size $\{OR\}$-BPs is equal to the class of functions representable by polynomial-size $\{AND\}$-BPs.*

We do not follow these complexity theoretical considerations and refer to Meinel (1988) for the investigation of nondeterministic BPs of bounded width.

In the rest of this section, we consider nondeterministic DTs. Alternating DTs correspond to Boolean B_2-formulas (see exercises). The main observation is that OR-DTs (we use this as short notation for {OR}-DTs) essentially are DNFs (Damm and Meinel (1992)).

Theorem 10.1.4. *A function $f \in B_n$ can be represented by an OR-DT with s 1-leaves iff it can be represented by a DNF with s monomials.*

Proof. The function which is represented by an OR-DT T is the disjunction of all monomials corresponding to paths from the root to a 1-leaf. OR-nodes do not contribute anything to these monomials. Hence, the simulation of OR-DTs by DNFs works as the simulation of DTs by DNFs described in the proof of Theorem 2.5.4.

For the other direction, let f be the disjunction of s monomials. Then we may start with an OR-gate with outdegree s. At these successors we represent the monomials of the given DNF, and each sub-DT contains one 1-leaf (and at most n 0-leaves). □

In the same way we obtain the following two results.

Theorem 10.1.5. *A function $f \in B_n$ can be represented by an AND-DT with s 0-leaves iff it can be represented by a CNF with s clauses. A function $f \in B_n$ can be represented by an EXOR-DT with s 1-leaves iff it can be represented by an EXORNF (parity of monomials) with s monomials.*

Corollary 10.1.6. *For OR-, AND-, or EXOR-DTs it is sufficient to have one nondeterministic node at the root.*

Proof. The result follows from the simulations in the proofs of Theorem 10.1.4 and Theorem 10.1.5. □

It is an open problem whether a similar result holds for nondeterministic OBDDs, FBDDs, or BPs.

By Theorem 10.1.4 and Theorem 10.1.5, it is easy to distinguish between the different nondeterministic DT models. The parity function has polynomial-size EXOR-DTs but exponential size for OR-DTs and AND-DTs. The function testing whether at least one row of a Boolean $n \times n$ matrix has at least two 1-entries has polynomial-size OR-DTs but exponential size for AND-DTs and

EXOR-DTs. For the negation of this function, the same results hold with interchanged roles of OR and AND (Damm and Meinel (1992)). Theorem 2.5.10 due to Jukna, Razborov, Savický, and Wegener (1997) gets a new interpretation. There is a function with polynomial OR-DT size and polynomial AND-DT size which has exponential DT size, i.e., P $\neq$ NP $\cap$ coNP for decision trees.

10.2 Upper Bound Techniques

For most problems in NP, the design of a polynomial-time nondeterministic algorithm is rather easy. With the usual guess-and-verify approach, we obtain nondeterministic π-OBDDs of polynomial size for many functions and often this even works independently from the variable ordering.

Theorem 10.2.1. *The following functions have polynomial-size OR-π-OBDDs for all variable orderings: HWB_n, $\overline{HWB}_n$, WS_n, $\overline{WS}_n$, ISA_n, $\overline{ISA}_n$, $\overline{excl}_n$, $\overline{PERM}_n$, ROW_n+COL_n, the test whether a graph is not regular, and the pointer-jumping function $PJ_{k,n}$ and its negation $\overline{PJ}_{k,n}$ for constant k.*

Proof. The functions HWB$_n$, WS$_n$, and ISA$_n$ are pointer functions with a single pointer (compare Definition 7.2.1). We guess the value of the pointer and verify that the guess is correct. More precisely, for HWB$_n$ we guess $i \in \{0, \dots, n\}$ and verify that $x_1 + \cdots + x_n = i$. This can be done, for each i, in quadratic size independently from the variable ordering. Furthermore, we store the value of x_i and output the value of x_i if the guess was correct. A similar approach works for $\overline{\text{HWB}}_n$, since the output equals $\overline{x}_i$ if $x_1 + \cdots + x_n = i$.

For WS$_n$ (and $\overline{\text{WS}}_n$) the pointer equals the sum of all ix_i mod p (see Theorem 6.2.10). Since $p \leq 2n$, this verification is possible with a width bounded by $2n$. For ISA$_n$ (and $\overline{\text{ISA}}_n$) we guess the value of $|y|$ and the value of $\alpha(x, y)$, i.e., the number represented by $(x_{|y|}, \dots, x_{|y|+k-1})$. For each of the n^2 possibilities, the verification is the computation of a monomial of length $2k$ and the output is $x_{\alpha(x,y)}$ if the guess was correct.

It is easy to verify that the function $\overline{\text{excl}}_n$ computes 1 for a graph $G(x)$ iff at least one of the following properties is fulfilled:

- $G(x)$ is empty,
- at least one vertex has a degree different from 0 or $\lceil n/2 \rceil - 1$,
- there exist two edges $\{v_1, v_2\}$ and $\{v_3, v_4\}$ such that $v_1 \neq v_3$ and $\{v_1, v_3\}$ is not an edge.

We guess one of the $1 + n + \binom{n}{3} + \binom{n}{4}$ possibilities and can verify each property easily.

For $\overline{\text{PERM}_n}$, we guess a row or a column and verify that this row or column contains no 1-entry or at least two 1-entries. A similar approach works for $\text{ROW}_n + \text{COL}_n$. To test whether a graph is not regular, it is sufficient to guess two nodes and to verify that they have different degree. For the pointer-jumping function, we may guess the whole path, since the number of possible paths is bounded by n^k. Then it is sufficient to check whether the pointers leaving nodes on the path have the right value. This is the computation of a monomial. Finally, the output for $\text{PJ}_{k,n}$ is the color of the last node. For $\overline{\text{PJ}_{k,n}}$, the output is the complement of this color. □

It is obvious that $\overline{f}_n$ has polynomial-size AND-π-OBDDs iff f_n has polynomial-size OR-π-OBDDs. Hence, Theorem 10.2.1 contains a lot of results on AND-π-OBDDs.

Corollary 10.2.2. *The following functions have polynomial-size EXOR-π-OBDDs for all variable orderings: HWB_n, WS_n, ISA_n, and the pointer-jumping function for constant k.*

Proof. This follows directly from the proof of Theorem 10.2.1. For the considered functions, there is at most one guess which leads to the output 1. In such a situation, OR-nondeterminism and EXOR-nondeterminism coincide. □

We add the obvious remark that a function $\overline{f}_n$ has polynomial-size EXOR-π-OBDDs iff f_n has polynomial-size EXOR-π-OBDDs. We may start with a binary EXOR-node. One successor is the 1-sink and the other one the source of an EXOR-π-OBDD for the negated function. This trick works for all EXOR-models.

Corollary 10.2.3. *The weighted sum function WS_n has exponential FBDD size but polynomial OR-OBDD and polynomial AND-OBDD size, i.e., $P \neq NP \cap coNP$ for OBDDs and FBDDs.*

Proof. We refer to Theorem 6.2.10 and Theorem 10.2.1. □

Jukna, Razborov, Savický, and Wegener (1997) have presented other functions with the same properties and the additional property that they are contained in $\Sigma_3^p \cap \Pi_3^p$, i.e., they can be computed by polynomial-size, unbounded fan-in $\{\text{AND}, \text{OR}, \text{NOT}\}$-circuits of depth 3 where negations only are allowed at the inputs and where we may choose whether the last gate is an OR- or an AND-gate (see exercises).

Although all previous results hold for all variable orderings, the variable-ordering problem does not become trivial for nondeterministic OBDDs.

Definition 10.2.4. The *equality test function* $\text{EQ}_n\colon \{0,1\}^{2n} \to \{0,1\}$ works on the variables $x_1, \ldots, x_n, y_1, \ldots, y_n$ and outputs 1 iff $x_i = y_i$ for all $i \in \{1, \ldots, n\}$.

It is obvious that EQ_n has linear OBDD size for the interleaved variable ordering $x_1, y_1, \ldots, x_n, y_n$. Let us investigate the variable ordering $x_1, \ldots, x_n$, $y_1, \ldots, y_n$. The corresponding OBDD size is exponential and it will be shown in Section 10.3 that this holds also for OR-OBDDs. The function $\overline{EQ}_n$ again has polynomial-size OR-π-OBDDs for all variable orderings. It is sufficient to guess i and to verify that $x_i \neq y_i$.

Proposition 10.2.5. *The functions $\overline{PERM_n}$ and $ROW_n + COL_n$ can be represented by linear-size OR-FBDDs with one nondeterministic node with fan-out 2 and by linear-size OR-OBDDs with one nondeterministic node with fan-out $2n$.*

Proof. For FBDDs, we guess whether the decisive event happens in a row or in a column and then construct a sub-OBDD with a rowwise, respectively, columnwise variable ordering. Using a rowwise variable ordering, it is easy to check whether some row does not contain exactly one 1-entry or to check whether some row contains only 1-entries. The same holds for the columns.

For OBDDs, we use an arbitrary variable ordering and guess the row or column where the decisive event happens. If we have to test for one specific row or column whether it does not contain exactly one 1-entry or whether it contains only 1-entries, this can be done for each variable ordering in linear size. □

The following example (Jukna (1995)) shows the power of nondeterminism. We have shown in Theorem 7.6.5 that the characteristic functions of some explicitly defined linear codes have exponential k-BP size even for slowly increasing k. The same holds for the negation of these functions.

Theorem 10.2.6. *The negations of the characteristic functions of linear codes have polynomial-size OR-π-OBDDs for each variable ordering.*

Proof. A linear code is a linear subspace of $\{0,1\}^n$ and can be described by some (at most n) linear equations. We guess the number j of the equation which is not fulfilled and, then, verify that this equation really is not satisfied. Since this is the check whether the parity of some set of variables is odd, it can be done with linear size independently from the chosen variable ordering. □

It will be noticed immediately that all considered OBDDs and FBDDs work in the guess-and-verify mode, i.e., no nondeterministic node follows a decision node. It is well known that this mode is no restriction for the time complexity of Turing machines. But BDDs model the space complexity of Turing machines. In the guess-and-verify mode, we may nondeterministically choose only among polynomially many possibilities if the size is polynomially bounded. This seems to be a significant restriction. But it is an open problem whether OBDDs or FBDDs with the restriction to the guess-and-verify mode can represent fewer functions in polynomial size than without this restriction. The following function is a candidate for such a separation. The input consists of n Boolean $n \times n$ matrices $X_1, \ldots, X_n$ and tests whether all these matrices are not permutation matrices. The following OR-OBDD for this function has polynomial size. We start with an OR-OBDD for $\overline{\text{PERM}}_n(X_1)$ (see Theorem 10.2.1), the 1-sink is replaced by the source of an OR-OBDD for $\overline{\text{PERM}}_n(X_2)$, and so on. There are inputs activating $(2n)^n$ computation paths and we see no way to get by with essentially less nondeterminism, even in the FBDD case.

We prove in the next section that PERM_n needs exponential OR-FBDD size. Therefore, the following result of Ablayev and Jukna (for the main idea see Jukna (1995)) is of interest. It proves again the power of semantic BDDs.

Theorem 10.2.7. *The function $PERM_n$ has polynomial-size semantic OR-$(1,+1)$-BPs.*

Proof. The BP consists of $2n$ layers, one for each row and one for each column of the matrix X_n. The layer for row i guesses an index $j \in \{1, \ldots, n\}$ and verifies that $x_{ij} = 1$. The 1-sink of each layer (except the last one) is replaced by the source of the next layer. We reach the 1-sink of the last row layer iff each row contains at least one 1-entry. Each path contains each variable at most once and all nodes on a path to the 1-sink are left via the 1-edge. The layer for column j guesses an index $i \in \{1, \ldots, n\}$ and verifies that $x_{lj} = 0$ for all $l \neq i$. The 1-sink of each layer (except the last one) is replaced by the source of the next layer. We reach the 1-sink of the last column layer iff each column contains at least $n-1$ entries with the value 0. This implies that the conjunction of the row part and the column part (replace the 1-sink of the row part by the source of the column part) represents PERM_n. Each path in the column part contains each variable at most once and all nodes on a path to the 1-sink are left via the 0-edge. This implies that all *computation paths* to the 1-sink of the whole BP are read once. If a path tests a variable for the second time, then it is tested in the row part with the value 1. Hence, on *computation paths*, variables are tested with the value 1 in the column part which implies that the 0-sink is reached immediately. This proves that the BP semantically is $(1,+1)$ (syntactically it is only $(1,+n)$). It is obvious that the size is $O(n^3)$. □

In summary, we have shown that nondeterminism is a powerful tool for OBDDs, FBDDs, and semantic $(1,+1)$-BPs.

Considering the construction of nondeterministic OBDDs leading to good upper bounds, one may conjecture that the choice of the variable ordering is not as important as for deterministic OBDDs. This conjecture is supported by the following result where we investigate nondeterministic OBDDs where all nodes are labeled by variables and may have an arbitrary number of outgoing edges.

Definition 10.2.8. Let π be a variable ordering and $x_{\pi^{-1}(1)}, \ldots, x_{\pi^{-1}(n)}$ the corresponding ordered list of variables. The *reversed variable ordering* π_R is described by the ordered list $x_{\pi^{-1}(n)}, \ldots, x_{\pi^{-1}(1)}$ of variables.

Theorem 10.2.9. *If $f \in B_n$ can be represented by OR-π-OBDDs of size s, then f can also be represented by OR-π_R-OBDDs of size $O(sn)$. The same holds for AND or EXOR instead of OR.*

Proof. Let G be an OR-π-OBDD representing f with size s. The result holds if f is a constant function. Otherwise, we eliminate the 0-sink and all edges leading to this sink. This does not change the function represented by the OR-π-OBDD. Moreover, we eliminate all inner nodes without an outgoing edge. In the next step, we add dummy nodes such that each path from the source to the 1-sink contains a node labeled by x_i for each variable x_i. This can be done as in the case of deterministic OBDDs. For each node and each x_i, it is sufficient to create one x_i-node. Hence, the size of the new OR-π-OBDD G' representing f is $O(sn)$. The graph G' is layered, it has one source and one sink, and the layers are labeled by $x_1, \ldots, x_n, 1$. We obtain G'' by reversing the direction of all edges and by relabeling the new labels by $x_n, \ldots, x_1, 1$. We claim that G'' represents f. Let us consider a c-edge of G' leading from the x_i-node v to the x_{i+1}-node w (or the 1-sink if $i = n$). This edge is activated iff $x_i = c$. The graph G'' contains the c-edge from w to v. The node w belongs to layer $i + 1$ of G' and, therefore, to layer $n + 1 - i$ of G''. Hence, w is labeled by x_i in G''. The edge (w, v) of G'' is activated by an input a iff the edge (v, w) of G' is activated by a. This implies that each path from the source of G' to the 1-sink of G' which is activated by a corresponds to a path from the source of G'' to the 1-sink of G'' which is activated by a. This proves that G'' represents f.

The same proof works for the EXOR-case. In the AND-case we remove the 1-sink and can argue in the same way with respect to the 0-sink. □

The following result of Savický (1998b) implies that there are no almost nice and no ambiguous functions (see Definition 5.1.3) for EXOR-OBDDs, while we have presented an almost nice function for OBDDs in Section 5.3. It is not known whether there exist ambiguous functions for OBDDs.

Theorem 10.2.10. *If the EXOR-π-OBDD size of $f_n \in B_n$ is bounded above by a polynomial $p(n)$ for at least a fraction $\varepsilon > 0$ of all variable orderings, the EXOR-π-OBDD size of f_n is bounded above by a polynomial $p^*(n)$ for all variable orderings.*

10.3 Lower Bound Techniques

It is not surprising that it is quite easy to design nondeterministic OBDDs or FBDDs of small size. But it will turn out that it is also not too difficult to obtain lower bounds of exponential size. The description of lower bound techniques in the previous chapters has been done in a way which supports the generalization to the nondeterministic case.

We investigate nondeterministic s-oblivious BDDs (including OBDDs, k-OBDDs, and k-IBDDs), FBDDs, and the more general case of k-BPs. For polynomial lower bounds in even more general models we refer to Razborov (1991) and for exponential lower bounds for BPs based on so-called corrupting Turing machines (a semantic restriction mentioned in Section 7.4) we refer to Jukna and Razborov (1998).

We already know that the theory of communication complexity is a powerful tool for proving lower bounds on the size of s-oblivious BDDs (see Section 7.5). A similar approach works in the nondeterministic case. Let G be a nondeterministic s-oblivious BDD which can be partitioned into k layers such that each layer belongs to Alice or to Bob. This partition leads to the following nondeterministic protocol of length $k\lceil \log |G| \rceil$. If Alice is the owner of the first layer, she chooses one of the paths activated by her partial input and her first message contains the number of the node in Bob's layer reached on the chosen path. Then Bob goes on in a similar way. Finally, a sink with some label c is reached and this information is contained in the last message. The interpretation of such a protocol is different from the interpretation of deterministic protocols.

If we consider OR-BDDs, the value of the function f represented by G on the chosen input a equals 1 iff at least one of the legal sequences of messages leads to the output 1. Instead of AND-BDDs for f we investigate OR-BDDs for $\overline{f}$. In the case of EXOR-BDDs, $f(a) = 1$ iff an odd number of sequences of messages leads to the output 1. This defines OR-protocols and EXOR-protocols and we obtain the same relations between communication complexity and the size of s-oblivious BDDs as in the deterministic case at the beginning of Section 7.5. The question is how we obtain lower bounds on the nondeterministic communication complexity.

We investigate a nondeterministic protocol for $f\colon \{0,1\}^n \times \{0,1\}^m \to \{0,1\}$ and the corresponding tree representing the protocol. This tree may contain nondeterministic nodes. It is still true that the set of inputs which may reach the same leaf is a rectangle. In the OR-case, the rectangle corresponding to a 0-leaf may contain inputs $a \in f^{-1}(0)$ and inputs $b \in f^{-1}(1)$ but rectangles corresponding to 1-leaves are monochromatic and have the color 1. Moreover, the union of the rectangles corresponding to the 1-leaves has to be equal to $f^{-1}(1)$. This leads to a lower bound based on *coverings*. The fooling-set method also works if we restrict ourselves to 1-fooling sets S (compare Definition 7.5.3), i.e., sets $S \subseteq \{0,1\}^n \times \{0,1\}^m$ such that $f(a,b) = 1$ for all $(a,b) \in S$ and $f(a_1,b_2) = 0$ or $f(a_2,b_1) = 0$ for different pairs (a_1,b_1), $(a_2,b_2) \in S$. We obtain the following lower bounds.

Theorem 10.3.1. *Let $f\colon \{0,1\}^n \times \{0,1\}^m \to \{0,1\}$. If a covering of $f^{-1}(1)$ by monochromatic rectangles requires t rectangles, the OR-nondeterministic communication complexity (with respect to the given partition of the variable set) is bounded below by $\lceil \log t \rceil$. The same lower bound holds if f has a 1-fooling set of size t.*

Definition 10.3.2. The $\mathbb{Z}_2$-*rank* of $f\colon \{0,1\}^n \times \{0,1\}^m \to \{0,1\}$ is denoted by $\text{rank}_{\mathbb{Z}_2}(f)$ and is defined as the rank of the communication matrix over the field $\mathbb{Z}_2$.

Theorem 10.3.3. *The EXOR-nondeterministic communication complexity of* $f\colon \{0,1\}^n \times \{0,1\}^m \to \{0,1\}$ *is bounded below by* $\lceil \log \text{rank}_{\mathbb{Z}_2}(f) \rceil$.

Proof. The proof is an adaptation of the proof of Theorem 7.5.6 which, as mentioned there, actually is a lower bound on the number of 1-leaves. The crucial observation is that we obtain the communication matrix as the EXOR-sum of the communication matrices $M(v)$ representing the rectangles $R(v)$ corresponding to the 1-leaves of the protocol tree. By the subadditivity of the rank function, we obtain

$$\text{rank}_{\mathbb{Z}_2}(f) \le \sum_{v \mid v \text{ is 1-leaf}} \text{rank}_{\mathbb{Z}_2}(M(v)) = \#\{v \mid v \text{ is 1-leaf}\}. \qquad \square$$

Our first applications of the lower bound techniques concern the simple functions EQ_n and IP_n.

Theorem 10.3.4. *Let the x-variables be given to Alice and the y-variables to Bob. Then the OR- and EXOR-communication complexity of EQ_n is bounded below by n and the OR- and AND-communication complexity of IP_n is bounded below by $\Omega(n/\log n)$.*

Proof. The communication matrix for EQ_n is the matrix with ones on the main diagonal and zeros elsewhere. This matrix has the full rank 2^n over $\mathbb{Z}_2$ and the inputs corresponding to the ones are a 1-fooling set of size 2^n. The communication matrix for IP_n is the Hadamard matrix, which is equal to the Sylvester matrix (see Definition 7.6.7) where $+1$ is replaced by 0 and -1 by 1. We have mentioned in Section 7.6 that the $\mathbb{Z}_3$-rank of $t \times u$ submatrices of the Sylvester matrix is bounded below by $ut/(2^{n+1}\ln(2^{n+1}/u))$. If the rank of a matrix is at least 2, the matrix is not monochromatic. This leads to the proposed bound. $\square$

The bound on the OR- and AND-complexity of IP_n can be improved to $\Omega(n)$ (see Kushilevitz and Nisan (1997)).

Krause (1992) and Krause and Waack (1991) have defined the functions EQ_n^* and IP_n^* in order to separate the different modes of nondeterminism.

Definition 10.3.5. For $x = (a_1, x_1, a_2, x_2, \ldots, a_n, x_n) \in \{0,1\}^{2n}$ let $x^* = (x_{i(1)}, \ldots, x_{i(m)})$ if $i(1) < \cdots < i(m)$, $a_{i(1)} = \cdots = a_{i(m)} = 1$, and $a_j = 0$ for all other j. Let y^* be defined similarly for $y = (b_1, y_1, \ldots, b_n, y_n)$. The functions EQ_n^*, $\text{IP}_n^* \in B_{4n}$ are defined on (x, y). The output of $\text{EQ}_n^*(x, y)$ equals 1 if x^* and y^* have the same length l and $\text{EQ}_l(x^*, y^*) = 1$. The output of $\text{IP}_n^*(x, y)$ equals 1 if x^* and y^* have the same length l and $\text{IP}_l(x^*, y^*) = 1$.

Theorem 10.3.6. *The function EQ_n^* has polynomial-size AND-π-OBDDs for all variable orderings, but OR- or EXOR-oblivious BDDs of length $O(n)$ need exponential size. The function IP_n^* has polynomial-size EXOR-π-OBDDs for all variable orderings, but OR- or AND-oblivious BDDs of length $O(n)$ need exponential size.*

Proof. The upper bounds are left as exercises. We have already presented all the tools for the lower bound proof. Let the length of the oblivious BDDs be bounded above by kn where k is a constant. There are at least $\lfloor n/2 \rfloor$ x-variables and $\lfloor n/2 \rfloor$ y-variables which are the labels of at most $2k$ levels each. We apply Lemma 7.5.1 to these sets of variables. Then we obtain $m = \Omega(n)$ x-variables and $m = \Omega(n)$ y-variables and the number of layers with respect to these sets of variables is bounded above by $4k+1$. We give these m x-variables to Alice and these m y-variables to Bob. All other variables are replaced by constants. We assign constants to the a- and b-variables in such a way that exactly the partners of the chosen x- and y-variables get the value 1. The given nondeterministic oblivious BDD G of length kn leads to a nondeterministic protocol with $4k+1$ rounds and protocol length $(4k+1)\lceil \log |G| \rceil$ for the function obtained by the replacements. This is the function EQ_m, respectively, IP_m. Because of the partition of the variables between Alice and Bob, Theorem 10.3.4 yields the desired bounds. □

Theorem 7.5.11 contains the lower bound of Gergov (1994) on the size of oblivious BDDs for the middle bit of multiplication. Actually, Gergov (1994) has also presented this bound for the nondeterministic case.

Theorem 10.3.7. *The size of OR-, AND-, or EXOR-oblivious BDDs representing $MUL_{n-1,n}$ is not polynomial if the length is $o(n \log n / \log\log n)$.*

Proof. By the proof of Theorem 7.5.11, it is sufficient to consider the nondeterministic communication complexity of the problem to decide for l-bit numbers x (given to Alice) and y (given to Bob) whether $|x|+|y| \geq 2^l$. The communication matrix (compare Fig. 7.5.1 for $l=3$) contains ones exactly below the diagonal from the lower-left corner to the upper-right corner. The rank of this matrix is $2^l - 1$ not only over the field $\mathbb{R}$ but also over $\mathbb{Z}_2$. This leads to the same lower bound for the EXOR-case as for the deterministic case in Theorem 7.5.11.

For the OR-case, we reformulate the problem. We have to decide whether $|x| \geq 2^l - |y|$. If we exclude the trivial cases $|x| = 0$ and $|y| = 0$, this is equivalent to the decision whether $|x| \geq |z| := 2^l - |y|$ where $|x|, |z| \in \{1, \dots, 2^l - 1\}$. If the OR-communication complexity of this problem is s, the function EQ_n has an OR-communication complexity of $O(s)$. We use the protocol for the test $|x| \geq |z|$ and, in the positive case, a similar protocol for the test $|x| \leq |z|$. Moreover, we use two bits for the information whether $|x| = 0$ and $|z| = 0$, respectively. This leads to the lower bound for the OR-case. The AND-case is similar, since we may consider OR-nondeterminism for the negated problem $|x| < |z|$ and this problem is equivalent to $|x| + 1 \leq |z|$. □

The remaining part of this section is devoted to OR-FBDDs (which are OR-1-BPs) and OR-k-BPs. The proof of exponential lower bounds on the EXOR-FBDD size of explicitly defined Boolean functions is still an open problem.

Theorem 10.3.8. *The OR-FBDD size of $PERM_n$ is bounded below by $\Omega(n^{-1/2}2^n)$.*

Proof. The proof of the lower bound on the OBDD size of PERM_n (see Theorem 4.12.3) actually works for OR-FBDDs. The only difference is that we do not choose *the* computation path for the input $M(\pi)$ describing the permutation matrix for π but one arbitrary path $p(\pi)$ activated by $M(\pi)$ and reaching the 1-sink. Then the cut-and-paste technique works. All variables are tested on $p(\pi)$. If $p(\pi)$ and $p(\pi')$ meet at v, the input described by the tests on $p(\pi)$ up to v and the tests on $p(\pi')$ starting at v has to be a permutation matrix. □

The reader is encouraged to study the proof of Theorem 4.12.3 and to work out that it is based on $(1, 2)$-rectangles.

The OR-case is much simpler than the EXOR-case, since each path leading to the 1-sink corresponds to an input a such that the OR-FBDD computes 1 on a. We remember (Theorem 10.2.1) that $\overline{\text{PERM}_n}$ has linear-size OR-OBDDs. All these properties of PERM_n have been recognized by Krause (1988), Jukna (1989), and Krause, Meinel, and Waack (1991).

In Section 7.6, we have presented lower bound techniques for k-BPs based on the rectangle method due to Borodin, Razborov, and Smolensky (1993) (see Definition 7.6.1). All the results of that section hold (and have been stated by the authors) also for OR-k-BPs. The reason is simply that in the proof of Theorem 7.6.2 all paths from the source to the 1-sink are considered and f is represented as a disjunction of the monomials corresponding to these paths. In order to obtain a more compact representation, traces are considered. But the essential observation is that the function represented by an OR-k-BP is also the disjunction of all monomials corresponding to the paths from the source to the 1-sink. We summarize the results.

Theorem 10.3.9 (cf. Theorem 7.6.2). *Let G be an OR-k-BP representing f with size s. For $r = (2s)^{ka}$, the function f can be represented as a disjunction of at most r (k, a)-rectangles.*

Theorem 10.3.10 (cf. Theorems 7.6.5, 7.6.8, 7.6.9, and Corollary 7.6.10).

(i) *The OR-k-BP size of some explicitly defined linear codes is bounded below by $2^{\Omega(n^{1/2}/k^k)}$.*

(ii) *The OR-k-BP size of the bilinear Sylvester function is bounded below by $2^{\Omega(n/4^k k^3)}$.*

(*iii*) *The OR-k-BP size of the hyperplanar sum-of-products predicate* HSP_q^{k+1} *(defined on N variables) is bounded below by* $2^{\Omega(N^{1/(k+1)}2^{-2k}k^{-4})}$.

(*iv*) *The OR-k-BP size of the conjunction of hyperplanar sum-of-products predicate* $CHSP_q^{k+1}$ *is bounded below by* $2^{\Omega(N^{1/(k+1)}2^{-2k}k^{-3})}$.

(*v*) *If* $k = o(\log^{1/2} n)$, *there are functions representable by polynomial-size k-IBDDs but not by polynomial-size OR-*$(k-1)$*-BPs.*

We present another application of Theorem 10.3.9 due to Borodin, Razborov, and Smolensky (1993). They have improved the FBDD lower bound for the clique function (Theorem 6.2.7) to OR-FBDDs.

Theorem 10.3.11. *If* $m(n) \leq n/3$, *the OR-FBDD size of* $cl_{n,m(n)}$ *is bounded below by* $2^{\Omega(m(n))}$.

Proof. We apply Theorem 10.3.9 for $k = 1$ and $a = 2$. Without loss of generality, we consider the case that $\binom{n}{2}$, the number of variables, is even. Then each $(1,2)$-rectangle can be described by a function $f_R = f_1(X_1) \wedge f_2(X_2)$ such that $|X_1| = |X_2| = \binom{n}{2}/2$ and $X_1 \cap X_2 = \emptyset$. If we consider the communication matrix of f_R with respect to the partition of the variables into X_1 and X_2, then $f_R^{-1}(1) = f_1^{-1}(1) \times f_2^{-1}(1)$, i.e., $f_R^{-1}(1)$ is a rectangle. In a communication game, the partition is fixed but in Theorem 10.3.9 each rectangle can be based on its own partition.

We prove that f_R can cover only an exponentially small fraction of all $\binom{n}{m(n)}$ graphs which consist of an $m(n)$-clique and $n - m(n)$ isolated vertices. Such graphs are called exactly $m(n)$-cliques. Let A_i, $i \in \{1, 2\}$, be the set of inputs a_i such that $f_i(X_i)$ computes 1 on a_i. In the communication matrix with respect to X_1 and X_2, we consider the rectangle $R = A_1 \times A_2$ and the subset $R' \subseteq R$ consisting of all exactly $m(n)$-cliques. Let (a_1, a_2), $(a_1', a_2') \in R'$. We claim that $a_1 = a_1'$ or $a_2 = a_2'$, i.e., all exactly $m(n)$-cliques in R belong to the same row or to the same column of the communication matrix. The graph belonging to (a_1, a_2) contains $\binom{m(n)}{2}$ edges, the same holds for (a_1', a_2'). Since R contains only graphs with $m(n)$-cliques, also (a_1, a_2') and (a_1', a_2) are graphs with $m(n)$-cliques. This is possible only if both inputs belong to exactly $m(n)$-cliques. Let us consider an input belonging to an exactly $m(n)$-clique, e.g., (a_1, a_2). If a_1 does not describe edges adjacent to $m(n)$ vertices, there is a vertex v of the $m(n)$-clique such that a_1 does not describe an edge adjacent to v and a_2 describes all $m(n) - 1$ edges adjacent to v. This implies that a_2 determines all $m(n)$ vertices of the exactly $m(n)$-clique. Since (a_1', a_2) is also an exactly $m(n)$-clique, it has to be the same clique and $a_1 = a_1'$. If a_1 describes edges adjacent to $m(n)$ vertices, we conclude that $a_2 = a_2'$.

Without loss of generality, we assume that R' is a subset of one row of the communication matrix. Let a_1 be the partial input describing the row. Let V_1

be the set of vertices v such that some edge adjacent to v is described by a_1. If $(a_1, a_2) \in R'$, the corresponding exactly $m(n)$-clique has $V_1 \cup V_2$ as vertex set and all $\binom{m(n)-|V_1|}{2}$ edges of the clique on V_2 belong to the $n(n-1)/4$ edges of X_2. Now we state a graph theoretical claim which we prove later.

Claim. *Each graph with n vertices and e edges contains at most $n\binom{\lfloor (2e)^{1/2} \rfloor}{s-1}$ cliques of size s.*

We apply this claim for $e = n(n-1)/4$ and $s = m(n) - |V_1|$. Then the number of cliques is bounded above by

$$n\binom{\lfloor (n(n-1)/2)^{1/2} \rfloor}{m(n)-|V_1|-1} \leq n\binom{\lfloor n/2^{1/2} \rfloor}{m(n)-|V_1|-1} \leq n\binom{\lfloor n/2^{1/2} \rfloor}{m(n)}.$$

The last inequality holds, since $m(n) \leq n/3 \leq (n/2^{1/2})/2$. Altogether, each rectangle covers at most $n\binom{\lfloor n/2^{1/2} \rfloor}{m(n)}$ of the $\binom{n}{m(n)}$ exactly $m(n)$-cliques and we need

$$\binom{n}{m(n)} \Big/ \left(n\binom{\lfloor n/2^{1/2} \rfloor}{m(n)} \right) \geq \frac{1}{n} \prod_{0 \leq i < m(n)} \frac{n-i}{\lfloor n/2^{1/2} \rfloor - i} = 2^{\Omega(m(n) - \log n)}$$

rectangles. The claimed bound on the OR-FBDD size follows from Theorem 10.3.9 and the simple lower bound $\binom{n}{2}$.

Proof of the claim. The claim is proved by a simple counting argument. Let the vertices $v_1, \ldots, v_n$ be numbered according to decreasing degree. We define the characteristic of a clique of size s as the maximum i such that v_i belongs to the clique. Now it is sufficient to prove an upper bound of $\binom{\lfloor (2e)^{1/2} \rfloor}{s-1}$ on the number of s-cliques with characteristic i. This is obvious if $i \leq \lfloor (2e)^{1/2} \rfloor$. Then we consider only cliques on $\lfloor (2e)^{1/2} \rfloor$ vertices with one fixed vertex. Now let $i > \lfloor (2e)^{1/2} \rfloor$. The sum of all degrees of the vertices equals $2e$. Since the degrees of $v_1, \ldots, v_n$ are decreasing, the degree of v_i is less than $\lfloor (2e)^{1/2} \rfloor$. The upper bound follows, since the s-cliques containing v_i can only contain neighbors of v_i besides v_i. □ □

10.4 Partitioned OBDDs

Nondeterminism is a powerful complexity theoretical concept and nondeterministic representations of Boolean functions can be much smaller than their deterministic counterparts. But we have also seen that the simple operation NOT may cause an exponential blow-up of the size. The function $\overline{\mathrm{PERM}_n}$ has polynomial OR-OBDD size but PERM_n has exponential size even for OR-FBDDs.

The satisfiability test for OR-OBDDs can be solved by a DFS traversal checking whether the 1-sink can be reached from the source, but the test whether the represented function is the constant 1 is coNP-complete. This is known for the disjunction of monomials and such functions can be represented (see Theorem 10.1.4) by ordered OR-DTs.

In order to obtain representation types with good algorithmic behavior, we have to consider restrictions where, in particular, negation is a simple operation, i.e., representation types where NP = coNP and NP $\neq$ P (otherwise we may use deterministic OBDDs). Partitioned BDDs introduced by Jain, Bitner, Fussell, and Abraham (1992) and more intensively studied by Narayan, Jain, Fujita, and Sangiovanni-Vincentelli (1996) have the desired properties. The first idea is to allow only one nondeterministic node with a fixed fan-out $k = k(n)$ at the source. The k edges leaving this OR-node lead to k OBDDs $G_1, \dots, G_k$ with fixed but possibly different variable orderings $\pi_1, \dots, \pi_k$. Since we have efficient OBDD synthesis algorithms only if the variable ordering is fixed, we assume that G_i and G_j do not share nodes if $i \neq j$. This has the further advantage that we may work with G_i in the main storage while all G_j, $j \neq i$, are in a secondary memory. We need a further restriction, since $\overline{\text{PERM}}_n$ is still representable in polynomial size. The essential idea which also leads to a canonical representation is to fix sets $P_i \subseteq \{0,1\}^n$ such that G_i has to represent f correctly on P_i and to compute 0 on all outputs outside P_i. The characteristic functions of P_i are called window functions w_i.

Definition 10.4.1. The vector $w = (w_1, \dots, w_k)$ of functions from B_n is called a vector of *window functions* for $\{0,1\}^n$ if $w_1 + \cdots + w_k = 1$. The window functions are called *disjoint* if $w_i \wedge w_j = 0$ for $i \neq j$.

Definition 10.4.2. A (k, w, π)-PBDD (*partitioned* BDD with k parts, the vector $w = (w_1, \dots, w_k)$ of window functions, and the vector $\pi = (\pi_1, \dots, \pi_k)$ of variable orderings) for $f \in B_n$ is an OR-FBDD G representing f with one nondeterministic node with fan-out k at the source such that the ith edge leaving the OR-node leads to the source of a π_i-OBDD G_i representing $f_i = f \wedge w_i$ and G_i and G_j do not share nodes if $i \neq j$.

Definition 10.4.3. A sequence of functions $f = (f_n)$ where $f_n \in B_n$ has polynomial-size PBDDs if f_n can be represented by polynomial-size (k_n, w_n, π_n)-PBDDs G_n, in particular the number of parts may depend on n.

As in the case of OBDDs and other BDD variants, we look for efficient algorithms, e.g., for the synthesis problem, only if the number of parts, the vector of variable orderings, and the vector of window functions are fixed. For many examples investigated in Section 10.2, it is easy to find appropriate window functions.

Theorem 10.4.4. *The following functions have polynomial-size PBDDs even under the restriction that all parts use the same variable ordering:* HWB_n, ISA_n, WS_n, *and the pointer-jumping function* $PJ_{k,n}$ *for constant* k.

Proof. The proof is left as an exercise. The design of appropriate window functions is easy. E.g., for HWB$_n$, let w_i, $0 \le i \le n$, compute 1 iff $x_1 + \cdots + x_n = i$. □

For these concrete functions, we easily find appropriate window functions because of our knowledge of the structural properties of the functions. Good heuristic algorithms for the automatic creation of appropriate window functions have been developed using methods known as *functional partitioning* (Jain, Bitner, Fussell, and Abraham (1992) and Lai, Pedram, and Vrudhula (1993)). In the following, we assume that window functions and variable orderings are given and fixed.

Theorem 10.4.5. *Let* G_f, G_g, *and* G_1 *be* (k, w, π)*-PBDDs representing* f, g, *and the constant* 1, *respectively:*

(*i*) *The evaluation problem on* G_f *can be solved in time* $O(k \cdot depth(G_f))$.

(*ii*) *A* (k, w, π)*-PBDD for* $f \wedge g$, $f + g$, *or* $f \oplus g$ *can be computed in time* $O(|G_f||G_g|)$.

(*iii*) *A* (k, w, π)*-PBDD for* $\overline{f}$ *can be computed in time* $O(|G_f||G_1|)$.

(*iv*) *The SAT problem on* G_f *can be solved in time* $O(|G_f|)$.

(*v*) *The SAT-COUNT problem on* G_f *can be solved in time* $O(|G_f|)$ *if the window functions are disjoint.*

(*vi*) *The representation by* (k, w, π)*-PBDDs is canonical and the reduction of* G_f *is possible in time* $O(|G_f|)$.

(*vii*) *The equivalence of* G_f *and* G_g *can be checked in time* $O(|G_f| + |G_g|)$.

Proof. For the evaluation problem, it is sufficient to follow the k paths activated by the input. For the binary synthesis problem, it is sufficient to verify that $(f \wedge w_i) \wedge (g \wedge w_i) = (f \wedge g) \wedge w_i$, $(f \wedge w_i) + (g \wedge w_i) = (f + g) \wedge w_i$, and $(f \wedge w_i) \oplus (g \wedge w_i) = (f \oplus g) \wedge w_i$. Hence, it is sufficient to apply the π_i-OBDD synthesis algorithm to the ith parts of G_f and G_g $(1 \le i \le k)$. The negation can be done as an EXOR-synthesis with the constant 1. The (k, w, π)-PBDD representing the constant 1 consists of π_i-OBDDs representing the window functions w_i, $1 \le i \le k$. The satisfiability test is easy even for general OR-FBDDs. If the window functions are disjoint, $|f^{-1}(1)|$ is the sum of all $|(f \wedge w_i)^{-1}(1)|$ and we can apply the OBDD SAT-COUNT algorithm to all parts of G_f. The representation by (k, w, π)-PBDDs is the representation by π_i-OBDDs for $f \wedge w_i$, $1 \le i \le k$. Each part can be reduced in linear time. The equivalence can also be checked individually for all the parts. □

All other operations listed in Section 1.3 are based on replacements by constants and this operation causes problems. Let g_n test whether the matrix X_n consists of rows with exactly one 1-entry and let h_n be the analogous function for the columns. The function $f_n = sg_n + \overline{s}h_n$ with an additional variable s can be represented by PBDDs with two parts and the window functions s and $\overline{s}$. For the first part we use a rowwise variable ordering and for the second part a columnwise variable ordering. For the replacement of s by 1, we obtain the function g_n. Then we have to represent in the second part $\overline{s}g_n$ by a columnwise variable ordering which needs exponential size. Moreover, $(\forall s)f_n = g_n \wedge h_n = \text{PERM}_n$ has exponential OR-FBDD size and, therefore, also exponential PBDD size.

Theorem 10.4.6. *The replacement and quantification problems may cause an exponential blow-up of the size for (k, w, π)-PBDDs.*

The heuristics for the generation of window functions often construct window functions which are the 2^m minterms with respect to a small set V' of m variables. Then the replacement of a variable $x_i \in X_n - V'$ by a constant is easy. This holds more generally if the window functions do not essentially depend on x_i. Then $f_{|x_i=c} \wedge w_j = (f \wedge w_j)_{|x_i=c}$ and the replacement can be done for each part in the usual way. Afterwards, quantification is a binary synthesis operation.

As the next step of our investigations of PBDDs, we compare the expressive power of PBDDs with other BDD variants (Bollig and Wegener (1997a, 1997b)).

Theorem 10.4.7. *There exist functions $f_n \in B_{n^2}$ which are representable by linear-size PBDDs with two disjoint parts and which need size $2^{\Omega(n^{1/2})}$ for FBDDs, size $2^{\Omega(n/k)}$ for k-OBDDs, and size $2^{\Omega(n)}$ for EXOR-OBDDs.*

Proof. The function f_n tests for $n \times n$ Boolean matrices X_n whether either the number of ones in X_n is odd and X_n contains a row consisting of ones only or the number of ones in X_n is even and X_n contains a column consisting of ones only.

The representation by PBDDs is easy. The window functions check whether the number of ones is odd or even. Then the first part uses a rowwise variable ordering and the second one a columnwise variable ordering.

In Theorem 6.2.13, we have proved a $2^{\Omega(n)}$ lower bound on the FBDD size of $\text{ROW}_n + \text{COL}_n$. A similar bound can be proved for f_n (see exercises).

For the two other bounds, we apply lower bound techniques based on communication complexity. Without loss of generality, n is even. For a given variable ordering, we consider the cut where for the first time $n - 2$ rows or columns have the property that at least one variable has been tested. Without loss of generality, these are the first $n-2$ rows. We give the variables which are tested before the cut to Alice and the other ones to Bob. Then we consider the following submatrix of the communication matrix. Exactly half of the variables tested as first variables in the first $n-2$ rows get the value 1 and the other half the value 0. All other variables given to Alice get the value 1. For each of these partial assignments (which have the same number of ones), we define a corresponding assignment to Bob's variables. The rows among the first $n-2$ consisting up to now of only ones are filled with zeros and the rows with a zero are filled with ones. The last but one row is filled with zeros ensuring that no column contains ones only and the last row is filled in a way ensuring that the number of ones is odd. Because of the definition of the cut, each assignment to Alice's variables together with the corresponding assignment to Bob's variables leads to the output 0. But the combination of an assignment to Alice's variables and a noncorresponding assignment to Bob's variables leads to the output 1. Hence, we obtain a submatrix of the communication matrix of size $N \times N$ where $N = \binom{n-2}{(n-2)/2} = 2^{\Omega(n)}$ such that the matrix has zeros only on the main diagonal. This matrix has rank N over $\mathbb{Z}_2$ and leads to a fooling set of size N. Hence, the lower bounds follow by the results of Section 10.3. □

Theorem 10.4.8.

(*i*) *There are functions representable by polynomial-size PBDDs (with two disjoint parts) which need exponential size for FBDDs, EXOR-OBDDs, and k-OBDDs if $k = O(n^{1-\varepsilon})$ (nonpolynomial size if $k = o(n/\log n)$).*

(*ii*) *Functions with polynomial-size k-OBDDs and $k = O(1)$ have polynomial-size PBDDs.*

(*iii*) *Functions with polynomial-size PBDDs with k parts have polynomial-size k-IBDDs.*

Proof. The first result is a corollary to Theorem 10.4.7.

The second result follows from the proof of Theorem 7.3.3, namely the SAT test for k-OBDDs G. The number of parts is bounded above by $|G|^{k-1}$, namely the different possibilities to reach the first nodes of the layers of the k-OBDD. This is a disjoint partition of the input space and each window function has a size bounded by $|G|^{k-1}$, since it is the conjunction of at most $k-1$ sub-OBDDs $G(v_i, v_{i+1})$ (for this notation see the proof of Theorem 7.3.3). The function represented in some part is the conjunction of the window function for that part and the function $G(v_r)$, where v_r depends on the part and describes the first node reached in the last layer. Hence, the size of each part is bounded by $|G|^k$.

The last result is easy. We obtain the k-IBDD in the following way. The first layer is the first part of the PBDD. Its 0-sink is replaced by the source of the second part of the PBDD and so on. If the PBDD uses only one common variable ordering, we even obtain a k-OBDD. □

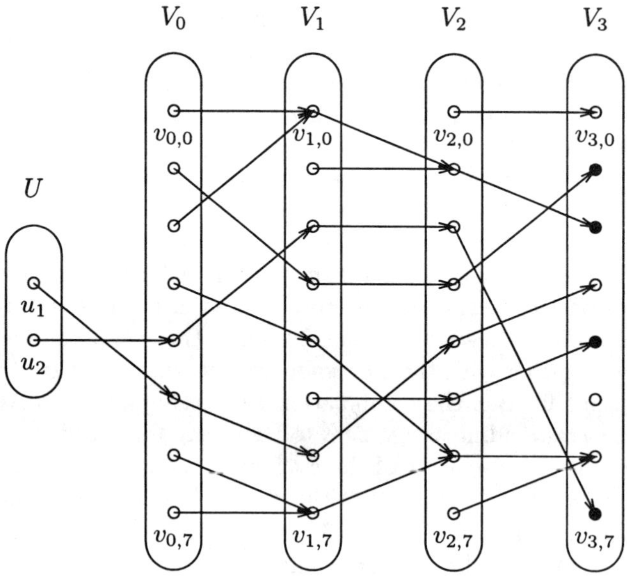

Figure 10.4.1: *The function* $P_{3,8}$ *for* $|s| = 0$*. The output equals* 1*, since an odd number of paths (exactly one) reaches a white vertex in* V_3*.*

The last result has the following implication. The function IP_n^* has polynomial-size EXOR-OBDDs (see Theorem 10.3.6) but no polynomial-size PBDDs with $k = O(1)$ parts. Otherwise, Theorem 10.4.8(iii) would imply the existence of polynomial-size k-IBDDs where $k = O(1)$ in contradiction to Theorem 10.3.6. This result can be extended to larger values of k (see Exercise 10.10).

The final question which we want to discuss in this section is how much nondeterminism is necessary to represent a function in small size. If a polynomial-size PBDD with k parts uses a common variable ordering, it is easy to obtain a polynomial-size PBDD with only $\lceil k/c \rceil$ parts if c is a constant. The OR-synthesis of c parts and also of c window functions can be done by the OBDD synthesis algorithm and the resulting size is at most the product of the sizes of the inputs. Bollig and Wegener (1997b) have proved, in the case of arbitrary variable orderings, a hierarchy result showing that there are functions $P_{k,n}$ representable by polynomial-size PBDDs with k parts but not representable by polynomial-size PBDDs with $k-1$ parts as long as $k = o(((\log n)/\log\log n)^{1/2})$. The definition of the functions $P_{k,n}$ is quite complicated (see Fig. 10.4.1).

Definition 10.4.9. The *path function* $P_{k,n}$ is defined on $(k-1+kn)\log n + \lceil \log k \rceil + n$ Boolean variables where n is chosen as a power of 2. The variables are denoted as follows:

- s_i, $0 \le i \le \lceil \log k \rceil - 1$, are selection variables describing a number $|s| \in \{0, \dots, 2^{\lceil \log k \rceil} - 1\}$,
- $z_{i,l}$, $1 \le i \le k-1$, $0 \le l \le \log n - 1$, describe pointers $|z_i| \in \{0, \dots, n-1\}$,
- $x_{i,j,l}$, $0 \le i \le k-1$, $0 \le j \le n-1$, $0 \le l \le \log n - 1$, describe pointers $|x_{i,j}| \in \{0, \dots, n-1\}$,
- c_j, $0 \le j \le n-1$, are color variables.

The value of $P_{k,n}(s, z, x, c)$ is defined in the following way. If $|s| \ge k$, the output equals 0. Otherwise, the input describes a graph on the vertex set $U \uplus V_0 \uplus \cdots \uplus V_k$ where $U = \{u_1, \dots, u_{k-1}\}$ and $V_i = \{v_{i,0}, \dots, v_{i,n-1}\}$. Each vertex of $U, V_0, \dots, V_{k-1}$ has outdegree 1 and each vertex of V_k is colored by a Boolean variable. We consider $k-1$ paths starting at $u_1, \dots, u_{k-1}$. The pointer starting at u_i leads to the vertex with number $|z_i|$ of V_0. The pointer starting at $v_{0,j}$ leads to the vertex with number $|x_{|s|,j}|$ of V_1 or, more generally, the pointer starting at $v_{r,j}$, $0 \le r \le k-1$, leads to the vertex with number $|x_{(|s|+r) \bmod k, j}|$ of V_{r+1}. The vertex with number j of V_k is colored white if $c_j = 1$ and black otherwise. For $i \in \{1, \dots, k-1\}$, we follow the path starting at u_i until after $k+1$ steps a vertex of V_k is reached. The function $P_{k,n}(s, z, x, c)$ outputs 1 if the number of paths reaching a white vertex is odd.

Lemma 10.4.10. *The function $P_{k,n}$ can be represented in size $O(2^k k^3 n^k)$ by PBDDs with k parts and disjoint window functions.*

Proof. The window function w_i, $0 \le i \le k-2$, tests whether $|s| = i$ and w_{k-1} tests whether $|s| \ge k-1$. These window functions are disjoint and each one has an OBDD size of $O(\log k)$ independently from the variable ordering. For the ith part, $0 \le i \le k-1$, we choose the following variable ordering: s-variables, z-variables, $x_{i,\cdot,\cdot}$-variables, $x_{(i+1) \bmod k,\cdot,\cdot}$-variables, $\dots$, $x_{(i+k-1) \bmod k,\cdot,\cdot}$-variables, c-variables.

We prove an upper bound of $O(2^k k^2 n^k)$ for the ith part. Without loss of generality, $i = 0$. First, it is checked whether $|s| = 0$. If $|s| \ne 0$, the 0-sink is reached. Otherwise, we follow all paths simultaneously. The z-variables are tested in a complete binary tree with n^{k-1} leaves. Then we know for each path which vertex in V_0 is reached. We prove an upper bound $O(2^k k n^k)$ for the size of the layer testing the $x_{0,\cdot,\cdot}$-variables. This layer has n^{k-1} leaves describing for each path which vertex in V_1 is reached. The following $k-1$ layers work similarly. Then we know for each path which vertex in V_k is reached. In a last step, we compute the parity of the colors of these vertices. Since the number of layers is k (plus the layer for the color variables) and the number of parts is k, we obtain the claimed bound.

Finally, we describe the layer testing the $x_{0,\cdot,\cdot}$-variables. We describe our current knowledge by $(w_1, \dots, w_{k-1}, m_1, \dots, m_{k-1})$ with the following meaning. If $m_l = 0$, the lth path reaches the w_lth vertex of V_0 and the variables describing the pointer which leaves this vertex are not yet tested. If $m_l = 1$, the lth path reaches the w_lth vertex of V_1. In the beginning, $m_1 = \cdots = m_{k-1} = 0$. The number of possible knowledge vectors is bounded above by $2^{k-1}n^{k-1}$. There are at most $2^{k-1}kn^{k-2}$ knowledge vectors where we have to test the $x_{0,j,\cdot}$-variables for some fixed value of $j \in \{0, \ldots, n-1\}$. For this, it is necessary that some w-value be equal to j. The $x_{0,j,\cdot}$-variables are then tested in a complete binary tree of size $O(n)$ and the knowledge vector is updated in the obvious way. The size of the jth sublayer is $O(2^k kn^{k-1})$ and the size of all n sublayers is $O(2^k kn^k)$. □

Theorem 10.4.11. *The size of PBDDs with $k-1$ parts representing $P_{k,n}$ is bounded below by 2^B where $B = \Omega(n^{1/(2k)}k^{-5}\log^{-1} n)$.*

Before we prove this lower bound, we describe the consequences.

Theorem 10.4.12. *There are Boolean functions representable by polynomial-size PBDDs with k parts and disjoint window functions but not by polynomial-size PBDDs with $k-1$ parts if $k = o(((\log n)/\log\log n)^{1/2})$.*

Proof. The result follows from Lemma 10.4.10 and Theorem 10.4.11. This is easy to see for constant k. If, in general, $k = (\log^{1/2} n)/\alpha(n)$, we restrict the vertex sets $V_0, \ldots, V_k$ to $r = 2^{(\log^{1/2} n)\alpha(n)}$ "active" vertices and $n-r$ "dummy" vertices (the output is 0 if a dummy vertex is reached). Then the upper bound of Lemma 10.4.10 is polynomially bounded while the lower bound of Theorem 10.4.11 is still superpolynomial. □

Proof of Theorem 10.4.11. Let G be a PBDD with $k-1$ parts representing $P_{k,n}$ and let the parts $G_1, \ldots, G_{k-1}$ be ordered according to the variable orderings $\pi_1, \ldots, \pi_{k-1}$. Without loss of generality, $k = o(\log n)$, since, otherwise, the bound is trivial.

The proof is technically involved. The main idea is as follows. The selection variables may distinguish between k essential situations, namely $|s| = 0, \ldots, |s| = k-1$. It is reasonable to conjecture that different situations need different variable orderings and that the number of available variable orderings is too small. By counting arguments, we prove the existence of an assignment of constants to some variables such that we essentially (after renaming) get the following situation (see Fig. 10.4.2).

The path starting at u_1 (similarly for the other paths) reaches $v_{0,0}$ and then has m possibilities to reach a node in V_1 where m is sufficiently large. The paths starting at these nodes are uniquely determined and disjoint until V_i, more precisely $V_{i(1)}$, is reached. Then, for each vertex which may be reached in V_i, there are two possible successors and there is one Boolean variable deciding between these possibilities. Moreover, all possible successors are distinct. Afterwards, the paths starting at the $2m$ possible vertices are uniquely determined and disjoint. For each possible vertex in V_i, there is exactly one path

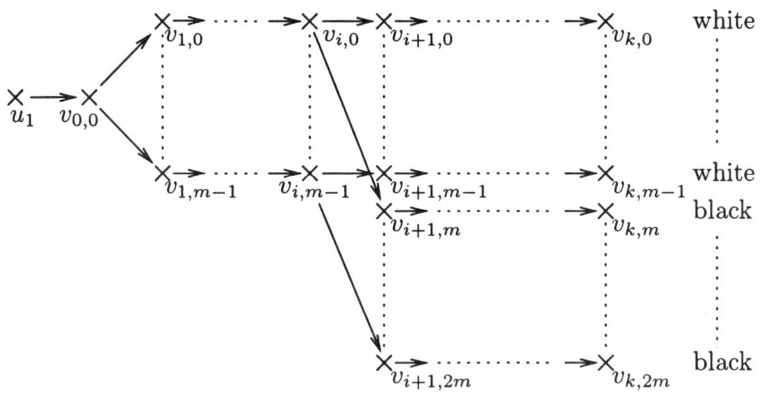

Figure 10.4.2: *Restrictions for the first path of $P_{k,n}$.*

reaching a white vertex. Moreover, all still possible paths starting at u_j are disjoint from all still possible paths starting at $u_{j'}$ if $j' \neq j$. The paths have disjoint "corridors." Finally, we ensure that, for each chosen vector of variable orderings $\pi = (\pi_1, \ldots, \pi_{k-1})$, there is one path where the variables deciding about the successors of $v_{i,0}, \ldots, v_{i,m-1}$ are tested before the variables deciding about the successor of the vertex $v_{0,0}$. Hence, each G_j is either large or has not enough information about one path and, therefore, cannot determine the output. In the following we make these ideas precise.

The number of x-variables equals $kn \log n$. Let A_j, $1 \leq j \leq k-1$, be the set of those $n \log n$ x-variables which are the first ones according to π_j. Hence, we can choose $n \log n$ variables not contained in any A_j. Moreover, we choose an index i such that at least $(n \log n)/k$ $x_{i,\cdot,\cdot}$-variables are not contained in any A_j. Without loss of generality, $i = 0$ and we fix the s-variables such that $|s| = 0$. Among the vertices in V_0, we choose those $k-1$ whose pointers contain the largest numbers of the chosen "late" $(n \log n)/k$ $x_{0,\cdot,\cdot}$-variables. It follows from a simple counting argument that, for each of these $k-1$ vertices, w.l.o.g. $V_0^* = \{v_{0,0}, \ldots, v_{0,k-2}\}$, there are at least $(\log n)/2k$ late variables. We fix the z-variables such that the pointer from u_i leads to $v_{0,i-1}$. With respect to π_j, there are $n \log n$ x-variables tested before the $(n \log n)/k$ late $x_{0,\cdot,\cdot}$-variables, among them at least $(n \log n)/k$ x-variables which are not $x_{0,\cdot,\cdot}$-variables. By the pigeonhole principle, we can choose some $j' \neq 0$ and a set B_j of at least $(n \log n)/k^2$ $x_{j',\cdot,\cdot}$-variables which are tested according to π_j before the late $x_{0,\cdot,\cdot}$-variables.

The sets B_j are not necessarily disjoint. There are at least n/k^2 indices h such that B_j contains an $x_{j',h,\cdot}$-variable. Hence, we can choose $C_j \subseteq B_j$ such that C_j contains at least n/k^3 variables belonging to different pointers and the property that C_i, $i \neq j$, does not contain an $x_{i',h,\cdot}$-variable if C_j contains an $x_{j',h,\cdot}$-variable. Let R_j be the set of vertices $v_{1,h}$ such that C_j

contains an $x_{j',h,\cdot}$-variable. For the pointer leaving $v_{0,j-1}$, we fix the at most $(1-1/(2k))\log n$ variables which are not late variables in such a way that the size of the set Q_j of vertices in R_j which are still reachable is maximized. Again by the pigeonhole principle,

$$|Q_j| \geq nk^{-3}2^{-(1-1/(2k))\log n} = n^{1/(2k)}k^{-3}.$$

In the following, we only consider inputs where the pointer from $v_{0,j-1}$ reaches a vertex in Q_j.

Now we ensure that we reach $v_{2,h}, \dots, v_{j',h}$ (remember that j' has been chosen above) starting at $v_{1,h} \in Q_j$. From $v_{j',h}$ we can reach two vertices $v_{j'+1,h}$ and $v_{j'+1,h'}$. We choose h' in such a way that h' differs from h in one bit position p and the variable $x_{j',h,p}$ belongs to C_j, i.e., to the set of variables tested before the late $x_{0,\cdot,\cdot}$-variables. In order to obtain different vertices in $V_{j'+1}$, we choose a set $Q'_j \subseteq Q_j$ such that $|Q'_j| \geq |Q_j|/\log n$ and the following property holds. There is one position p such that for $v_{1,h} \in Q'_j$ the variable $x_{j',h,p}$ belongs to C_j.

The sets $W_j \subseteq \{0, \dots, n-1\}$, $1 \leq j \leq k-1$, containing all i such that the path starting at u_j may reach $v_{j'+1,i}$ are not disjoint. The successors $v_{j'+1,h'}$ may cause difficulties. We construct sets $Q''_j \subseteq Q'_j$ such that the corresponding W_j-sets are disjoint. If we choose $v_{1,h}$ as an element of Q''_j, we include h and h' in W_j and forbid at most two vertices for each Q'_l, $l \neq j$. Hence, it is possible to choose appropriate sets $Q''_j \subseteq Q'_j$ such that

$$m := |Q''_j| \geq |Q'_j|/(2(k-1)) \geq |Q_j|/(2(k-1)\log n) \geq n^{1/(2k)}/(2k^4 \log n).$$

Finally, we fix pointers in such a way that, for $v_{1,h} \in Q''_j$, the path from $v_{j'+1,h}$ reaches $v_{j'+2,h}, \dots, v_{k,h}$ and the path from $v_{j'+1,h'}$ reaches $v_{j'+2,h'}, \dots, v_{k,h'}$. The vertex $v_{k,h}$ is colored white and the vertex $v_{k,h'}$ is colored black. We obtain the situation described in Fig. 10.4.2 and, for different u-vertices, the path systems are disjoint.

We consider the PBDD after all these replacements of variables by constants. The colors of the different paths (more precisely their last vertices) depend on disjoint sets of variables. This PBDD computes 1 for exactly half the remaining inputs. Therefore, there is one part, w.l.o.g., G_1, computing 1 on a fraction of at least $1/2(k-1)$ of the remaining inputs. Now we fix all variables not belonging to the path system starting at u_1 in such a way that the fraction of inputs where G_1 computes 1 is maximal. Let G_1^* denote the OBDD resulting from G_1 by this replacement. By the pigeonhole principle, the fraction of 1-inputs for G_1^* is still at least $1/2(k-1)$. Without loss of generality, the number of white paths among the paths starting at $u_2, \dots, u_{k-1}$ is even. Then G_1^* has to compute 1 for a fraction of at least $1/2(k-1)$ of the inputs such that the path (starting at u_1) is white and G_1^* has to compute 0 for all inputs where the path is black. The OBDD G_1^* tests the m variables deciding about the color of the path before the variables deciding about the successor of $v_{0,0}$. We have $m2^m$ inputs and G_1^* has to compute 1 on at least $m2^m/2(k-1)$ of them. We consider the cut after the test of all variables deciding about the color. Let s^* be the number of nodes of G_1^* below the cut. We may think of these nodes as nodes with m outgoing edges, one for each value of the variables describing the successor of $v_{0,0}$. If less than $m/4(k-1)$ edges leaving such a node w lead to the 1-sink, the node w is called "poor." The other nodes are called "rich." The number of paths reaching the 1-sink via poor nodes can be bounded by $m2^m/4(k-1)$. Hence, at least $m2^m/4(k-1)$ paths have to reach the 1-sink via rich nodes. For the rich node w, we assume, w.l.o.g., that the first $m' \geq m/4(k-1)$ edges reach the 1-sink. This is only possible if the paths for the corresponding m' vertices reach a white vertex. This implies that w is reached for at most $2^{m(1-1/4(k-1))}$ inputs. Hence, at most $m2^{m(1-1/4(k-1))}$ paths can reach the 1-sink via w and

$$s^* \geq \frac{m2^m/4(k-1)}{m2^{m(1-1/4(k-1))}} = 2^{m/4(k-1)}/4(k-1) = 2^{\Omega(m/k)}.$$

Combining this with the lower bound on m, we have proved the theorem. □

10.5 Algorithms for EXOR-OBDDs

In this section, we investigate EXOR-π-OBDDs with a fixed variable ordering π, w.l.o.g., $\pi = id$. We use the model where all inner nodes are labeled by Boolean variables and may have an unbounded number of outgoing 0-edges and 1-edges. This model has been discussed in Section 10.1. The function f_v represented at v computes 1 on input a iff the number of paths activated by a and leading from v to the 1-sink is odd. If an x_i-node v has 0-edges leading to $u_1, \ldots, u_k$ and 1-edges leading to $w_1, \ldots, w_l$, we conclude that

$$f_v = \overline{x}_i \wedge (f_{u_1} \oplus \cdots \oplus f_{u_k}) \oplus x_i \wedge (f_{w_1} \oplus \cdots \oplus f_{w_l}).$$

It is obvious that edges to the 0-sink can be eliminated and the constant 0 is represented by an empty BDD.

In the following, we describe efficient algorithms for the important operations on EXOR-π-OBDDs. The above remarks lead to a linear-time evaluation algorithm using a bottom-up approach. The following simple operations will be used in later algorithms.

Double edges are always eliminated. More precisely, if r edges with the same label lead from v to w, they are replaced by (r mod 2) edges of the same kind. This does not change the function computed at v, since $g \oplus g = 0$. EXOR-OBDDs may contain inner nodes without outgoing edges. These nodes represent the constant 0 and are always eliminated together with their incoming edges.

The index $ind(w)$ of a node w is defined as k if w is labeled by x_k and as $n+1$ if w is a sink. Let $v_1, \ldots, v_m$ be nodes such that $ind(v_1) = \cdots = ind(v_r) = k$ and $ind(v_i) > k$ if $i > r$. We create an x_k-node v representing $f_{v_1} \oplus \cdots \oplus f_{v_m}$. The node v gets c-edges leading to all c-successors of $v_1, \ldots, v_r$ and to $v_{r+1}, \ldots, v_m$.

It follows easily that $f_v = f_{v_1} \oplus \cdots \oplus f_{v_m}$ and that we still respect the given variable ordering. This operation is called *creation of linear combinations.*

The following operation, called *elimination of linear combinations,* is the inverse operation to the above one. Let $f_v = f_{v_1} \oplus \cdots \oplus f_{v_m}$ and $ind(v) \le ind(v_i)$, $1 \le i \le m$. The aim is to eliminate v without changing the functions represented at other nodes. This is possible by replacing edges to v by edges to $v_1, \ldots, v_m$. This does not imply that all nodes besides the node representing f can be eliminated. The reason is that the function f_v represented at the node v usually is not the EXOR-sum of functions represented at other nodes.

Now it is quite easy to obtain efficient algorithms for the problems binary synthesis and replacement by constants (Gergov and Meinel (1996)). The size of an EXOR-OBDD G is again denoted by $|G|$ but, in the case of nondeterministic OBDDs, we have to measure the size as the number of edges. The number of edges $|G|$ can be as large as $\Theta(|V(G)|^2)$ for the node set $V(G)$ of G.

Theorem 10.5.1.

(*i*) *The operation replacement by constants on EXOR-π-OBDDs G can be performed in time $O(|G|)$ and does not increase the number of nodes.*

(*ii*) *The EXOR-synthesis of EXOR-π-OBDDs G_f and G_g can be performed in time $O(|G_f| + |G_g|)$. The result G_h has $|V(G_f)| + |V(G_g)| + 1$ nodes.*

(*iii*) *The negation of an EXOR-π-OBDD can be performed in time $O(1)$.*

(*iv*) *The AND-synthesis of EXOR-π-OBDDs G_f and G_g can be performed in time $O(|G_f| \cdot |G_g|)$. The result G_h has $|V(G_f)| \cdot |V(G_g)|$ nodes.*

Proof. Without loss of generality, we consider the replacement of x_i by 1. For each node v labeled by x_i, we eliminate all outgoing 0-edges and for each outgoing 1-edge we create a 0-edge leading from v to the same node as the considered 1-edge. Afterwards, it is possible to eliminate all x_i-nodes but this may increase the number of edges considerably.

For the EXOR-synthesis, let v_f and v_g be the nodes representing f, respectively, g. It is sufficient to create an x_1-node representing $f \oplus g$.

The negation of f is the EXOR-synthesis of f and 1. This special case can be handled even more directly. If f is represented at an inner node v, this node gets two more outgoing edges, a 0-edge leading to the 1-sink and a 1-edge leading to the 1-sink. The case that f is represented by a sink or an empty diagram is trivial.

The EXOR-π-OBDD G_h for $h = f \wedge g$ is defined on $V(G_f) \times V(G_g)$ as in the basic π-OBDD synthesis algorithm. Let $(v, w) \in V(G_f) \times V(G_g)$, where the x_i-node v represents

$$f_v = \overline{x}_i \wedge (f_{v_1} \oplus \cdots \oplus f_{v_k}) \oplus x_i \wedge (f_{v_{k+1}} \oplus \cdots \oplus f_{v_{k+l}})$$

and the x_j-node w represents

$$g_w = \overline{x}_j \wedge (g_{w_1} \oplus \cdots \oplus g_{w_m}) \oplus x_j \wedge (g_{w_{m+1}} \oplus \cdots \oplus g_{w_{m+r}}).$$

Our construction is done in such a way that we represent $f_v \wedge g_w$ at (v, w). If $i = j$, then

$$\begin{aligned} f_v \wedge g_w &= \overline{x}_i \wedge (f_{v_1} g_{w_1} \oplus f_{v_1} g_{w_2} \oplus \cdots \oplus f_{v_k} g_{w_m}) \\ &\quad \oplus x_i \wedge (f_{v_{k+1}} g_{w_{m+1}} \oplus f_{v_{k+1}} g_{w_{m+2}} \oplus \cdots \oplus f_{v_{k+l}} g_{w_{m+r}}). \end{aligned}$$

Hence, (v, w) gets the label x_i, 0-edges leading to $(v_1, w_1), (v_1, w_2), \ldots, (v_k, w_m)$, and 1-edges leading to $(v_{k+1}, w_{m+1}), \ldots, (v_{k+l}, w_{m+r})$. If (v_a, w_b) represents $f_{v_a} \wedge g_{w_b}$, we can conclude that (v, w) represents $f_v \wedge g_w$. If $i < j$ (the case $i > j$ is handled similarly) and also in the case that w is the 1-sink,

$$f_v \wedge g_w = \overline{x}_i \wedge (f_{v_1} g_w \oplus \cdots \oplus f_{v_k} g_w) \oplus x_i \wedge (f_{v_{k+1}} g_w \oplus \cdots \oplus f_{v_{k+l}} g_w).$$

In this case, (v, w) gets the label x_i, 0-edges leading to $(v_1, w), \ldots, (v_k, w)$, and 1-edges leading to $(v_{k+1}, w), \ldots, (v_{k+l}, w)$. If (v_a, w) represents $f_{v_a} \wedge g_w$, the node (v, w) represents $f_v \wedge g_w$. Finally, the node (v^*, w^*) for the 1-sinks v^* and w^* is defined as the 1-sink of G_h. In this case, (v^*, w^*) represents $f_{v^*} \wedge g_{w^*}$ by definition. Now it follows by backward induction on the index that (v, w) represents $f_v \wedge g_w$. □

It is sufficient to have synthesis algorithms for AND, EXOR, and NOT. Each binary operator can be expressed by an AND- and at most three NOT-operations or by an EXOR- and at most one NOT-operation or as a NOT-operation.

Moreover, we remember that replacement by functions is a combination of replacement by constants and an ite synthesis step. Quantification consists of two replacements by constants followed by a binary synthesis step.

All these results are quite useless as long as we have no method to control the size of the created EXOR-π-OBDDs. Even for linear-size circuits of functions with small OBDD size, it is possible to obtain EXOR-π-OBDDs of exponential size using the synthesis algorithms described above.

Nobody knows how to minimize the size of EXOR-π-OBDDs in polynomial time. Waack (1997) has applied methods from linear algebra to describe which functions have to be represented in EXOR-π-OBDDs for f. This has led to a polynomial-time algorithm to minimize the *node size* of EXOR-π-OBDDs. Before presenting these results, we discuss some conclusions. Satisfiability can be easily tested after a node minimization. The unique node minimal EXOR-π-OBDD for the constant 0 is the empty diagram. The equivalence test is performed as a nonsatisfiability test for $f \oplus g$.

In π-OBDDs, we have to represent all subfunctions $f_{|x_1=a_1,\ldots,x_{i-1}=a_{i-1}}$. The essential observation of Waack (1997) is that, in EXOR-π-OBDDs, it is necessary and sufficient to represent a basis of the vector space spanned by the subfunctions represented in π-OBDDs for the same function. To be more precise, we consider the representation of a Boolean function f by its value table as an element of $(\mathbb{Z}_2)^{2^n}$. This set is a $\mathbb{Z}_2$ vector space where addition is a componentwise EXOR and scalar multiplication by 0 or 1 is defined in the obvious way. Different Boolean functions $f_1, \ldots, f_k$ span a subspace whose dimension is at least $\lceil \log k \rceil$ and at most k. It will turn out that EXOR-π-OBDDs for f are much smaller than π-OBDDs for f if the dimension of the subspace of all $f_{|x_1=a_1,\ldots,x_{i-1}=a_{i-1}}$, $1 \le i \le n+1$, is much smaller than the number of these subfunctions.

Definition 10.5.2.

(i) Let G be an EXOR-π-OBDD where $\pi = id$. The subspace $V_{G,k}$, $1 \le k \le n+1$, of $(\mathbb{Z}_2)^{2^n}$ is defined as the vector space spanned by the functions f_v represented at the nodes v of G whose index is at least k.

(ii) The subspace $V_{f,k}$, $1 \le k \le n+1$ and $f \in B_n$, is defined as the vector space spanned by the subfunctions $f_{|x_1=a_1,\ldots,x_m=a_m}$ where $k-1 \le m \le n$ and $a_i \in \{0,1\}$.

It follows from the definition that $V_{G,k+1} \subseteq V_{G,k}$ and $V_{f,k+1} \subseteq V_{f,k}$. Furthermore, we obtain a relation between $V_{G,k}$ and $V_{f,k}$ if G represents f.

Lemma 10.5.3. *Let G be an EXOR-π-OBDD representing f. Then $V_{f,k} \subseteq V_{G,k}$ for all k.*

Proof. By definition, it is sufficient to prove that $f' = f_{|x_1=a_1,\ldots,x_m=a_m}$, $m \ge k-1$, is contained in $V_{G,k}$. We consider the partial paths activated by the partial assignment $x_1 = a_1, \ldots, x_m = a_m$. Then f' is the EXOR-sum of all f_v such that v is reached by an odd number of activated partial paths. Since $ind(v) > m$, also $ind(v) \ge k$ and $f_v \in V_{G,k}$. This implies that f', the EXOR-sum of these f_v, is also contained in $V_{G,k}$. □

Theorem 10.5.4. *The EXOR-π-OBDD representing $f \in B_n$ with the minimal number of nodes contains $\dim(V_{f,1})$ nodes if $\pi = id$. The number of x_k-nodes equals $\dim(V_{f,k}) - \dim(V_{f,k+1})$.*

Proof. Lemma 10.5.3 implies that the number of nodes v where $ind(v) \ge k$ is at least $\dim(V_{f,k})$. Hence, the lower bounds of the theorem are proved.

We prove the upper bounds by the bottom-up construction of an appropriate EXOR-π-OBDD representing f. The upper bounds hold for the constant functions. In all other cases, it is easy to see that $V_{f,n+1}$ contains the constants 0 and 1 and has dimension 1 and the EXOR-OBDD G consisting of the 1-sink fulfills the property $V_{f,n+1} = V_{G,n+1}$.

Let $k \le n$. We assume that the levels $k+1, \ldots, n+1$ of G contain exactly $\dim(V_{f,k+1})$ nodes and that $V_{f,k+1} = V_{G,k+1}$. Our aim is to add $\dim(V_{f,k}) - \dim(V_{f,k+1})$ x_k-nodes to G such that $V_{f,k} = V_{G,k}$. Finally, we obtain an EXOR-π-OBDD G with $\dim(V_{f,1})$ nodes where $V_{f,1} = V_{G,1}$. Since $V_{f,k} \supseteq V_{f,k+1}$, the theorem from linear algebra that any linearly independent set can be extended to a basis implies the existence of $m = \dim(V_{f,k}) - \dim(V_{f,k+1})$ vectors, i.e., functions $g_1, \ldots, g_m$ such that these functions together with the functions represented at the levels $k+1, \ldots, n+1$ of G are a basis of $V_{f,k}$. We create m x_k-nodes $v_1, \ldots, v_m$ and choose their successors in such a way that v_i represents g_i. By definition, the function $g_{i|x_k=c}$ is contained in $V_{f,k+1}$. Hence, it is the EXOR-sum of functions represented at the levels $k+1, \ldots, n+1$ of G. If we create c-edges from v_i to the nodes representing these functions, we guarantee that g_i is represented at v_i. Since $f \in V_{f,1}$, $f \in V_{f,k} - V_{f,k+1}$ for some k. During the construction of the x_k-level, we choose f as a function represented by an x_k-node. This ensures that G represents f. □

There is no unique node-minimal EXOR-π-OBDD representing f, since the basis of vector spaces is not unique. The proof of Theorem 10.5.4 does not include an efficient algorithm for the minimization of the node size of an EXOR-π-OBDD G. In particular, we should not work with functions represented explicitly as elements of $(\mathbb{Z}_2)^{2^n}$.

Theorem 10.5.5. *Let G be an EXOR-π-OBDD representing f. A node minimal EXOR-π-OBDD G' representing f can be computed in time $O(n|V(G)|^3)$ using $O(|V(G)|^2)$ space.*

Proof. In a first DFS traversal, we eliminate all nodes not reachable from the node representing f. The resulting graph is again denoted by G. We cannot be sure that the functions represented in G are linearly independent. Hence, by Theorem 10.5.4, G is perhaps not node minimal. In the first phase, we use a bottom-up approach to eliminate nodes such that the resulting BDD represents functions which are linearly independent. In a second DFS traversal, nodes not reachable from the node representing f are eliminated. In the resulting graph G, the vector space $V_{G,1}$ may be larger than $V_{f,1}$ and, therefore, larger than necessary. In the second phase, we use a top-down approach to replace nodes such that the resulting EXOR-π-OBDD is, by Theorem 10.5.4, a node-minimal one for f.

We describe the first phase of the algorithm. If G does not contain a 1-sink, we are done by constructing the empty EXOR-OBDD representing the constant 0. Otherwise, we merge all 1-sinks and eliminate all 0-sinks. The functions (there is only one) represented at level $n+1$ are linearly independent. For the inductive step, we assume that the functions represented at the levels $k+1, \ldots, n+1$ are linearly independent and the functions represented at other levels have not been changed. We want to change the level k in such a way that the functions on the levels $k, \ldots, n+1$ span the same space as before and are linearly independent. Let $v_1, \ldots, v_l$ be the nodes on level k representing $g_1, \ldots, g_l$ and $w_1, \ldots, w_m$ the nodes on the levels $k+1, \ldots, n+1$, representing $h_1, \ldots, h_m$.

The first step is to obtain a short representation of the functions $g_1, \ldots, g_l, h_1, \ldots, h_m$, namely by vectors of length $2m$. We interpret the vector

$(a_1, \ldots, a_m, b_1, \ldots, b_m)$ as the following function F on $x_k, \ldots, x_n$. The function $F_{|x_k=0}$ is the EXOR-sum of all h_i such that $a_i = 1$ and $F_{|x_k=1}$ is the EXOR-sum of all h_i such that $b_i = 1$. Since h_j itself cannot essentially depend on x_k, this function is represented by $(a_1, \ldots, a_m, b_1, \ldots, b_m)$, where $a_j = b_j = 1$ and $a_r = b_r = 0$ if $r \neq j$. Let $S(i)$ contain the indices j such that a 0-edge leads from v_i to w_j. Then we choose $a_j = 1$ iff $j \in S(i)$ for the representation of g_i, similarly for b_j and the 1-edges. We obtain a $(2m) \times (m + l)$ matrix containing the representations of $h_1, \ldots, h_m, g_1, \ldots, g_l$ in this order as columns. This representation still carries all the information about linear dependencies. If g_i is the linear combination of some of the functions $h_1, \ldots, h_m, g_1, \ldots, g_{i-1}$, the column corresponding to g_i is the linear combination of the corresponding columns. The first m columns are linearly independent. By Gaussian elimination (consisting of row operations only), we determine whether g_i is a linear combination of some of the functions $h_1, \ldots, h_m, g_1, \ldots, g_{i-1}$. In the positive case, we eliminate v_i by the operation elimination of linear combinations.

There is one special case which needs more care. We cannot eliminate the source. If the function represented at the source, namely f, is the linear combination of the functions represented at $u_1, \ldots, u_r$, let w.l.o.g. u_1 be one node with the smallest index among these nodes. With the operation creation of linear combinations we create a node u with $ind(u) = ind(u_1)$ representing f and, afterwards, we can eliminate u_1.

We remark that the following property holds at the end of the first phase. Each x_k-node represents a function essentially depending on x_k. Otherwise, the x_k-node v represents g where $g_{|x_k=0} = g_{|x_k=1}$. Since the functions represented on the levels $k+1, \ldots, n+1$ are linearly independent, the functions $g_{|x_1=0}$ and $g_{|x_k=1}$ are represented in the same way, i.e., the 0-edges leaving v lead to the same nodes as the 1-edges leaving v. Then g is the linear combination of the functions represented at the direct successors of v and v is eliminated in the first phase.

Now we describe the second phase. The aim is to obtain an EXOR-π-OBDD representing only linearly independent functions from $V_{f,1}$, among them f. Let x_r be the label of the source representing f. Then the desired properties hold for the levels $1, \ldots, r$. We assume that these properties hold for the levels $1, \ldots, k-1$ where $k - 1 \geq r$ and want to establish the property for the levels $1, \ldots, k$.

For each node v on the levels $1, \ldots, k-1$ and each $c \in \{0, 1\}$, we check whether at least one c-edge leaving v reaches a node at level k. In the positive case, we apply the operation creation of linear combinations to create an x_k-node computing the EXOR-sum of all functions which are represented at c-successors of v on the levels $k, \ldots, n+1$. We replace these c-edges leaving v by a c-edge to the new node. This does not change the function represented at v. Afterwards, all old nodes at level k can be eliminated, since they are no longer reachable from the source. We claim that the function f_w represented at a new x_k-node w belongs to $V_{f,k}$. If w is a c-successor of v, $f_{v|x_i=c} = f_w \oplus g$ where g is the

EXOR-sum of some functions represented on the levels $i+1, \dots, k-1$. Hence, by the induction hypothesis, $g \in V_{f,1}$, $f_v \in V_{f,1}$, and also $f_{v|x_i=c} \in V_{f,1}$. This implies that $f_w = f_{v|x_i=c} \oplus g \in V_{f,1}$. Since f_w does not essentially depend on $x_1, \dots, x_{k-1}$, we even know that $f_w \in V_{f,k}$.

The functions now represented on level k are not necessarily linearly independent from each other and from the functions on the later levels. With the same approach as in the first phase, we obtain an x_k-level representing functions from $V_{f,k}$ such that the functions represented on the levels $k, \dots, n+1$ are linearly independent. Since the EXOR-π-OBDD still represents f, the x_k-level contains exactly $\dim(V_{f,k}) - \dim(V_{f,k+1})$ nodes. This implies by Theorem 10.5.4 that, at the end of the second phase, we obtain a node-minimal EXOR-π-OBDD representing f.

The runtime of the algorithm is dominated by the $2n$ Gaussian elimination steps and the result on the storage space is obvious. □

We have used the straightforward cubic-time Gaussian elimination algorithm. This reflects what is done in applications. From a complexity theoretical point of view, we may use the asymptotically best-known algorithm which decreases the exponent from 3 to 2.38 (Coppersmith and Winograd (1990)). In any case, the algorithms on EXOR-π-OBDDs are too slow for applications. One reason is that the node minimization cannot be integrated into the synthesis and the other reason is that Gaussian elimination takes a lot of time, not only in the worst case but also in the "usual case." Löbbing, Sieling, and Wegener (1998) have shown that we cannot avoid such messy computations, i.e., EXOR-π-OBDD can be handled in polynomial time but not "efficiently enough." We present their result without proof.

Theorem 10.5.6. *If the node minimization of EXOR-π-OBDDs with n nodes can be performed in $t(n)$ steps, it is possible to compute the $\mathbb{Z}_2$-rank of $n \times n$ matrices in time $O(t(n))$.*

This result holds even if we restrict the assumption to EXOR-π-OBDDs with $2n+1$ nodes which result from the EXOR-synthesis of two node-minimal EXOR-π-OBDDs with n nodes each. Hence, Theorem 10.5.6 holds in situations which occur in the synthesis process.

We finish this section with some comments. Theorem 10.5.4 contains a lower bound technique for EXOR-π-OBDDs which has the same structure as the lower bound technique presented in Theorem 3.1.4 for π-OBDDs. This method is not explicitly based on communication complexity and has been used by Jukna (1999) to prove exponential lower bounds on the EXOR-OBDD size of characteristic functions of linear codes where the Hamming distance of two codewords is large and the same holds for the dual code.

Another observation is that π-OFDDs and even π-OKFDDs (see Chapter 8) are special variants of EXOR-π-OBDDs. We obtain an EXOR-π-OBDD for f from a π-OFDD for f by adding to each inner node v an outgoing

1-edge which leads to the 0-successor of v. This result follows directly from Reed-Muller's decomposition rule. For π-OFDDs or π-OKFDDs, it is possible that the AND-synthesis leads to an exponential blow-up. If f and g are represented by small-size π-OFDDs or π-OKFDDs and if we ask whether $f \wedge g \equiv 0$, we may convert the given DDs to EXOR-π-OBDDs and perform the AND-synthesis, a node minimization, and the check whether the resulting diagram is empty. Many other problems on OFDDs and OKFDDs can be answered efficiently via EXOR-OBDDs (see, e.g., Exercises 8.15 and 8.18).

In Chapter 9, we discussed word-level representations, namely MTBDDs, BMDs, HDDs, EVBDDs, *BMDs, and Kronecker *BMDs. Scholl, Becker, and Weis (1998) (for similar results see also Thathachar (1998b)) introduced *word-level linear combination diagrams* (WLCDs) as a common generalization where nodes may have many outgoing 0-edges and 1-edges and additive and multiplicative weights. The value computed at some edge is computed from the value computed at the node reached by the edge and the weights corresponding to these edges. If $x_i = c$, the value computed at an x_i-node v is the sum of the values computed at the c-edges leaving v. Generalizing the approach of Waack (1997), it is possible to characterize the π-WLCD size of functions by the dimension of appropriate vector spaces. The function word-level division computes the value of $\lfloor |x|/|y| \rfloor$ from $(x_{n-1}, \ldots, x_0, y_{n-1}, \ldots, y_0)$ if $|y| \neq 0$. Scholl, Becker, and Weis (1998) have proved that the WLCD size of word-level division is bounded below by $2^{\Omega(n)}$ and this implies the same lower bound for all word-level representations mentioned above. Hence, there are representation types where multiplication is easy but division is hard.

10.6 Exercises and Open Problems

10.1.E Consider Ω-BPs for (a) $\Omega = \{\text{NOT}, \text{OR}\}$ and (b) $\Omega = \{\rightarrow\}$, where $(x \rightarrow y) = 1$ iff $x = 0$ or $x = y = 1$. Decide which of the five essential models is polynomially related to the considered model.

10.2.M Prove that $\{\text{AND}, \text{OR}\}$-DTs are polynomially related to formulas over the basis B_2.

10.3.M (See Damm and Meinel (1992).) Describe a function with polynomial formula size and exponential-size OR-DTs, AND-DTs, and EXOR-DTs.

10.4.O Is it possible to simulate polynomial-size OR-OBDDs by polynomial-size OR-OBDDs where no nondeterministic node follows a decision node?

10.5.O Is it possible to simulate polynomial-size OR-FBDDs by polynomial-size OR-FBDDs where no nondeterministic node follows a decision node?

10.6.M Prove that EQ_n^* has polynomial-size AND-π-OBDDs for all variable orderings.

10.7.M Prove that IP_n^* has polynomial-size EXOR-π-OBDDs for all variable orderings.

10.8.D (See Jukna, Razborov, Savický, and Wegener (1997).) Let $f_n \in B_n$, $n = 2^k$, be the following pointer function. The variables are partitioned into k $s \times s$ matrices, where $s = \lfloor (n/k)^{1/2} \rfloor$, and the set of the remaining variables. Each matrix is responsible for one bit of the pointer. This bit is 1 iff the majority of the rows contain more ones than zeros. If the pointer has the value p, the output of f_n equals x_p. Use this function and padding arguments to prove the existence of Boolean functions f_n^* contained in $\Sigma_3^p \cap \Pi_3^p$ and representable by polynomial-size OR- and AND-OBDDs but not by polynomial-size FBDDs.

10.9.E Prove that the conjunction of hyperplanar sum-of-products predicate CHSP_q^k has polynomial-size AND-FBDDs.

10.10.E Improve the lower bounds of Theorem 10.3.6. For which length can we obtain nonpolynomial lower bounds?

10.11.M Generalize the proof of Theorem 6.2.6 to prove an exponential lower bound on the OR-FBDD size of excl_n.

10.12.O Prove an exponential lower bound on the EXOR-FBDD size of an explicitly defined Boolean function.

10.13.E Prove Theorem 10.4.4.

10.14.M Prove that the function defined in Exercise 10.8 has polynomial-size PBDDs.

10.15.E Determine the size of (k, w, π)-PBDDs for the constant 1 and the variable x_i.

10.16.M Prove the FBDD lower bound stated in Theorem 10.4.7.

10.17.M (See Bollig and Wegener (1997b).) Prove that the path function $P_{k,n}$ can be represented in polynomial size by FBDDs, 2-OBDDs, and EXOR-OBDDs.

10.18.M Prove that functions representable by polynomial-size k-OBDDs for constant k are also representable by polynomial-size EXOR-OBDDs.

10.19.E Determine the complexity of the SAT-COUNT problem for EXOR-π-OBDDs.

10.20.M Design an efficient redundancy test for EXOR-π-OBDDs.

10.21.E Prove that the equality test EQ_n is almost ugly for EXOR-OBDDs.

10.22.O Is it an NP-hard problem to minimize the (edge) size of an EXOR-π-OBDD?

Chapter 11

Randomized BDDs and Algorithms

Probabilistic methods have turned out to be useful in almost all areas of computer science (see Motwani and Raghavan (1995)). We distinguish between randomized algorithms working on deterministic BDD variants (Section 11.1) and the use of randomized BDD variants (Sections 11.2–11.8).

11.1 Randomized Equivalence Tests

Based on the so-called fingerprinting technique, Blum, Chandra, and Wegman (1980) have proposed a randomized algorithm checking the equivalence of FBDDs. Let G_f and G_g be FBDDs representing $f \in B_n$ and $g \in B_n$, respectively. A naive approach is to check whether $f(a) = g(a)$ for a random input $a \in \{0,1\}^n$. If $f(a) \neq g(a)$, we can conclude that f and g are not equivalent. If $f(a) = g(a)$, we still have to believe that f and g may be equivalent. The error probability of such a test is close to 1 for functions f and g differing only on a polynomial number of inputs. A better idea is based on the algebraic version of Shannon's decomposition rule. For an x_i-node representing h and having successors representing h_0 and h_1, Shannon's decomposition rule can be written as $h(a) = \overline{a}_i \wedge h_0(a) \oplus a_i \wedge h_1(a)$. In a BDD, this formula has to be evaluated in the field $\mathbb{Z}_2$. We embed $\mathbb{Z}_2$ into a larger field $\mathcal{F}$ and interpret Shannon's decomposition rule in $\mathcal{F}$. This leads to $\mathcal{F}$-BDDs, which we introduce formally for FBDDs.

Definition 11.1.1. Let G be an FBDD over $X_n = \{x_1, \ldots, x_n\}$ and $\mathcal{F}$ be a field. The function $f\colon \mathcal{F}^n \to \mathcal{F}$ represented by G as an $\mathcal{F}$-FBDD is defined as follows. A sink with label c represents the constant function c. Let v be an

x_i-node whose direct successors are v_0 and v_1. Then

$$f_v(a) := (1 - a_i) * f_{v_0}(a) + a_i * f_{v_1}(a),$$

where the operations are performed in $\mathcal{F}$.

For $\mathcal{F} = \mathbb{Z}_2$, we obtain the usual interpretation of FBDDs.

Lemma 11.1.2. *If the FBDDs G_1 and G_2 represent the same Boolean function f as $\mathbb{Z}_2$-FBDDs, they represent the same function $f_{\mathcal{F}}$ as $\mathcal{F}$-FBDDs.*

Proof. First, we consider complete FBDDs, i.e., FBDDs where all paths from the source to a sink have length n and, therefore, contain tests of all n variables. Then, G_1 and G_2 contain for each $b \in f^{-1}(1)$ a unique path from the source to the 1-sink, namely the computation path for b. For $c \in \{0, 1\}$, let $I_c(b)$ be the set of all i where $b_i = c$. The complete $\mathcal{F}$-FBDDs G_1 and G_2 represent the same function

$$f_{\mathcal{F}}(x) = \sum_{b \in f^{-1}(1)} \prod_{i \in I_0(b)} (1 - x_i) * \prod_{i \in I_1(b)} x_i.$$

In the general case, it is easy to obtain complete FBDDs from G_1 and G_2 by including dummy tests, i.e., x_i-nodes v which have the same node w as 0-successor and 1-successor. Then $f_v(a) = (1 - a_i)f_w(a) + a_i f_w(a) = f_w(a)$. This implies that the inclusion of dummy nodes does not change the function represented by the $\mathcal{F}$-FBDD. Hence, the lemma holds for all FBDDs. □

Lemma 11.1.3. *Let $S \subseteq \mathcal{F}$ have size s. If the FBDDs G_f and G_g represent different Boolean functions f and g as $\mathbb{Z}_2$-FBDDs, they differ as $\mathcal{F}$-FBDDs on at least $(s-1)^n$ inputs $a \in S^n$.*

Proof. We only investigate inputs $a \in S^n$ and prove the claim by induction on n. The claim is obvious for $n = 0$. For $n > 0$, we apply the evaluation rule for $\mathcal{F}$-FBDDs. Hence,

$$f_{\mathcal{F}}(x) = (1 - x_1) * f_{\mathcal{F}|x_1=0}(x) + x_1 * f_{\mathcal{F}|x_1=1}(x)$$

or briefly $f = (1 - x_1)f_0 + x_1 f_1$ and, analogously, $g = (1 - x_1)g_0 + x_1 g_1$. Since f and g are different, f_0 and g_0 are different or f_1 and g_1 are different (or both). Without loss of generality, we assume that f_0 and g_0 are different. Since these functions essentially depend on at most $n - 1$ variables, we conclude by the induction hypothesis that they differ on at least $(s-1)^{n-1}$ assignments to $x_2, \ldots, x_n$. In order to have $f(a) = g(a)$ for some $a = (a_1, \ldots, a_n)$ where $f_0(a') \neq g_0(a')$ for $a' = (a_2, \ldots, a_n)$, it is necessary that

$$(1 - a_1)f_0(a') + a_1 f_1(a') = (1 - a_1)g_0(a') + a_1 g_1(a').$$

This is equivalent to

$$a_1(f_1(a') - f_0(a') + g_0(a') - g_1(a')) = g_0(a') - f_0(a').$$

Since $g_0(a') - f_0(a') \neq 0$, this is impossible if $d := f_1(a') - f_0(a') + g_0(a') - g_1(a') = 0$. If $d \neq 0$, we get the unique solution $a_1 = (g_0(a') - f_0(a')) * d^{-1}$. For fixed a' there is one a_1 such that $f(a) = g(a)$ and it may even happen that $a_1 \notin S$. Hence, we obtain at least $(s-1)^{n-1}(s-1) = (s-1)^n$ inputs $a \in S^n$ where f and g differ. □

Algorithm 11.1.4. Let $\mathcal{F}$ be a field and $S \subseteq \mathcal{F}$.
Input: FBDDs G_f and G_g representing $f \in B_n$ and $g \in B_n$, respectively,

(1) Choose $a \in S^n$ randomly.

(2) Evaluate G_f and G_g as $\mathcal{F}$-FBDDs on a. Call the results $r(f)$ and $r(g)$, respectively.

(3) Output "$f \neq g$" if $r(f) \neq r(g)$ and "presumably $f = g$" otherwise.

Theorem 11.1.5. *Let $s = |S|$. The runtime of Algorithm 11.1.4 is dominated by $O(|G_f| + |G_g|)$ field operations. The algorithm is a randomized equivalence test for FBDDs with one-sided error whose error probability is bounded above by $1 - (1 - \frac{1}{s})^n$.*

Proof. The result on the runtime follows if we apply the usual evaluation algorithm taking $O(1)$ time per node. If $f = g$, the algorithm answers "presumably $f = g$" by Lemma 11.1.2. If $f \neq g$, Lemma 11.1.3 implies that the algorithm gives the right answer with a probability of at least $(s-1)^n/s^n = (1 - \frac{1}{s})^n$. □

Corollary 11.1.6. *The equivalence test for FBDDs is contained in co-RP.*

Proof. Choose a field $\mathcal{F} = \mathbb{Z}_p$ for some small prime $p \geq 2n$. Then the field operations can be performed efficiently. Choosing $S = \mathcal{F}$, the error probability is bounded above by $1-(1-\frac{1}{2n})^n$ which is at most $1/2$ by Bernoulli's inequality. □

Using larger fields, we may decrease the error probability but the bit complexity of the field operations increases. We also may choose $\mathcal{F} = \mathbb{R}$ and $S = \{1, \ldots, 2n\}$. Then $r(f)$ may be exponentially large and we have to work with numbers of bit length $\Theta(n)$.

EXOR-FBDDs and, in particular, EXOR-OBDDs combine Shannon's decomposition rule with the EXOR-sum for the combination of the results computed at the a_i-edges leaving x_i-nodes (see Section 10.5). If the characteristic of the chosen field equals 2, addition coincides with EXOR on $\{0,1\}$ and we can directly generalize our results on FBDDs to EXOR-FBDDs. This has been observed by Gergov and Meinel (1996), who suggested GF(2^m), where $2^m \geq 2n$, as a field. Hence, the equivalence test for EXOR-FBDDs can also be performed efficiently with the fingerprinting algorithm. This algorithm is much faster than the deterministic equivalence test for EXOR-OBDDs based on the results of Section 10.5. Savický (1998a) has used the fingerprinting technique to develop a randomized equivalence test for syntactic $(1,+k)$-BPs G which runs in time $|G|^{O(k)}$, i.e., in polynomial time for constant k.

In applications, the value of an $\mathcal{F}$-FBDD or $\mathcal{F}$-OBDD on a random input $a \in S^n$, $S \subseteq \mathcal{F}$, is called the *signature.* Jain, Abraham, Bitner, and Fussell (1992) and Shen, Devadas, and Ghosh (1995) have presented heuristic ideas to compute the signature without constructing the whole FBDD or OBDD from the given circuit. Now we know that EXOR-gates and, therefore, NOT-gates do not cause problems if we choose the appropriate field. Disjunctions like $f_1 + \cdots + f_k$ can be replaced by $f_1 \oplus \cdots \oplus f_k$ if $f_i \wedge f_j = 0$ for $i \neq j$. This is called an orthogonal partition. In such a situation, the signature of the disjunction can also be computed as the sum of the signatures of the f_i. In general, the Boolean sum $f_1 + f_2$ is equal to the $\mathcal{F}$-expression $f_1 + f_2 - f_1 * f_2$ and, in the case that $f_1 * f_2 = 0$ on the Boolean inputs, Lemma 11.1.2 implies that $f_1 * f_2$ has the signature 0. It can also be shown that the signature of $g \wedge h$ is the product of the signatures of g and h if g and h are defined on disjoint sets of variables.

The common disadvantage of the randomized equivalence test algorithms is that they make the error on the "wrong side." They may accept two functions as equivalent if they are not while the decision that functions are not equivalent is free of errors. Even if the error probability is reduced by a sequence of independent runs, we cannot *verify* in a strong sense that specification and realization are equivalent.

11.2 Randomized BDD Variants

The usual OR-nondeterminism can be interpreted as randomization with one-sided error and the weak restriction that the error probability is less than 1. The following definition of randomized BPs is inspired by this point of view (compare Definition 10.1.1).

Definition 11.2.1. A *randomized* BP (or BDD) G is a directed acyclic graph with decision nodes for Boolean variables and *randomized nodes.* A randomized node is an unlabeled node with two outgoing edges. Sinks can be labeled by 0, 1, or "?". The *random computation path* for a is defined as follows. At decision nodes labeled by x_i, the outgoing a_i-edge is chosen. At randomized nodes, each outgoing edge is chosen independently from all other random decisions with probability 1/2. The *acceptance probability* $acc_G(a)$ or $Prob(G(a) = 1)$ of G on a is the probability that the random computation path reaches a 1-sink. The *rejection probability* $rej_G(a)$ or $Prob(G(a) = 0)$ of G on a is the probability that the random computation path reaches a 0-sink.

A randomized BP is a nondeterministic representation of f if $acc_G(a) > 0$ for $a \in f^{-1}(1)$ and $rej_G(a) = 1$ for $a \in f^{-1}(0)$. Here we look for randomized representations of Boolean functions defined like randomized algorithms for decision problems.

Definition 11.2.2. Let G be a randomized BP on n variables.

(i) G represents $f \in B_n$ with *unbounded error* if $Prob(G(a) = f(a)) > 1/2$ for all inputs a.

(ii) G represents $f \in B_n$ with *two-sided ε-bounded error*, $0 \le \varepsilon < 1/2$, if $Prob(G(a) \neq f(a)) \le \varepsilon$ for all inputs a.

(iii) G represents $f \in B_n$ with *one-sided ε-bounded error*, $0 \le \varepsilon < 1$, if $Prob(G(a) \neq 1) \le \varepsilon$ for all $a \in f^{-1}(1)$ and $Prob(G(a) = 0) = 1$ for all $a \in f^{-1}(0)$.

(iv) G represents $f \in B_n$ with *zero error and ε-failure*, $0 \le \varepsilon < 1$, if $Prob(G(a) = \overline{f}(a)) = 0$ and $Prob(G(a) = \ ?) \le \varepsilon$ for all inputs a.

Definition 11.2.3.

(i) A function $f = (f_n)$ is contained in ZPP$_\varepsilon$-BP, RP$_\varepsilon$-BP, or BPP$_\varepsilon$-BP if it can be represented by polynomial-size randomized BPs with zero error and ε-failure, one-sided ε-bounded error, or two-sided ε-bounded error, respectively, where $0 \le \varepsilon < 1$ in the first two cases and $0 \le \varepsilon < 1/2$ in the last case.

(ii) A function $f = (f_n)$ is contained in ZPP-BP, RP-BP, or BPP-BP if it is contained in some ZPP$_\varepsilon$-BP, $0 \le \varepsilon < 1$; RP$_\varepsilon$-BP, $0 \le \varepsilon < 1$; or BPP$_\varepsilon$-BP, $0 \le \varepsilon < 1/2$, respectively.

(iii) A function $f = (f_n)$ is contained in PP-BP if it can be represented by polynomial-size randomized BPs with unbounded error.

We are mainly interested in restricted BPs like π-OBDDs, OBDDs, FBDDs, k-OBDDs, and k-BPs. It is obvious how to define the randomized counterparts of these BDD variants and the corresponding complexity classes are denoted in the obvious way, e.g., RP-OBDD or BPP-FBDD. The complexity classes for deterministic BPs are denoted by P-BP, P-OBDD, ... , and the complexity classes for nondeterministic BDD variants by, e.g., NP-OBDD or coNP-FBDD.

Another approach to defining randomized BPs is the introduction of *probabilistic variables* $z_1, \dots, z_r$ in addition to the usual variables $x_1, \dots, x_n$ (Ablayev and Karpinski (1996)). The input is an assignment to the usual variables and the probabilistic variables independently take the values 0 and 1 with probability 1/2. As long as the probabilistic variables obey the read-once property, it follows directly that the acceptance and rejection probability is equal to the corresponding probability if we treat a node labeled by a probabilistic variable as a randomized node. In this case, we obtain an alternative equivalent definition of randomized BPs. The situation may change if we can test probabilistic variables more than once. This allows the possibility of "storing" information with the help of probabilistic variables. Since BPs correspond to space-restricted computations (see Theorem 2.1.9), this possibility may increase the representational power of randomized BPs. In Section 11.4, we present some results on this generalized variant of randomized BPs. Our restricted variant is more natural, since it allows an efficient computation of the acceptance and rejection probability.

Proposition 11.2.4. *Let G be a randomized BP and a be an input. The acceptance probability $acc_G(a)$ and the rejection probability $rej_G(a)$ can be computed within $O(|G|)$ arithmetic steps.*

Proof. The acceptance probability is computed for each node v where we use the notation $acc_v(a)$. This probability is 1 for the 1-sink and 0 for the 0-sink and the ?-sink. If v is an x_i-node and w is the a_i-successor of v, $acc_v(a)$ equals $acc_w(a)$. If v is a randomized node with successors w' and w'', $acc_v(a) = (acc_{w'}(a) + acc_{w''}(a))/2$. A similar approach works for the rejection probability. □

If d is an upper bound on the number of randomized nodes on a path, it is sufficient to work with numbers of bit length d. The complexity of the problem of computing the acceptance probability changes if we allow probabilistic variables to be read twice. In Theorem 7.3.1, it is shown that the satisfiability test for 2-IBDDs G is NP-complete. We may interpret a 2-IBDD as a randomized BP without any usual variable, i.e., all variables are considered as probabilistic variables. The acceptance probability for an arbitrary input a is positive iff the 2-IBDD is satisfiable. Hence, the test whether the acceptance probability for an arbitrary input is positive is NP-complete for randomized BPs with probabilistic variables which may be tested at least twice. This result justifies the choice of our definition.

We do not investigate randomized DTs, since the known results often concern their depth (Heiman and Wigderson (1991), Heiman, Newman, and Wigderson (1993)) and do not fit into the scope of our investigations.

11.3 Probability Amplification

Probability amplification is one of the most important techniques in the design of randomized algorithms. For algorithms with zero error and ε-failure, it is easy to decrease the failure probability by independent repetitions. The same holds for the reduction of the error probability of algorithms with one-sided error and two-sided ε-bounded error. These results can be proved similarly for BPs. We state the results more carefully by pointing out the relation between the failure or error probability and the number of read accesses to the variables. Without loss of generality, we assume that only BPs with zero error have a ?-sink.

Proposition 11.3.1. *Let G_n be a randomized k-BP representing $f_n \in B_n$ with zero error and ε-failure (or one-sided ε-bounded error). Then f_n can be represented by a randomized (mk)-BP G_n^m with zero error and ε^m-failure (or one-sided ε^m-bounded error) such that the size of G_n^m is bounded by $m|G_n|$.*

Proof. We use m copies of G_n. By our definition of randomized nodes, this implies that the random decisions are independent. In the case of zero error and ε-failure, we replace the ?-sink of the ith copy, $1 \le i \le m-1$, by the source of the $(i+1)$th copy. This implies that the ?-sink of G_n^m is reached with a probability bounded by ε^m. Moreover, G_n^m works with zero error. In the case of one-sided error, the 0-sink of the ith copy is replaced by the source of the $(i+1)$th copy of G_n which obviously decreases the error probability to ε^m. □

Proposition 11.3.2. *Let G_n be a randomized k-BP representing $f_n \in B_n$ with two-sided ε-bounded error. Let $0 < \varepsilon' \le \varepsilon < 1/2$. Then f_n can be represented by a randomized (mk)-BP G_n^m with two-sided ε'-bounded error where $m = O((\frac{1}{2} - \varepsilon)^{-2} \log((\varepsilon')^{-1}))$ and $|G_n^m| = O(m^2|G_n|)$.*

Proof. The idea is to repeat the computation of G_n m times and to take a majority vote. More precisely, we start with an OBDD for the majority function MAJ_m. The size of such an OBDD is $O(m^2)$. Then we replace each inner node v with a copy of G_n whose c-sinks, $c \in \{0,1\}$, are replaced with the c-successor of v. The resulting randomized BP G_n^m is an (mk)-BP of size $O(m^2|G_n|)$. This BP accepts an input a iff at least $\lceil m/2 \rceil$ out of m independent copies of G_n accept a. If $a \in f_n^{-1}(1)$ (the case $a \in f_n^{-1}(0)$ can be handled similarly), $rej_{G_n}(a) \le \varepsilon$. By Chernoff's bounds, the probability that G_n^m rejects a can be bounded above by

$$2\exp\left(-\left(\frac{1}{2}-\varepsilon\right)^2 m/(4\varepsilon(1-\varepsilon))\right) \le 2\exp\left(-\left(\frac{1}{2}-\varepsilon\right)^2 m\right).$$

This probability is bounded by ε' if we set $m = \lceil (\frac{1}{2}-\varepsilon)^{-2}\ln(2/\varepsilon')\rceil$. □

Theorem 11.3.3. *P-BP = ZPP-BP = RP-BP = BPP-BP.*

Proof. For each constant $\varepsilon \in (0,1)$ (or $\varepsilon \in (0,1/2)$ for two-sided error), the failure or error probability of a randomized BP can be decreased from ε to $\varepsilon^* < 2^{-n}$ using Proposition 11.3.1 or Proposition 11.3.2, respectively. The size increase of the randomized BP is polynomially bounded. Afterwards, a well-known nonuniform derandomization technique can be applied. Let $p(n)$ be the number of randomized nodes. Each outcome of the random decisions corresponds to a vector $r \in \{0,1\}^{p(n)}$. For each $a \in \{0,1\}^n$, the number of vectors r leading to a bad event is bounded by $\varepsilon^* 2^{p(n)}$. Hence, the number of pairs (a, r) such that r is bad for a is bounded above by $\varepsilon^* 2^{p(n)} 2^n < 2^{p(n)}$. This implies the existence of some r^* which does not lead to a bad event for any input a. We obtain a deterministic BP for the function represented by the given randomized BP by replacing edges to the ith randomized node w_i with edges to the r_i^*-successor of w_i. □

The situation may change if we consider depth-restricted BPs like k-BPs, FBDDs, or OBDDs. For these representation types R, only the following relations are obvious:

- P-$R \subseteq$ ZPP-$R =$ coZPP-R,
- BPP-$R =$ coBPP-$R \subseteq$ PP-$R =$ coPP-R,
- RP-$R \subseteq$ NP-$R \subseteq$ PP-R.

Sauerhoff (1999a) has shown that the usual property RP $\subseteq$ BPP also holds for k-BPs, FBDDs, and OBDDs. We explicitly mention FBDDs, although they are 1-BPs.

Theorem 11.3.4. *Let R be one of the representation types k-BP, FBDD, or OBDD. Then RP-$R \subseteq$ BPP-R.*

Proof. Let G_n be a polynomial-size randomized BP of type R representing $f_n \in B_n$ with one-sided ε-bounded error where $\varepsilon < 1$ is a constant. We construct a randomized BP G'_n of type R in the following way. For some r chosen later, we start with a complete binary tree of depth r consisting of randomized nodes. We replace r^* of the 2^r leaves with the 1-sink and the remaining leaves with the source of G_n. (For later purposes, we mention that the tree can be reduced to at most r inner nodes.)

If $a \in f_n^{-1}(1)$, the new BP G'_n accepts a with a probability of at least $r^* 2^{-r} + (1 - r^* 2^{-r})(1 - \varepsilon)$, since G_n accepts a with a probability of at least $1 - \varepsilon$. If $a \in f_n^{-1}(0)$, G'_n rejects a with a probability of $1 - r^* 2^{-r}$, since G_n rejects a. We obtain the minimal error probability if

$$r^* 2^{-r} + (1 - r^* 2^{-r})(1 - \varepsilon) = 1 - r^* 2^{-r},$$

which is equivalent to $r^* = \frac{\varepsilon}{1+\varepsilon} 2^r$. Then the error is two-sided but the error probability is bounded by $\varepsilon/(1 + \varepsilon) < 1/2$. We have to choose r^* as an integer which increases the error probability at most by 2^{-r}. Hence, we choose r as a constant such that

$$\frac{\varepsilon}{1+\varepsilon}+2^{-r}<\frac{1}{2}. \qquad \square$$

In the following, we restrict ourselves to π-OBDDs. For this representation type, we have efficient synthesis algorithms. Instead of independently repeating a randomized BP, we can apply synthesis algorithms to obtain a parallel version of the repetition technique (Agrawal and Thierauf (1998), Sauerhoff (1999a)).

Theorem 11.3.5. *Let G_n be a randomized π-OBDD representing $f_n \in B_n$ with one-sided ε-bounded error, $0 \le \varepsilon < 1$. Then f_n can be represented by a randomized π-OBDD G'_n with one-sided ε^m-bounded error in size $|G_n|^m$. In the case of two-sided ε-bounded error, $0 \le \varepsilon < 1/2$, and some $\varepsilon' \le \varepsilon$, f_n can be represented by a randomized π-OBDD G''_n with two-sided ε'-bounded error in size $|G_n|^m$ where $m = O\left(\left(\frac{1}{2}-\varepsilon\right)^{-2}\log\left((\varepsilon')^{-1}\right)\right)$.*

Proof. Here it is convenient to use randomized π-OBDDs where the randomized nodes are labeled with different probabilistic variables and independent copies get different probabilistic variables. All these copies can be understood as π^*-OBDDs with a common variable ordering π^* of all usual and all probabilistic variables. In the case of one-sided error, we compute the disjunction of m independent copies of G_n. The result follows as in the proof of Proposition 11.3.1, where we have computed the disjunction by replacing the 0-sink of one copy by the source of the next copy. The size bound follows from the size bound for the synthesis of π-OBDDs.

In the case of two-sided error, we consider the π-OBDD as a π-MTBDD (see Section 9.2) where the addition of integers is a legal synthesis operation. We apply the corresponding synthesis algorithm to m copies of G_n to obtain G''_n with the proposed size bound. Then we replace the sinks labeled by some $i < \lceil m/2 \rceil$ by a 0-sink and the other sinks with a 1-sink. The result on the error probability follows in the same way as in the proof of Proposition 11.3.2. $\square$

With the parallel execution of independent randomized BPs we save depth at the cost of a larger increase of size. Only if m is a constant can we guarantee that the size of the resulting randomized π-OBDD is polynomial if the given π-OBDD is.

It is not surprising that results on randomized one-way communication complexity lead to results on randomized OBDDs. Ďuriš, Hromkovič, Rolim, and Schnitger (1997) have proved that Las Vegas one-way communication, i.e., zero error and ε-failure one-way communication, can save at most half the bits of deterministic protocols for one-way communication. Karpinski and Mubarakzjanov (1999) have observed that this directly implies the following result.

Theorem 11.3.6. *P-π-OBDD = ZPP-π-OBDD and P-OBDD = ZPP-OBDD.*

In Section 11.8, it will be proved that even a weak probability amplification result like Theorem 11.3.5 is not possible for FBDDs.

11.4 Throw the Coins First

It is well known (Garey and Johnson (1979)) that nondeterministic Turing machine computations can be efficiently simulated by nondeterministic computations using the guess-and-verify mode. First, bits are generated nondeterministically and then a deterministic computation reading each nondeterministic bit once starts. It is an open problem whether similar results hold for nondeterministic OBDDs or FBDDs (see the open problems 10.4 and 10.5).

Here we are faced with a similar problem with respect to the randomized decisions.

Definition 11.4.1.

(i) A *generalized randomized BP* works with probabilistic variables which can be read more than once.

(ii) A randomized BP works in the *throw-and-decide mode* if no randomized node follows a decision node.

Newman (1991) has shown how communication protocols with public coins can be simulated by communication protocols with private coins. Sauerhoff (1999a) has applied Newman's technique to randomized BPs.

Theorem 11.4.2. *Let G be a generalized randomized BP representing $f \in B_n$ with two-sided ε-bounded error, $0 \le \varepsilon < 1/2$, and let $0 < \delta < 1/2 - \varepsilon$. Then there exists a randomized BP G' working in the throw-and-decide mode and representing f in size $O(n\delta^{-2}|G|)$ with two-sided $(\varepsilon + \delta)$-bounded error. Moreover, the depth of G' with respect to the randomized nodes is bounded by $\lceil \log 2n\delta^{-2} \rceil$. The same result holds for k-BPs, FBDDs, OBDDs, and π-OBDDs.*

Proof. We regard G as a deterministic BP G^* working on the n variables of f and the r variables which are probabilistic variables of G. The BP G^* represents a function $f^* \in B_{n+r}$. Let $Z(a,b)$ be the function taking the value 1 if $f^*(a,b) \neq f(a)$ and the value 0 if $f^*(a,b) = f(a)$.

For fixed a and random b, $Z(a,b)$ is a random variable describing whether G errs on input a and, therefore, $E(Z(a,b)) \le \varepsilon$. Let G_b be the deterministic BP obtained from G (or G^*) by replacing the probabilistic variables by the constant vector b. For all considered BP variants, it is obvious that $|G_b| \le |G|$. We prove the theorem in the following way. The BP G' starts with a complete binary randomized tree of depth $d := \lceil \log 2n\delta^{-2} \rceil$. Each leaf is replaced with the source of some G_b where the vectors b are chosen independently according to the uniform distribution. It is sufficient to prove the existence of vectors $b_1, \ldots, b_D$, where $D = 2^d$, such that the error probability is bounded by $\varepsilon + \delta$. Therefore, it is sufficient to bound the average worst-case error probability (the average is taken over all choices of $b_1, \ldots, b_D$) by $\varepsilon + \delta$.

For this purpose, we fix some $a \in \{0,1\}^n$. The error probability of G' on a if $b_1, \ldots, b_D$ are chosen is equal to $(Z(a, b_1) + \cdots + Z(a, b_D))/D$. For each i, $E(Z(a, b_i)) \leq \varepsilon$, and the random variables $Z(a, b_i)$ are independent. Therefore, we can apply Chernoff's bounds and obtain

$$\begin{aligned}\text{Prob}\left(D^{-1} \sum_{1 \leq i \leq D} Z(a, b_i) - \varepsilon > \delta\right) &\leq 2\exp\Big(-\delta^2 D / \big(4\varepsilon(1-\varepsilon)\big)\Big) \\ &\leq 2\exp(-\delta^2 D) \leq 2\exp(-2n) < 2^{-n}.\end{aligned}$$

Hence, the probability that for random $b_1, \ldots, b_D$ there exists some $a \in \{0,1\}^n$ such that $(Z(a, b_1) + \cdots + Z(a, b_D))/D > \varepsilon + \delta$ is smaller than 1. This finally implies the existence of some $b_1, \ldots, b_D$ such that the error probability of the resulting BP G' is bounded by $\varepsilon + \delta$. □

A similar result holds for each BDD variant where a sequence of operations replacement by constants cannot lead to a superpolynomial increase of the size. Usually, the size decreases by replacements by constants. For ZBDDs, the size may increase but not by more than a factor of $(n + 1)$ (see Section 8.1) while for G-FBDDs and OFDDs the size may increase exponentially.

In Theorem 11.4.2, the error probability increases from ε to $\varepsilon + \delta$. For BPs, OBDDs, and π-OBDDs, we can apply the probability amplification results from Section 11.3 and can decrease the error probability from $\varepsilon + \delta$ to ε while preserving the polynomial size of the DDs. Moreover, the proof of Theorem 11.4.2 also works for one-sided error and zero error. This leads to the following corollary.

Corollary 11.4.3. *If f is represented by polynomial-size (generalized) randomized BPs, OBDDs, or π-OBDDs with constant error probability $\varepsilon < 1$ ($\varepsilon < 1/2$ for two-sided error), then f can be represented by polynomial-size randomized BPs, OBDDs, or π-OBDDs, resp., which work in the throw-and-decide mode and guarantee the same error bound.*

Sauerhoff (1999a) has also shown that, for OBDDs, it is not possible to significantly decrease the depth of the randomized tree in Theorem 11.4.2.

11.5 Upper Bound Results

The design of small-size randomized OBDDs and FBDDs uses the throw-and-decide mode established in the last section and the fingerprinting technique originally introduced by Freivalds (1979) and first applied to randomized OBDDs by Ablayev and Karpinski (1996). We introduce the main ideas with a typical example, the design of a polynomial-size randomized π-OBDD representing the equality test function EQ_n for an arbitrary variable ordering π.

Proposition 11.5.1. *$EQ_n \in coRP_{\varepsilon(n)}$-$\pi$-OBDD for each variable ordering π as long as $\varepsilon(n)^{-1}$ is polynomially bounded.*

Proof. We regard EQ_n as the equality test of $x = (x_0, \dots, x_{n-1})$ and $y = (y_0, \dots, y_{n-1})$ which we interpret as binary numbers $|x|$ and $|y|$. The vectors x and y are equal iff $|x| = |y|$. If $|x| = |y|$, also $|x| - |y| \equiv 0 \bmod p$ for each number p. If $|x| \neq |y|$, there are at most n different prime numbers p such that $|x| - |y| \equiv 0 \bmod p$. This follows from the fact that the product of the smallest n primes is at least 2^n and $0 \leq |x|, |y| < 2^n$.

The idea is to choose a random prime number p among the s smallest primes and to check whether $|x| - |y| \equiv 0 \bmod p$. In the negative case, we know that $|x| \neq |y|$ and reject the input. In the positive case, we accept the input. An input (x, y), where $|x| \neq |y|$, is accepted with a probability which is bounded above by n/s, which is at most $\varepsilon(n)$ if $s \geq n\varepsilon(n)^{-1}$.

In our case, we choose s as the smallest power of 2 which is at least $n\varepsilon(n)^{-1}$. Then $s = O(n\varepsilon(n)^{-1})$ and the size of all considered primes is $O(s \log s) = O(n\varepsilon(n)^{-1} \log n)$, since $\varepsilon(n)^{-1}$ is polynomially bounded.

The randomized part of the OBDD has depth $\log s$. This part simulates the random choice of one of the s smallest primes. The π-OBDD G_p (for the prime p) checks whether $|x| - |y| \equiv 0 \bmod p$. This is obviously possible in size $O(np) = O(ns \log s)$. The estimate $p = O(s \log s)$ follows from the prime number theorem. Hence, the total size is $O(ns^2 \log s) = O(n^3\varepsilon(n)^{-2} \log n)$, which is polynomially bounded. □

It is easy to apply this technique to some other functions. The following result on EQ_n^* (see Definition 10.3.5) is similar to a result of Ablayev and Karpinski (1996) and the result on multgraph, the graph of multiplication (defined in Section 1.4), has been proved independently by Ablayev and Karpinski (1998) and Agrawal and Thierauf (1998).

Theorem 11.5.2. *$EQ_n^* \in coRP_{\varepsilon(n)}$-OBDD as long as $\varepsilon(n)^{-1}$ is polynomially bounded.*

Proof. We use the variable ordering $a_1, x_1, \dots, a_n, x_n, b_1, y_1, \dots, b_n, y_n$. $\mathrm{EQ}_n^*(a, x, b, y) = 1$ iff $x^* = y^*$, where x^* is the subvector of all x_i where $a_i = 1$, and y^* is defined similarly with respect to b. We use the same approach as in the proof of Proposition 11.5.1. If the prime number p is chosen, we additionally count the number of indices i where $a_i = 1$ and of j where $b_j = 1$. If $a_i = 1$ and a_i is the kth a-variable with value 1, then x_i contributes $x_i 2^{k-1}$ to $|x^*|$, similarly for b and y. Altogether, we get an additional factor of n^2 for the OBDD size, since we count the i where $a_i = 1$ and the j where $b_j = 1$. The input is accepted only if $a_1 + \cdots + a_n = b_1 + \cdots + b_n$ and $|x^*| - |y^*| \equiv 0 \bmod p$. The total size is $O(n^5\varepsilon(n)^{-2}\log n)$. □

Theorem 11.5.3. *The graph of multiplication is contained in coRP*$_{\varepsilon(n)}$*-OBDD as long as* $\varepsilon(n)^{-1}$ *is polynomially bounded.*

Proof. For $x = (x_{n-1}, \ldots, x_0)$, $y = (y_{n-1}, \ldots, y_0)$, and $z = (z_{2n-1}, \ldots, z_0)$, it has to be tested whether $|x|\cdot|y| = |z|$. We use the variable ordering $x_0, \ldots, x_{n-1}$, $y_0, \ldots, y_{n-1}, z_0, \ldots, z_{2n-1}$. Again, we use an approach similar to the proof of Proposition 11.5.1. Here we choose $s \geq 2n\varepsilon(n)^{-1}$, since $0 \leq |x| \cdot |y|, |z| < 2^{2n}$.

We only describe the OBDD G_p testing whether $|x| \cdot |y| - |z| \equiv 0 \bmod p$. While testing the x-variables, we compute $|x| \bmod p$ (width p is sufficient for this purpose). Our aim during the tests of the y-variables is to store the value val of $|x| \bmod p$ and, after the test of y_i, the intermediate result $res_i \equiv val \cdot (y_0 + \cdots + y_i 2^i) \bmod p$. Obviously, $res_{i+1} \equiv res_i + val \cdot y_{i+1} 2^{i+1} \bmod p$ and width p^2 is sufficient for this phase. At the end of this phase, we know $|x|\cdot|y| \bmod p$ and we may forget $|x| \bmod p$. The third phase checks whether $|x|\cdot|y| - |z| \equiv 0 \bmod p$ in the same way as we have tested $|x| - |y| \equiv 0 \bmod p$ in the proof of Proposition 11.5.1. The total size is $O(ns^3 \log^2 s) = O(n^4\varepsilon(n)^{-3}\log^2 n)$. □

The next result (Sauerhoff (1999a)) on the permutation matrix test function PERM$_n$ is interesting, since we know that this function needs exponential size for OBDDs, FBDDs, k-OBDDs, OR-OBDDs, and OR-FBDDs.

Theorem 11.5.4. *The permutation matrix test function PERM*$_n$ *is contained in coRP*$_{\varepsilon(n)}$*-OBDD as long as* $\varepsilon(n)^{-1}$ *is polynomially bounded.*

Proof. The key idea is to reformulate PERM$_n$ as a function which mainly tests the equality of simple arithmetic expressions. Let $x_i = (x_{i,n-1}, \ldots, x_{i,0})$ be the ith row of X. Then PERM$_n(X) = 1$ iff each row vector x_i contains exactly one 1-entry and the sum of all $|x_i|$ equals $2^n - 1$. Now we may use the approach of the proof of Proposition 11.5.1. Since $0 \leq |x_1| + \cdots + |x_n| < n2^n$, we choose $s \geq (n + \log n)\varepsilon(n)^{-1}$. We use a rowwise ordering of the variables.

We only describe the OBDD G_p testing whether each row contains exactly one 1-entry and $|x_1| + \cdots + |x_n| - (2^n - 1) \equiv 0 \bmod p$. We start with the intermediate result $-(2^n - 1) \bmod p$ and add (in $\mathbb{Z}_p$) $x_{ij}2^j$ after reading x_{ij}. The test whether each row contains exactly one 1-entry increases the OBDD size by a factor of at most 2. The total size is $O(n^2s^2 \log s) = O(n^4\varepsilon(n)^{-2}\log n)$. □

For many representation types, $\overline{\text{PERM}_n}$ and ROW$_n$ + COL$_n$ have similar behavior. Theorem 11.5.4 implies that $\overline{\text{PERM}_n} \in$ RP$_{\varepsilon(n)}$-OBDD. It will be shown in Section 11.7 that $\text{ROW}_n + \text{COL}_n \notin$ BPP-k-OBDD, implying that $\text{ROW}_n + \text{COL}_n \notin$ RP-OBDD. We only know the following positive results on $\text{ROW}_n + \text{COL}_n$. Since $\text{ROW}_n + \text{COL}_n \in$ NP-OBDD (Proposition 10.2.5), also $\text{ROW}_n + \text{COL}_n \in$ PP-OBDD. Moreover, $\text{ROW}_n + \text{COL}_n \in$ RP-FBDD. We use one randomized node. With probability 1/2, we check with a rowwise ordering whether $\text{ROW}_n(X) = 1$ and, with probability 1/2, we check with a columnwise ordering whether $\text{COL}_n(X) = 1$. If $\text{ROW}_n(X) + \text{COL}_n(X) = 0$, we reject X. If $\text{ROW}_n(X) + \text{COL}_n(X) = 1$, we accept X with a probability of at least 1/2.

The previous applications of the fingerprinting technique are more or less straightforward. The following result (Sauerhoff (1999a)) is a sophisticated application of this technique.

Theorem 11.5.5. *The exactly half clique function $excl_n$ is contained in coRP$_{\varepsilon(n)}$-OBDD as long as $\varepsilon(n)^{-1}$ is polynomially bounded.*

Proof. The input $X = (x_{ij})_{1 \le i < j \le n}$ describes an undirected graph and it has to be tested whether this graph consists of a clique of size $t = \lceil n/2 \rceil$ and $n - t$ isolated vertices. The adjacency vector $y_i = (y_{i1}, \ldots, y_{in})$ is defined by $y_{ij} = x_{ji}$ if $i > j$, $y_{ij} = x_{ij}$ if $i < j$, and $y_{ii} = 1$. The problem is that we cannot read the adjacency vectors in rowwise order, since each variable is contained in two adjacency vectors. Otherwise, the following observation would lead to a straightforward application of the fingerprinting technique. The graph described by X is an exactly half clique iff $n - t$ adjacency vectors contain exactly one 1-entry and the other t adjacency vectors contain more than one 1-entry and are equal.

We look for another necessary and sufficient condition for the fact $excl_n(X) = 1$. This condition should allow a randomized check for a columnwise variable ordering, i.e., we test the vectors $c_2, \ldots, c_n$ one after the other where $c_j = (x_{1j}, \ldots, x_{j-1,j})$. Obviously, $excl_n(X) = 0$ if all column vectors are zero vectors. Otherwise, let $j_{\min}$ and $j_{\max}$ be the smallest, respectively, largest index j such that c_j is not a zero vector. If $excl_n(X) = 1$, $j_{\max}$ is the last vertex of the clique and $j_{\min}$ is the second-smallest vertex of the clique. Let $i_{\min}$ be the smallest i such that $x_{i,j_{\min}} = 1$. Then $x_{i_{\min}+1,j_{\min}} = \cdots = x_{j_{\min}-1,j_{\min}} = 0$ and $i_{\min}$ is the first vertex of the clique. A vertex $j > j_{\min}$ belongs to the clique iff c_j is not the zero vector and a vertex $i < j_{\max}$ belongs to the clique iff $x_{i,j_{\max}} = 1$. Moreover, if j and $j' > j$ belong to the clique, $(x_{1j}, \ldots, x_{j-1,j}) = (x_{1j'}, \ldots, x_{j-1,j'})$, i.e., c_j is a prefix of $c_{j'}$. We introduce the following notation. Let $a_j = 1$ iff $j = i_{\min}$ or c_j contains a 1-entry. Let $b_i = 1$ iff $i = j_{\max}$ or $x_{i,j_{\max}} = 1$.

Claim. *The property $excl_n(X) = 1$ holds iff X contains a 1-entry, $a = b$, a and b contain t 1-entries each, and c_j is a prefix of $c_{j'}$ if $j < j'$ and c_j and $c_{j'}$ are not zero vectors.*

Proof of the claim. The "only if part" has been proved above. We prove the "if part." We claim that X contains exactly the clique on the vertices i where

$a_i = b_i = 1$. Let $a_i = b_i = 0$. Then $i \neq i_{\min}$, c_i is the zero vector, $i \neq j_{\max}$, and $x_{i,j_{\max}} = 0$. Since c_i is the zero vector, $x_{1i} = \cdots = x_{i-1,i} = 0$. If $j > i$ and $a_j = b_j = 0$, c_j is the zero vector and $x_{i,j} = 0$. If $j > i$, $a_j = b_j = 1$, and $x_{ij} = 1$, we conclude that c_j is a prefix of $c_{j_{\max}}$ and $x_{i,j_{\max}} = 1$ in contradiction to $b_i = 0$. Hence, $x_{ij} = 0$ also in this case. Now let $a_i = b_i = 1$. We have already shown that the vertices k where $a_k = b_k = 0$ are isolated. Hence, it is sufficient to prove that $x_{ij} = 1$ if $a_j = b_j = 1$ and $j > i$. Since $a_j = 1$, we have $j \leq j_{\max}$. Since $i < j$, also $i < j_{\max}$. Therefore, $b_i = 1$ implies $x_{i,j_{\max}} = 1$. Since $j > i \geq i_{\min}$, the assumption $a_j = 1$ implies that c_j is not a zero vector and, therefore, a prefix of $c_{j_{\max}}$. This implies $x_{ij} = 1$. □

All considered vectors, namely a, b, and c_j, $2 \leq j \leq n$, have a length which is bounded by n. For a randomized equality check it is sufficient to choose $s \geq n\varepsilon(n)^{-1}$ (compare the proof of Proposition 11.5.1). We only describe the OBDD G_p testing whether X contains a 1-entry, a and b contain exactly t 1-entries, $|a| - |b| \equiv 0 \bmod p$, and c_j is a prefix of $c_{j'}$ (if $j < j'$ and c_j and $c_{j'}$ are not 0-vectors). All these checks are performed simultaneously. The first one is obvious. If we read the first 1-entry, we know $i_{\min}$ and $j_{\min}$. We also know that $a_{i_{\min}} = 1$, $a_{j_{\min}} = 1$, and $a_i = 0$ for all other $i \leq j_{\min}$. Hence, we can compute the partial sum to compute $|a| \bmod p$ (p different possible values). Later, we know the value of a_j after having read c_j. Reading c_j, we assume that $j = j_{\max}$ and compute $|b| \bmod p$ under this assumption (p different values). Whenever we later find another 1-entry, we forget the wrong $|b| \bmod p$ value and try another one. During all these computations, we count the number of i where $a_i = 1$ and the number of j where $b_j = 1$ (n^2 different partial results). At the end, we know $|a| \bmod p$ and $|b| \bmod p$ and can compare them. We always store the index j of the last column with a 1-entry (n possibilities) and $|c_j| \bmod p$ (p different results). If we find a new 1-entry in $c_{j'}$, we compute $|c_{j'}| \bmod p$ (p different results). After having read the first $j-1$ entries of $c_{j'}$, we compare $|c_j| \bmod p$ and the partial result for $|c_{j'}| \bmod p$. This is an equality test for the property that c_j is a prefix of $c_{j'}$. Finally, we accept the input if it has passed all checks. An input X with $excl_n(X) = 1$ passes all checks. If $excl_n(X) = 0$, at least one condition is not fulfilled and the probability of accepting such an input is bounded by n/s. The total size of G_p can be estimated by $O(n^2 \cdot p \cdot p \cdot n^2 \cdot n \cdot p \cdot p) = O(n^5 p^4)$. The size altogether is bounded by $O(n^5 s^5 \log^4 s) = O(n^{10} \varepsilon(n)^{-5} \log^4 n)$. □

We finish this section with the design of two randomized FBDDs (Sauerhoff (1999a)).

Definition 11.5.6. The *matrix storage access* function $\mathrm{MSA}_n \in B_n$, $n = 2^k$, is defined on $x = (x_0, \ldots, x_{n-1})$. The variables are partitioned into k $s \times s$ matrices $M_0, \ldots, M_{k-1}$, where $s = \lfloor (n/k)^{1/2} \rfloor$, and the set of the remaining variables. The matrix M_i contains the variables $x_{is^2}, \ldots, x_{(i+1)s^2-1}$. Let

$a_i = 1$ iff M_i contains a row consisting of ones only, i.e., $a_i = \mathrm{ROW}_s(M_i)$. Then $\mathrm{MSA}_n(x) = x_{|a|}$.

Proposition 11.5.7. *The matrix storage access function has exponential FBDD size but polynomial OR-OBDD size and AND-OBDD size.*

Proof. The proof is left to the reader (see exercises). Jukna, Razborov, Savický, and Wegener (1997) have proved such a result for a related function. □

Theorem 11.5.8. *The matrix storage access function MSA_n can be represented by polynomial-size randomized FBDDs with zero error and 1/2-failure.*

Proof. The FBDD G starts with a single randomized node deciding whether we read G_1 or G_2. In order to simplify the notation we assume that $k = 2^l$ and $(n/k)^{1/2} = 2^{(k-l)/2}$ is an integer. Then the set of the remaining variables is empty. We distinguish two cases.

Case 1. The variable $x_{|a|}$ is contained in $M_0, \ldots, M_{k-l-1}$. Then G_1 computes the correct output and G_2 outputs "?".

Case 2. The variable $x_{|a|}$ is contained in $M_{k-l}, \ldots, M_{k-1}$. Then G_2 computes the correct output and G_1 outputs "?".

Now we describe G_1 and G_2. In G_1, the variables of $M_{k-l}, \ldots, M_{k-1}$ are read in an order where the variables of each matrix are read blockwise in rowwise order. It is obvious that polynomial size is sufficient to compute $(a_{k-1}, \ldots, a_{k-l})$. These are the high-order bits of $|a|$. Since $2^{k-l} = s^2$, $r := |(a_{k-1}, \ldots, a_{k-l})|$ is the index such that M_r contains $x_{|a|}$. If $r \geq k - l$, the ?-sink is reached. If $r < k - l$, the variables of the matrices M_i, $i < k - l$ and $i \neq r$, are read in an order where the variables of each matrix are read blockwise in rowwise order. It is obvious that polynomial size is sufficient to compute all a_j except a_r. Then there are two possible values of $|a|$ left. We read M_r in rowwise order and compute a_r and $x_{|a|}$ in polynomial size. The sub-FBDD G_1 has the desired properties.

In G_2, the variables of $M_0, \ldots, M_{k-l-1}$ are read to compute $(a_{k-l-1}, \ldots, a_0)$ in polynomial size. The key observation is that only l address bits are missing and only $2^l = \log n$ addresses are possible. Each matrix $M_0, \ldots, M_{k-1}$ contains exactly one of the variables x_i such that $i = |a|$ is possible. Now the variables of $M_{k-l}, \ldots, M_{k-1}$ are read to compute $|a|$ and to store the $l = \log \log n$ variables x_i such that $i = |a|$ is possible. This can be done in polynomial size. If $x_{|a|}$ is contained in one of the matrices $M_0, \ldots, M_{k-l-1}$, the ?-sink is reached. Otherwise, the sink with label $x_{|a|}$ is reached. The sub-FBDD G_2 also has the desired properties. □

Proposition 11.5.7 and Theorem 11.5.8 imply the following corollary which is in contrast to the statement P-OBDD = ZPP-OBDD in Theorem 11.3.6.

Corollary 11.5.9. *P-FBDD* $\neq$ *ZPP-FBDD.*

The following function introduced by Sauerhoff (1998) will play an important role in Section 11.8, where we prove that probability amplification is not possible for randomized FBDDs.

Definition 11.5.10. The *row mod sum function* $\mathrm{RMS}_n \in B_{n^2}$ is defined on an $n \times n$ matrix X and outputs 1 iff the number of rows of X where the number of ones is a multiple of 3 is even. The *column mod sum function* $\mathrm{CMS}_n \in B_{n^2}$ is similarly defined for the columns of X. The *mod sum function* $\mathrm{MS}_n \in B_{n^2}$ is defined as the conjunction of RMS_n and CMS_n.

Proposition 11.5.11. *MS* $\in$ *coRP*$_{1/2}$*-FBDD and MS* $\in$ *BPP*$_{1/3+\varepsilon(n)}$*-FBDD as long as* $\log(\varepsilon(n)^{-1})$ *is polynomially bounded.*

Proof. Let G be the randomized FBDD which starts with a single randomized node leading to G_1 and G_2. The FBDD G_1 is an OBDD using a rowwise variable ordering and representing RMS_n in polynomial size and G_2 is a polynomial-size OBDD representing CMS_n with a columnwise variable ordering. If $\mathrm{RMS}_n(X) \wedge \mathrm{CMS}_n(X) = 1$, G accepts X with probability 1. If $\mathrm{RMS}_n(X) = 0$ or $\mathrm{CMS}_n(X) = 0$, G_1 or G_2 rejects X and the error probability is bounded by 1/2. This proves $\mathrm{MS} \in \mathrm{coRP}_{1/2}$-FBDD.

For the proof of the second claim, we analyze the proof of Theorem 11.3.4. A randomized FBDD G with one-sided ε-bounded error can be used to construct a randomized FBDD G' with size $|G| + r$ and two-sided error where the error probability is bounded by $\varepsilon/(1+\varepsilon) + 2^{-r}$. This result is applied for $\varepsilon = 1/2$ and polynomially bounded r. □

Randomized BDDs and OBDDs have turned out to be quite powerful representations.

11.6 Efficient Algorithms and Hardness Results

We restrict ourselves to the discussion of randomized π-OBDDs G. In Proposition 11.2.4, we have proved that the acceptance probability $acc_G(a)$ for a given input a can be computed efficiently. This implies that the evaluation problem can be solved efficiently if the error bound is fixed.

The most important problem is the synthesis problem. Let G_f and G_g be randomized π-OBDDs representing f and g, respectively. We may label the randomized nodes by different variables and may regard G_f and G_g as π^*-OBDDs for a variable ordering including the probabilistic variables. Then we may apply the usual synthesis algorithm for Boolean operators $\otimes \in B_2$ to construct a DD G^* and we may replace the nodes labeled by probabilistic variables by randomized nodes. Does G^* represent $h := f \otimes g$? First, we consider randomized

π-OBDDs with two-sided ε-bounded error where $\varepsilon < 1/2$ is a constant. Using the probability amplification technique for OBDDs (Theorem 11.3.5), we can decrease the error probability to 1/4. If the synthesis algorithm is applied after this reduction of the error probability, G^* is a randomized π-OBDD representing h with two-sided 7/16-bounded error and, by another probability amplification, the error probability can be decreased to 1/4. The reason is that G_f and G_g work independently and each one works correctly with probability at least 3/4. Hence, the probability that both work correctly is at least 9/16. The same approach works for randomized π-OBDDs with zero error and ε-failure.

Theorem 11.6.1. *Let $0 < \varepsilon < 1/2$ be a constant. The synthesis problem for randomized π-OBDDs with two-sided ε-bounded error or zero error and ε-failure can be solved in polynomial time.*

The approach fails for randomized π-OBDDs with unbounded error, since we have no probability amplification technique which is strong enough (e.g., HWB $\in$ PP-OBDD (see exercises), but HWB $\notin$ BPP-OBDD (see Section 11.7)). It also fails for one-sided ε-bounded error, since we can conclude from known results that RP-OBDD $\neq$ coRP-OBDD and negation can cause an exponential blow-up.

Proposition 11.6.2. *RP-OBDD $\neq$ coRP-OBDD.*

Proof. The permutation matrix test function PERM is contained in coRP-OBDD (see Theorem 11.5.4) but not in NP-FBDD (see Theorem 10.3.8). We know that RP-OBDD $\subseteq$ NP-OBDD $\subseteq$ NP-FBDD (see Section 11.3) and, therefore, PERM $\notin$ RP-OBDD. □

The proposed synthesis algorithms for the ZPP- and BPP-models run in polynomial time. The problem is that the frequent application of the probability amplification technique during a synthesis process typically leads to large randomized π-OBDDs and there is no idea for an efficient minimization technique. Moreover, we have no idea how to introduce randomization into π-OBDDs. We start with π-OBDDs for the variables and it does not seem to be useful to work with randomized nodes. But this implies that randomization should be introduced somewhere in the synthesis process. We have to admit that randomized OBDDs do not seem to have applications.

Finally, we present a result of Agrawal and Thierauf (1998) proving that the satisfiability problem and, therefore, also the equivalence problem is hard for randomized π-OBDDs with one-sided ε-bounded error.

Theorem 11.6.3. *Let $0 < \varepsilon < 1$ be a constant. It is NP-complete to decide for a randomized OBDD which is known to work with one-sided ε-bounded error whether some input is accepted (with probability at least $1 - \varepsilon$).*

Proof. Manders and Adleman (1978) have proved that it is NP-complete to decide the language Q of all triples (a, b, c) of natural numbers such that $ax^2 + by = c$ for some natural numbers x and y. If a, b, and c have bit length n, we can bound the bit length of x and y by n. Using the fingerprinting technique (as for the graph of multiplication in Theorem 11.5.3), we can design a polynomial-size randomized OBDD G with one-sided ε-bounded error deciding for n-bit numbers a, b, c, x, and y whether $ax^2 + by = c$. Let G_{a^*,b^*,c^*} be the randomized OBDD obtained from G by replacing the variables a, b, and c by the constants a^*, b^*, and c^*. We may decide whether $(a^*, b^*, c^*) \in Q$ by checking whether G_{a^*,b^*,c^*} accepts some input with a probability of at least $1 - \varepsilon$. □

Theorem 11.6.3 implies the same result for randomized OBDDs with two-sided ε-bounded error and randomized OBDDs with unbounded error and all generalizations of randomized OBDDs.

11.7 Lower Bounds for Randomized OBDDs and k-OBDDs

Randomized communication complexity or, more precisely, communication with *randomized protocols* is the main tool to prove lower bounds on the size of randomized OBDDs, k-OBDDs, k-IBDDs, and s-oblivious BDDs. We only cite the main results on lower bounds for the length of randomized protocols and refer the reader to the monographs of Hromkovič (1997) and Kushilevitz and Nisan (1997). We need a result on one-round communication complexity for the direct storage access function DSA or multiplexer MUX which, in the context of communication complexity, is often called the *index function* INDEX. Kremer, Nisan, and Ron (1995) have proved that this function is difficult if Alice gets the data variables and Bob the address variables, which corresponds to a bad variable ordering. Nisan and Wigderson (1993) have considered the pointer-jumping function PJ (see Definition 7.5.12) and communications restricted to k rounds. Lower bounds without restriction on the number of communication rounds have been obtained by Kalyanasundaram and Schnitger (1992) and Razborov (1992) for a function called *set disjointness* DISJ. The inputs $x \in \{0,1\}^n$ given to Alice and $y \in \{0,1\}^n$ given to Bob are interpreted as characteristic vectors of subsets A and B of $\{1, \ldots, n\}$. The output is 1 iff $A \cap B = \emptyset$. The set disjointness function is nothing else than the negation of the disjoint quadratic function DQF. Again, the partition of the inputs between Alice and Bob corresponds to a bad variable ordering.

Theorem 11.7.1.

(*i*) *The data variables of the index function* $INDEX_n$ (DSA_n, MUX_n) *are given to Alice and the address variables to Bob. Each randomized one-*

round protocol where Alice sends a message and Bob has to decide about the output and whose two-sided error probability is bounded by $1/8$ has a length of $\Omega(n)$.

(*ii*) *Each randomized $2k$-round protocol for the pointer-jumping scenario (Bob gets the pointers starting in W and has to start the communication) whose two-sided error probability is bounded by $1/3$ has a length of $\Omega(n/k^2 - k \log n)$.*

(*iii*) *The x-variables of the set disjointness function $DISJ_n$ or the disjoint quadratic function DQF_n are given to Alice and the y-variables to Bob. Each randomized protocol for $DISJ_n$ or DQF_n whose two-sided error probability is bounded by a constant $\varepsilon < 1/2$ has a length of $\Omega(n)$.*

We have to generalize the relation between the size of oblivious BDDs and the length of communication protocols derived in Section 7.5 to the randomized case. Moreover, we need an appropriate reduction concept. We have to reduce functions $f = (f_n)$ to functions $g = (g_n)$ preserving the chosen partitions of the variables. The partition of the variable set is crucial, since DSA and DQF have linear OBDD size. The concept of *rectangular reductions* is an appropriate analogue of many-one reductions for Turing machines. We use the notation $f(x, y) \in B_{n+m}$ for Boolean functions $f\colon \{0,1\}^n \times \{0,1\}^m \to \{0,1\}$ where the first n variables are given to Alice and the other m variables are given to Bob.

Definition 11.7.2. Let $f(x, y) \in B_{n+m}$ and $g(x, y) \in B_{k+l}$. A pair (φ_A, φ_B) of functions $\varphi_A\colon \{0,1\}^n \to \{0,1\}^k$ and $\varphi_B\colon \{0,1\}^m \to \{0,1\}^l$ is called a rectangular reduction from f to g if $f(a, b) = g(\varphi_A(a), \varphi_B(b))$ for all $(a, b) \in \{0,1\}^n \times \{0,1\}^m$.

The reduction concept is called rectangular since Alice can compute $\varphi_A(a)$ and Bob can compute $\varphi_B(b)$ and, afterwards, they can apply communication protocols for g to evaluate f. Rectangles of the communication matrix for f are mapped to rectangles of the communication matrix for g by (φ_A, φ_B). The obvious and important consequence of a rectangular reduction from f to g is that the communication complexity of f is bounded above by the communication complexity of g. This holds for each type of protocol.

Let $g(x, y) \in B_{n+m}$ and let G be a randomized k-OBDD representing g and respecting the variable ordering π where the x-variables are tested before the y-variables. In the same way as discussed in Section 7.5, we may partition G into $2k$ layers such that each layer contains only randomized nodes and nodes labeled by variables given to the same player. This player may choose a random computation path through her or his layer and can communicate the first node of the next layer. This leads to a randomized protocol with the same error or failure guarantee as the given randomized k-OBDD. The length of the protocol is bounded by $(2k - 1)\lceil \log |G| \rceil$ if it is sufficient that one player "knows" the result. This implies

$$|G| > 2^{R(g)/(2k-1)-1}$$

for the randomized communication complexity $R(g)$ of the given type of $(2k-1)$-round protocols. If there is a rectangular reduction from $f(x', y')$ to $g(x, y)$, we can replace $R(g)$ by $R(f)$ in the lower bound.

Sauerhoff (1999b) has applied this lower bound technique to the class of k-stable functions.

Definition 11.7.3. A function $f \in B_n$ is called *k-stable* if for all sets $V \subseteq X_n$ containing at most k variables and all $x_i \in V$ there exists some assignment to the variables in $X_n - V$ such that the resulting subfunction of f is x_i or $\overline{x}_i$.

It is easy to see that each k-stable function is k-mixed. By Lemma 6.2.4, k-stable functions have an FBDD size of at least $2^k - 1$.

Theorem 11.7.4. *Let $f \in B_n$ be k-stable. The size of randomized OBDDs representing f with two-sided ε-bounded error, where $\varepsilon < 1/2$ is a constant, is bounded below by $2^{\Omega(k)}$.*

Proof. Since f is k'-stable for $k' \leq k$ if it is k-stable, we may assume that $k = 2^m$ for an integer m. Without loss of generality, we can investigate π-OBDDs for $\pi = id$. We regard f as a function $f(y, z) \in B_{k+(n-k)}$, i.e., Alice obtains $y_1, \dots, y_k$ and Bob $z_1, \dots, z_{n-k}$. Let $i \in \{1, \dots, k\}$. Since f is k-stable, there is an assignment $b(i)$ to Bob's variables such that the resulting subfunction of f is y_i or $\overline{y}_i$. Let I be the set of indices i where the resulting subfunction is y_i.

We consider the function $\text{INDEX}_k^*(x, a)$ which outputs x_i if $|a| = i$ and $i \in I$ and $\overline{x}_i$ if $|a| = i$ and $i \notin I$. It is obvious that Theorem 11.7.1(i) also holds for INDEX_k^*. We describe a rectangular reduction from $\text{INDEX}_k^*(x, a)$ to $f(y, z)$. The function $\varphi_A\colon \{0,1\}^k \to \{0,1\}^k$ is the identity. The function $\varphi_B\colon \{0,1\}^m \to \{0,1\}^{n-k}$ maps the address vector a of INDEX_k^* to the assignment $b(|a|)$. Then $f(\varphi_A(x), \varphi_B(a)) = x_i$ if $|a| = i$ and $i \in I$ and $f(\varphi_A(x), \varphi_B(a)) = \overline{x}_i$ if $|a| = i$ and $i \notin I$. Hence, $f(\varphi_A(x), \varphi_B(a)) = \text{INDEX}_k^*(x, a)$. Now we can apply the generalization of Theorem 11.7.1(i) and our lower bound technique for randomized OBDDs. This implies the result for error probabilities bounded by $1/8$. For constants $\varepsilon > 1/8$ (but $\varepsilon < 1/2$) we obtain the lower bound with the probability amplification result of Theorem 11.3.5. □

A lot of functions are known to be k-stable for large k:

- the clique function $cl_{n,s}$ is k-stable for $k = \min\{\binom{s}{2} - 1, (n - s + 2)/2\}$ (this is implicitly proved in the proof of Theorem 6.2.7),
- the *Hamiltonian circuit function* HAM_n deciding whether an undirected graph contains a Hamiltonian circuit is $\Omega(n^{\varepsilon})$-stable for some $\varepsilon > 0$ (Dunne (1985) or Exercise 6.5),

- the determinant DET_n is $(n-1)$-stable (Dunne (1985) or Exercise 6.4),
- the matrix storage access function MSA_n, $n = 2^k$, is $\lfloor (n/k)^{1/2} \rfloor$-stable (see Jukna, Razborov, Savický, and Wegener (1997) or Exercise 11.9).

Corollary 11.7.5. *The functions $cl_{n,\lceil (2n/3)^{1/2} \rceil}$, HAM_n, DET_n, and MSA_n are not contained in BPP-OBDD.*

From the results on MSA_n in Corollary 11.7.5 and Proposition 11.5.7, we obtain another corollary.

Corollary 11.7.6. *NP-OBDD $\cap$ coNP-OBDD $\not\subseteq$ BPP-OBDD.*

For functions which are not known to be k-stable for large k, we may directly use the reduction technique. This has been done by Ablayev (1996, 1997) for the weighted sum function WS_n and by Ablayev and Karpinski (1998) for the middle bit of multiplication. We cite the last result without proof.

Theorem 11.7.7. *The size of randomized OBDDs representing the middle bit of multiplication $MUL_{n-1,n}$ with two-sided ε-bounded error, where $\varepsilon < 1/2$ is a constant, is bounded below by $2^{\Omega(n/\log n)}$.*

Functions with polynomial-size FBDDs cannot be k-stable for $k = \omega(\log n)$. Hence, it is interesting to prove exponential-size lower bounds for the size of randomized OBDDs representing such functions. Bollig, Löbbing, Sauerhoff, and Wegener (1999) prove such a result for the hidden weighted bit function HWB which is known to be contained in PP-OBDD (see exercises).

Theorem 11.7.8. *The size of randomized OBDDs representing HWB_n with two-sided ε-bounded error, where $\varepsilon < 1/2$ is a constant, is bounded below by $2^{\Omega(n)}$.*

Proof. We adapt the lower bound proof technique for the deterministic case (see Theorem 4.10.2). Without loss of generality, $n = 10m$ and $m = 2^k$ for some integer k. We reorder the variables $x_1, \ldots, x_n$ according to an arbitrarily fixed variable ordering π. The first $6m$ variables according to π are given to Alice and the remaining $4m$ to Bob. As in the proof of Theorem 4.10.2, we may choose $s = m$ or $s = 5m$ such that at least $2m$ variables x_i, $s \le i \le s + 4m$, are given to Alice. We choose m of these variables and call them $x_{i(1)}, \ldots, x_{i(m)}$. We complete the proof by designing a rectangular reduction from $\mathrm{INDEX}_m(y, a)$ to $\mathrm{HWB}_n(x)$ with the given ordering and partition of the variables. The function $\varphi_A\colon \{0,1\}^m \to \{0,1\}^{6m}$ is defined in the following way. Let $b \in \{0,1\}^m$. Then $\varphi_A(b)$ contains b_j at that position where $x_{i(j)}$ stands in the variable ordering. The other positions are filled with zeros and ones such that the number of ones equals s. This is possible, since the number of ones in b is bounded by m, $s = m$ or $s = 5m$, and we have $5m$ open positions. The function $\varphi_B\colon \{0,1\}^k \to$

$\{0,1\}^{4m}$ is defined in the following way. If $|c| = j$, $\varphi_B(c)$ is a vector containing exactly $i(j) - s$ ones. This is possible since $s \le i(j) \le s + 4m$. The input $(\varphi_A(b), \varphi_B(c))$ has to be interpreted according to the given variable ordering π. By definition, $(\varphi_A(b), \varphi_B(c))$ contains $i(|c|)$ ones and the output of HWB_n is the value of $x_{i(|c|)}$ which is, by definition, equal to $b_{|c|}$ and, therefore, equal to $\mathrm{INDEX}_m(b, c)$. □

A similar proof leads to a $2^{\Omega(n/\log n)}$ lower bound on the size of randomized OBDDs representing the indirect storage access function ISA_n with two-sided ε-bounded error (see exercises). We finish this section with a lower bound for randomized k-OBDDs (Sauerhoff (1999a)).

Theorem 11.7.9. *The size of randomized k-OBDDs representing $ROW_n + COL_n$ with two-sided ε-bounded error, where $\varepsilon < 1/2$ is a constant, is bounded below by $2^{\Omega(n/k)}$.*

Proof. Without loss of generality, n is even. We fix a variable ordering π and give the first $n^2/2$ variables to Alice and the other $n^2/2$ variables to Bob. We claim that there are at least $n/2$ rows or at least $n/2$ columns such that the variables of each row or column are not given to the same player. If Alice gets all variables of one column and Bob gets all variables of another column, then the property holds for all rows. If only Alice (or Bob) owns some columns, the number of these columns is bounded by $n/2$ and at least $n/2$ columns are "distributed." Otherwise, all columns are distributed. Hence, we may assume w.l.o.g. that the first $n/2$ rows are distributed between Alice and Bob. We describe a rectangular reduction from $\mathrm{DQF}_{n/2} = \overline{\mathrm{DISJ}}_{n/2}$ to $\mathrm{ROW}_n + \mathrm{COL}_n$ with the given partition of the variables and apply Theorem 11.7.1(iii). The function $\varphi_A\colon \{0,1\}^{n/2} \to \{0,1\}^{n^2/2}$ assigns for $a \in \{0,1\}^{n/2}$ the value a_i to all variables of the ith row given to Alice and the value 0 to all variables of the last $n/2$ rows given to Alice. The function $\varphi_B\colon \{0,1\}^{n/2} \to \{0,1\}^{n^2/2}$ assigns for $b \in \{0,1\}^{n/2}$ the value b_j to all variables of the jth row given to Bob and the value 0 to all variables of the last $n/2$ rows given to Bob. It is obvious that $(\mathrm{ROW}_n + \mathrm{COL}_n)(\varphi_A(a), \varphi_B(b)) = \mathrm{DQF}_{n/2}(a, b)$. □

11.8 Lower Bounds for Randomized FBDDs and k-BPs

As known from the deterministic and nondeterministic cases, communication complexity cannot be used directly for lower bound proofs on the size of FBDDs and k-BPs. Moreover, randomized FBDDs do not obey a graph ordering. For example, we may use two different variable orderings if we start with a randomized node. The only remaining lower bound technique we know of is the rectangle method based on (k, a)-rectangles (see Definition 7.6.1). In Section 7.6, we have already mentioned that this technique can be understood as an appropriate generalization of communication complexity. Sauerhoff (1998) has generalized this technique to the randomized case. In this context, it is useful to identify a (k, a)-rectangle R, which is defined as a Boolean function, with the set $R^{-1}(1)$. Such a rectangle R is called f-monochromatic if $R \subseteq f^{-1}(0)$ or $R \subseteq f^{-1}(1)$.

Definition 11.8.1. A function $f \in B_n$ is called an (s, k, a)-*step function* if $\{0,1\}^n$ can be partitioned into $(2s)^{ka}$ f-monochromatic (k, a)-rectangles.

Theorem 11.8.2. *Functions $f \in B_n$ whose k-BP size equals s are (s, k, a)-step functions.*

Proof. This result is only a corollary to Theorem 7.6.2. Theorem 7.6.2 directly implies that $f^{-1}(1)$ can be partitioned into at most $r = (2s)^{ka}$ f-monochromatic (k, a)-rectangles. The same holds for $f^{-1}(0)$. Hence, we obtain a bound of $2r$ for the number of f-monochromatic (k, a)-rectangles in the partition of $\{0,1\}^n$. But the bound r in the proof of Theorem 7.6.2 is a bound on the number of traces and each trace either leads to a (k, a)-rectangle in the partition of $f^{-1}(1)$ or to a (k, a)-rectangle in the partition of $f^{-1}(0)$. □

Applying a method due to Yao (1983), we prove that functions with small randomized k-BP size can be approximated with small error by (s, k, a)-step functions.

Theorem 11.8.3. *Let μ be a probability measure on $\{0,1\}^n$. If $f \in B_n$ can be represented by a generalized randomized k-BP of size s which has two-sided ε-bounded error, there exists an (s, k, a)-step function φ such that φ μ-approximates f with ε-bounded error, i.e., the set of all a where $f(a) \neq \varphi(a)$ has a μ-measure bounded above by ε.*

Proof. Let G be a generalized randomized k-BP representing f in size s with two-sided ε-bounded error and let r be the number of probabilistic variables used by G. Let $f^* \in B_{n+r}$ be the function represented by G as a deterministic BP on $n+r$ variables. Let $g \in B_{n+r}$ be defined by $g(a, b) = 1$ iff $f^*(a, b) \neq f(a)$. Since the error probability of G is bounded by ε, we conclude that

$$\sum_{b \in \{0,1\}^r} g(a, b) \leq \varepsilon 2^r$$

for each $a \in \{0,1\}^n$. Hence,

$$\sum_{b \in \{0,1\}^r} \sum_{a \in \{0,1\}^n} \mu(a) g(a, b) = \sum_{a \in \{0,1\}^n} \mu(a) \sum_{b \in \{0,1\}^r} g(a, b) \leq \varepsilon 2^r$$

and there exists some $b^* \in \{0,1\}^r$ such that

$$\sum_{a \in \{0,1\}^n} \mu(a) g(a, b^*) \leq \varepsilon.$$

We replace the probabilistic variables of G with the assignment b^* and obtain a deterministic k-BP G_{b^*} of size s representing the function $\varphi \in B_n$ defined by $\varphi(a) = f^*(a, b^*)$. We have proved that φ is a μ-approximation of f with ε-bounded error. By Theorem 11.8.2, φ is an (s, k, a)-step function. □

Theorem 11.8.3 implies that functions are hard for randomized k-BPs with ε-bounded error if they are hard to approximate by step functions. The following property is a sufficient condition that a function f where $f^{-1}(1)$ is not small is hard to approximate by step functions. The property says that each (k, a)-rectangle covers a certain fraction of inputs from $f^{-1}(0)$ or is quite small.

Definition 11.8.4. Let $f \in B_n$; μ be a probability measure on $\{0,1\}^n$; and k, a, $\alpha > 0$, and $\delta(n)$ be constants. The function f has the *rectangle balance property* with respect to $(\mu, k, a, \alpha, \delta(n))$ if

$$\mu(R \cap f^{-1}(0)) \geq \alpha\mu(R \cap f^{-1}(1)) - \delta(n)$$

holds for each (k, a)-rectangle R.

Theorem 11.8.5. *If $f \in B_n$ has the rectangle balance property with respect to $(\mu, k, a, \alpha, \delta(n))$, the size of each generalized randomized k-BP representing f with two-sided ε-bounded error is bounded below by*

$$\frac{1}{2}\left[\left(\alpha\mu(f^{-1}(1)) - \varepsilon \cdot \max\{\alpha, 1\}\right)/\delta(n)\right]^{1/ka}.$$

Proof. It is sufficient to combine all our information. By Theorem 11.8.3, there exists an (s, k, a)-step function φ which μ-approximates f with ε-bounded error. Let $R_{i,0}$, $1 \leq i \leq r_0$, be the (k, a)-rectangles where φ computes 0 and $R_{i,1}$, $1 \leq i \leq r_1$, be the (k, a)-rectangles where φ computes 1. The rectangles can be chosen such that $r_0 + r_1 \leq (2s)^{ka}$. For our purpose, it is sufficient to prove that

$$r_1 \geq \left(\alpha\mu(f^{-1}(1)) - \varepsilon \cdot \max\{\alpha, 1\}\right)/\delta(n).$$

All $R_{i,1}$, $1 \leq i \leq r_1$, are (k, a)-rectangles. We sum up the inequalities obtained for $R_{i,1}$ by the assumption that f has the rectangle balance property. This implies

$$\sum_{1 \leq i \leq r_1} \mu(R_{i,1} \cap f^{-1}(0)) \geq \alpha \sum_{1 \leq i \leq r_1} \mu(R_{i,1} \cap f^{-1}(1)) - r_1\delta(n).$$

Now we use the fact that the rectangles $R_{i,0}$, $1 \leq i \leq r_0$, and $R_{i,1}$, $1 \leq i \leq r_1$, partition $\{0,1\}^n$. Hence,

$$\mu(f^{-1}(1)) = \sum_{1 \leq i \leq r_0} \mu(R_{i,0} \cap f^{-1}(1)) + \sum_{1 \leq i \leq r_1} \mu(R_{i,1} \cap f^{-1}(1))$$

and

$$\begin{aligned} r_1\delta(n) &\geq \alpha \sum_{1\leq i\leq r_1} \mu(R_{i,1}\cap f^{-1}(1)) - \sum_{1\leq i\leq r_1} \mu(R_{i,1}\cap f^{-1}(0)) \\ &= \alpha\mu(f^{-1}(1)) - \alpha \sum_{1\leq i\leq r_0} \mu(R_{i,0}\cap f^{-1}(1)) \\ &\quad - \sum_{1\leq i\leq r_1} \mu(R_{i,1}\cap f^{-1}(0)). \end{aligned}$$

For $c \in \{0,1\}$, let

$$e_c = \sum_{1\leq i\leq r_c} \mu(R_{i,c}\cap f^{-1}(\bar{c})).$$

Then we have proved that

$$r_1\delta(n) \geq \alpha\mu(f^{-1}(1)) - (\alpha e_0 + e_1)$$

and it is sufficient to prove

$$\alpha e_0 + e_1 \leq \varepsilon \cdot \max\{\alpha, 1\}.$$

Finally, we make use of the property that φ μ-approximates f with ε-bounded error, i.e.,

$$\sum_{1\leq i\leq r_0} \mu(R_{i,0}\cap f^{-1}(1)) + \sum_{1\leq i\leq r_1} \mu(R_{i,1}\cap f^{-1}(0)) \leq \varepsilon,$$

which can be rewritten as

$$e_0 + e_1 \leq \varepsilon.$$

Hence, $\alpha e_0 + e_1 \leq (e_0 + e_1)\cdot\max\{\alpha, 1\} \leq \varepsilon\cdot\max\{\alpha, 1\}$ and we have proved the theorem. □

It is still a major step to apply this technique to a concrete function. Sauerhoff (1998) has obtained the following result on the mod sum function MS_n.

Theorem 11.8.6. *Let $\varepsilon < 1/4$ be a constant. The size of a generalized randomized FBDD representing MS_n with two-sided ε-bounded error is bounded below by $2^{\Omega(n)}$.*

Sketch of Proof. Without loss of generality, $n = 4m$. We try to apply Theorem 11.8.5 and choose μ as the uniform distribution on $\{0,1\}^{n^2}$. The row mod sum function RMS_n tests whether the input matrix X contains an even number of rows where the number of ones is a multiple of 3. Let us consider a random input. The number of ones in a row is binomially distributed with parameters n and $1/2$ and approximately a third of the inputs lead to a row where the number of ones is a multiple of 3 (the fraction is $\frac{1}{3} \pm o(1)$). This holds for the rows independently from each other. Hence, the number of rows where the number of ones is a multiple of 3 is approximately binomially distributed with parameters n and $1/3$ and the fraction of inputs with an even number of rows whose number of ones is a multiple of 3 seems to be approximately $1/2$. The same holds for the column mod sum function CMS_n. Because the rows and columns tend to contain many ones and zeros, RMS_n and CMS_n take their values almost independently and we expect that $\mathrm{MS}_n = \mathrm{RMS}_n \wedge \mathrm{CMS}_n$ takes the value 1 for approximately a quarter of the inputs. This is expressed in the following claim, whose proof is omitted.

Claim 1. $\mu(MS_n^{-1}(1)) = \frac{1}{4} + o(1)$.

Now we set $\alpha = 1$ and $a = 2$. Moreover, $k = 1$, since we investigate FBDDs. The theorem follows from Theorem 11.8.5 if we can prove that

$$\mu(R \cap \mathrm{MS}_n^{-1}(0)) \geq \mu(R \cap \mathrm{MS}_n^{-1}(1)) - \delta(n)$$

for each $(1, 2)$-rectangle R and some function $\delta(n) = \gamma^n$ where $\gamma < 1$. Here we see why our proof only works if $\varepsilon < 1/4$. We want to ensure that $\mu(\mathrm{MS}_n^{-1}(1)) - \varepsilon$ is bounded below by a positive constant.

A $(1, 2)$-rectangle R corresponds to a balanced partition of the input matrix, i.e., Alice gets a variable set X_A of size $n^2/2$ and Bob gets the set X_B of the remaining variables. Then R can be described as rectangle $A \times B$ where A is a set of assignments to X_A and B a set of assignments to X_B. A row or column of the input matrix X is called (X_A, X_B)-distributed if it contains at least two X_A-variables and at least two X_B-variables. It is not too difficult to prove the following claim (we omit the proof).

Claim 2. *Each balanced partition (X_A, X_B) of X leads to at least $n/4$ (X_A, X_B)-distributed rows or at least $n/4$ (X_A, X_B)-distributed columns.*

The following results are obtained for an arbitrary balanced partition (X_A, X_B) which, w.l.o.g., leads to at least $n/4$ (X_A, X_B)-distributed rows. We choose $n/4$ rows which are (X_A, X_B)-distributed and choose a set $X'_A \subseteq X_A$ containing exactly two variables of each of the chosen rows and a set $X'_B \subseteq X_B$ containing exactly two variables of each of the chosen rows. Each partial assignment to the variables outside $X'_A \cup X'_B$ restricts R to some rectangle $R' = A' \times B'$ where A' contains assignments to the variables in X'_A and B' contains assignments to the variables in X'_B. Moreover, MS_n is restricted to a subfunction MS'_n. Finally, the uniform distribution μ is restricted to the uniform distribution μ' on $\{0,1\}^n$, the input set for the variables in $X'_A \cup X'_B$. We prove the rectangle balance property for the restricted version of the problem, i.e., we prove that

$$\mu'(R' \cap \mathrm{MS'}_n^{-1}(0)) \geq \mu'(R' \cap \mathrm{MS'}_n^{-1}(1)) - \delta(n).$$

Then we easily obtain the rectangle balance property for R and MS_n by averaging over all partial assignments.

We know that $\mathrm{MS}'_n = \mathrm{RMS}'_n \wedge \mathrm{CMS}'_n$, where RMS'_n and CMS'_n are the subfunctions of RMS_n and CMS_n according to the same assignment as has been used to obtain MS'_n from MS_n. Since we have many (X_A, X_B)-distributed rows, we believe that RMS'_n alone is responsible for the rectangle balance property. More precisely, we want to prove that

$$\mu'(R' \cap \mathrm{RMS'}_n^{-1}(0)) \geq \mu'(R' \cap \mathrm{RMS'}_n^{-1}(1)) - \delta(n).$$

This implies the above property, since

$$\mu'(R' \cap \mathrm{MS'}_n^{-1}(0)) \geq \mu'(R' \cap \mathrm{RMS'}_n^{-1}(0))$$

and

$$\mu'(R' \cap \mathrm{MS'}_n^{-1}(1)) \leq \mu'(R' \cap \mathrm{RMS'}_n^{-1}(1))$$

follow from $\mathrm{MS}'_n = \mathrm{RMS}'_n \wedge \mathrm{CMS}'_n$.

Finally, we investigate RMS'_n. There are $n/2$ rows which have been replaced with constants. The only influence of this partial assignment is the distinction whether we have to look for an even or an odd number of rows where the number of ones is a multiple of 3 among the remaining $n/2$ rows. Let us consider one of the remaining rows. All but four variables are replaced with constants. Even if additionally the two variables given to Alice (or Bob) are fixed, there is still an assignment to the remaining variables which makes the number of ones in this row a multiple of 3. There is also an assignment with the opposite property. The different rows are independent. Hence, $\mu'(\mathrm{RMS'}_n^{-1}(1))$ is close to $1/2$. Moreover, if R' is not too small, there are enough rows which independently influence the output. Using methods from linear algebra (spectral norm of matrices and size of eigenvalues), the proof of the theorem can be finished by the proof of the following claim (which we omit).

Claim 3.

$\mu'(R' \cap \mathit{RMS'}_n^{-1}(0)) \geq \mu'(R' \cap \mathit{RMS'}_n^{-1}(1)) - \delta(n)$ *for* $\delta(n) = (\sqrt{14}/4)^{n/4}$. □

Combining this theorem with Proposition 11.5.11, we obtain the following results.

Theorem 11.8.7.

(i) *NP-FBDD* $\not\subseteq$ *BPP*$_\varepsilon$*-FBDD if* $\varepsilon < 1/4$ *is a constant.*

(ii) *BPP*$_\delta$*-FBDD* $\subsetneq$ *BPP*$_\varepsilon$*-FBDD if* $\delta < 1/4$ *and* $\varepsilon > 1/3$ *are constants.*

(iii) *RP*$_\delta$*-FBDD* $\subsetneq$ *RP*$_{1/2}$*-FBDD* $\subseteq$ *NP-FBDD if* $\delta < 1/3$ *is a constant.*

Proof. For the first claim, consider $\overline{\text{MS}}_n$, which is not contained in BPP_ε-FBDD (Theorem 11.8.6) but is contained in $\text{RP}_{1/2}$-FBDD (Proposition 11.5.11) and, therefore, also in NP-FBDD. The second claim follows from the properties proved for MS_n. For the last claim, we again consider $\overline{\text{MS}}_n$, which is contained in $\text{RP}_{1/2}\text{-FBDD} \subseteq \text{NP-FBDD}$. If $\overline{\text{MS}}_n \in \text{RP}_\delta$-FBDD, also $\overline{\text{MS}}_n \in \text{BPP}_{\delta/(1+\delta)+\alpha}$-FBDD for each constant $\alpha > 0$ (see the proof of Theorem 11.3.4). If $\delta < 1/3$, then $\delta/(1+\delta) < 1/4$ and we may conclude that $\overline{\text{MS}}_n \in \text{BPP}_\varepsilon$-FBDD for some $\varepsilon < 1/4$ in contradiction to Theorem 11.8.6. □

Sauerhoff (1998) also has applied his lower bound technique to prove exponential-size lower bounds for randomized k-BPs and $k > 1$. We only cite his result.

Theorem 11.8.8.

(*i*) *The size of generalized randomized k-BPs with three-valued decision nodes representing the bilinear Sylvester function* SYL_n *with two-sided* ε*-bounded error is bounded below by* $2^{\Omega(n/4^k k^3)}$ *as long as* ε *is a constant smaller than* $1/3$.

(*ii*) *There is an explicitly defined Boolean variant* SYL_n^* *of the bilinear Sylvester function such that for each constant* $\varepsilon < 1/2$ *there is a constant* $c(\varepsilon) > 0$ *such that the following holds. The size of generalized randomized k-BPs representing* SYL_n^* *with two-sided* ε*-bounded error is bounded below by* $2^{\Omega(n/c(\varepsilon)^k k^3)}$.

We remark that the probability amplification result of Proposition 11.3.2 is applied for the proof of the second claim of this theorem. Thathachar (1998a) has applied Sauerhoff's technique to obtain lower bounds for his conjunction of the hyperplanar sum-of-products predicate CHSP_q^k. Sauerhoff (1999a) has improved the bound on the error probability.

Theorem 11.8.9. *The size of generalized randomized* $(k-1)$*-BPs representing* CHSP_q^k *(which is defined on N variables) with two-sided* ε*-bounded error has a size bounded below by* $2^{\Omega((\alpha(k)-\varepsilon)N^{1/k}2^{-2k}k^{-3})}$ *where* $\alpha(k) = q^{-(k+1)}(1+o(1))$ *(for* $N \to \infty$*). (The parameters k and* ε *may both depend on N.)*

11.9 Exercises and Open Problems

11.1.E If f and g essentially depend on disjoint sets of variables, prove that the signature of $f \wedge g$ is equal to the product of the signatures of f and g.

11.2.M (See Sauerhoff (1999a).) Let G be a randomized OBDD working on n Boolean variables ordered with respect to π and on r probabilistic variables. Prove that there exists a graph theoretically isomorphic randomized

OBDD G' working on n Boolean and at most $(n+1)r$ probabilistic variables such that the acceptance and rejection probabilities of G and G' are the same and G' is ordered with respect to a variable ordering of all variables.

11.3.M Discuss the consequences if we allow randomized BPs to contain randomized nodes with an arbitrary number of outgoing edges which are chosen with equal probability.

11.4.E Prove a probability amplification result for randomized s-oblivious BDDs.

11.5.D Efficient synthesis algorithms are known for G-FBDDs. Nevertheless, a corresponding result to Theorem 11.3.5 cannot be proved. Describe why the proof of Theorem 11.3.5 does not work in this situation.

11.6.O Is it possible to generalize Corollary 11.4.3 to FBDDs (or k-BPs for some fixed k)?

11.7.M (See Sauerhoff (1999a).) The function *shifted equality test* $\text{SEQ}_n \in B_{2n+k}$, where $n = 2^k$, is defined on $a = (a_{k-1}, \ldots, a_0)$, $x = (x_0, \ldots, x_{n-1})$, and $y = (y_0, \ldots, y_{n-1})$. It tests whether x is equal to $y(a)$, which is the vector resulting from a cyclic shift of y by $|a|$ positions. Prove that $\text{SEQ}_n \in \text{coRP}_{\varepsilon(n)}\text{-OBDD}$ as long as $\varepsilon(n)^{-1}$ is polynomially bounded.

11.8.M (See Sauerhoff (1999a).) The function *equal adjacent rows* $\text{EAR}_n \in B_{n^2}$ checks whether a Boolean $n \times n$ matrix X contains two adjacent rows which are equal. Prove that $\text{EAR}_n \in \text{coRP}_{\varepsilon(n)}\text{-OBDD}$ as long as $\varepsilon(n)^{-1}$ is polynomially bounded.

11.9.M Prove Proposition 11.5.7.

11.10.M Prove Theorem 11.5.8 without the simplifying assumptions.

11.11.E (See Bollig, Löbbing, Sauerhoff, and Wegener (1999).) Prove that HWB $\in$ PP-OBDD.

11.12.D (See Agrawal and Thierauf (1998).) Let G be a randomized OBDD. Prove that it is coNP-complete to decide whether for each input a either $acc_G(a) \geq 3/4$ or $rej_G(a) \geq 3/4$ holds.

11.13.M Describe lower bound techniques for randomized k-IBDDs and s-oblivious BDDs using the results from Section 7.5 and Section 11.7.

11.14.D (See Sauerhoff (1999a).) Prove that the size of randomized OBDDs representing the indirect storage access function ISA_n with two-sided ε-bounded error, where $\varepsilon < 1/2$ is a constant, is bounded below by $2^{\Omega(n/\log n)}$.

11.15.M Prove exponential lower bounds on the size of randomized $(k-1)$-OBDDs representing the pointer-jumping function $\mathrm{PJ}_{k,n}$ with two-sided 1/3-bounded error where k is a constant. For which $k = k(n)$ do you obtain nonpolynomial lower bounds?

11.16.E Improve the results of Exercise 11.15 to ε-bounded error and constants $\varepsilon < 1/2$.

11.17.D Solve Exercises 11.15 and 11.16 for randomized $(k-1)$-IBDDs.

11.18.M Prove that ZPP-FBDD = RP-FBDD $\cap$ coRP-FBDD.

11.19.O Prove or disprove that ZPP-OBDD = RP-OBDD $\cap$ coRP-OBDD.

11.20.E Prove that RP-OBDD $\cap$ coRP-OBDD $\subsetneqq$ NP-OBDD $\cap$ coNP-OBDD.

11.21.O Prove for some explicitly defined Boolean function $f = (f_n)$ that $f \notin$ PP-OBDD.

11.22.O Prove or disprove that NP-FBDD $\subseteq$ BPP-FBDD.

11.23.O Prove or disprove that $\mathrm{RP}_{\varepsilon}$-FBDD $\subsetneqq$ $\mathrm{RP}_{\varepsilon'}$-FBDD for all $0 \le \varepsilon < \varepsilon' < 1$.

11.24.O Prove or disprove that $\mathrm{BPP}_{\varepsilon}$-FBDD $\subsetneqq$ $\mathrm{BPP}_{\varepsilon'}$-FBDD for all $0 \le \varepsilon < \varepsilon' < 1/2$.

11.25.O Prove or disprove that RP-FBDD $\cap$ coRP-FBDD = NP-FBDD $\cap$ coNP-FBDD.

11.26.O Consider the last five open problems for k-BPs, $k \neq 1$, instead of FBDDs.

(All mentioned open problems are from Sauerhoff (1999a).)

Chapter 12

Summary of the Theoretical Results

12.1 Algorithmic Properties

We have investigated a lot of BP and BDD variants. These variants have been introduced for various reasons. Here we discuss their algorithmic properties and weigh which representations are useful for applications. It is necessary to consider time-space trade-offs. A quadratic runtime for the representation type R_1 is "more efficient" than a linear runtime for the representation type R_2 and the same operation if the representations of type R_1 typically are "much smaller" than representations of type R_2. Hence, the results summarized in the next two subsections, namely the size of selected functions for various representation types and the complexity landscapes, have also to be taken into account. Moreover, the worst-case estimates have to be compared with the behavior on "typical inputs."

We start our résumé with the bit-level representations. General BPs and BDDs do not allow an efficient minimization nor efficient equivalence tests and are not useful for applications. No efficient minimization algorithm for DTs is known and DTs are too large for many functions. Some representation types like k-BPs and $(1,+k)$-BPs have been introduced for complexity theoretical reasons only. Since nobody knows how to introduce randomized nodes during a synthesis process, there is no idea how to use randomized BDDs in applications. The situation is different for nondeterministic BDDs. EXOR-nondeterminism allows a lot of polynomial-time algorithms which nevertheless seem to be too inefficient. In order to allow an efficient implementation of the negation operation, OR- and AND-nondeterminism have been restricted to partitioned OBDDs with fixed window functions. General oblivious BDDs have some nice properties but, in

	EVAL	SYN	SAT	EQU	RBC	MIN	RED
π-OBDD	n	$s_f s_g$	s_f	$s_f + s_g$	s_f	s_f	yes
G-FBDD	n	$s_f s_g s_G$	s_f	$s_f + s_g$	exp	s_f	yes
π-k-OBDD	kn	$s_f s_g$	s_f^{2k-1}	$(s_f s_g)^{2k-1}$	s_f	?	no
$\pi(k)$-IBDD	kn	$s_f s_g$	NPC	coNPC	s_f	NPH	no
π-ZBDD	n	$ns_f s_g$	s_f	$s_f + s_g$	ns_f	s_f	yes
π-OFDD	s_f	exp	s_f	$s_f + s_g$	exp	s_f	yes
π-d-OKFDD	s_f	exp	s_f	$s_f + s_g$	exp	s_f	yes
$\pi(k)$-PBDD	kn	$s_f s_g$	s_f	$s_f + s_g$	exp	s_f	yes
$\oplus$-π-OBDD	s_f	$s_f s_g$	ns_f^3	$n(s_f + s_g)^3$	s_f	ns_f^3	no

Table 12.1.1: *Algorithmic properties of selected BDD variants.*

applications, only k-OBDDs and k-IBDDs are discussed. The remaining models are OBDDs, FBDDs, k-OBDDs, k-IBDDs, ZBDDs, OFDDs, OKFDDs, PBDDs, and EXOR-OBDDs (also denoted as $\oplus$-OBDDs). They all allow efficient synthesis algorithms only if a variable ordering π, a sequence of variable orderings $\pi(k) = (\pi_1, \ldots, \pi_k)$, a graph ordering G, and/or a decomposition type list d is fixed. This restriction can be relaxed by allowing reordering techniques like the sifting algorithm. Table 12.1.1 contains the best-known asymptotic runtimes of algorithms for the operations evaluation EVAL, synthesis SYN, satisfiability test SAT, equivalence test EQU, replacement by constants RBC, minimization MIN, and the information whether a reduction RED is possible, i.e., whether the result of minimization is unique up to isomorphism. The other operations are less important or can be performed as combinations of the considered operations. The table uses n for the number of variables, s_f and s_g for the size of G_f and G_g, respectively, s_G for the size of the graph ordering G, NPC, coNPC, and NPH for NP-complete, coNP-complete, and NP-hard, respectively, and exp for a possible exponential blow-up of the size. The results for RBC refer to a replacement of a set of variables by constants.

We add some comments to Table 12.1.1. A synthesis process is only efficient if the size increase due to the single synthesis steps is in most steps very moderate. The factor s_G for G-FBDDs occurs only as long as G_f and G_g do not reflect the full structure of G. The additional factor n for the synthesis of π-ZBDDs is only necessary for non-0-preserving operations. The exponential blow-up of the size of π-OFDDs and π-d-OKFDDs is not possible for EXOR-synthesis steps and is not observed very often in applications for the other synthesis operations. It has been discussed in previous chapters how the exponential blow-up of the size for RBC can be prevented for G-FBDDs and $\pi(k)$-PBDDs. For the variants based on Shannon's decomposition rule (OBDDs, FBDDs, k-OBDDs, k-IBDDs, PBDDs, and $\oplus$-OBDDs), it is no problem to replace Boolean variables by multivalued ones.

The considered word-level DDs have similar properties. MTBDDs (also called ADDs) are the word-level counterpart of OBDDs, while BMDs are the counterpart of OFDDs and share their problems. Replacement by constants may cause a blow-up of the size and some synthesis operations like multiplication may lead to an exponential blow-up of the size. The introduction of edge weights results in EVBDDs and *BMDs. The size of representations decreases but the algorithmic problems become more difficult. OKFDDs are a combination of OBDDs, OFDDs, and a third type of DDs based on the negative Reed-Muller decomposition rule. In a similar way, we obtain HDDs from MTBDDs and BMDs and Kronecker *BMDs from EVBDDs and *BMDs.

12.2 Bounds for Selected Functions

This monograph is organized in such a way that different chapters refer to different types of BDDs. Therefore, the results for a selected function are scattered. Here we list the known upper and lower bounds for many functions and refer to the corresponding theorem (marked by T), proposition (P), corollary (C), or exercise (E).

Each symmetric function can be represented in size $O(n^2)$ by OBDDs (T2.3.4) and OFDDs (T8.2.8). We have more precise bounds for some special symmetric functions:

- threshold function $T_{k,n}$: $k(n-k+1)+2$ for OBDDs (T4.7.2) and $O\big((n\log^3 n)/\log\log n\log\log\log n\big)$ for BPs (T2.3.10).
- majority function $\mathrm{MAJ}_n = T_{\lceil n/2\rceil,n}$: $\Omega((n\log n)/\log\log n)$ for BPs (T2.2.8).
- exact counting $E_{k,n}$: $nk-k^2+n+2$ for OBDDs (T4.7.2) and $O\big((n\log^2 n)/\log\log n\big)$ for BPs (T2.3.9).
- modular counting $\mathrm{MOD}_{k,n}$: $nk-k^2+n+2$ for OBDDs (T4.7.2) and $O\big((n\log^4 n)/(\log\log n)^2\big)$ for BPs (T2.3.10).

The next group contains the arithmetic operations:

- addition ADD_n: $9n - 5$ for OBDDs (T4.4.3), similarly for subtraction.
- multiple addition $MULTADD_n$: $O(n^4)$ for OBDDs (T4.4.4).
- multiplication MUL_n: $O(n^4)$ for BPs (T2.3.5).
- middle bit of multiplication $MUL_{n-1,n}$: at least $2^{n/8}$ for OBDDs (T4.5.2), similarly for OFDDs (see the remark in Section 8.2); $2^{\Omega(n^{1/3})}$ for FBDDs (T6.2.14); $2^{\Omega(nk^{-3}2^{-4k})}$ for oblivious BDDs of length $2kn$, i.e., not polynomial for a length $l = o(n \log n / \log \log n)$ (T7.5.11), similarly for oblivious OR-OBDDs, AND-OBDDs, and EXOR-OBDDs (T10.3.7); $2^{\Omega(n/\log n)}$ for randomized OBDDs with two-sided ε-bounded error if $\varepsilon < 1/2$ (T11.7.7).
- graph of multiplication multgraph_n: $O(n^3)$ for BPs (E.2.11), contained in $\text{coRP}_{\varepsilon(n)}$-OBDD if $\varepsilon(n)^{-1} = \text{poly}(n)$ (T11.5.3).
- word-level multiplication: $O(n^2)$ for BMDs (T9.3.2), at least 2^n for EVBDDs (T9.5.6), and $O(n)$ for $*$BMDs (T9.6.2).
- squaring SQU_n: similar lower bounds as for $MUL_{n-1,n}$ (e.g., C4.6.3 and C6.2.15).
- multiplicative inverse INV_n: similar lower bounds as for $MUL_{n-1,n}$ (e.g., C4.6.3 and C6.2.15).
- division DIV_n: similar lower bounds as for $MUL_{n-1,n}$ (e.g., C4.6.3 and C6.2.15).

A further class with interesting properties is the class of storage access or pointer functions:

- direct storage access DSA_n or multiplexer MUX_n: $2n + 1$ for OBDDs (T4.3.2).
- indirect storage access ISA_n: $\Omega(n^2/\log^2 n)$ and $O(n^2)$ for BPs (T2.2.6, E.2.5); $\Omega(2^{n/\log n})$ for OBDDs (T4.3.3); $O(n^2)$ for FBDDs (T6.1.3) and 2-OBDDs (T7.2.2); $O(n^2 \log n)$ for OR- and AND-OBDDs (T10.2.1), EXOR-OBDDs (C10.2.2), and π-PBDDs (T10.4.4); $2^{\Omega(n/\log n)}$ for randomized OBDDs with two-sided ε-bounded error if $\varepsilon < 1/2$ (E.11.14).
- hidden weighted bit HWB_n: $O(n^2)$ for BPs (T2.3.3); $\Omega(2^{n/5})$ and $O(2^{0.2029n})$ for OBDDs (T4.10.2, T4.10.5); $\Omega(n^{-3/2}2^{n/5})$ for OFDDs (T8.2.10); $O(n^2)$ for FBDDs (T6.1.4), 2-OBDDs (T7.2.2), and $(1,+1)$-BPs (T7.2.2); $O(n^3)$ for OR- and AND-OBDDs (T10.2.1), EXOR-OBDDs (C10.2.2), and π-PBDDs (T10.4.4); $2^{\Omega(n)}$ for randomized OBDDs with two-sided ε-bounded error if $\varepsilon < 1/2$ (T11.7.8).

- weighted sum WS_n: $2^{n-O(n^{1/2})}$ for FBDDs (T6.2.10); $O(n^2)$ for 2-OBDDs (T7.2.2) and $(1,+1)$-BPs (T7.2.2); $O(n^3)$ for OR- and AND-OBDDs (T10.2.1), EXOR-OBDDs (C10.2.2), and π-PBDDs (T10.4.4); $2^{\Omega(n)}$ for randomized OBDDs with two-sided ε-bounded error if $\varepsilon < 1/2$ (Ablayev (1997)).

- matrix storage access MSA_n: exponential for FBDDs (P11.5.7 and E.11.9), polynomial for OR- and AND-OBDDs (P11.5.7) contained in $\mathrm{ZPP}_{1/2}$-FBDD (T11.5.8), $2^{\Omega(n^{1/2}/\log^{1/2} n)}$ for randomized OBDDs with two-sided ε-bounded error if $\varepsilon < 1/2$ (C11.7.5).

The function determinant DET_n is an important one from linear algebra and has the following properties where $N = n^2$ is the number of variables: $\Omega(N^{3/2}/\log N)$ for BPs (T2.2.6), $2^{\Omega(N^{1/2})}$ for FBDDs (E.6.4), $2^{\Omega(N^{1/2})}$ for randomized OBDDs with two-sided ε-bounded error if $\varepsilon < 1/2$ (C11.7.5).

Functions combining properties of rows and columns of matrices cause difficulties for read-once DDs. We have obtained the following results where $N = n^2$ is the number of variables:

- permutation matrix test PERM_n: $\Omega\left(N^{-1/4}2^{N^{1/2}}\right)$ for OBDDs (T4.12.3), FBDDs (T6.2.12), and OR-FBDDs (T10.3.8); $2^{\Omega(N^{1/2}/k)}$ for k-OBDDs (T7.5.7); $O(N)$ for 2-IBDDs (T7.2.7) and AND-OBDDs (T10.2.1); polynomial-size for semantic OR-$(1,+1)$-BPs (T10.2.7); contained in $\mathrm{coRP}_{\varepsilon(n)}$-OBDD if $\varepsilon(n)^{-1} = \mathrm{poly}(n)$ (T11.5.4), but not in RP-OBDD (P11.6.2).

- test for a 1-row or 1-column ROW_n+COL_n: $2^{\Omega(N^{1/4})}$ for OBDDs (T4.12.2), improved to $\Omega(N^{-7/4}2^{N^{1/2}})$ for FBDDs (T6.2.13); $N^{-3/2}2^{\Omega(N^{1/2}/k)}$ for k-OBDDs (T7.5.8); $O(N)$ for 2-IBDDs (T7.2.7) and OR-OBDDs (T10.2.1); $2^{\Omega(N^{1/2}/k)}$ for randomized k-OBDDs with two-sided ε-bounded error if $\varepsilon < 1/2$ (T11.7.9). The last bound also holds for $s\mathrm{ROW}_n + \bar{s}\mathrm{COL}_n$, which can be represented by FBDDs of size $O(N)$.

- mod sum function MS_n: contained in $\mathrm{coRP}_{1/2}$-FBDD (P11.5.11) and $\mathrm{BPP}_{1/3+\varepsilon(n)}$-FBDD if $\log \varepsilon(n)^{-1} = \mathrm{poly}(n)$ (P11.5.11), $2^{\Omega(n)}$ for randomized FBDDs with two-sided ε-bounded error if $\varepsilon < 1/4$ (T11.8.6).

Various types of clique functions have been considered as typical functions describing graph properties. The following results have been obtained where $N = \binom{n}{2}$ is the number of variables:

- clique function $cl_{k,n}$: $2^{\Omega(N^{1/2})}$ for certain $k(n)$ and FBDDs (T6.2.7), $2^{\Omega(k(n))}$ for $k(n) \le n/3$ and OR-FBDDs (T10.3.11), $2^{\Omega(N^{1/2})}$ for certain

$k(n)$ and randomized OBDDs with two-sided ε-bounded error if $\varepsilon < 1/2$ (C11.7.5).

- exactly half clique function excl_n: $2^{\Omega(N^{1/2})}$ for FBDDs (T6.2.6), $O(N^2)$ for oblivious BDDs of linear length $2N$ and 2-BPs (T7.2.6), polynomial size for AND-OBDDs (T10.2.1), and contained in $\text{coRP}_{\varepsilon(n)}$-OBDD if $\varepsilon(n)^{-1} = \text{poly}(n)$ (T11.5.5).

- odd number of triangles $\oplus cl_{n,3}$: $2^{\Omega(N)}$ for FBDDs (T8.2.12) and $O(N^2)$ for OFDDs (T8.2.12).

- isolated triangle function $1cl_{n,3}$: $2^{\Omega(N)}$ for FFDDs (T8.2.12) and $O(N^2)$ for OBDDs (T8.2.12).

Characteristic functions of linear codes and the bilinear Sylvester function have been considered, since their clear algebraical structure simplifies lower bound proofs:

- characteristic functions of certain linear codes: superpolynomial lower bounds for semantic $(1,+k)$-BPs and $k = o(n/\log n)$ (see the remark in Section 7.4), $2^{\Omega(n^{1/2}k^{-k})}$ for k-BPs (T7.6.5), polynomial upper bounds for AND-OBDDs (T10.2.6), $2^{\Omega(n^{1/2}k^{-k})}$ for OR-k-BPs (T10.3.10).

- bilinear Sylvester function SYL_n: $2^{\Omega(n4^{-k}k^{-3})}$ for k-BPs (T7.6.8), OR-k-BPs (T10.3.10), and randomized k-BPs with two-sided ε-bounded error if $\varepsilon < 1/3$ (T11.8.8). These results hold for BPs with three-valued variables. The bound $2^{\Omega(nc(\varepsilon)^{-k}k^{-3})}$ holds for a Boolean variant SYL_n^*, $\varepsilon < 1/2$, and some $c(\varepsilon) > 0$ (T11.8.8).

The last group of functions is investigated for hierarchy results:

- pointer-jumping function $\text{PJ}_{k,n}$ on $N = \Theta(n \log n)$ variables: $O(kn^2)$ for k-OBDDs (P7.5.13), $2^{\Omega(n^{1/2}/k)}$ for $(k-1)$-OBDDs (T7.5.15), i.e., not polynomial size for $k = o(n^{1/2}/\log n)$, and not polynomial size for $(k-1)$-IBDDs if $k \le (1-\delta)\log\log n$ (T7.5.17), related results for the randomized case (E.11.15–E.11.17), polynomial size for constant k and OR-OBDDs (T10.2.1) and EXOR-OBDDs (C10.2.2).

- hyperplanar sum-of-products predicate HSP_q^k on $N = n^k$ variables: $O(kN)$ for k-IBDDs (P7.2.9), $2^{\Omega(N^{1/k}2^{-2k}k^{-4})}$ for $(k-1)$-BPs (T7.6.9) and OR-$(k-1)$-BPs (T10.3.10), which are superpolynomial if $k = o(\log^{1/2} n)$.

- conjunction of hyperplanar sum-of-products predicate CHSP_q^k on $N = n^k$ variables: $O(kN)$ for k-IBDDs (P7.2.9), $2^{\Omega(N^{1/k}2^{-2k}k^{-3})}$ for $(k-1)$-BPs (T7.6.9) and OR-$(k-1)$-BPs (T10.3.10), polynomial size for AND-OBDDs (E.10.9), not contained in BPP_ε-$(k-1)$-BP for some $\varepsilon > 0$ and k (T11.8.8).

This list contains a lot of separation results, i.e., results that functions have polynomial size for one representation type and not polynomial size for another representation type. We explicitly mention some hierarchy results:

- P-$(k-1)$-OBDD $\subsetneq$ P-k-OBDD as long as $k(n){=}o(n^{1/2}/\log^{3/2} n)$. This is proved for the pointer-jumping function $\mathrm{PJ}_{k,n}$ (C7.5.16).
- P-k-OBDD $\not\subseteq$ P-$(k-1)$-IBDD as long as $k(n) \le (1-\delta)\log\log n$ for some $\delta > 0$. Again, $\mathrm{PJ}_{k,n}$ is an example (T7.5.17).
- P-$(k-1)$-IBDD $\subsetneq$ P-k-IBDD, P-$(k-1)$-BP $\subsetneq$ P-k-BP, and P-OR-$(k-1)$-BP $\subsetneq$ P-OR-k-BP as long as $k(n) = o(\log^{1/3} n)$. This follows from the results on HSP_q^k and CHSP_q^k. Moreover, P-k-IBDD $\not\subseteq$ P-OR-$(k-1)$-BP for these k.
- P-BPP_ε-$(k-1)$-BP $\subsetneq$ P-BPP_ε-k-BP for certain $\varepsilon > 0$ and constant k. This follows from the results on CHSP_q^k. Moreover, P-k-IBDD $\not\subseteq$ P-BPP_ε-$(k-1)$-BP for certain $\varepsilon > 0$ and constant k.
- P-$(1,+(k-1))$-BP $\subsetneq$ P-$(1,+k)$-BP as long as $k \le n^{1/2}/(2\log n)$ (T7.4.1).
- P-$(1,+k)$-BP $\not\subseteq$ P-$(1,+(k-1))$-semantic-BP as long as $k \le n^{1/6}/(2\log^{1/3} n)$ (T7.4.1).
- P-$(k-1)$-PBDD $\subsetneq$ P-k-PBDD (partitioned BDDs with k parts) as long as $k(n) = o\big(((\log n)/\log\log n)^{1/2}\big)$ (T10.4.12).

12.3 Complexity Landscapes

It would be confusing (and even impossible because of the limited number of pages) to draw a figure with the landscape of all considered complexity classes. We show the partial landscape for deterministic, nondeterministic, and randomized OBDDs, as well as the corresponding landscape for FBDDs (Sauerhoff (1999a)). The reader is asked to produce larger landscapes.

Figure 12.3.1 contains the complexity landscape for OBDDs, where $\nearrow$ stands for $\subseteq$, $\nearrow$ for $\subsetneq$, and $\nearrow$ (dotted) for $\not\subseteq$.

We discuss the described relations. (1) has been proved (T11.3.6). The other inclusion properties are obvious and we look for separating examples. For (4), (6), (7), and (14), consider HWB_n, which is contained in NP-OBDD and coNP-OBDD (T10.2.1) and, therefore, also in PP-OBDD but not in P-OBDD (T4.10.2) or BPP-OBDD (T11.7.8) and, therefore, not in RP-OBDD (T11.3.4). For (3), (5), (9), (10), (11), and (12), consider PERM_n and $\overline{\mathrm{PERM}}_n$. We know that $\mathrm{PERM}_n \in$ coRP-OBDD (T11.5.4) and, therefore, $\mathrm{PERM}_n \in$ BPP-OBDD (T11.3.4) but $\mathrm{PERM}_n \notin$ RP-OBDD (P11.6.2) and we know that

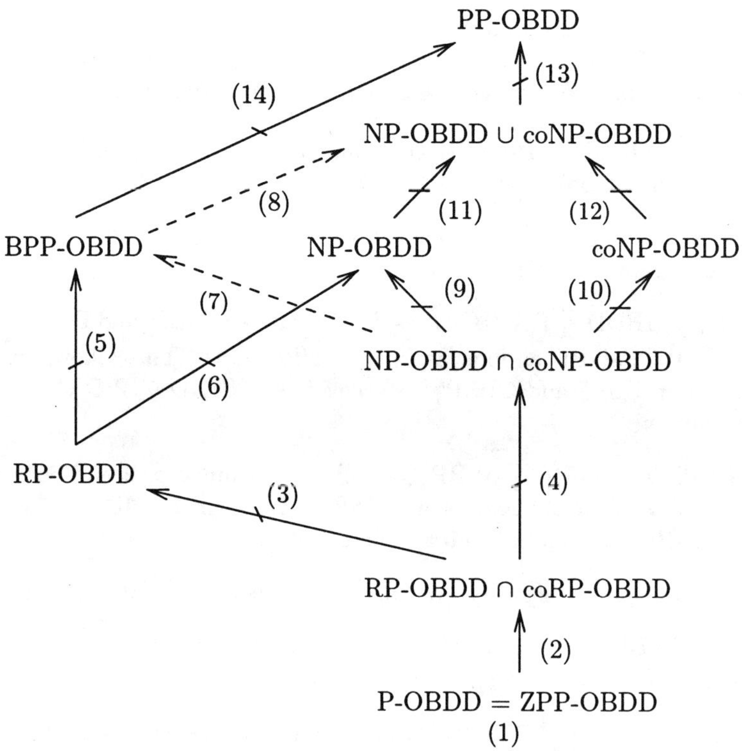

Figure 12.3.1: *The complexity landscape for OBDDs.*

$\text{PERM}_n \in$ coNP-OBDD (T10.2.1) but $\text{PERM}_n \notin$ NP-OBDD (T10.3.8). Finally, let $2\text{PERM}_n(X, Y) = \text{PERM}_n(X) \wedge \overline{\text{PERM}}_n(Y)$. Then $2\text{PERM}_n \notin$ NP-OBDD ∪ coNP-OBDD (T10.3.8 for appropriate subfunctions) and $2\text{PERM}_n \in$ BPP-OBDD ⊆ PP-OBDD. Let G_1 be a polynomial-size randomized OBDD representing $\text{PERM}_n(X)$ with one-sided ε-bounded error and making errors only if $\text{PERM}_n(X) = 0$. Let G_2 be a polynomial-size randomized OBDD representing $\overline{\text{PERM}}_n(Y)$ with one-sided ε-bounded error and making errors only if $\overline{\text{PERM}}_n(Y) = 1$. If we replace the 1-sink of G_1 with the source of G_2, we obtain a polynomial-size OBDD representing 2PERM_n with two-sided ε-bounded error. This proves (8) and (13).

Let ⊕P-OBDD be the class of functions representable by polynomial-size EXOR-OBDDs. Then $\overline{\text{EQ}^*_n} \in$ NP-OBDD, $\overline{\text{EQ}^*_n} \notin$ coNP-OBDD, and $\overline{\text{EQ}^*_n} \in$ ⊕P-OBDD (T10.3.6). For EQ^*_n, we may interchange the roles of NP-OBDD and coNP-OBDD. Finally, $\text{IP}^*_n \in$ ⊕P-OBDD, $\text{IP}^*_n \notin$ NP-OBDD, and $\text{IP}^*_n \notin$ coNP-OBDD (T10.3.6).

Figure 12.3.2 contains the complexity landscape for FBDDs. This landscape can be refined, since we have no probability amplification method and we may

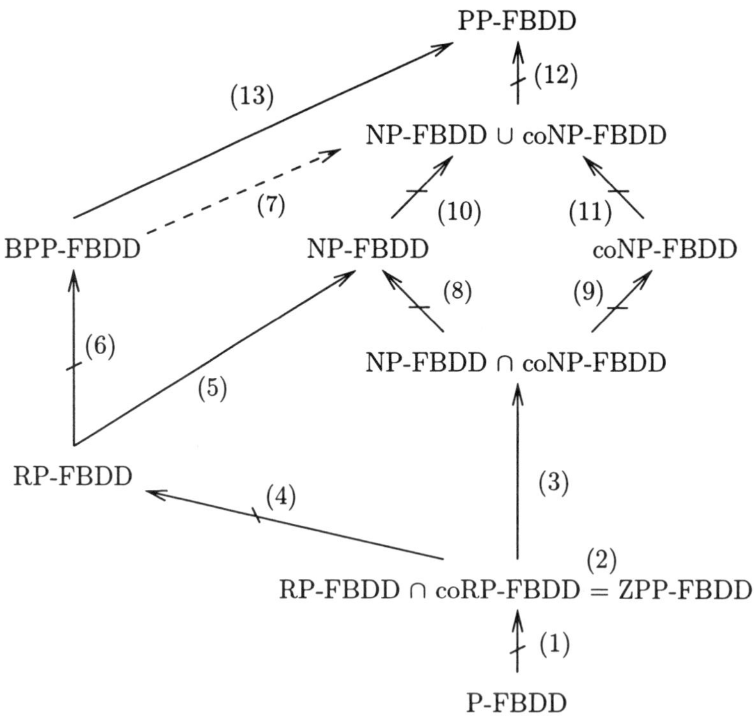

Figure 12.3.2: *The complexity landscape for FBDDs.*

obtain different complexity classes for different error probabilities $\varepsilon > 0$. Here, (2) is Exercise 11.18 and (3), (5), and (13) are obvious. Property (1) is proved by considering MSA_n (P11.5.7, T11.5.8). (4) follows since $\overline{\mathrm{PERM}_n} \in$ RP-FBDD (T11.5.4) and $\mathrm{PERM}_n \notin$ NP-FBDD (T10.3.8), implying that $\overline{\mathrm{PERM}_n} \notin$ coRP-FBDD. RP-FBDD $\subseteq$ BPP-FBDD is proved in Theorem 11.3.4. Moreover, RP-FBDD $\neq$ BPP-FBDD, since RP-FBDD $\neq$ coRP-FBDD (this follows from (4)) but, obviously, BPP-FBDD = coBPP-FBDD. This proves (6). We obtain (8)–(11) by considering PERM_n (T10.2.1, T10.3.8). Finally, (7) and (12) follow in the same way as in the OBDD case by considering $2\mathrm{PERM}_n$.

Chapter 13

Applications in Verification and Model Checking

Several BDD models, in particular OBDDs, are used for many different problems. The early OBDD package due to Brace, Rudell, and Bryant (1990) has been superseded by a lot of successors, e.g., the Long package (Long (1993)) and the Boulder package CUDD (Somenzi (1998)). The successful Berkeley package HSIS for formal verification (Aziz et al. (1994)) uses OBDDs. It is nowadays impossible to discuss all relevant applications in three chapters of a monograph. The idea is to present a representative selection of typical applications. This implies that in some places the most advanced applications are not considered. Instead of this we try to build a bridge between our knowledge of BDDs and the applications. Our main emphasis is the description of the operations needed for some applications and how these operations can be performed with BDDs. The organization is as follows. We start in this chapter with verification and model checking, still the main areas of BDD applications. In Chapter 14, we focus on applications in other CAD areas and, in Chapter 15, we show that BDDs can have applications in various further areas.

The sections of this chapter deal with the verification of combinational circuits and sequential circuits and with model checking, respectively. We omit other subjects about verification like, e.g., the verification of protocols (Hu and Dill (1993)). For more intensive descriptions of the applications of BDDs in verification and model checking, we refer to Hachtel and Somenzi (1996), McFarland (1993), McMillan (1994), and Minato (1996).

13.1 Verification of Combinational Circuits

The verification of combinational circuits is the most classical area of application of BDDs. The general idea can be described as follows. Let S be the specification of a Boolean function f and R be a circuit representing f'. The task is to verify that $f = f'$. Often S is given as a circuit or can be translated into a circuit-like description. In order to represent f and f' by OBDDs, one starts with the variables or primary inputs of the circuits. A circuit is a (topologically sorted) list of gates and we use the synthesis algorithm to construct the OBDDs representing the functions computed at the gates of the circuit. This implies even in the case where f has a single output (one primary output) that we have to represent a lot of functions simultaneously. One usually works with SBDDs allowing complemented edges. At the end, it is sufficient to check whether the pointers to the nodes representing the outputs of f coincide with the corresponding pointers for f'.

This simple approach causes a lot of important decisions. First, an initial variable ordering has to be chosen (see Section 5.6). It may happen that the different stages of the synthesis process require different variable orderings. Usually, only local reordering techniques are applied (see Section 5.8) but, if necessary, we also know how to perform a global reordering efficiently (see Section 5.7).

OBDDs with a fixed variable ordering have the best algorithmic properties as long as the OBDD size does not explode. In such a situation, one may switch to other BDD models like ZBDDs, FBDDs with a fixed graph ordering, OFDDs, OKFDDs, partitioned BDDs with fixed window functions, or to some word-level representation. OKFDDs have the advantage that we may save the advantages of OBDDs on certain layers. Nevertheless, we are faced with a trade-off between simplicity and efficiency of algorithms, the number and complexity of free parameters (one or more variable orderings, variable ordering vs. graph ordering, window functions, decomposition type list), the difficulty of a good choice for the free parameters, and the size of the resulting representations. Without some additional knowledge, one should start with the most efficient method for simple functions and if this method needs too much time and/or space, one should switch to another method. Such a process is called *filtering* (Mukherjee, Jain, Takayama, Fujita, Abraham, and Fussell (1997)) and is realized in the system FLOVER. Such a filter-based approach contains, besides BDD techniques, also techniques not using BDDs.

Several ideas have been suggested to avoid the peak of the OBDD size during the sequence of synthesis steps to obtain an OBDD for f or $f \oplus f'$. It has been observed that most often the OBDD representing f is much smaller than the largest OBDD during the synthesis process. In particular, we hope that $f \oplus f'$ is the constant 0 and has an OBDD size of 1. Ashar, Ghosh, and Devadas (1992) have replaced a variable x_i with large fan-out in the circuit by independent variables $x_{i,1}, \ldots, x_{i,d}$. This destroys the canonicity of the representation and even the uniqueness of the considered function, e.g., 0 may be replaced by $x_{i,1} \oplus x_{i,2}$. We may reconstruct the old function by the application of the *smoothing operator* due to Touati, Savoj, Lin, Brayton, and Sangiovanni-Vincentelli (1990) where we replace $x_{i,2}, \ldots, x_{i,d}$ by $x_{i,1}$ to obtain the function $f_{|x_{i,1}=\cdots=x_{i,d}}$.

If the specification and the realization share some subcircuit, it is possible to replace the output (or the outputs) of the common subcircuit by a new variable y called the *auxiliary variable*. This also may lead to *false negatives*, since the specification and the realization may be equal but there may be some assignments a to x and b to y such that $y = b$ is impossible if $x = a$. Such "impossible" inputs may satisfy the constructed OBDD and, in that case, are called false negatives. Therefore, at the end we have to replace y by the function represented by the common subcircuit. Such an approach has been presented by van Eijk (1997). Brand (1993) has used methods known from testing to simplify the verification of large combinational circuits by OBDDs. The approach of Shin and Hachtel (1997) has a similar flavor. They cut the circuit for $f \oplus f'$ into two parts and compute the implications caused by the logic relations between the inputs of the circuit and the functions computed at the cut. These implications are decomposed to a conjunction of functions. Then an OBDD for the second part, namely the part whose input variables are the signals at the cut, is constructed and this OBDD typically is smaller than the OBDD for the whole circuit. Again, we obtain a representation with false negatives. Finally, the parts of the conjunctive decomposition can be applied subsequently to destroy these false negatives.

Lai and Sastry (1992) have proposed the idea of *hierarchical verification* with word-level representations. This idea has been used by Lai, Pedram, and Vrudhula (1996) with EVBDDs and by Bryant and Chen (1995) with *BMDs. The approach works if the circuit contains subcircuits which have their own specification and if there is a specification of the whole circuit based on the results of the subcircuits. The subcircuits are verified in the traditional way. The verification that the interaction of the subcircuits leads to a correct realization may use word-level representations, in particular, if the inputs and outputs of the subcircuits have a natural interpretation as integers. Hence, this approach is applied to the verification of arithmetic circuits.

In the remainder of this section, we consider the verification of multiplier and divider circuits. It is obvious that the typical adder circuits are easy to verify. The verification of multipliers has been discussed quite often, since multiplication has no polynomial-size bit-level representation for BDD variants with good algorithmic properties and $c6288$, the most difficult International Symposium on Circuits and Systems (ISCAS) benchmark circuit for OBDDs, is a 16-bit multiplier (each factor has length 16). The bug in the Pentium floating-point divider has motivated research on the verification of circuits for division.

Circuits for functions with a well-studied structure like multiplication or division follow one of a limited number of design ideas. Hence, it is worthwhile to look for verification techniques for special design techniques. Burch (1991) observed that it is much easier to verify multipliers where in a first phase the pairwise conjunctions $x_i y_j$ are computed (a trivial verification task) and where afterwards the numbers $x_i y_j 2^{i+j}$ are added with adders which are not restricted to the special form of the considered numbers. Here we may use the new input variables $z_{ij} = x_i y_j$ and it is obvious to "verify by mathematics" that multiplication is reduced to multiple addition, which can be represented by polynomial-size OBDDs (see Theorem 4.4.4). Bryant and Chen (1998) have used the above-mentioned hierarchical verification technique for multipliers which are called $(n+m)$-add-steppers. The partial product p is initialized as 0 and, in general, p is a binary number $(h_{n-1}, \ldots, h_0, l_{m-1}, \ldots, l_0)$ where it is known that $(l_{m-1}, \ldots, l_0)$ belongs to the final product. Then y_m is multiplied by x and, if $(h'_n, \ldots, h'_0)$ is the sum of $(h_{n-1}, \ldots, h_0)$ and $y_m \cdot |x|$, we set $p = (h'_n, \ldots, h'_1, l_m, l_{m-1}, \ldots, l_0)$ where $l_m = h'_0$.

Jain, Bitner, Fussell, and Abraham (1992) have used partitioned OBDDs with disjoint window functions and have combined this approach with randomized equivalence tests (see Section 11.1). They have used the variable ordering $x_0, y_0, x_1, y_1, \ldots, x_{n-1}, y_{n-1}$ and have partitioned the input space by replacing some variables with constants. Since the replacement of a variable by 0 implies a larger simplification of a multiplier than a replacement by 1, they have chosen some k and have considered subcircuits where at most k variables are replaced with 0. The idea is to obtain a partitioning of the input space where all functions have been simplified in a comparable way. This approach is not restricted to special multipliers and, therefore, it is slower than verification methods which are tuned to multipliers of a given type.

Hamaguchi, Morita, and Yajima (1995) have considered the word-level verification of multipliers by *BMDs. The usual algorithm to transform the given circuit gate-by-gate into a *BMD is not efficient, since we know that the middle bit of multiplication has exponential OFDD size and, therefore, exponential *BMD size. We may only take advantage of *BMDs if we do not try to represent the single output bits of multiplication separately. The whole output $z = (z_{2n-1}, \ldots, z_0)$ of a multiplier is interpreted as a binary number. There is a linear-size *BMD representation of $|z| = z_{2n-1}2^{2n-1} + \cdots + z_0$. We start with this *BMD and work backward toward the inputs, i.e., we work on a reversed topologically sorted list of the gates. We always represent the result of the circuit with respect to the "inputs" of that circuit whose gates we have considered. In the beginning, no gate has been considered and $z_{2n-1}, \ldots, z_0$ are the inputs. At the end, all gates have been considered and $x_{n-1}, \ldots, x_0, y_{n-1}, \ldots, y_0$ are the inputs. Let us discuss the situation where we consider the binary gate G. Its output signal s is an input to the circuit part already considered. Its input signals s_1 and s_2 of G are considered as new variables. Then s is replaced by $s_1 \otimes s_2$ if G realizes $\otimes$. It may happen that s_1 (similarly for s_2) is the output of a gate G' such that another edge leading from G' to some gate G'' has been considered before as signal s_3. Then we have to apply the smoothing operator and have to set $s_1 = s_3$. At the end, we obtain a *BMD for multiplication working on the primary inputs of the circuit.

Although not all details of the erroneous Pentium floating-point divider have been published, we know enough about the underlying design to discuss verification methods which might have found the error (for theorem-proving techniques

$x_i + y_i$	-6	-5	-4	-3	-2	-1	0	1	2	3	4	5	6
v_i	-2	-1	0	1	-2	-1	0	1	2	-1	0	1	2
c_i	-1	-1	-1	-1	0	0	0	0	0	1	1	1	1

Table 13.1.1: *Radix-4 addition.*

or word-level model checking we refer to Clarke, German, and Zhao (1999) and Clarke, Khaira, and Zhao (1996), respectively). Here we discuss methods based mainly on OBDDs. First, we describe some features of radix-4 representations which are used in the Pentium floating-point divider.

Definition 13.1.1. A *radix-4 representation* of an integer a is a vector $x = (x_{n-1}, \ldots, x_0)$ where $-3 \le x_i \le 3$ and

$$a = |x|_4 := x_{n-1} \cdot 4^{n-1} + x_{n-2} \cdot 4^{n-2} + \cdots + x_0.$$

Radix-4 representations of integers are redundant—numbers may have many representations of the same length. This redundancy can be used to perform arithmetic operations in small depth. We describe the addition of two numbers x and y given by radix-4 representations of length n. We choose v_i and c_i such that $x_i + y_i = 4c_i + v_i$. Table 13.1.1 shows our special choice. Only the representation of -3 and 3 is unusual; we represent 3 as $1 \cdot 4 + (-1) \cdot 1$ and not as $0 \cdot 4 + 3 \cdot 1$. But this implies $-2 \le v_i \le 2$ and $-1 \le c_i \le 1$. We have guaranteed that $|x|_4 + |y|_4 = |v|_4 + 4|c|_4$. Let $s_0 = v_0$ and $s_i = v_i + c_{i-1}$ if $1 \le i \le n$. Then $-3 \le s_i \le 3$ and it is easy to obtain the radix-4 representation s for $|x|_4 + |y|_4$. Hence, addition can be performed in linear size and constant depth.

If we know that the result of an arithmetic circuit has an absolute value bounded by r, the computation can be performed in $\mathbb{Z}_m$ for some $m \ge 2r + 1$. For radix-4 representation it is useful to choose $m = 4^k + 1$ for some integer k. This implies that a carry c stands for $c \cdot 4^k \equiv -c \bmod m$ and it is easy to obtain $-c$ from c. This choice of m also simplifies shifts, i.e., the multiplication of a radix-4 number $x = (x_k, \ldots, x_0)$ by 4^l. We conclude that

$$|x|_4 \cdot 4^l = x_k 4^{k+l} + \cdots + x_{k-l+1} 4^{k+1} + x_{k-l} 4^k + \cdots + x_0 \cdot 4^l$$

$$\equiv (-x_k 4^l - \cdots - x_{k-l+1} \cdot 4) + (x_{k-l} 4^k + \cdots + x_0 4^l) \bmod m.$$

The multiplication with 4^l can be implemented as an addition of the numbers $(0, \ldots, 0, -x_k, \ldots, -x_{k-l+1}, 0)$ and $(x_{k-l}, \ldots, x_0, 0, \ldots, 0)$. A multiplication by 2 is considered as an addition.

The SRT division method is an iterative procedure like the school method for division. In each round of the algorithm, one further part of the quotient (from left to right) is produced. The school method for division works with an irredundant representation of numbers. In order to compute the first bit of the quotient, the whole dividend and the whole divisor have to be considered. In general, we have a remainder r (initialized as dividend) and the divisor. Then the next part q_i of the quotient q is produced and the new remainder is obtained by subtracting the product of q_i and d shifted in an appropriate way from the old remainder.

Bryant (1996) has described the design of a radix-4 SRT divider which, as the Pentium divider, is based on the design of Atkins (1968). The main idea is the following. Since a redundant representation of numbers is used, q_i (now not only a bit but an integer with small absolute value) is not uniquely determined and we hope that a legal q_i can be computed by considering only a short part of the remainder and a short part of the divisor. The computation of q_i will be done using a precomputed PD-table (PD $\hat{=}$ partial division) which contains for each partial remainder and each partial divisor a legal part of the quotient. It has to be verified that such a legal output always exists.

Now we specify the details of one round of the divider. The divisor d is a number in the interval $[1, 2)$, its binary representation has a 1 to the left of the binary point and up to n bits to the right of the binary point. The remainder r is initialized as dividend d^*. The remainder is described by two's complement numbers r_1 and r_2 such that $r = r_1 + r_2$. In the beginning, we may set $r_1 = d^*$ and $r_2 = 0$. The numbers r_1 and r_2 have the sign bit and three further bits to the left of the binary point and up to n bits to the right of the binary point. Hence, $r_1, r_2 \in [-8, 8)$. The divider requires that

$$-8d \leq 3(r_1 + r_2) \leq 8d$$

holds. If $|d^*| \leq 2$, this assumption is fulfilled in the beginning. For arbitrary numbers, we need a precomputation to ensure that the numbers are shifted in the right way. This shift is reversed in a postcomputation. For a specific design, n is fixed ($n = 64$ might be a reasonable choice). We discuss the design for general n in order to argue that one iterative step of the divider constructs one position of a radix-4 representation of the quotient in linear size and constant depth.

Before we describe one iterative step of the divider, we describe the PD-table. It is a 128×16 matrix, one row for each 7-bit number (the sign bit and three further bits to the left of the binary point) and one column for each 5-bit number with a one to the left of the binary point and four bits to the right. The entries are numbers from $\{-2, -1, 0, 1, 2\}$.

The following computations are performed. Let $r_{1,\text{high}}$ and $r_{2,\text{high}}$ consist of the seven most significant bits of r_1 and r_2 respectively (the sign bit and three bits to the left and three bits to the right of the binary point). We compute $r_{\text{high}} = r_{1,\text{high}} + r_{2,\text{high}}$ and consider the corresponding row of the PD-table. Let d_{high} consist of the five most significant bits of d (including the one to the left of the binary point). We consider the corresponding column of the PD-table.

The $(r_{\text{high}}, d_{\text{high}})$-entry q^* of the PD-table is promised to be a legal coefficient of the quotient for all numbers r (or r_1 and r_2) and d leading to r_{high} and d_{high}. This part of the circuit has constant size, since we work with $O(1)$ bits of r_1, r_2, and d and since the PD-table has constant size.

The new remainder r_{new} is defined as $4(r - q^*d) = 4(r_1 + r_2 - q^*d)$. Since $q^* \in \{-2, -1, 0, 1, 2\}$, it is possible to compute $-q^*d$ in linear size and constant depth. Then we use a carry-save adder (CSA) as an adder of r_1, r_2, and $-q^*d$ and multiply the results by 4 to obtain the numbers $r_{1,\text{new}}$ and $r_{2,\text{new}}$ such that $r_{\text{new}} = r_{1,\text{new}} + r_{2,\text{new}}$. This step is also possible in linear size and constant depth.

Given the PD-table and a circuit for the computations described above, the problem is to verify that the assumption $-8d \leq 3(r_1 + r_2) \leq 8d$ implies that $-8d \leq 3(r_{1,\text{new}} + r_{2,\text{new}}) \leq 8d$ and that the new remainder fulfills the equation $r_{1,\text{new}} + r_{2,\text{new}} = 4(r_1 + r_2 - q^*d)$ (in particular, there is no overflow). This property can be transformed into a circuit and the realization of the divider can then be checked against this "specification." Using this approach, we cannot find errors which are contained in the specification *and* in the realization. To reduce the chance of such an error, one can use a conservative design style for the specification circuit. Then we verify that the behavior of the radix-4 SRT divider coincides with the behavior of a quite different circuit. Bryant (1996) has performed this type of verification for words of bit length 70. The peak size of the constructed OBDDs was $4.2 \cdot 10^6$.

We discuss some features of the radix-4 SRT divider which are the reasons for its correctness. The restriction of the entries of the PD-table to the set $\{-2, -1, 0, 1, 2\}$ has been done for efficiency reasons. The largest number representable with this restriction and starting with q_0 (the position multiplied by 4^0) is less than

$$\sum_{-\infty < i \leq 0} 2 \cdot 4^i = 8/3$$

and this is the reason that we have to ensure the invariant $-8d \leq 3(r_1 + r_2) \leq 8d$. If $q_0 = 2$, the quotient is less than 8/3 and larger than

$$2 - \sum_{-\infty < i \leq -1} 2 \cdot 4^i = 4/3.$$

This implies that the value $q_0 = 2$ is only allowed if

$$4d \leq 3(r_1 + r_2) \leq 8d.$$

Similar calculations lead to the following implications:

- $q_0 = 1$ is only allowed if $d \leq 3(r_1 + r_2) \leq 5d$,
- $q_0 = 0$ is only allowed if $-2d \leq 3(r_1 + r_2) \leq 2d$,

- $q_0 = -1$ is only allowed if $-5d \le 3(r_1 + r_2) \le -d$,
- $q_0 = -2$ is only allowed if $-8d \le 3(r_1 + r_2) \le -4d$.

The intervals overlap. In certain situations, two results are allowed. All these considerations are based on the complete knowledge of r_1, r_2, and d. The efficiency of the divider relies on the fact that only the first $O(1)$ bits of these numbers are used. Hence, only intervals R_1, R_2, and D are known such that $r_1 \in R_1$, $r_2 \in R_2$, and $d \in D$. The entry of the PD-table has to be correct for all these triples (r_1, r_2, d). We have seen above that the precise knowledge of r_1, r_2, and d is not necessary to predict q^*. The claim is that the chosen truncations of r_1, r_2, and d are not too short and that the entries of the PD-table are chosen appropriately. This is exactly what has been verified by Bryant's approach. It has been reported that the erroneous PD-table of the Pentium divider contains some entries 0 instead of 2 for numbers where approximately $3(r_1 + r_2) = 8d$.

It is hard work to design the PD-table by hand. Using OBDDs, this computation can be automated as has been shown by Bryant (1996). Hence, the bug of the Pentium divider could have been avoided (and not only found) by such an OBDD-based approach. For $q^* \in \{-2, -1, 0, 1, 2\}$ let $\text{PD}(r_{\text{high}}, d_{\text{high}}, q^*)$ be the predicate which is true iff for all r_1, r_2, and d such that the high part of d equals d_{high}, the sum of the high parts of r_1 and r_2 equals r_{high}, and $-8d \le 3(r_1 + r_2) \le 8d$, the conclusion that $-8d \le 3(r_{1,\text{new}} + r_{2,\text{new}}) \le 8d$ and $r_{1,\text{new}} + r_{2,\text{new}} = 4(r_1 + r_2 - q^*d)$ is true. We may design a circuit testing this property depending on r_1, r_2, d, r_{high}, d_{high}, and q^*. This circuit is transformed into an OBDD and, finally, a universal quantification operation is performed for r_1, r_2, and d. Hence, the OBDD representing $\text{PD}(r_{\text{high}}, d_{\text{high}}, q^*)$ works on the small number of 14 variables (7 variables for r_{high}, 4 variables for d_{high}, and 3 variables for q^*). Hence, the resulting OBDD is guaranteed to be small. Bryant (1996) has implemented his approach and has constructed OBDDs with a maximum of 4.5 million BDD nodes. Using the resulting small OBDD, it is easy to construct the PD-table by listing all satisfying inputs. The PD-table has no entries remaining empty and this is an automatic proof that the truncation of the numbers used by the radix-4 SRT divider allows a correct division.

We have seen that it is an important substep to check linear inequalities like $3r \le 8d$, which is equivalent to the test whether $8d - 3r \ge 0$. Clarke, Fujita, and Zhao (1995a) have considered general linear functions $f(x) = c_1x_1 + \cdots + c_mx_m$ where $x_i = x_{i,n-1}2^{n-1} + \cdots + x_{i,0}2^0$. They have used the variable ordering $x_{1,0}, \ldots, x_{1,n-1}, \ldots, x_{m,0}, \ldots, x_{m,n-1}$ and the obvious linear-size BMD representing f. Let $g(x) = g(x_{1,0}, \ldots, x_{m,n-1})$ be the Boolean function which takes the value 1 iff $f(x) \ge 0$. Then it can be proved that the OBDD representing g can be constructed in time $O(n^2(c_1 + \cdots + c_m))$ from the BMD for f. For the parameters of the linear functions considered for the radix-4 SRT divider, we obtain the bound $O(n^2)$.

13.2 Verification of Sequential Circuits

The underlying structure of finite automata, finite state machines, and sequential circuits is a finite state transition structure (FST).

Definition 13.2.1. An FST $T = (S, S_0, X, \delta)$ consists of a finite state set S, a nonempty set $S_0 \subseteq S$ of initial states, a finite input alphabet X, and a next state transition function δ defined on $S \times X$ and describing by $\delta(s,x) \subseteq S$ the set of possible successor states if input x is read in state s. The FST is called complete if $\delta(s,x) \neq \emptyset$ for all (s,x) and deterministic if $|\delta(s,x)| = 1$ for all (s,x). It is called strongly deterministic if it is deterministic and $|S_0| = 1$.

We do not define formally how an FST reacts on a sequence $x_1, \ldots, x_n$ of inputs. We use the notation $\delta(s, (x_1, \ldots, x_n))$ to describe the set of reachable states if one starts in s and reads $x_1, \ldots, x_n$. Moreover, for $S' \subseteq S$, $\delta(S', (x_1, \ldots, x_n))$ is the union of all $\delta(s, (x_1, \ldots, x_n))$ and $s \in S'$.

Definition 13.2.2. A *finite state machine* (FSM) consists of an FST $T = (S, S_0, X, \delta)$, a finite output alphabet X_{out}, and an output function λ defined on $S \times X$ and describing the output $\lambda(s,x) \in X_{\text{out}}$ produced in state s after having read x.

In complexity theory, optimization and search problems are reduced to decision problems, since these problems contain the core of many complexity theoretical problems. In the same way, we obtain the model of a *finite automaton* (FA). We either may restrict ourselves to the output alphabet $X_{\text{out}} = \{0,1\}$ or to the partition of S into accepting and rejecting states. FSMs model the behavior of simple real-world automata like drink machines but also the behavior of complicated control systems. In particular, each *sequential circuit* can be understood as an FSM. A sequential circuit consists of a finite memory whose contents may be interpreted as the state and a combinational circuit which computes from the present state and the input the next state and an output. The memory is implemented in physical systems by latches (flip-flops, registers). Hence, there is a direct simulation of a sequential circuit by an FSM. It is also possible to simulate an FSM by a sequential circuit. Then it is necessary to use binary encodings of S, X, and X_{out}. If, e.g., $|S|$ is not a power of two, this leads to dummy states which should not be reachable. Afterwards, we realize δ and λ by a combinational circuit. Since we may have dummy states and inputs, the functions δ and λ may be incompletely specified (compare Section 3.6 for problems and options arising from this fact). Because of these simple simulations, we do not distinguish between FSMs and sequential circuits and, as in most of the literature, we prefer to talk about FSMs.

The verification of a deterministic FSM M against a specification M' also given as an FSM is an equivalence check of M and M'. We have to check whether the FSM M starting at some $s_0 \in S_0$ produces for each finite sequence $x = (x_1, \ldots, x_n)$ of inputs the same sequence of outputs as the FSM M' starting at some $s'_0 \in S'_0$. For this purpose, we may consider the product $M^* = M \times M'$ of M and M'. The FSM M^* has the state set $S^* = S \times S'$, the set $S^*_0 = S_0 \times S'_0$ of initial states, the same input alphabet as M and M' (FSMs with different input alphabets cannot be equivalent), the output alphabet $X^*_{\text{out}} = X_{\text{out}} \times X'_{\text{out}}$, the state transition function δ^* defined by

$$\delta^*((s,s'),x) = (\delta(s,x), \delta'(s',x)),$$

and the output function λ^* defined by

$$\lambda^*((s,s'),x) = (\lambda(s,x), \lambda'(s',x)).$$

The FSMs M and M' are equivalent iff for M^* only states (s,s') are reachable where $\lambda(s,x) = \lambda(s',x)$ holds for each $x \in X$. We have reduced the verification problem to the problem of computing the set of reachable states of an FSM M^*. The problem is called the *reachability problem* and the process to compute or to estimate the set R of reachable states is called reachability analysis. For the reachability problem, it is sufficient to consider the FST behind the FSM.

The reachability problem also has applications for the design of sequential systems. The behavior of an FSM on a nonreachable state does not matter. Knowing that a set S' of states is not reachable, we may arbitrarily change $\delta(s,x)$ and $\lambda(s,x)$ for $s \in S'$. This may lead to a simplification of these functions. Moreover, if we interconnect FSMs, the nonreachable states of one FSM may be used as don't cares for the following FSMs.

In classical automata theory (Hopcroft and Ullman (1979)), FSMs are described by complete function tables for δ and λ. The reachability problem can be solved with a linear-time DFS traversal and causes no problem. This is similar to the verification of combinational circuits if the specification consists of a complete function table. The systems we try to verify are too large to admit the description of a complete function table, e.g., if the sequential circuit has 100 or 200 latches. Hence, we have to assume that δ is given by some compact logical description like a combinational circuit.

The OBDD approach for reachability analysis assumes that OBDDs for the functions and sets (represented as characteristic functions) describing the FST of the FSM M are given. The OBDDs work on the variable vectors s, s', and x where s describes the present state, s' the next state, and x the input letter. The next state function δ can be decomposed to $\delta_1, \ldots, \delta_m$ if the FSM works with m latches and δ_k describes the next state of the kth latch. Then the following functions are given by OBDDs:

- $I(s)$, the characteristic function of the set S_0 of initial states,
- $T_k(s,s',x)$, which takes the value 1 if $s'_k \in \delta_k(s,x)$,
- $T(s,s',x)$, the conjunction of all $T_k(s,s',x)$, $1 \le k \le m$.

The basic step of a BFS traversal is the *image computation*. Let $R(s)$ be the characteristic function describing the set of states known to be reachable. We always start with $R(s) = I(s)$. An OBDD for

$$N(s') = \exists s\ \exists x\ \left(T(s, s', x) \wedge R(s)\right)$$

can be computed by the usual OBDD operations and describes the set of states reachable in one step from the set of states described by $R(s)$. The OBDD for $N(s')$ essentially depends only on the s'-variables. Let $N^*(s)$ be the OBDD obtained from the OBDD for $N(s')$ by renaming the variables in such a way that a variable s'_i is replaced by the corresponding s_i. For this purpose, we assume that the chosen variable ordering arranges the variables in s and in s' in an analogous way. The function $NEW(s) = N^*(s) \wedge \overline{R}(s)$ describes the set of states whose reachability has been proved during the last step. The function $R(s) + N^*(s) = R(s) + NEW(s)$ describes the set of states known to be reachable. This function can be used as $R(s)$ for the following image computation. We may use the test $NEW(s) \equiv 0$ as a stopping criterion.

We discuss some of the ideas proposed to improve this standard reachability algorithm, which is based on ideas in the early paper of Supowit and Friedman (1986).

One idea is *early quantification*. Since $R(s)$ does not essentially depend on x, we may rewrite the equation for $N(s')$ as

$$N(s') = \exists s\ \left(\exists x\ (T(s, s', x)) \wedge R(s)\right).$$

Sometimes, it may help to construct the OBDD for $T(s, s') = \exists x\ (T(s, s', x))$. Early quantification can be used for each subset of the x-variables.

The relation $T(s, s')$ describes the pairs (s, s') of states such that s' is reachable from s in one step. Let

$$U_0(s, s') = T(s, s') + \text{EQ}(s, s')$$

describe the set of pairs (s, s') such that s' is reachable in at most 2^0 steps. We know that it is necessary to use an interleaved variable ordering to obtain a small-size OBDD for the equality test EQ. Starting with U_0, we may use the technique of *iterative squaring* to compute

$$U_k(s, s') = \exists s''\ \left(U_{k-1}(s, s'') \wedge U_{k-1}(s'', s')\right)$$

describing the set of pairs (s, s') such that s' is reachable from s within at most 2^k steps. This method together with the stopping criterion $U_k(s, s') = U_{k-1}(s, s')$ can be used for a reachability analysis. The number of iterations is logarithmically smaller than for the standard algorithm. Nevertheless, this approach often leads to very large intermediate OBDDs, which are caused by the existential quantification and the encoding of three states instead of two.

The standard algorithm uses $R(s) + NEW(s)$ as the description of the set of states whose successors are computed in the next iteration as $N^*(s)$. Let $R^*(s)$ be any function such that

$$NEW(s) \leq R^*(s) \leq NEW(s) + R(s).$$

If we run the standard algorithm with $R^*(s)$ instead of $R(s)$, we obtain the same function $N^*(s) \wedge \overline{R}(s)$ as a result of the step. The reason is that a state is reached for the first time iff it is reached from a state which was reached for the first time during the last iteration. Hence, we may consider $R^*(s)$ as an incompletely specified function with on-set $NEW(s)$ and don't care set $R(s) \wedge \overline{NEW}(s)$. This may be used to obtain some $R^*(s)$ with small OBDD size. This basic idea due to Coudert, Berthet, and Madre (1989a) has been the starting point for large successes in reachability analysis with OBDDs.

Coudert, Berthet, and Madre (1989b) have considered alternative representations of characteristic functions like $R(s)$ by so-called functional vectors $f = (f_1, \ldots, f_r)$ such that $R(s) = 1$ iff $f(a) = s$ for some a. They also present algorithms to switch between both types of representation.

We have based our considerations on the joint translation relation $T(s, s', x)$, which is the conjunction of all $T_k(s, s', x)$. By definition, $T_k(s, s', x)$ does not essentially depend on s'_j if $j \neq k$. Burch, Clarke, and Long (1991) and Burch, Clarke, Long, McMillan, and Dill (1994) have proposed working with OBDDs for all the functions $T_k(s, s', x)$ or conjunctions of small sets of them (corresponding to natural groups of latches). This typically allows us to work with smaller OBDDs. It may also lead to better and earlier applications of the early quantification technique (Touati, Savoj, Lin, Brayton, and Sangiovanni-Vincentelli (1990)). Hu, York, and Dill (1994) have described techniques to work with functions $T = T_1 \wedge \cdots \wedge T_m$ given by OBDDs for $T_1, \ldots, T_m$, so-called implicitly conjoined OBDDs. Even with these techniques, the OBDDs for the transition relation may be too large. Cabodi, Camurati, and Quer (1994) have used the idea of auxiliary variables to overcome this problem. How to deal with auxiliary variables has already been described in Section 13.1.

The main idea of Coudert, Berthet, and Madre (1989a) has been generalized in different ways. The function $R^*(s)$ may be decomposed into a disjunction $R_1^*(s) + \cdots + R_l^*(s)$ and we may search for new reachable states from the different sets $R_i^*(s)$ separately and may later compute the union of the computed sets of reachable states (Cabodi, Camurati, and Quer (1996)). The approach of Ravi and Somenzi (1995) introduces the notion of *density* of an OBDD defined as the number of states represented by the OBDD divided by the size of the OBDD. The OBDD for $R(s)$ is replaced by an OBDD for $R'(s) \leq R(s)$ which is more dense. One hopes to find quickly a lot of reachable states by searching from $R'(s)$ and this may help to avoid the peak size of OBDDs. At the end, the OBDDs may be smaller and one searches from those states which have not been considered in $R'(s)$.

Since reachability analysis is such a difficult task, one may ask whether generalizations of OBDDs help. Narayan, Isles, Jain, Brayton, and Sangiovanni-Vincentelli (1997) were able to improve the considered methods by using partitioned OBDDs with disjoint window functions. The window functions $w_1, \ldots, w_k$ where $k = 2^m$ only depend on m of the present state variables and represent the different minterms on these variables. Also the number k is fixed in advance. Obviously, there is room for more sophisticated ideas to construct window functions. The choice of the m variables which partition the input space is done by the following heuristic which is based only on the transition relation $T(s, s', x)$. A variable s_j seems to be a good choice if it leads to a balanced partition, more precisely if $p_j(T)$, the maximum of the OBDD size of $T_{|s_j=0}(s, s', x)$ and the OBDD size of $T_{|s_j=1}(s, s', x)$, is small. Moreover, $r_j(T)$, the sum of the OBDD size of $T_{|s_j=0}(s, s', x)$ and the OBDD size of $T_{|s_j=1}(s, s', x)$, should be small. The cost of s_j is defined as the weighted sum of $p_j(T)$ and $r_j(T)$, i.e., for some parameters $\alpha, \beta > 0$,

$$\text{cost}_j(T) = \alpha \cdot p_j(T) + \beta \cdot r_j(T)$$

and the m variables with the smallest cost are chosen. If $T = T_1 \wedge \cdots \wedge T_l$ is given by OBDDs for $T_1, \ldots, T_l$, the cost of each variable is computed with respect to each T_i. Then the weighted sum of these cost factors is taken. Here, functions T_i with small OBDD size should have small influence.

These simple window functions have the property that the computation of $R_j(s) = w_j(s) \wedge R(s)$ and $w_j(s) \wedge T(s, s', x)$ is always easy. We first consider transitions within the jth part and define

$$\begin{aligned} N_j(s') &= \exists s \, \exists x \, (w_j(s) \wedge w_j(s') \wedge T(s, s', x) \wedge R_j(s)) \\ &= w_j(s') \wedge \left(\exists s \, \exists x \, (w_j(s) \wedge T(s, s', x) \wedge R_j(s))\right). \end{aligned}$$

Hence, we may use all considered techniques to compute the set $R_j^*(s)$ of states within the jth part which are reachable from $R_j(s)$ if only states in the same part are allowed as intermediate states. During these computations, we may treat $T(s, s', x)$ and $R_j(s)$ as incompletely specified functions where the don't care set is described by $\overline{w}_j(s) + \overline{w}_j(s')$.

In a next step, we compute for each part which states are reachable directly from a state outside this part and known as reachable. Let

$$N_{ij}(s') = w_j(s') \wedge \left(\exists s \, \exists x \, (w_i(s) \wedge T(s, s', x) \wedge R_i^*(s))\right).$$

Let $N_{ij}^*(s)$ be the OBDD obtained from the OBDD for $N_{ij}^*(s')$ by a renaming of the variables as described before. The disjunction of $R_j^*(s)$ and all $N_{ij}^*(s)$, $i \neq j$, describes the states of the jth part known to be reachable. With this new state set described as $R_j(s)$, we may restart the above-mentioned algorithm to find reachable states in the jth part. This procedure can be iterated until no new reachable states are found.

Matsunaga, McGeer, and Brayton (1993) have presented a recursive procedure to compute the set of reachable states. Their approach is based on the classical transitive closure algorithm with adjacency matrices and is adapted to BDD techniques. Hachtel, Macii, Pardo, and Somenzi (1994a, 1994b) have gone a step further and have used MTBDDs (or ADDs) to perform a probabilistic analysis of FSMs and to compute the steady-state probabilities of an FSM considered as a homogeneous discrete Markoff chain.

13.3 Symbolic Model Checking

Burch, Clarke, McMillan, Dill, and Hwang (1992) define *model checking* as the process of determining whether a given formula is true in a given model. For our purpose, it is not necessary to give a formal definition of models. Model checking allows quite general types of formulas as specification. In Section 13.1, we have assumed that a Boolean function is specified as a combinational circuit and that a realization, another combinational circuit, has to be checked against this specification. This assumption is quite natural for Boolean functions. The reachability analysis for FSMs considered in Section 13.2 is already an abstraction, since we do not consider specific inputs and only look for the existence of an input leading to the considered state. Nevertheless, we have implicitly assumed that everything we are talking about is expressible by FSMs. Model checking is based on a complete abstraction of all details of an implementation. The formula is expressed in the framework of some appropriate logical system. In a multi-user system, we may like to verify that each user has infinite access to the CPU and this property has to hold for all actions of all other users. Such systems are still finite but very large and we are again faced with the "state explosion problem." It is impossible to describe the state set explicitly by enumeration and, as in the previous section, the state set is described implicitly or symbolically. Model checking with symbolically described state sets and transition relations is called *symbolic* model checking.

We consider computation paths of sequential systems. These paths can be described as sequences of states where the state s_{i+1} is a legal successor of s_i if the transition relation allows us to switch from s_i directly to s_{i+1}. It should be clear that we abstract from specific inputs. Considering computation paths, we discuss temporal aspects. If we switch from s_i to s_{i+1}, we reach s_{i+1} one unit of time later than s_i. Hence, we should choose a logical system admitting operators describing the temporal behavior of a system.

A *linear-time temporal logic* only considers single computation paths $s = (s_0, s_1, s_2, \dots)$. On such a computation path, s_i is reached at time i, the relation $i \to s_i$ leads to the notion of *linear-time* temporal logic. Let $s = (s_0, s_1, s_2, \dots)$ be a computation path and let f be a Boolean formula. The temporal operators are defined in the following way:

- Xf: the formula f is fulfilled for s_1, the next state. The operator X is called the *nexttime operator*.
- Gf: the formula f is fulfilled for all s_i, $i \geq 0$. The operator G is called the *global operator*.
- Ff: there exists some i such that f is fulfilled for s_i. The operator F is called the (sometime in the) *future operator*.
- f U g: there exists some i such that g is fulfilled for s_i and f is fulfilled for all s_j, $j < i$. The operator U is called the *until operator*.

It is easy to see that the future operator is not necessary. It is always possible to replace Ff by *true* U f where *true* is the formula which always is true.

We do not consider specific inputs. A deterministic FSM where we identify all input letters is a nondeterministic FSM with a one-letter input alphabet. Starting in s_0, we have a lot of possible computation paths and we want to express properties of the set of all these computation paths. It is possible to describe all computation paths as a tree with root s_0. Each node labelled by a state s has as many successors as it has possible successors on computation paths. The computation path "branches" at the state s. The tree describes the possible behavior of a finite-state system. If we know the present node in the tree, we know the past and the present, since the way back to the root is unique. We do not know the future. Linear-time temporal logics cannot express properties of the unknown future. A logical system allowing us to express properties of the set of possible computation paths is called *branching-time temporal logic*. Symbolic model checking with OBDDs is most often based on the logical system called *computation tree logic* (CTL).

CTL allows two path quantifiers and requires that each of the so-called forward-time operators X, G, F, and U is directly preceded by a path quantifier. The path quantifier E is the existential quantifier. E.g., E Ff means that for at least one of the possible computation paths Ff is true. The other path quantifier A is the universal quantifier, i.e., the property following A has to hold for all possible computation paths. Since the Boolean negation $\neg$ is available, we can replace the universal quantifier A in the following way by existential quantifiers:

- $\mathrm{A\,X}f = \neg\mathrm{E\,X}(\neg f)$: a property holds on all computation paths in the next state iff there is no computation path where it does not hold in the next state (the usual deMorgan law),
- $\mathrm{A\,G}f = \neg\mathrm{E\,F}(\neg f) = \neg\mathrm{E}\,(\mathit{true}\mathrm{U}\neg f)$: a property always holds on all computation paths iff there is no computation path and no point of time where the property does not hold,
- $\mathrm{A}\,(f\,\mathrm{U}\,g) = \neg[\mathrm{E\,G}(\neg g) + \mathrm{E}\,(\neg g\,\mathrm{U}\,\neg f \wedge \neg g)]$: it is true for all computation paths that f holds until g is fulfilled iff it is not true that there is one

computation path such that g is never true and no computation path where at some point in time f and g are not fulfilled and before that point in time g is not fulfilled.

Hence, we may use all operators X, G, F, and U and quantifiers E and A to express properties in a concise way. But in order to prove theorems about CTL formulas it is sufficient to consider X, G, U, and E. We list some examples of natural CTL formulas:

- The properties f and g will not occur at the same time (mutual exclusion): $\mathrm{A\,G}(\neg(f \wedge g))$.
- Each request *req* will be acknowledged (sooner or later): $\mathrm{A\,G}(\mathit{req} \Rightarrow \mathrm{A\,F}\mathit{ack})$.
- Each request will be stored until it has been acknowledged: $\mathrm{A\,G}(\mathit{req} \Rightarrow \mathrm{A}\,(\mathit{req}\,\mathrm{U}\,\mathit{ack}))$.
- There is no deadlock, i.e., it is always possible to drive the system to its initial state q_0: $\mathrm{A\,G}(\mathrm{E\,F}q_0)$.

Since we investigate finite systems, it is quite easy to prove that the problem of checking the validity of a CTL formula is decidable. We are interested in OBDD-based algorithms which solve this problem efficiently. It is sufficient to describe algorithms for $\mathrm{E\,X}f$, $\mathrm{E}\,(f\ \mathrm{U}\ g)$, and $\mathrm{E\,G}f$ where f and g are characteristic functions of subsets of the state set and are given by OBDDs. The result is the characteristic function of the subset of all states where the given CTL formula is true. Our aim is to compute an OBDD representing this characteristic function.

The nexttime operator X is easy to handle, since we only have to consider one unit of time. We describe the transition relation as the function $T(s, s')$. Then

$$\mathrm{E\,X}f(s) = \exists s'\ \big(T(s,s') \wedge f(s')\big)$$

can be computed with the known OBDD operations. The equation looks similar to the equation

$$N(s') = \exists s\ \big(T(s,s') \wedge R(s)\big)$$

for the image computation (Section 13.2) but there is one difference. The roles of the present state and the next state are interchanged. Hence, we perform a kind of inverse image computation or *backward analysis*. All methods for an efficient image computation can be adapted to this situation.

In order to motivate the following algorithms for $\mathrm{E\,G}f$ and $\mathrm{E}\,(f\ \mathrm{U}\ g)$, we describe the algorithms for the reachability analysis (Section 13.2) in another way. Let

$$A_{\text{new}}(s') = I(s) + \exists s\ \big(T(s,s') \wedge A(s)\big)$$

be an equation. A function $A(s)$ is a fixed point of this equation if $A_{\text{new}}(s) = A(s)$. It is easy to verify that $R(s)$, the function describing the set of reachable states, is the smallest or least fixed point of this equation. The algorithms for the reachability analysis start with $A(s) = I(s)$ and compute new A-sets until a fixed point is reached. Let $B = f \wedge \mathrm{E\,X}(B)$ or in more detail

$$B(s) = f(s) \wedge \exists s' \left(T(s, s') \wedge B(s')\right).$$

It is straightforward to show that $\mathrm{E\,G}f$ is a fixed point of this equation and even the greatest fixed point. Hence, in order to compute $\mathrm{E\,G}f$ we may start with $B(s) = f(s)$ and then iteratively compute

$$B_{\text{new}}(s) = f(s) \wedge \exists s' \left(T(s, s') \wedge B(s')\right)$$

until a fixed point is reached. Finally, let $C = g + (f \wedge \mathrm{E\,X}(C))$ or

$$C(s) = g(s) + \left(f(s) \wedge \exists s' \left(T(s, s') \wedge C(s')\right)\right).$$

Again, it is easy to see that $E(f\,\mathrm{U}\,g)$ is the least fixed point of this equation. In order to compute $\mathrm{E}\,(f\;\mathrm{U}\;g)$, we may initialize $C(s) = g(s)$ and may iteratively compute

$$C_{\text{new}}(s) = g(s) + \left(f(s) \wedge \exists s' \left(T(s, s') \wedge C(s')\right)\right)$$

until a fixed point is reached. The theory behind this algorithm is called fixed-point semantics and we apply Kleene's fixed-point theorem to prove the correctness of the proposed algorithms. Altogether, we have seen that the evaluation of one operator of a CTL formula is a fixed-point computation comparable to a reachability analysis. A CTL formula may contain a lot of operators which makes model checking much more expensive than reachability analysis. This approach has been described by Burch, Clarke, McMillan, and Dill (1990) and has been continued by Burch, Clarke, McMillan, Dill and Hwang (1992). They and Burch, Clarke, Long, McMillan, and Dill (1994) have also investigated how *fairness constraints* can be integrated into symbolic model checking with CTL.

A computation path is called *fair* with respect to a set of fairness constraints if each constraint holds infinitely often along the path. A fairness constraint can be an arbitrary CTL formula. The path quantifiers in CTL formulas are now restricted to fair paths. A typical example is the nondeterministic access to the CPU in a multi-user system. Then a set of fairness constraints can express that every user eventually gets access to the CPU.

Let $C = \{c_1, \ldots, c_n\}$ be a finite set of fairness constraints each expressed as a CTL formula. We use E_C as the existential path quantifier which asks for the existence of a fair computation path with the required properties. We cannot express, e.g., $\mathrm{E}_C\mathrm{G}f$ directly in CTL. By definition, $\mathrm{E}_C\mathrm{G}f(s)$ is true iff there exists a computation path starting at s such that f is true on all states of the path and each formula $c_i \in C$ holds infinitely often. This leads to the following characterization of the set Z of all states fulfilling $\mathrm{E}_C\mathrm{G}f$. It is the largest set of states s such that $f(s)$ is true and for all $c_i \in C$ there is a path of positive length starting at s and leading through states s' where f is true to a state s'' in Z where $c_i(s'')$ is true. Hence, $\mathrm{E}_C\mathrm{G}f$ is the largest fixed point of the equation

$$Z(s) = f(s) \wedge \bigwedge_{1 \le i \le n} \mathrm{E\,X}[\mathrm{E}\,(f\,\mathrm{U}\,(Z \wedge c_i))]$$

and we may initialize $Z(s)$ as $f(s)$ and may perform a fixed-point computation with respect to the given equation. We have to perform a nested fixed-point computation, since $\mathrm{E}\,(f\,\mathrm{U}\,(Z \wedge c_i))$, $1 \le i \le n$, is contained within the equation and has to be computed by a fixed-point computation. With the global operator we can describe the set of states which are lying on some fair computation path as $h := \mathrm{E}_C\mathrm{G}\mathit{true}$. Now it is easy to conclude that

$$\mathrm{E}_C\mathrm{X}f = \mathrm{E\,X}(f \wedge h)$$

and

$$\mathrm{E}_C(f\,\mathrm{U}\,g) = \mathrm{E}\,(f\,\mathrm{U}\,(g \wedge h)).$$

Enders, Filkorn, and Taubner (1993) have extended the considered approach to systems with several parallel components. They avoid the consideration of the whole system with general transition relations, since this would lead to systems with a very large state space. They restrict the set of operations to parallel composition, restriction, and relabelling and prove that this ensures that the OBDD size only grows linearly with respect to the number of parallel components.

A different approach for the verification that a system satisfies a given property is testing *language containment*. One specifies both the system and the property (or the task) and then verifies that the language of the system is contained in the language of the property. Touati, Brayton, and Kurshan (1995) have shown how OBDDs can be used for the language containment problem of so-called ω-automata working on infinite input sequences. Hojati, Shiple, Brayton, and Kurshan (1993) have argued that language containment is a complementary approach to CTL-based formal verification of finite state systems. Moreover, they have shown how algorithms for the language containment problem can be used as subroutines in model-checking algorithms for CTL-based model checking.

Chapter 14

Further CAD Applications

14.1 Two-Level Logic Minimization

The design of optimal circuits is a fundamental problem. A lot of optimization criteria can be considered, among them time, depth, testability, hazard freeness, wire length, power consumption, and size or area. Most of these optimization problems are too hard to be solved exactly in reasonable time. *Two-level logic minimization* is a subproblem which can be attacked algorithmically and has applications in logic synthesis, reliability analysis, and automated reasoning. We restrict circuits to two logical levels, i.e., the circuits work on the literals $x_1, \overline{x}_1, \ldots, x_n, \overline{x}_n$ which are connected, w.l.o.g., to AND-gates on the first level where monomials or products of literals are computed. The results of the first level lead to an OR-gate on the second level computing the considered function as a polynomial or sum of products. We investigate the problem of minimizing the number of gates (in a similar way we may minimize the number of wires). Hence, the circuit is restricted to small depth (assuming unbounded fan-in) and, under this restriction, the size is minimized. The problem of two-level logic minimization has been investigated since the early fifties. We list the most important concepts.

Definition 14.1.1.

(i) A *monomial* or *product* is a conjunction of literals.

(ii) A *polynomial* or *sum of products* is a disjunction of monomials.

(iii) An *implicant* of a Boolean function f is a monomial $m \leq f$, i.e., $m(a) = 1$ implies $f(a) = 1$. The set of all implicants of f is denoted by $\mathrm{I}(f)$.

(iv) A *prime implicant* of a Boolean function f is an implicant m of f such that no proper shortening of m is an implicant of f. The set of all prime implicants of f is denoted by $\mathrm{PI}(f)$.

(v) An *essential prime implicant* of a Boolean function f is a prime implicant $m \in \mathrm{PI}(f)$ such that there exists some $a \in f^{-1}(1)$ where $m(a) = 1$ and $m'(a) = 0$ for all other $m' \in \mathrm{PI}(f)$. The set of all essential prime implicants of f is denoted by $\mathrm{EPI}(f)$.

For each reasonable cost measure, it is easy to prove that some optimal polynomial for a function f only consists of prime implicants and contains all essential prime implicants. Hence, two-level minimization can be reduced to the following two problems.

(1) Compute the set $\mathrm{PI}(f)$ of all prime implicants.

(2) Solve the set-covering problem where prime implicants are considered as sets covering elements from $f^{-1}(1)$.

For a long time, only explicit descriptions of the function (a list of all inputs from $f^{-1}(1)$ or a polynomial representing f) have been worked with and the list of all prime implicants leading to very large inputs for the NP-hard set-covering problem has been constructed (see Coudert (1994)). Nowadays, one considers such complex functions that it is necessary to represent f and $\mathrm{PI}(f)$ implicitly, e.g., by some BDD variant.

Coudert and Madre (1992) have described an OBDD-based algorithm computing the set of prime implicants of a Boolean function f. We only consider the case of completely specified functions, although the algorithm can easily be generalized to work also for incompletely specified functions.

The number of different monomials equals 3^n (a monomial may contain x_i or $\overline{x}_i$ or none of them). We use the redundant representation by $2n$ bits $y_1, z_1, \ldots, y_n, z_n$ with the following meaning. The variable y_i decides whether an x_i-literal is contained in the monomial and the variable z_i decides whether x_i or $\overline{x}_i$ is contained in the monomial, i.e., the monomial contains x_i if $(y_i, z_i) = (1, 1)$, $\overline{x}_i$ if $(y_i, z_i) = (1, 0)$, and none of them if $y_i = 0$. Often we are working with OBDDs describing properties of two monomials described by $y_1, z_1, \ldots, y_n, z_n$ and $y'_1, z'_1, \ldots, y'_n, z'_n$ and an input described by $x_1, \ldots, x_n$. Then a variable ordering is used where each group of variables with the same index is tested one directly after the other. The ordering within a group is $y_i, y'_i, z_i, z'_i, x_i$. The ordering of the groups can be chosen with respect to the situation.

The function g_1 checks whether the monomial m described by (y, z) computes 1 on the input described by x. This can be described by the formula $((y_i = 1) \Rightarrow (z_i \equiv x_i))$ for all i and it is obvious that the chosen variable ordering ensures a linear-size OBDD representation. Since the encoding of monomials is redundant, we need the function g_2 checking whether (y, z)

and (y', z') describe the same monomial. This is equivalent to the formula $(y_i \equiv y_i') \wedge \big((y_i = 1) \Rightarrow (z_i \equiv z_i')\big)$ for all i where the OBDD size is linear. Prime implicants are shortest implicants and g_3 checks whether the monomial described by (y, z) is a shortening of the monomial described by (y', z'). This is equivalent to the formula $(y_i = 1) \Rightarrow \big((y_i' = 1) \wedge (z_i \equiv z_i')\big)$ for all i, where again the OBDD size is linear. Now we are able to describe the characteristic functions of the sets I(f), PI(f), and EPI(f) by formulas which can be used to create OBDDs. Here we identify sets and their characteristic functions:

- $\mathrm{I}(f)(y,z) = 1 \Leftrightarrow \Big(\forall x\colon \big(g_1(y,z,x) = 1 \Rightarrow f(x) = 1\big)\Big).$
- $\mathrm{PI}(f)(y,z) = 1 \Leftrightarrow \Big(I(f)(y,z) = 1\Big) \wedge \Big(\forall (y',z')\colon \big(\mathrm{I}(f)(y',z') = 1 \wedge g_3(y',z',y,z) = 1\big) \Rightarrow g_2(y,z,y',z') = 1\Big).$
- $\mathrm{EPI}(f)(y,z) = 1 \Leftrightarrow \Big(PI(f)(y,z) = 1\Big) \wedge \Big(\exists x\colon \big(f(x) = 1\big) \wedge \big(\forall (y',z')\colon (\mathrm{PI}(f)(y',z') = 1 \wedge g_1(y',z',x) = 1\big) \Rightarrow g_2(y,z,y',z') = 1)\Big).$

It has been observed that the OBDD representing I(f) is often much larger than the OBDD representing PI(f) or EPI(f). Hence, we try to avoid the computation of an OBDD for I(f). Let $h = f_{|x_1=0} \wedge f_{|x_1=1}$, $h_0 = f_{|x_1=0}$, and $h_1 = f_{|x_1=1}$ and let us assume that we have representations of PI(h), PI(h_0), and PI(h_1). The following claims are well known and easy to prove:

- $m \in \mathrm{PI}(f),\ x_1, \overline{x}_1 \notin m \Leftrightarrow m = m_0 m_1, m_0 \in \mathrm{PI}(h_0),\ m_1 \in \mathrm{PI}(h_1)$
  $\Leftrightarrow m \in \mathrm{PI}(h),$
- $x_1 m \in \mathrm{PI}(f),\ x_1 \notin m \Leftrightarrow m \in \mathrm{PI}(h_1) - \mathrm{I}(h_0) \Leftrightarrow m \in \mathrm{PI}(h_1) - \mathrm{PI}(h),$
- $\overline{x}_1 m \in \mathrm{PI}(f),\ \overline{x}_1 \notin m \Leftrightarrow m \in \mathrm{PI}(h_0) - \mathrm{I}(h_1) \Leftrightarrow m \in \mathrm{PI}(h_0) - \mathrm{PI}(h).$

We define $\mathrm{PI}(h_0)(y,z) \otimes \mathrm{PI}(h_1)(y,z)$ as the set of monomials $m = m_0 m_1$ where m_0 belongs to $\mathrm{PI}(h_0)(y,z)$ and m_1 belongs to $\mathrm{PI}(h_1)(y,z)$. Then

$$\begin{aligned}
\mathrm{PI}(f)(y,z) &= \overline{y}_1 \wedge \big(\mathrm{PI}(h_0)(y,z) \otimes \mathrm{PI}(h_1)(y,z)\big) + y_1 \overline{z}_1 \wedge \mathrm{PI}(h_0)(y,z) \\
&\quad \wedge \overline{\mathrm{I}(h_1)}(y,z) + y_1 z_1 \wedge \mathrm{PI}(h_1)(y,z) \wedge \overline{\mathrm{I}(h_0)}(y,z) \\
&= \overline{y}_1 \wedge \mathrm{PI}(h)(y,z) + y_1 \overline{z}_1 \wedge \mathrm{PI}(h_0)(y,z) \wedge \overline{\mathrm{PI}(h)}(y,z) \\
&\quad + y_1 z_1 \wedge \mathrm{PI}(h_1)(y,z) \wedge \overline{\mathrm{PI}(h)}(y,z).
\end{aligned}$$

These equations are the basis of two algorithms for the computation of an OBDD representation of $\mathrm{PI}(f)$. Experiments have shown that the second approach is often more efficient, since the realization of the $\otimes$-operator is time-consuming. We stop the recursion whenever we obtain a subproblem already solved.

Altogether, we have OBDDs for f and $\mathrm{PI}(f)$ leading to an implicit description of the PI-table which is the input of a set-covering problem. The table entry at position (x, y, z) equals 1 iff $f(x) = 1$, $\mathrm{PI}(f)(y, z) = 1$, and $g_1(y, z, x) = 1$, i.e., the prime implicant (y, z) covers the input x.

It will turn out that it is more convenient to work with a table R where the rows and the columns are indexed by monomials. We denote the rows by x and the columns by p. The value of $R(x,p)$ is initialized by 1 iff x is a minterm covered by the prime implicant p of f. Otherwise, $R(x,p) = 0$. It is easy to obtain the new table, called the transformed PI-table, from the given one. We denote the set of nonzero rows by X and the set of nonzero columns by P.

Set-covering problems can be simplified by the application of dominance relations. The x-row is dominated by the x'-row (denoted by $x \leq_X x'$) iff $R(x',p) \leq R(x,p)$ for all p. Then we can remove the x-row, which is done by setting $R(x,p) = 0$ for all p. We still have to cover x'. Whenever we cover x', we also cover x. The case $x \equiv_X x'$, i.e., $x \leq_X x'$ and $x' \leq_X x$, occurs if the x-row and the x'-row are identical. Hence, $\equiv_X$ is an equivalence relation and we can remove all but one element of each equivalence class. The p-column is dominated by the p'-column (denoted by $p \leq_P p'$) iff $R(x,p) \leq R(x,p')$ for all x. Then we can remove the p-column, since p' covers at least all rows covered by p and, by our assumption, all implicants (or columns) have the same cost. We obtain equivalence classes based on $\equiv_P$ and it is sufficient to keep one column of each equivalence class. It is easy to see that we have to choose all columns which are the only ones to cover some x (in the beginning these are the essential prime implicants). If column p is chosen, all rows x where $R(x,p) = 1$ can be eliminated. In an implicit description, eliminated rows and columns are replaced by zeros. We may cyclically apply the described three reduction techniques until a fixed point is obtained, i.e., none of the three techniques eliminates some row or column. The result of this reduction process is called the *cyclic core* of the set-covering problem. Our description leads directly to an OBDD-based algorithm for the computation of the cyclic core of the problem. Experiments have shown that this approach is too inefficient. The reason is that we have to work with quantified formulas, since $x \leq_X x'$ leads to the formula

$$\forall p \left(R(x',p) = 1 \Rightarrow R(x,p) = 1\right).$$

Coudert, Madre, and Fraisse (1993) have chosen an approach which leads to a more efficient algorithm. The efficiency cannot be proved formally but is observed in experiments. The approach is based on so-called transposing functions which map monomials x to monomials $\tau(x)$ and monomials p to monomials $\rho(p)$ such that the following properties hold:

- $x \leq_X x' \Leftrightarrow \tau(x) \leq \tau(x')$,
- $p \leq_P p' \Leftrightarrow \rho(p) \leq \rho(p')$,
- p covers $x \Leftrightarrow \rho(p)$ covers $\tau(x)$,
- $\rho(p) \leq p$, and
- $\tau(x) \geq x$.

These properties ensure that it is sufficient to solve the new covering problem which has ones only at positions $(\tau(x), \rho(p))$ where $R(x,p) = 1$. Moreover, each cover can be used as a cover for the original problem. The property $\rho(p) \leq p$ ensures that we only choose implicants of the given function. If $\rho(p)$ covers $\tau(x)$, it also covers x, since $\tau(x) \geq x$. The main advantage of this approach is that the quantified formulas for the properties $x \leq_X x'$ and $p \leq_P p'$ are replaced by the properties $\tau(x) \leq \tau(x')$ and $\rho(p) \leq \rho(p')$ checking whether one monomial is a lengthening of another one.

In the following, we define the transposing functions chosen by Coudert, Madre, and Fraisse (1993) and we present an algorithm for the computation of the cyclic core of the new problem.

The transposing function τ is defined in the following way. The monomial $\tau(x)$ describes the intersection of all $p^{-1}(1)$ such that $p \in P$ covers x. This is equivalent to the description of $\tau(x)$ as the conjunction of all $p \in P$ covering x. For example, the monomial $x_1x_2x_3$ may be covered by the prime implicants x_1 and x_2 leading to $\tau(x_1x_2x_3) = x_1x_2$. Now it is easy to prove that $x \leq_X x' \Leftrightarrow \tau(x) \leq \tau(x')$. If x' dominates x, then x is at least covered by the same $p \in P$ as x' and, therefore, the conjunction of all $p \in P$ covering x' is a (not necessarily proper) shortening of the conjunction of all $p \in P$ covering x, i.e., $\tau(x) \leq \tau(x')$. If p covers x, then $p \geq x$. Moreover $\tau(x) \geq x$, since $\tau(x)$ is the conjunction of some p such that $p \geq x$. The transposing function ρ is defined in the following way. The monomial $\rho(p)$ is the longest monomial covering all $x \in X$ covered by p. This definition needs some explanation, since for the original PI-table always $\rho(p) = p$. This reflects that no prime implicant dominates another one in the beginning. Let us consider the prime implicant x_1x_2 covering $x_1x_2\overline{x}_3$ and $x_1x_2x_3$. If an implicant is already chosen which covers $x_1x_2x_3$, the row for $x_1x_2x_3$ is no longer contained in X and $\rho(x_1x_2) = x_1x_2\overline{x}_3$. It is also easy to prove that $p \leq_P p' \Leftrightarrow \rho(p) \leq \rho(p')$. Since p covers a subset of the inputs covered by p', $\rho(p')$ covers all inputs covered by p and $\rho(p)$ may be longer, which means smaller as a function. Obviously, p itself is a monomial covering all $x \in X$ covered by p. This implies that $\rho(p) \leq p$. If $\rho(p)$ covers $\tau(x)$, then $p \geq \rho(p)$ covers $x \leq \tau(x)$. Let us now assume that p covers x. By definition,

$\tau(x)^{-1}(1) \subseteq p^{-1}(1)$ and p covers $\tau(x)$. Then $\tau(x) \in X$ and $\rho(p)$ is defined in such a way that it covers the same monomials from X as p. Hence, $\rho(p)$ covers $\tau(x)$. Altogether, the chosen transposing functions fulfill all properties listed above.

This approach also leads to a simple detection of essential columns, i.e., columns $p \in P$ covering a row $x \in X$ which is only covered by p. By definition, this is equivalent to $\tau(x) = p$. Hence, we first apply the transposing function τ. It is easy to obtain OBDDs describing the sets X (after the application of τ) and P. Then the OBDD for $P \cap X$ describes the set of essential columns.

Altogether, we obtain the following algorithm for the computation of the cyclic core. We start with the transformed PI-table and then apply the following three steps until none of them causes a change. Remember that X and P are sets of monomials.

(1) X is replaced with the set of maximal elements of all $\tau(x)$, $x \in X$.

(2) The set of essential columns ESS is computed as the intersection of X and P. We add ESS to the set of already chosen implicants, remove the columns from ESS in the set-covering problem, and also remove all rows covered by columns from ESS.

(3) P is replaced with the set of maximal elements of all $\rho(p)$, $p \in P$.

The efficiency of this approach relies on the efficiency of the computation of the set of maximal elements of all $\tau(x)$, $x \in X$, and of all $\rho(p)$, $p \in P$. For $k \in \{1, \ldots, n\}$, let $X(1_k)$, $X(x_k)$, and $X(\overline{x}_k)$ be sets of monomials defined on the set $\{x_1 \ldots, x_n\} - \{x_k\}$ of variables. The set $X(1_k)$ contains all monomials $x \in X$ not containing an x_k-literal, the set $X(x_k)$ contains all x not containing x_k such that $xx_k \in X$, and the set $X(\overline{x}_k)$ contains all x not containing $\overline{x}_k$ such that $x\overline{x}_k \in X$. The sets $P(1_k)$, $P(x_k)$, and $P(\overline{x}_k)$ are defined analogously. These sets can be computed from OBDDs representing X or P. E.g., $X(x_k)$ is computed in the following way. We compute the intersection of X and the set of all monomials containing x_k. Then we replace the variables responsible for x_k by the appropriate constants. Let $\mathrm{Sup}(X, P)$ contain the monomials $x \in X$ such that $x \geq p$ for some $p \in P$ and let $\mathrm{Sub}(X, P)$ contain the monomials $x \in X$ such that $x \leq p$ for some $p \in P$. These sets can be computed by the following recursive approach. The terminal cases are

$$\mathrm{Sup}(X, P) = \mathrm{Sub}(X, P) = \emptyset \text{ if } X = \emptyset \text{ or } P = \emptyset,$$
$$\mathrm{Sup}(X, P) = X \text{ if } P \text{ is the set of all monomials,}$$
$$\mathrm{Sub}(X, P) = X \text{ if } P \text{ contains the constant } 1.$$

Moreover,

$$\begin{aligned}\mathrm{Sup}(X,P)(x_k) &= \mathrm{Sup}\big(X(x_k), P(x_k)\big),\\ \mathrm{Sup}(X,P)(\overline{x}_k) &= \mathrm{Sup}\big(X(\overline{x}_k), P(\overline{x}_k)\big),\\ \mathrm{Sup}(X,P)(1_k) &= \mathrm{Sup}\big(X(1_k), P(1_k)\cup P(x_k)\cup P(\overline{x}_k)\big).\end{aligned}$$

If $x \geq p$ and x contains x_k, p has to contain x_k, similarly for $\overline{x}_k$. If x contains neither x_k nor $\overline{x}_k$, p may contain x_k, $\overline{x}_k$, or none of them. In order to obtain $\mathrm{Sup}(X,P)$ we compute the union of $\mathrm{Sup}(X,P)(1_k)$, $\mathrm{Sup}^*(X,P)(x_k)$, and $\mathrm{Sup}^*(X,P)(\overline{x}_k)$. The set $\mathrm{Sup}^*(X,P)(x_k)$ contains xx_k if $\mathrm{Sup}(X,P)(x_k)$ contains x. The set $\mathrm{Sup}^*(X,P)(\overline{x}_k)$ is defined analogously. Similarly, we obtain

$$\begin{aligned}\mathrm{Sub}(X,P)(1_k) &= \mathrm{Sub}\big(X(1_k), P(1_k)\big),\\ \mathrm{Sub}(X,P)(x_k) &= \mathrm{Sub}\big(X(x_k), P(1_k)\cup P(x_k)\big),\\ \mathrm{Sub}(X,P)(\overline{x}_k) &= \mathrm{Sub}\big(X(\overline{x}_k), P(1_k)\cup P(\overline{x}_k)\big).\end{aligned}$$

Now, the set $\mathrm{Sub}(X,P)$ can be computed in a similar way as $\mathrm{Sup}(X,P)$. The next step is to describe a recursive approach for the computation of the set $\mathrm{MT}(X,P)$ of maximal elements of all $\tau(x)$, $x \in X$, with respect to $\leq$. Here MT abbreviates MaxTau. The terminal cases are

$$\begin{aligned}\mathrm{MT}(X,P) &= \emptyset \text{ if } X=\emptyset \text{ or } P=\emptyset,\\ \mathrm{MT}(X,P) &= \{1\} \text{ if } P=\{1\}.\end{aligned}$$

To prepare the recursive calls, we choose a variable x_k. In the first step, we compute the following sets:

$$\begin{aligned}A_1 &= \mathrm{Sub}\big(X(x_k), P(x_k)\big),\\ A_0 &= \mathrm{Sub}\big(X(\overline{x}_k), P(\overline{x}_k)\big), \text{ and}\\ A &= X(1_k)\cup\big(X(x_k)-A_1\big)\cup\big(X(\overline{x}_k)-A_0\big).\end{aligned}$$

A monomial x belongs to A_1 if it does not contain x_k, $xx_k \in X$, and there is some $p \in P(x_k)$ covering x, i.e., $px_k \in P$ and px_k covers xx_k. A monomial x belongs to A if it does not contain an x_k-literal and at least one of the following properties is fulfilled: $x \in X$, or $xx_k \in X$ but no $p \in P(x_k)$ covers x, or $x\overline{x}_k \in X$ but no $p \in P(\overline{x}_k)$ covers x. These sets are used for the following recursive calls:

$$\begin{aligned}B &= \mathrm{MT}\big(A, P(1_k)\big),\\ B_1 &= \mathrm{MT}\big(A_1, P(1_k)\cup P(x_k)\big),\\ B_0 &= \mathrm{MT}\big(A_0, P(1_k)\cup P(\overline{x}_k)\big).\end{aligned}$$

It follows from our description of A above that it is sufficient to consider $P(1_k)$ in the computation of the maximal elements $\tau(x)$, $x \in A$. If $x \in A_1$, it is not necessary to consider some $p \in P(\overline{x}_k)$. We have to be careful with the interpretation of B, B_1, and B_0, which are defined on a variable set not containing x_k. In order to obtain $\mathrm{MT}(X, P)$ we have to consider the influence of the x_k-literals. Let C, C_1, and C_0 be the sets of all monomials containing neither x_k nor $\overline{x}_k$, containing x_k, and containing $\overline{x}_k$, respectively. These sets can be represented by OBDDs of constant size. If we consider B as a set of monomials syntactically defined on the variable set including x_k, the set B contains the monomials xx_k and $x\overline{x}_k$ if it contains x. The set $B \cap C$ contains the same monomials as B but the monomials are defined on the set of all variables $x_1, \ldots, x_n$. This is an operation which is not necessary for ZBDDs. A monomial $x \in B_1$ can be thrown away if $x \le x'$ and $x' \in B$. The reason is that x' is also contained in $B \cap C$ and we would append x_k to x. Obviously, $xx_k \le x'$ and xx_k is not a maximal element. Hence, the result of $\mathrm{MT}(X, P)$ can be computed in the following way:

$$\mathrm{MT}(X, P) = (B \cap C) \cup \big((B_1 - \mathrm{Sub}(B_1, B)) \cap C_1\big) \cup \big((B_0 - \mathrm{Sub}(B_0, B)) \cap C_0\big).$$

The recursive approach for the computation of the set $\mathrm{MR}(X, P)$ (MR abbreviates MaxRho) of maximal elements of all $\rho(p)$, $p \in P$, with respect to $\le$ works in a similar way. The terminal cases are

$$\begin{aligned} \mathrm{MR}(X, P) &= \emptyset \text{ if } X = \emptyset \text{ or } P = \emptyset, \\ \mathrm{MR}(X, P) &= P \text{ if } X \text{ contains all monomials.} \end{aligned}$$

Remember that $\rho(p)$ is the longest monomial covering all $x \in X$ covered by p. Hence, we keep all $p \in P(1_k)$ covering some $x \in X(1_k)$ or some $x \in X(x_k) \cap X(\overline{x}_k)$. These are the monomials where we do not have to append x_k or $\overline{x}_k$. We compute

$$D = \mathrm{Sup}\big(P(1_k), X(1_k)\big) \cup \big(\mathrm{Sup}\big(P(1_k), X(x_k)\big) \cap \mathrm{Sup}\big(P(1_k), X(\overline{x}_k)\big)\big).$$

Now we consider those monomials where we later append x_k. These are monomials $p \in P(1_k) \cup P(x_k)$ covering monomials from $X(x_k)$. Since we are interested in maximal elements, we can throw away elements contained in D. Let

$$\begin{aligned} D_1 &= \mathrm{Sup}\big(P(1_k) \cup P(x_k), X(x_k)\big) - D, \\ D_0 &= \mathrm{Sup}\big(P(1_k) \cup P(\overline{x}_k), X(\overline{x}_k)\big) - D. \end{aligned}$$

The situation for the recursive calls is similar to the recursive computation of Sup. We obtain

$$\begin{aligned} E &= \mathrm{MR}\big(X(1_k) \cup X(x_k) \cup X(\overline{x}_k), D\big), \\ E_1 &= \mathrm{MR}\big(X(x_k), D_1\big), \\ E_0 &= \mathrm{MR}\big(X(\overline{x}_k), D_0\big). \end{aligned}$$

At the end, we have to "reinsert" the variable x_k in a similar way as in the procedure MT, namely,

$$\mathrm{MR}(X,P) = (E \cap C) \cup \big((E_1 - \mathrm{Sub}(E_1, E)) \cap C_1\big) \cup \big((E_0 - \mathrm{Sub}(E_0, E)) \cap C_0\big).$$

As in most applications of OBDD techniques, we cannot guarantee that the OBDD size does not explode during all the steps where OBDDs are involved. Since we are considering sets which typically are small compared to the set of all monomials, ZBDDs may perform better than OBDDs.

After having computed the cyclic core of the set-covering problem, we have still to solve a set-covering problem (which hopefully is smaller and easier than the given one). Coudert (1995) has described a branch-and-bound algorithm with improved lower bound techniques for the solution of this problem. We omit the details, since they are independent of BDD techniques. After having decided to choose or to eliminate some column, a new set-covering problem is obtained and the technique of computing the cyclic core can be applied again. If, e.g., some column is eliminated, another column may turn into an essential one.

14.2 Multilevel Logic Synthesis

No efficient method for multilevel logic minimization is known and we have to be satisfied with methods for *multilevel logic synthesis* leading to combinational circuits with some nice properties. BDD techniques may support approaches for multilevel logic synthesis. Those results are not presented. Here we concentrate on the idea of using BDDs as combinational circuits. This is always easily possible, since BDD nodes can be simulated by small circuits for the ite operation. The same holds for DD nodes for the positive or negative Reed-Muller decomposition (see Section 8.3). Even nondeterministic nodes do not cause problems, since they can easily be simulated by OR-, AND-, or EXOR-gates. In order to obtain small combinational circuits, preferably of small depth, we need BDDs of small size. The depth of OBDDs, OFDDs (also OKFDDs), and FBDDs is bounded by the number n of variables. Moreover, OBDDs and OFDDs with a fixed variable ordering can be minimized efficiently.

Before OFDDs were introduced as a representation of Boolean functions (Kebschull and Rosenstiel (1993)), they were used as a tool for multilevel logic synthesis (Kebschull, Schubert, and Rosenstiel (1992)). The idea behind this approach was the improvement of circuits simulating fixed-polarity or mixed-polarity Reed-Muller expansions.

Becker (1992) has investigated combinational circuits simulating OBDDs and FBDDs. He has proved that such circuits are easily testable.

Ishiura (1992) has tried to avoid a major disadvantage of combinational circuits simulating OBDDs. The depth is bounded above by $O(n)$ for the number

n of variables but in most cases it is bounded below by $\Omega(n)$. To reduce the depth, Ishiura (1992) has started with quasi-reduced OBDDs for Boolean functions f. Without loss of generality, we assume that $\pi = id$, i.e., level i is labeled by x_i, $1 \le i \le n$, and level $n+1$ consists of the sinks. Let s_i be the size of level i and let $R[i,j]$ be the $s_i \times s_j$ matrix containing at position (k,l) the Boolean function depending on $x_i, \ldots, x_{j-1}$ and accepting those inputs a where the path activated by a and starting at the kth node of level i reaches the lth node on level j. It is easy to describe the matrix $R[i,i+1]$ which only consists of entries from $\{0,1,x_i,\overline{x}_i\}$. For two matrices A and B containing Boolean functions as entries, we define the Boolean matrix product $C = A \cdot B$ by

$$C_{ij} = A_{i1}B_{1j} + \cdots + A_{im}B_{mj}$$

where the product is AND and the sum equals OR. The matrix product is defined, as usual, only if the number m of columns of A is equal to the number of rows of B. Since paths from the ith level to the jth level pass through the kth level, we can conclude that

$$R[i,j] = R[i,k] \cdot R[k,j]$$

for $i < k < j$. Finally, $R[1,n+1]$ describes $f_1, \ldots, f_m, \overline{f}_1, \ldots, \overline{f}_m$ if the OBDD represents $f = (f_1, \ldots, f_m)$. In order to obtain small depth, we use a balanced tree to compute $R[1,n+1]$, i.e., we recursively compute $R[1,n+1]$ as $R[1,\lfloor n/2 \rfloor+1] \cdot R[\lfloor n/2 \rfloor+1, n+1]$. How can we estimate the size and depth of the resulting circuit? We denote by w the width of the given OBDD, which is defined as the size of the largest level. The computation of each entry of $R[i,j]$ given a circuit computing the entries of $R[i,k]$ and $R[k,j]$ is performed along the definition of the Boolean matrix product. Hence, w binary AND-gates and $w-1$ binary OR-gates are sufficient. The AND-gates work in parallel and the OR-gates are organized as a balanced binary tree. All entries of $R[i,j]$ can be computed with $(2w-1)w^2 = O(w^3)$ binary gates in depth $\lceil \log w \rceil + 1$.

Altogether, we have to perform n such matrix multiplications and the depth with respect to matrix multiplications is $\lceil \log n \rceil$. We have proved the following result.

Theorem 14.2.1. *If f is defined on n variables and represented by a quasi-reduced OBDD of width w, it is possible to construct a combinational fan-in 2 circuit representing f in depth $O((\log n) \cdot \log w)$ and size $O(nw^3)$. The construction takes time $O(nw^3)$.*

Ishiura (1992) has proved that this circuit is also easily testable.

Testability is one issue in logic synthesis, hazard freeness another one. In order to define hazards, we denote by $[a,b]$ the subcube of $\{0,1\}^n$ consisting of all c such that $a_i = b_i$ implies $a_i = c_i$. If the input switches from a to b and if the switching bits may switch in an arbitrary order, exactly the inputs from $[a,b]$ are possible as intermediate inputs.

Definition 14.2.2. The Boolean function f contains a *static function hazard* for the input transition from a to b if $f(a) = f(b)$ and $f(a) \neq f(c)$ for some $c \in [a, b]$. The Boolean function f contains a *dynamic function hazard* for the input transition from a to b if $f(a) \neq f(b)$ and there exist some $c \in [a, b]$ and $d \in [c, b]$ where $f(a) \neq f(c)$ and $f(d) \neq f(b)$.

A static function hazard implies that the output may change during a transition from a to b, although $f(a) = f(b)$. A dynamic function hazard implies that a transition from a to b can be performed in a way that we switch from a to c, then to d, and finally to b. The output changes for this transition at least three times. Hazards describe output changes which are not necessary. No implementation can avoid a glitch on a transition which contains a function hazard if we assume arbitrary gate and wire delays. The only problems one can avoid are logic hazards.

Definition 14.2.3. Let $a \to b$ be a transition which is not a function hazard for f and let C be a combinational circuit representing f. The circuit C contains a *static logic hazard* for the input transition from a to b if $f(a) = f(b)$ and some delay assignment causes the circuit to output the value different from $f(a)$ during the transition. The circuit C contains a *dynamic logic hazard* for the input transition from a to b if $f(a) \neq f(b)$ and some delay assignment causes the circuit to change its output during the transition at least three times.

It is a classical result that a two-level realization of f by prime implicants is free of logic hazards iff it contains all prime implicants. Such a realization is often much more expensive than a minimal two-level realization. Lin and Devadas (1995) have investigated combinational circuits based on OBDDs (and also on FBDDs). Each BDD node is replaced by a hazard-free circuit for the ite operation.

Theorem 14.2.4. *Let G be an FBDD representing f such that no OBDD reduction rule is applicable. Let C be the combinational circuit obtained from G by replacing the BDD nodes by hazard-free realizations. The circuit C is free of static logic hazards.*

Proof. The proof is done by induction on the number n of variables x_i such that the FBDD contains an x_i-node. For $n = 0$, G represents a constant function by a sink and C is hazard-free. For $n \geq 1$, we assume w.l.o.g. that the first node of the FBDD is labeled by x_1. Let the transition from a to b be a static one, w.l.o.g. $f(a) = f(b) = 1$, without static function hazard, i.e., $f(c) = 1$ for all $c \in [a, b]$.

In the first case, we assume that $a_1 = b_1$, w.l.o.g. $a_1 = 1$. During the transition from a to b, the first FBDD node v is only influenced by the sub-FBDD reached by the 1-edge leaving v. This sub-FBDD represents $f_{|x_1=1}$,

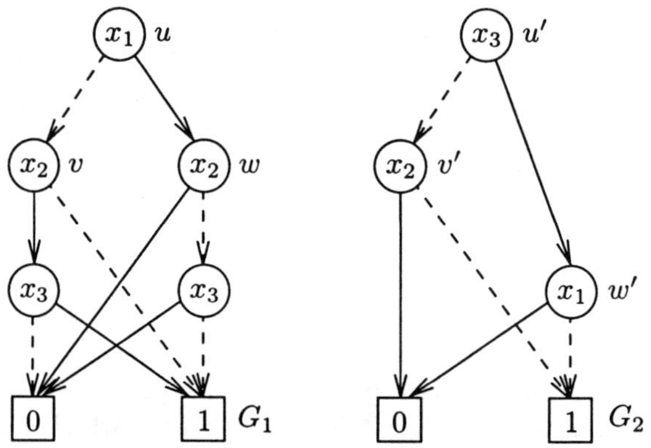

Figure 14.2.1: *OBDDs representing* $\overline{x}_3\overline{x}_2 + x_3\overline{x}_1$.

which equals f for all $c \in [a, b]$. We conclude by the induction hypothesis that we have no static logic hazard in this case.

In the second case, we assume that $a_1 \neq b_1$, w.l.o.g. $a_1 = 1$ and $b_1 = 0$. Since $f(c) = 1$ for all $c \in [a, b]$, also $f_{|x_1=0}(c) = f_{|x_1=1}(c) = 1$ for all $c \in [a, b]$. We conclude by the induction hypothesis that the subcircuits obtained from the FBDDs whose sources are the direct successors of v are free of static logic hazards, i.e., they constantly produce 1 during the transition from a to b. Since, furthermore, the subcircuit replacing the BDD node v is hazard-free, the whole circuit is free of static logic hazards. □

The situation changes if we investigate dynamic logic hazards.

Example 14.2.5. Let $f(x_1, x_2, x_3) = \overline{x}_3\overline{x}_2 + x_3\overline{x}_1$, $a = (0, 0, 0)$, and $b = (1, 1, 0)$. Then $f(0, 0, 0) = 1$ and $f(1, 1, 0) = 0$ and the transition from a to b does not lead to a dynamic function hazard. Let G_1 and G_2 be reduced OBDDs representing f for the variable orderings (x_1, x_2, x_3) and (x_3, x_2, x_1), respectively (see Fig. 14.2.1).

For the input $a = (0, 0, 0)$, u "computes" 1 denoted by $u \to 1$. Moreover, $v \to 1$ and $w \to 1$. We assume that x_2 switches from 0 to 1 and we obtain the input $c = (0, 1, 0)$. Both v and w are supposed to change but we assume that there is a large delay at w. Then v switches to 0 and causes u to switch to 0. We assume that x_1 switches from 0 to 1 before w has reacted on the first switch. This causes u to switch to the value of w which is still 1. Finally, w reacts on the switch of x_2 and switches to 0 causing u to switch to 0. This identifies a dynamic logic hazard. The situation for G_2 is different. For $a = (0, 0, 0)$, $u' \to 1$, $v' \to 1$, and $w' \to 1$. We know that x_3 does not change its value during the transition from a to b. Hence, the switch of x_1 does not have influence on the circuit simulating G_2. Only the switch of x_2 causes a switch but this does not lead to a dynamic logic hazard.

Lin and Devadas (1995) have derived conditions on the chosen variable ordering π to prevent a dynamic logic hazard for some transition and the circuit based on reduced π-OBDDs. Moreover, they have described how one can decide whether a variable ordering exists which fulfills the conditions for all transitions and how to construct such a variable ordering if it exists. The conditions may require different variable orderings for different inputs. Then the conditions cannot be fulfilled by OBDDs but perhaps by FBDDs. We refer to Lin and Devadas (1995) for the details.

14.3 Functional Simulation

Verification is better than validation by simulation but also much more expensive. Hence, design validation by *functional simulation* is still a key step in the design of digital systems. The task is to compute for a sequence of inputs $a_1, \ldots, a_r$ the outputs $b_1, \ldots, b_r$ realized by the system on $a_1, \ldots, a_r$. Hence, we are faced with the evaluation problem and it is sufficient to consider a single input vector a. The evaluation of a combinational circuit with s binary gates can be performed in time $\Theta(s)$. In typical situations, s is much larger than the number n of input variables. If a table of all 2^n function values is available, the output $f(a)$ can be found by a simple table lookup. Circuits are small but evaluation takes time while evaluation is trivial for function tables which are typically much too large to be stored. OBDDs (or FBDDs) might be a good compromise, since they are usually much smaller than function tables and evaluation is possible in time $O(n)$. We postpone the discussion why we should perform functional simulation if verification is possible by constructing OBDDs. First, we discuss other problems with this approach. Circuits encountered in practice have more than one output—the number m of outputs may be even larger than n. Using OBDDs (more precisely SBDDs) we cannot get in general a better bound than $O(nm)$ for the time to evaluate $f(a)$ and then it may be better to evaluate the circuit directly, which still takes time $O(s)$.

Definition 14.3.1. The *characteristic function* of a Boolean function $f\colon \{0,1\}^n \to \{0,1\}^m$ is the Boolean function $F\colon \{0,1\}^{n+m} \to \{0,1\}$ where $F(a,b) = 1$ iff $f(a) = b$.

Ashar and Malik (1995) suggest constructing an OBDD for $F(x,y)$ and evaluating it in time $O(n+m)$. The problem is that we have to know the output $b = f(a)$ in order to evaluate F at (a,b). The solution is quite easy by using variable orderings where all x-variables are tested before all y-variables. Having read the input $x = a$, only one assignment to the y-variables, namely

$b = f(a)$, leads to the 1-sink and we can determine $b = f(a)$ by following the unique path to the 1-sink. The new disadvantage is that the considered variable ordering seems to be quite bad. Let, e.g., $f(a) = a$ be the identity. Then $F(x, y) = \mathrm{EQ}(x, y)$ is the equality test whose OBDD size is exponential if all x-variables are tested before all y-variables. It is sufficient to test all x-variables which influence the output y_i before y_i is tested. In our example, this leads to an interleaved variable ordering with linear OBDD size for the equality test. There are heuristics to compute good variable orderings which fulfill the mentioned restriction. The sifting algorithm can be restricted such that x_i and y_j are never swapped if x_i may influence y_j.

The whole approach is efficient if the OBDD is not too large. But then OBDD-based verification is better than functional simulation. Hence, we look for a solution in the situation where the OBDD size explodes. Then we choose a set of gates such that they are not connected by paths and such that the OBDD for the characteristic function with these gates as outputs is not too large. In a first step, we determine the values computed at the chosen gates. Afterwards, we replace the chosen gates by new input variables and the task is to evaluate this circuit on the given input and the resulting values at the chosen gates. This process can be iterated with a given threshold for the OBDD size. The approach is still an improvement of the direct evaluation of the circuit if the number of considered characteristic functions remains small.

Another issue of functional simulation is *fault simulation*. For a given test input t and a set of possible faults, we have to determine the set of faults where the faulty circuit and the good circuit differ. We consider stuck-at faults where wires are replaced by constant values. The number of possible single faults grows linearly with the circuit size. If we also want to consider multiple faults consisting of up to m single faults, the size of the fault set grows exponentially with m. Let $f_0, \ldots, f_{N-1}$ be the list of multiple faults we allow. If N is too large, it is impossible to simulate the circuit on t and all types of faults. Takahashi, Ishiura, and Yajima (1994) have encoded the multiple faults by vectors of length $n \geq \lceil \log N \rceil$. A function $g \in B_n$ is interpreted as the characteristic function of a subset F' of $F = \{f_0, \ldots, f_{N-1}\}$ where $f_i \in F'$ iff $g(a^i) = 1$ for the encoding a^i of f_i. The aim is to compute the sets of multiple faults observable at the outputs of the circuit. It is easy to describe the sets of faults which are observable at the inputs. Then we use a topologically sorted list of the wires and gates of the circuit and compute the sets of observable multiple faults for all wires and gates with respect to this ordering.

Let w be a wire transporting the signal 0 in a good circuit with respect to the input t. Let L_0 be the set of faults containing the stuck-at-0 fault at w and let L_1 be defined similarly for the stuck-at-1 fault. Finally, let L be the set of faults observable at the source of w and L' the similar set for the sink of w. The characteristic function of $L' = (L \cup L_1) \cap \overline{L_0}$ can be computed with the usual OBDD techniques. A similar formula can be derived if the good signal is 1.

For the fault set propagation at gates, we investigate the example of a binary AND-gate whose first input is 0 and whose second input is 1 if the circuit works correctly. Let A and B be the corresponding sets of observable faults for the sinks of the input signals and let C be the set of faults observable at the source of the wires leaving the gate. Then $C = A \cap \overline{B}$, since the output differs from the good output 0 only if both inputs are 1. As long as the OBDD size of the characteristic functions does not explode, we may handle a large set of faults simultaneously.

It has turned out that the best-known coding technique is fault number tuple (FNT) coding. Each single fault gets a unique number. If the number of different faults is bounded by m, the fault numbers are concatenated with respect to the usual ordering of numbers. In order to encode fault sets with $m' < m$ faults, the encoding of the first fault is repeated $m - m'$ times to obtain codewords of fixed length. There are words belonging to illegal fault sets, i.e., if a wire has a stuck-at-0 and a stuck-at-1 fault at the same time. These words cause problems if we compute the complement of a set. We always have to remove illegal codewords by computing the intersection with the set of all possible fault sets. The recommended variable orderings test first the first bits of all m fault numbers, then their second bits, and so on.

14.4 Test Generation

Verification can be performed on the design level but verification cannot detect production faults. The only possibility of detecting such faults is to simulate (see Section 14.3) the circuit on a set of test patterns and to compare the result with the corresponding results of the verified design. One may assume that the produced circuit is not totally different from the design. Otherwise, we can detect all possible faults only by checking the circuit on all inputs. Hence, the number of faults and their types are restricted. We work with a fault model describing all faults we want to detect. Faults can be redundant, i.e., these faults do not change the input-output behavior of the circuit and cannot be detected. The aim is to compute a small or even minimal set of test patterns such that all nonredundant faults are detected by an investigation of the outputs of the circuit on the test patterns.

Let f be the function the circuit should compute and g the function computed if a certain fault occurs. The set of test patterns detecting this fault is equal to the set of inputs satisfying $f \oplus g$. A fault is redundant iff $f \equiv g$ or $f \oplus g \equiv 0$. In any case, we may use our synthesis techniques to obtain an OBDD representation of the set of test patterns for the fixed fault. If h_i describes the set of test patterns for the ith fault, the intersection of some h_i describes the set of test patterns which cover all faults involved in the intersection.

It is interesting to note that BDDs had already been used for test generation before OBDDs were defined. Akers (1978b) and also Abadir and Reghbati (1986) use BDDs for a compact representation of the function computed by a circuit or subcircuit. These BDDs are constructed from descriptions of the circuit and not by a synthesis process. Nevertheless, all examples are indeed OBDDs. These OBDDs support the classical D-algorithm for the computation of test patterns. Paths in the OBDDs describe so-called experiments, i.e., partial assignments to the variables which are used by the D-algorithm. Here it is interesting to see that in these early papers BDDs are used as a static representation of Boolean functions. Bryant (1985) has recognized that people using BDDs indeed use OBDDs and that OBDDs can be applied dynamically, since many operations can be performed efficiently.

Jóźwiak and Mijland (1992) remark that OBDDs can describe the same function as a circuit but there is no counterpart to the possible faults of the circuit. Different circuits for f lead to the same reduced π-OBDD for f. OBDDs only describe a function and nothing more. Jóźwiak and Mijland (1992) have looked for BDDs where circuit faults have a counterpart and they have restricted circuits to two-level representations. We have shown (Theorem 10.1.4) that polynomial-size two-level representations can be simulated by OR-DTs and vice versa. Jóźwiak and Mijland (1992) represent two-level representations by OR-OBDDs (all examples are indeed OR-DTs). They describe some heuristic ideas for the choice of a good variable ordering. In order to minimize the size it is not the best choice to use nondeterministic nodes only at the top. Finally, they show how stuck-at faults of the two-level circuit can be described in the corresponding OR-DT. This is another example of applications of nondeterministic BDDs. Many other authors use OBDDs in their algorithms for test generation but they only need the standard OBDD techniques.

14.5 Timing Analysis

A fundamental issue of integrated circuit design is to meet given time constraints. *Timing analysis* is the task of analyzing the time behavior for all inputs under some given information about the delay of inputs, gates, and wires. Timing analysis determines the set of critical input vectors (leading to the largest delay), critical gates, and critical paths. This information may be used for resynthesizing, e.g., to reduce the maximal delay or to lower the power consumption.

One may think that the *delay* of a circuit is easy to compute. We start at the input variables which have a given delay. Wires contribute some delay and at a gate we have to take the maximum of the delays on the incoming wires and have to add the delay of the gate. This would lead to a linear-time DFS algorithm. But this point of view is too pessimistic. Gates may have *controlling inputs* and *sensitizing inputs*. Controlling inputs determine the output of a gate like 0 for AND-gates or 1 for OR-gates while sensitizing or noncontrolling inputs imply that we have to know other inputs of the gate before we know the output. An AND-gate has a controlling and a sensitizing input while an EXOR-gate only has sensitizing inputs. A path from an input to an output of the circuit is called *sensitizable* if it is possible that an event at the input, i.e., changing its value, may change the value at all gates of the path. Paths which are not sensitizable are called *false paths*. The maximal delay of the circuit is the maximal delay along all sensitizable paths which may be smaller than the maximal delay along all paths. Our simple linear-time DFS algorithm cannot distinguish between sensitizable and false paths.

We present the approach of Bahar, Cho, Hachtel, Macii, and Somenzi (1994) to compute an MTBDD or ADD (see Section 9.2) representing for a combinational circuit the delay for each input vector. The algorithm works under the following assumptions. The delay does not depend on the previous input (floating mode of the circuit), gate delays are included in the delay of the incoming wires, and the wire delay may depend on the input x, i.e., we may distinguish rising and falling input transitions.

These assumptions lead to the following formalization of the problem. Let v be a gate of the circuit having m inputs. We denote by $d(v_j, x)$ the delay on the jth incoming wire for input x. The parameter $d(v_j, x)$ depends on x only via the property whether the wire carries the value 0 or 1. We denote by $\mathrm{AT}(v, x)$ the *arrival time* at v on input x, i.e., the earliest point of time where the final output signal of v is fixed. The aim is to produce MTBDDs which represent for all output gates the arrival times for all input vectors x. We obtain the following recursive equations for $\mathrm{AT}(v, x)$, where $\mathrm{AT}(v_j, x)$ describes the arrival time at that gate which is the starting point of the jth incoming wire of v. If v has no controlling input wire for input x, then

$$\mathrm{AT}(v, x) = \max\{\mathrm{AT}(v_j, x) + d(v_j, x) \mid 1 \le j \le m\}.$$

Otherwise, we denote the set of controlling input wires by $C(v, x)$ and obtain

$$\mathrm{AT}(v, x) = \min\{\mathrm{AT}(v_j, x) + d(v_j, x) \mid j \in C(v, x)\}.$$

These formulas can be evaluated in linear time for each x but we want to get a solution for all x. In order to construct the MTBDD for $\mathrm{AT}(v)$, we need the MTBDDs M_j for $\mathrm{AT}(v_j)$, $1 \le j \le m$, and the OBDDs G_j representing the Boolean functions f_j, $1 \le j \le m$, which are computed at the wires arriving at v, and the OBDD G representing the Boolean function f_v computed at v. Furthermore, we need the information about the delay values. Before describing an algorithm for the construction of the MTBDD for $\mathrm{AT}(v)$, we present a small example.

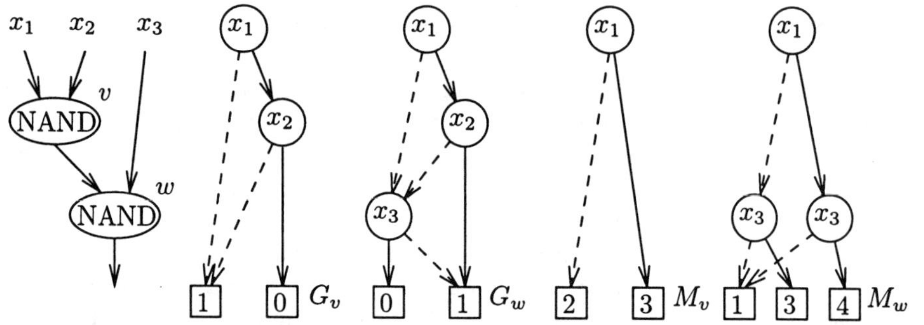

Figure 14.5.1: *A circuit, OBDDs for the functions, and MTBDDs for the arrival times.*

Example 14.5.1. Figure 14.5.1 contains a circuit, OBDDs G_v and G_w for the functions computed at the gates v and w, respectively, and MTBDDs M_v and M_w for AT(v) and AT(w), respectively, where we assume that the inputs have the arrival times AT(x_1) = 1, AT(x_2) = 2, and AT(x_3) = 0, and all delay values are equal to 1. We use the variable ordering x_1, x_2, x_3.

The value of x_2 does not influence the arrival times. If $x_1 = x_3 = 1$, the corresponding subfunction of f_v equals $\overline{x}_2$ and the subfunction of f_w equals x_2. The path (x_2, v, w) is sensitizable and determines the maximal delay. We obtain an OBDD representing the set of critical inputs, namely those leading to the largest arrival time, by replacing in the MTBDD for AT(w) the largest sink by 1 and all other sinks by 0. (The resulting OBDD is not guaranteed to be reduced.) We look for an algorithm computing M_w from G_v, G_w, and M_v.

The main idea is to generalize the synthesis algorithm for OBDDs to a synthesis of $m+1$ OBDDs $G, G_1, \ldots, G_m$ and m MTBDDs $M_1, \ldots, M_m$. Without loss of generality, we assume that we use the variable ordering $\pi = id$. Each situation is described by the vector of nodes $(v, v_1, \ldots, v_m, w_1, \ldots, w_m)$ such that we have reached v in G, v_i in G_i, and w_j in M_j. We start at the sources of all BDDs. If $(v, v_1, \ldots, v_m, w_1, \ldots, w_m)$ is contained in the computed-table, we return the corresponding result. The terminal cases are discussed later. Otherwise, we are prepared to construct a new node v^* of the MTBDD M representing the arrival times for the considered partial assignment. Its label is x_i if i is the minimal index of all labels of the nodes $v, v_1, \ldots, v_m, w_1, \ldots, w_m$. Then we recursively apply our synthesis algorithm by setting at first $x_i = 0$ and then $x_i = 1$, i.e., we replace those of the nodes $v, v_1, \ldots, v_m, w_1, \ldots, w_m$ which are labeled by x_i with their 0-successors or 1-successors, respectively. At the end of these recursive calls, we obtain the successors of v^* and know whether v^* can be eliminated. Using the unique-table, we may decide whether v^* can be merged with some already computed node. We still have to describe the terminal cases. It is necessary to reach a sink of G to know whether the gate has controlling inputs and to know which of the two equations for $\mathrm{AT}(v, x)$ has to be applied. If there are no controlling inputs, it is necessary and sufficient to know the arrival times of all incoming wires and the delay on these wires. If the delay does not depend on the value of the wires, it is not necessary to know the value on these wires. If there are controlling inputs, it is necessary to know the values on all incoming wires in order to decide which of them are controlling. Furthermore, it is necessary to know the arrival times of the controlling inputs. This information is sufficient to compute the arrival time. The runtime of this algorithm is bounded above by the product of the sizes of the input BDDs (assuming constant-time operations on the hash tables).

14.6 Technology Mapping

Logic design uses libraries of standard cells realizing Boolean functions. If a circuit has been partitioned into subcircuits (usually with a single output), we have to look for a cell realizing the function $g \in B_n$ computed by the subcircuit. This is not a simple equivalence test whether $f \equiv g$ for a function $f \in B_n$ realized by a cell in the library. The question whether $f \equiv g$ is indeed pointless. There is no relation between the set of variables of f and g. We may use an arbitrary renaming of the variables of f. Let $x_1, \ldots, x_n$ be the variables of g and $y_1, \ldots, y_n$ the variables of f. For each permutation ρ on $\{1, \ldots, n\}$, we denote by ρf the function realized by the circuit for f after renaming y_i by $x_{\rho(i)}$. The problem is to decide whether $g \equiv \rho f$ for some permutation ρ. It is easy to see that this problem is NP-hard. We may decide the satisfiability of a formula g by looking for a permutation ρ such that $g \equiv \rho h$ for the function $h \equiv 0$. Standard cells are used in an even more general form. We may negate the inputs and the outputs. Hence, the *technology-mapping* problem for $f \in B_n$ and $g \in B_n$ is to decide whether there exist $a, a_1, \ldots, a_n \in \{0, 1\}$ and a permutation ρ on $\{1, \ldots, n\}$ such that $g \oplus a \equiv \rho_{a_1,\ldots,a_n} f$, where $\rho_{a_1,\ldots,a_n} f$ is the function realized by the circuit for f after replacing y_i with $x_{\rho(i)} \oplus a_i$. In the positive case, we are interested in a solution.

Most of the cells of a library will not match the given function g. Therefore, efficient tests proving that a cell is not appropriate are very useful. We have shown in Section 5.5 how we can check whether the variables x_i or x_j (positive symmetry) or x_i or $\overline{x}_j$ (negative symmetry) are symmetric and how to compute the equivalence classes of symmetric variables. If we consider positive and negative symmetries, the spectrum of the sizes of the equivalence classes does not change if we permute the variables and negate some of the variables and perhaps the output. If f and g have different spectra, the cell for f cannot realize g.

Clarke, McMillan, Zhao, Fujita, and Yang (1997) present another test based on the *Walsh transform* of Boolean functions. The Walsh matrix (see Section

9.2) is defined by

$$W_0 = [1] \quad \text{and} \quad W_n = \begin{bmatrix} W_{n-1} & W_{n-1} \\ W_{n-1} & -W_{n-1} \end{bmatrix}$$

and has a linear-size MTBDD representation. The matrix W_n is of size $2^n \times 2^n$. Boolean functions f can be represented by vectors of length 2^n describing the value table. The vector $\widehat{f}$ contains the values $1 - 2f(x)$, i.e., 0 is replaced with 1 and 1 with -1 (compare Section 2.5), in the lexicographical ordering of the x-values.

Definition 14.6.1. The Walsh transform of $f \in B_n$ is equal to the vector $W \cdot \widehat{f}$. The kth-order *Walsh spectrum* of f is denoted by $W_k(f)$ and is equal to the subvector of the Walsh transform corresponding to the inputs x where $x_1 + \cdots + x_n = k$.

We investigate how negations of the inputs or the output and the permutation of variables change the Walsh transform. Because of the symmetry of the Walsh matrix, a permutation of the variables leads to the corresponding permutation of the Walsh transform and also to a permutation of the kth-order Walsh spectrum. The last claim relies on the fact that a permutation of the variables does not change the number of ones in the input. If f is negated, $\widehat{f}$ and $W\widehat{f}$ are also negated. A simple calculation shows that the absolute values of the Walsh transform do not change if some inputs are negated. More precisely, the Walsh transform at position a is negated iff the number of i, where $a_i = 1$ and x_i is negated, is odd. Altogether, we obtain the following conclusion. Let $|W_k(f)|$ denote the sequence obtained by sorting the absolute values of $W_k(f)$. In order to match f and g, it is necessary that $|W_k(f)| = |W_k(g)|$ for all k.

The Walsh transform can be computed as a matrix-vector product (see Section 9.2). This may lead to an MTBDD with many different sinks. If we are only interested in $W_k(f)$, we may simplify the computation. We construct an OBDD for $E_{k,n}$ checking whether x contains exactly k ones. The MTBDD representing the Walsh matrix and OBDDs representing f and $E_{k,n}$ are inputs for a synthesis step producing an MTBDD which outputs $W\widehat{f}(x)$ if $E_{k,n}(x) = 1$ and the default value $+\infty$ otherwise. For small k, it is easy to determine all values of $W_k(f)$ and to compute $|W_k(f)|$. Clarke, McMillan, Zhao, Fujita, and Yang (1997) propose to check only whether $|W_0(f)| = |W_0(g)|$ and $|W_1(f)| = |W_1(g)|$.

Kukimoto, Fujita, and Brayton (1994) show how to use OBDDs for the *redesign* of combinational circuits, a problem which is a generalization of the problem of checking whether a cell can realize a given function. We assume that a combinational circuit is given and that the gates have to remain unchanged but we may reconnect the gates, i.e., we may choose how the gates are connected. The problem is motivated by gate-array or standard-cell based designs where modifying the routing without changing the placement of the gates is allowed. This is possible as long as enough space is left for the rerouting.

The given circuit works on $x_1, \ldots, x_n$ and contains the gates $v_1, \ldots, v_k$ in this topological order. The output is computed at v_k. We introduce the connection variables c_{ij} where $1 \le i < n+k$, $1 \le j \le k$, and $i - n < j$. The variable c_{ij}, where $i \le n$, describes whether x_i is an input of v_j and the variable c_{ij}, where $i > n$, describes whether a wire leads from v_{i-n} to v_j. The variable y_j, $1 \le j \le k$, describes the output of gate v_j. The function $F(x, y, c)$ outputs 1 iff the c-variables describe a legal connection and y describes the values computed at the gates if the gate connections are described by c and the input is given by x. The function F can be described as the conjunction of the following conditions.

- If the fan-in of v_j is restricted by m, it is necessary that the negative threshold function $\overline{T}_{m+1,n+j-1}(c_{1j}, \ldots, c_{n+j-1,j})$ outputs 1.

- If v_j is an AND-gate, it is necessary that the value of y_j is equal to the conjunction of all $x_i c_{ij}$, $1 \le i \le n$, and all $y_l c_{n+l,j}$. For other types of gates, we get similar conditions.

We are only interested in the input-output behavior of the circuit. Let $G(x, y_k, c)$ output 1 iff the c-variables describe a legal connection and y_k is the value computed at v_k if the gate connections are described by c and the input is given by x. Then

$$G(x, y_k, c) = (\exists y_1, \ldots, y_{k-1}) F(x, y, c).$$

Let $f(x)$ be the function we want to compute with a redesigned circuit and let $g(x, y_k) = 1$ iff $f(x) = y_k$. Finally, the characteristic function of all gate connections c resulting in the representation of f is given by

$$h(c) = (\forall x, y_k)\big(g(x, y_k) \equiv G(x, y_k, c)\big).$$

We may use the known techniques to construct an OBDD representing h. Because of the quantification of a lot of variables, we are faced with the problem of an explosion of the OBDD size. To prevent this, techniques described in Section 13.2 like early quantification should be applied.

If $h \equiv 0$, no redesign is successful. Otherwise, we are interested in a redesign with the minimal number of reconnections. An edge activated by $c_{ij} = a \in \{0, 1\}$ is associated with the cost 1 if $c_{ij} = \overline{a}$ for the given design and with the cost 0 otherwise. Then we look for a shortest path from the source to the 1-sink which can be computed by one of the well-known shortest-path algorithms. This path describes a partial redesign. The variables not tested on this path may have an arbitrary value and we use the value realized in the given design.

14.7 Synchronizing Sequences

In Section 13.2, we have investigated sequential circuits or equivalently FSMs. The underlying FST structure is called strongly deterministic if the transition function is deterministic and if there is a unique initial state s_0. In order to work with such an FSM, we have to ensure that the system always starts in s_0. A computation may stop in some state s. Either there is an external reset forcing the machine to switch to state s_0 or we need a *synchronizing sequence* $x = (x_1, \ldots, x_k)$ of input vectors such that $\delta(s, (x_1, \ldots, x_k)) = s_0$ for all states $s \in S$. The problem is to decide whether an FSM has a *reset state* s_0 such that a synchronizing sequence forces the FSM to state s_0. We distinguish the problem where s_0 is fixed and we try to compute a corresponding synchronizing sequence x and the problem where we look for a pair (s_0, x) of a reset state and a synchronizing sequence.

As a simple example we consider a sequential adder where $S = \{0,1\}$ and the present state represents the last carry, $X = \{0,1\}^2$, the output function λ computes the sum bit $c \oplus a \oplus b$ from the carry c and the current input $(a, b) \in X$, and the transition function δ computes the new carry $c' = T_{2,3}(a, b, c)$ ($T_{2,3}$ is a threshold function). Both states can be used as reset states and the corresponding synchronizing sequences, namely $(0,0)$ for $s_0 = 0$ and $(1,1)$ for $s_0 = 1$, have length 1. Because of the interpretation of the FSM as a sequential adder, only the reset state $s_0 = 0$ is useful.

Algorithms working on an explicit representation of the FSM have been known for a long time. Pixley, Jeong, and Hachtel (1994) have applied results of automata theory and have reduced the problem of the computation of a synchronizing sequence and the decision problem whether a synchronizing sequence exists to an image computation problem (see Section 13.2). The image computation has to be performed for the product machine $M \times M$ which is the product of two disjoint copies of the given FSM M. We have seen in Section 13.2 that we are faced with the problem of exploding OBDD size.

Rho, Somenzi, and Pixley (1993) base their algorithm on the observation that most FSMs considered in applications have a synchronizing sequence and even a very short one. For $k = 1, 2, 3, \ldots$, they decide whether a synchronizing sequence of length k exists and in the positive case they compute an OBDD representing the set of all synchronizing sequences of length k. As in Section 13.2, we denote by $T(s, s', x)$ the characteristic function of the transition function which outputs 1 iff $\delta(s, x) = s'$. The characteristic function T_k of the transition function for inputs $x = (x_1, \ldots, x_k)$ of length k can be described by

$$\begin{aligned} T_k(s, s', x) \quad = \quad & (\exists s^{(2)}, \ldots, s^{(k-1)}) \\ & (T(s, s^{(2)}, x_1) \wedge T(s^{(2)}, s^{(3)}, x_2) \wedge \cdots \wedge T(s^{(k-1)}, s', x_k)). \end{aligned}$$

The function r_k is the characteristic function of all pairs of inputs $x = (x_1, \ldots, x_k)$ and states s such that s is a reset state and x a corresponding synchronizing sequence. Then

$$r_k(x,s) = \forall s' \, \big(T_k(s',s,x)\big).$$

If we are only interested in a synchronizing sequence for an arbitrary reset state, we obtain the characteristic function r_k^* defined by

$$r_k^*(x) = \exists s \, \big(r_k(x,s)\big).$$

Rho, Somenzi, and Pixley (1993) suggest using the following expansion before applying OBDDs. Let $\delta_i(s',x)$ denote the ith bit of $s = (s_1,\ldots,s_n) = \delta(s',x)$:

$$\begin{aligned} r_k^*(x) &= \exists s \, \forall s' \, \big(T_k(s',s,x)\big) \\ &= \big(\exists s_1,\ldots,s_n\big) \, \forall s' \bigwedge_{1\le i\le n} \big(s_i \equiv \delta_i(s',x)\big) \\ &= \big(\exists s_1,\ldots,s_n\big) \, \forall s' \bigwedge_{1\le i\le n} \big((s_i \wedge \delta_i(s',x)) + (\overline{s}_i \wedge \overline{\delta}_i(s',x))\big) \\ &= \bigwedge_{1\le i\le n} \exists s_i \, \forall s' \, \big((s_i \wedge \delta_i(s',x)) + (\overline{s}_i \wedge \overline{\delta}_i(s',x))\big) \\ &= \bigwedge_{1\le i\le n} \Big[\forall s' \, \big(\delta_i(s',x)\big) + \forall s' \, \big(\overline{\delta}_i(s',x)\big)\Big]. \end{aligned}$$

In the last step, we have applied the definition of the existential quantifier $(\exists s_i)\, f = f_{|s_i=1} + f_{|s_i=0}$. This simple trick eliminates the quantification of s.

14.8 Boolean Unification

The behavior of some systems can be described by systems of Boolean equations based on independent variables and dependent variables. Given an arbitrary assignment to $x = (x_1,\ldots,x_n)$, we look for an assignment to $y = (y_1,\ldots,y_m)$ to fulfill the set of equations $h_i(x,y) = h_i'(x,y)$, $1 \le i \le k$. The vector of functions $f(x) = \big(f_1(x),\ldots,f_m(x)\big)$ describes a solution of this problem if

$$h_i\big(a, f(a)\big) = h_i'\big(a, f(a)\big), \quad 1 \le i \le k,$$

holds for all $a \in \{0,1\}^n$. A set of equations can have many solutions. For applications, it is useful to obtain a representation of all solutions. This postpones the choice of a specific solution to a point of time where one knows criteria to distinguish which solution is "better" than another one. The vector of functions $F(x,u) = \big(F_1(x,u),\ldots,F_m(x,u)\big)$ describes all solutions of the problem if each solution $f(x) = \big(f_1(x),\ldots,f_m(x)\big)$ can be obtained from $F(x,u)$ by replacing the vector u of variables by appropriate constants and if each replacement of u by constants leads to a solution of the problem. We describe the construction of an OBDD representing such a function $F(x,u)$ where the length of $u=(u_1,\ldots,u_m)$ is equal to the length of $y=(y_1,\ldots,y_m)$ and $F_i(x,u)$ does not essentially depend on $u_1,\ldots,u_{i-1}$.

The first step of our algorithm is to replace the equations $h_i(x,y)=h'_i(x,y)$ with $h''_i(x,y):=h_i(x,y)\oplus h'_i(x,y)=0$ and then with

$$g(x,y):=h''_1(x,y)+\cdots+h''_k(x,y)=0.$$

Hence, we only have one equation whose right side equals 0. If $h_1,\ldots,h_k$, $h'_1,\ldots,h'_k$ are given by OBDDs, the OBDD for g may be obtained by a usual synthesis process.

Boolean unification (Büttner and Simonis (1987)) is a powerful tool to compute the solution of $g(x,y)=0$ by successive variable elimination. The following theorem contains the theoretical background.

Theorem 14.8.1. *Let $g(x_1,\ldots,x_n,y_1,\ldots,y_m)$ be a Boolean function, $g_0=g_{|y_1=0}$, and $g_1=g_{|y_1=1}$. Let $f(x)=(f_1(x),\ldots,f_m(x))\in B_{n,m}$.*

(*i*) $g_0\wedge g_1(x,f(x))=0$
$\Rightarrow\ g(x,g_0(x,f(x))+u\overline{g}_1(x,f(x)),f_2(x),\ldots,f_m(x))=0.$

(*ii*) $g(x,f(x))=0\ \Rightarrow\ \exists\alpha\in B_n\colon f_1(x)=g_0(x,f(x))+\alpha(x)\overline{g}_1(x,f(x)).$

Proof. For the first claim, we use the Shannon decomposition of g with respect to $y_1=g_0(x,f(x))+u\overline{g}_1(x,f(x))$. Our aim is to prove that the equations $\overline{y}_1g_0(x,0,f_2(x),\ldots,f_m(x))=0$ and $y_1g_1(x,0,f_2(x),\ldots,f_m(x))=0$ hold. Since g_0 and g_1 do not essentially depend on y_1, this is equivalent to

$$[\neg(g_0(x,f(x))+u\overline{g}_1(x,f(x)))]\wedge g_0(x,f(x))=0$$

and

$$(g_0(x,f(x))+u\overline{g}_1(x,f(x)))\wedge g_1(x,f(x))=0.$$

The first equality follows by an application of de Morgan's law. In the second expression, we obtain on the left-hand side $g_0(x,f(x))g_1(x,f(x))$, which is equal to 0 by assumption, and $u\overline{g}_1(x,f(x))g_1(x,f(x))$, which obviously is equal to 0.

For the second claim, let

$$\alpha(x):=g_0(x,f(x))+f_1(x).$$

We claim that

$$f_1(x)=g_0(x,f(x))+g_0(x,f(x))\overline{g}_1(x,f(x))+f_1(x)\overline{g}_1(x,f(x)).$$

The assumption $g(x,f(x))=0$ can be rewritten as

$$\overline{f}_1(x)g_0(x,f(x))+f_1(x)g_1(x,f(x))=0.$$

If $f_1(x)=1$, this implies $g_1(x,f(x))=0$ and $f_1(x)\overline{g}_1(x,f(x))=1$. Hence, the claim is fulfilled. Now we assume that $f_1(x)=0$ which implies $g_0(x,f(x))=0$. Again, the claim is fulfilled. □

Theorem 14.8.1 shows how we can obtain the set of all solutions for $g(x,y) = 0$ from the set of all solutions for $g_0g_1(x,y) = 0$. This leads to a recursive algorithm, since g_0g_1 cannot depend essentially on y_1. Let $F_i(x,u)$, $2 \le i \le m$, describe the set of all solutions of $g_0g_1(x,y) = 0$ where F_i does not essentially depend on $u_1, \dots, u_{i-1}$. Together with

$$\begin{aligned} F_1(x_1,\dots,x_n,u_1,\dots,u_m) :=& g_0(x_1,\dots,x_n,0,F_2(x,u),\dots,F_m(x,u)) \\ &+ u_1\overline{g}_1(x_1,\dots,x_n,0,F_2(x,u),\dots,F_m(x,u)), \end{aligned}$$

we obtain the set of all solutions of $g(x,y) = 0$. We still have to describe the terminal case of this recursive approach, namely an equation with one y-variable. The equation $g(x,y_1) = 0$ can be rewritten as

$$\overline{y}_1h_0(x) + y_1h_1(x) + h_2(x) = 0.$$

This equation has a solution iff $h_2 \equiv 0$ and $h_0h_1 \equiv 0$. In the positive case, the set of all solutions is described (according to Theorem 14.8.1) by $y_1 = h_0(x) + u_1\overline{h}_1(x)$.

This approach can be performed with OBDDs. We include new u-variables and the y-variables have to be replaced with those functions describing the set of all solutions.

Example 14.8.2. We consider the following set of equations:

$$\begin{aligned} y_1y_2 &= 0, \\ y_2 + x_1\overline{y}_1 &= x_2(\overline{x}_1\overline{x}_3 + x_1x_3). \end{aligned}$$

The reader may verify that the equations describe an abstraction of an RS-flip-flop. The variables y_1 and y_2 describe the R-wire and S-wire, respectively. The first equation describes the necessary condition that at least one of R and S has to be equal to 0. The second equation describes the new state of the RS-flip-flop depending on the old state x_1 and the inputs x_2 and x_3. This equation is equivalent to

$$(y_2 + x_1\overline{y}_1) \oplus x_2(\overline{x}_1\overline{x}_3 + x_1x_3) = 0.$$

Altogether, we have to solve the equation

$$g(x_1,x_2,x_3,y_1,y_2) = y_1y_2 + [(y_2 + x_1\overline{y}_1) \oplus x_2(\overline{x}_1\overline{x}_3 + x_1x_3)] = 0.$$

We obtain that

$$\begin{aligned} g_0(x,y) &= (y_2 + x_1) \oplus x_2(\overline{x}_1\overline{x}_3 + x_1x_3) \\ &= \overline{y}_2x_2\overline{x}_3 + y_2\overline{x}_2 + y_2\overline{x}_1x_3 + x_1\overline{x}_3 + x_1\overline{x}_2, \\ g_1(x,y) &= y_2 + [y_2 \oplus x_2(\overline{x}_1\overline{x}_3 + x_1x_3)] \\ &= y_2 + \overline{x}_1x_2\overline{x}_3 + x_1x_2x_3, \end{aligned}$$

implying that

$$\begin{aligned} g_0 g_1(x,y) &= y_2\overline{x}_2 + y_2\overline{x}_1x_3 + y_2x_1\overline{x}_3 + \overline{y}_2\overline{x}_1x_2\overline{x}_3 \\ &= \overline{y}_2h_0(x) + y_2h_1(x) + h_2(x), \end{aligned}$$

where $h_0(x) = \overline{x}_1x_2\overline{x}_3$, $h_1(x) = \overline{x}_2 + \overline{x}_1x_3 + x_1\overline{x}_3$, and $h_2(x) = 0$. Hence, $h_2 \equiv 0$ and also $h_0h_1 \equiv 0$. This implies that we obtain a description of all solutions by

$$F_2(x_1,x_2,x_3,u_1,u_2) = \overline{x}_1x_2\overline{x}_3 + u_2 \wedge [\neg(\overline{x}_2 + \overline{x}_1x_3 + x_1\overline{x}_3)] = \overline{x}_1x_2\overline{x}_3 + u_2x_1x_2x_3.$$

Applying Theorem 14.8.1, we obtain the function

$$F_1(x_1,x_2,x_3,u_1,u_2) = g_0\big(x_1,x_2,x_3,0,F_2(x,u)\big) + u_1\overline{g}_1\big(x_1,x_2,x_3,0,F_2(x,u)\big)$$

describing the set of all solutions. In this equation, we have to replace $F_2(x,u)$ with the solution computed above. Finally, we obtain

$$F_1(x_1,x_2,x_3,u_1,u_2) = x_1\overline{x}_2 + x_1\overline{x}_3 + u_1(\overline{x}_2 + \overline{x}_1x_3),$$

which in this example is independent of u_2.

Chapter 15

Applications in Optimization, Counting, and Genetic Programming

Representation types or data structures for Boolean functions with good algorithmic properties should be applicable to all types of problems concerned with Boolean functions. In this chapter, we present applications of BDDs in areas quite different from those discussed in Chapters 13 and 14. In Section 15.1, it is shown how a branch-and-bound based integer-programming solver may work with EVBDDs. A lot of very efficient algorithms for optimization problems on graphs are known. They work on explicit graph descriptions. If the graphs are very large and can only be described implicitly, BDD techniques are promising. In Section 15.2, an implicit network flow algorithm is presented. OBDDs may describe all solutions of a problem and then it is easy to determine the number of solutions. In Section 15.3, an OBDD-based computation of the number of knight's tours on an 8×8 chessboard is described. Finally, it is discussed in Section 15.4 how OBDDs can be used as a data structure for Boolean functions in genetic programming.

15.1 Integer Programming

Integer programming is one of the most important NP-equivalent optimization problems. Many problems have a direct representation as an integer-programming problem. The aim is to minimize (or maximize) a linear function under conditions given by linear inequalities and the condition that the variables have to take integer values. Without the last condition, we obtain an instance of linear programming which can be solved efficiently. Without loss of generality, we assume that the problem is given in the following form:

$$c_1x_1 + \cdots + c_nx_n \to min$$

where

$$a_{i1}x_1 + \cdots + a_{in}x_n + b_i \le 0, \quad 1 \le i \le m,$$

and

$$x_1, \ldots, x_n \in \mathbb{Z}.$$

We have no BDD variant which works with variables which may take arbitrary integer values. In most practical problems, we have bounds like $d_i \le x_i \le e_i$. Then we replace x_i with $x_i - d_i$ and obtain $0 \le x_i \le e_i - d_i$. If $e_i - d_i \le 2^k - 1$, we replace x_i with k Boolean variables $x_{i,0}, \ldots, x_{i,k-1}$ with the interpretation that $x_i = x_{i,0} + 2x_{i,1} + \cdots + 2^{k-1}x_{i,k-1}$. We have to add the inequality $x_i - (e_i - d_i) \le 0$. This leads to the special case of *binary programming* where the variables have to take Boolean values. We restrict ourselves to binary programming but use the more familiar notation integer programming.

First, we describe an MTBDD-based integer-programming solver. This algorithm will lead in most interesting cases to exponential-size representations but the main ideas are easy to describe. Afterwards, we discuss improvements.

The functions $f(x_1, \ldots, x_n) = c_1x_1 + \cdots + c_nx_n$ and $g_i(x_1, \ldots, x_n) = a_{i1}x_1 + \cdots + a_{in}x_n + b_i$, $1 \le i \le m$, are represented by MTBDDs. Then it is easy to obtain OBDDs for the functions h_i where $h_i(x_1, \ldots, x_n) = 1$ iff $g_i(x_1, \ldots, x_n) \le 0$. It is sufficient to replace sinks whose labels are positive by zeros and the other sinks by ones. The function $h = h_1 \wedge \cdots \wedge h_m$ describes the set of admissible inputs and an OBDD representing h can be computed by synthesis. The OBDD for h is interpreted as an MTBDD. In a last step, we apply the synthesis algorithm to the MTBDDs G_f and G_h representing f and h, resp., and the operator $\downarrow : \mathbb{Z} \times \mathbb{Z} \to \mathbb{Z} \cup \{\infty\}$ defined by

$$z_1 \downarrow z_2 = \begin{cases} z_1 & \text{if } z_2 = 1, \\ \infty & \text{if } z_2 = 0. \end{cases}$$

This leads to an extended MTBDD (the sink label ∞ is also allowed) describing the value of the goal function on the admissible inputs. We obtain an OBDD describing *all* optimal solutions by replacing the sink with the minimal label with 1 and all other sinks with 0.

The problem with this approach is that MTBDDs may need exponential size to represent linear functions, e.g., for $x_0 + 2x_1 + \cdots + 2^{n-1}x_{n-1}$. EVBDDs represent affine functions essentially depending on n variables with $n+1$ nodes and this holds for all variable orderings (Theorem 9.5.5). In Section 9.5, we have argued that MTBDDs can be understood as an unfolding of EVBDDs.

Lai, Pedram, and Vrudhula (1994) have the opposite point of view and describe EVBDDs as flattened MTBDDs. EVBDDs have the advantage that the first step, the representation of the linear goal function and the affine functions describing the constraints, does not cause problems.

Lai, Pedram, and Vrudhula (1994) describe an integer-programming solver working with EVBDDs. Using EVBDDs, we are faced with two problems. Given an EVBDD for an affine function g, we have to return an OBDD describing h where $h(x) = 1$ iff $g(x) \leq 0$. The other problem is the $\downarrow$-synthesis problem described above.

For both problems we provide the EVBDD nodes v for f and g_i, $1 \leq i \leq m$, with the additional information $\max(v)$ and $\min(v)$ describing the maximal and minimal, respectively, value of the function represented at v. This is a trivial task for affine functions.

Let G_g be such an EVBDD and let us consider the problem of computing an OBDD (with the same variable ordering) to represent h where $h(x) = 1$ iff $g(x) \leq 0$. Let e be an edge to v and let w_v be the total weight already seen. We obtain the following terminal cases. If $w_v + \max(v) \leq 0$, return the result 1. If $w_v + \min(v) > 0$, return the result 0. Moreover, we use a computed-table. If we cannot find the result by these checks, we perform recursive calls for the 0- and the 1-successor. For the 1-successor we have to add the weight on this edge to the weight already seen. After having obtained the results of both recursive calls, we can decide whether the OBDD node for the initial call (w_v, v) can be eliminated directly or whether it can be merged with an already constructed node (this information is contained in the unique-table). During all these computations, we manage the additive weights as described in Section 9.5. It is not guaranteed that this algorithm runs in polynomial time with respect to the size of G_g.

Let G_f be the EVBDD representing the goal function f and containing the additional min- and max-information and let G_h be the OBDD representing the set of admissible inputs. The EVBDD for $f \downarrow h$ often is too large to be represented. Therefore, we only try to compute the value of an optimal solution (it is easy to store the best solution ever found which, finally, is an optimal solution). The algorithm works with a local bound lb with the interpretation that starting in the described situation it is the goal to find an admissible input whose value is less than lb. We initialize $lb := \max(v)+1$ for the source v of G_f. Having found a solution with value $lb^* < lb$, we set $lb := lb^*$ and again look for a better solution. The situation is described as usual by a pair (v, w) of nodes v of G_f and w of G_h. This implies that some variables have been replaced with constants and that we have adapted the local bound correctly. Fixing $x_i = 1$, we have a contribution of c_i and the local bound of the subproblem is $lb - c_i$. Although we do not construct an EVBDD for $f \downarrow h$, we follow the idea of the corresponding synthesis algorithm. Since we do not create an EVBDD, we do not need a unique-table.

The computed-table contains tuples (v, w, bou, val) such that the integer *val* is the result of a call of the "synthesis algorithm" with the pair (v, w) of nodes and the local bound *bou*. If no solution whose value is smaller than *bou* has been found, *val* is set to *bou*. Otherwise, $val < bou$ and *val* is the value of an optimal solution of the considered subproblem. Such an entry of the computed-table can be used in the following way. If $val < lb$ and $val < bou$, we know that *val* is also the solution of our present problem (v, w, lb). If $val \geq lb$ and $val < bou$, we know that, at (v, w), it is not possible to obtain a solution with a better value than *lb*. If $lb \leq bou = val$, the same consequence is correct. Only if $lb > bou = val$, can we not stop. Starting at (v, w) we have looked for a solution whose value is less than *bou* and we know that such a solution does not exist. Now we start again at (v, w) but we look for a solution whose value is less than $lb > bou$ and such a solution is still possible.

Now we explain the modified synthesis algorithm. We have the following terminal cases:

- If w is the 0-sink of G_h, no input is admissible and we cannot find a solution with the desired properties.
- If $\min(v) \geq lb$, no input, whether it is admissible or not, has a value which is small enough, i.e., we cannot find a solution with the desired properties.
- If v is the sink of G_f, $\min(v) = 0$, since an EVBDD has only a 0-sink. If $\min(v) < lb$ and w is not the 0-sink, an optimal solution has the value 0 and is good enough.
- If w is the 1-sink of G_h, all inputs are admissible and an optimal solution has the value $\min(v)$. Either $\min(v) \geq lb$ (see above) or $\min(v) < lb$ and an optimal solution is good enough.

If we are not in a terminal case, we look for a solution in the computed-table (see above).

If we have not found a solution of our problem, we create in the usual way two subproblems. We solve the subproblem with the smaller min-value first (ties can be broken arbitrarily), since, in the positive case, we start the other subproblem with a smaller local bound and this may lead earlier to a negative answer. Finally, we have solved the considered subproblem. In any case, we store the result in the computed-table. If the result is better than the local bound, the local bound is updated. Afterwards, we look for solutions which are better than the new best-known solution. Altogether, we have obtained an EVBDD-based integer-programming solver.

This pure approach will often lead to large EVBDDs. Therefore, Lai, Pedram, and Vrudhula (1994) have integrated their algorithm into a branch-and-bound procedure. This procedure works with the following modules. Lower bounds are computed with the linear-programming relaxation where $x_i \in \{0, 1\}$

is replaced by $0 \le x_i \le 1$. The search strategy is a DFS strategy, i.e., one of the successors of the current node is always used for branching. The successor with the smaller lower bound is chosen first. For the branching, we select one variable x_i and fix it to 0 and 1, respectively. The branching rule follows the variable ordering chosen for the EVBDDs.

Finally, we discuss the integration of the EVBDD approach and the branch-and-bound procedure. We start with EVBDDs for the goal function f and the functions $g_1, \ldots, g_m$ describing the constraints. The user chooses two parameters n^* and s^*. Remember that $h_i(x) = 1$ iff $g_i(x) \le 0$. An EVBDD (or OBDD) for h_i is called the Boolean form for g_i. We convert a constraint into its Boolean form only if it essentially depends on at most n^* variables. Otherwise, we apply the branch-and-bound procedure which replaces variables by constants and reduces the number of variables in the EVBDDs. The conjunction of the Boolean forms of the constraints may lead to an increase of the size. Such conjunctions are only performed if the size of the considered OBDDs is bounded by c^*. Otherwise, the branch-and-bound procedure is used. Moreover, we do not perform the $\downarrow$-synthesis of EVBDDs for subfunctions of f and h. As described above, we only decide whether the subproblem contributes to a better solution and, in the positive case, we compute an optimal solution.

Putting all these ideas together, we obtain an integer-programming solver based on branch-and-bound which takes advantage of the representation of the goal function and the constraints by EVBDDs.

15.2 Network Flow

The network flow problem is a maximization problem on directed graphs $G = (V, E)$ with a source $s \in V$ (indegree 0) and a terminal $t \in V$ (outdegree 0). Given some *capacity constraint* function $c \colon E \to \mathbb{N}$, we look for a flow $f \colon E \to \mathbb{N}$ such that f respects the capacity constraint, i.e., $0 \le f(e) \le c(e)$ for all $e \in E$; f respects the *flow conservation constraint* at all vertices $v \in V - \{s, t\}$, i.e.,

$$\sum_{u|(u,v)\in E} f(u,v) = \sum_{w|(v,w)\in E} f(v,w);$$

and f has among all admissible flows the largest flow value defined by

$$val(f) = \sum_{v|(s,v)\in E} f(s,v).$$

Moreover, it is presupposed that $(y, x) \notin E$ if $(x, y) \in E$.

We assume that the reader is familiar with classical maxflow algorithms based on *augmenting paths* (see Cormen, Leiserson, and Rivest (1990)). The classical algorithms work on an explicit description of the graph, i.e., G is given by a list of its vertices and adjacency lists. There are well-known maxflow algorithms whose runtime can be bounded by $O(|V|^3)$ or $O(|V||E|\log|V|)$. Here we investigate the situation where the graph size does not allow an explicit representation. The vertices get n-bit binary numbers (allowing 2^n nodes) and the graph is represented by an OBDD for the function $E(x,y)$ which takes the value 1 iff there is an edge from the vertex with number x to the vertex with number y. As discussed for the reachability analysis problem in Section 13.2, one should use an interleaved variable ordering where x_i and y_i are tested one after the other. The source is described by the function $s(x)$, which only takes the value 1 if x equals the source, and the terminal by the function $t(x)$. We simplify the problem and assume that $c \equiv 1$. This special case is the so-called maximum flow problem for 0-1 networks.

Hachtel and Somenzi (1997) have presented an implicit network flow algorithm for 0-1 networks which works with OBDDs. Their algorithm is based on the $O(|V|^3)$ algorithm due to Malhotra, Pramodh Kumar, and Maheshwary (1978). The original runtime $O(|V|^3)$ is meaningless for an OBDD approach. We hope to obtain good runtimes for graphs which are regular enough to allow a small-size OBDD representation even for very large $|V|$ and $|E|$. Hachtel and Somenzi (1997) have computed a maximum flow for a graph with more than 10^{27} vertices and more than 10^{36} edges in less than one CPU minute. Such a result is only possible if the problem instance is a somewhat simple one.

The algorithm starts with the initial flow $F \equiv 0$ which is admissible. Then it constructs a layered network containing all shortest augmenting paths. This is done by a sequential computation of the layers and, therefore, the algorithm can only be efficient if during all phases the augmenting paths are short. If, e.g., the network consists of one directed path from the source through all vertices to the terminal and $|V|$ is large, the algorithm will not work efficiently. Since all capacities are equal to 1, we look for a maximal set of shortest augmenting paths and improve the flow with these augmenting paths. Afterwards, we start the next phase with the new flow. The algorithm stops if we do not find an augmenting path. The value of the maximum flow is equal to the number of edges leaving the source s and carrying the flow 1. We obtain this value easily by constructing an OBDD representing $F(x,y) \wedge s(x)$ and by applying the SAT-COUNT algorithm.

An augmenting path may contain forward edges (x,y) (more precisely $E(x,y)=1$) without flow and backward edges (y,x) with flow. Hence,

$$A(x,y) := (E(x,y) \wedge \overline{F}(x,y)) + (E(y,x) \wedge F(y,x)) = (E(x,y) \wedge \overline{F}(x,y)) + F(y,x)$$

describes the set of edges which can lie on augmenting paths. The equality holds, since $F(y,x)=1$ implies $E(y,x)=1$. We perform a reachability analysis with source s and the edge set described by $A(x,y)$ (see Section 13.2) until we find the terminal t. If we store the set of vertices found in the ith step, we obtain a layered network of all vertices reachable from s and whose distance from s is not larger than the distance from s to t. For the convenience of the reader, we explicitly describe a similar approach which is closer to classical maxflow algorithms. Let $\text{NEW}_0(x) = s(x)$, let $F(x, y)$ describe the current flow, and let $E(x, y)$ describe the edge set. If we have computed the layers $0, \ldots, m$, $\text{NEW}_i(x)$ describes the vertex set of the ith layer and $R_m(x) = \text{NEW}_0(x) + \cdots + \text{NEW}_m(x)$ describes the set of all already reached vertices. The algorithm stops if $R_m(x) \wedge t(x) \not\equiv 0$. Otherwise, the set of backward edges between layer m and layer $m+1$ is represented by

$$B_m(x, y) = \overline{R}_m(x) \wedge F(x, y) \wedge \text{NEW}_m(y),$$

which is a preimage computation. The set of forward edges between layer m and layer $m+1$ is represented by

$$F_m^*(x, y) = \text{NEW}_m(x) \wedge \overline{F}(x, y) \wedge E(x, y) \wedge \overline{R}_m(y),$$

which is an image computation. The edges between layer m and layer $m+1$ are represented by

$$U_m(x, y) = B_m(y, x) + F_m^*(x, y).$$

Using an interleaved variable ordering, it is easy to obtain $B_m(y, x)$ from $B_m(x, y)$. Finally,

$$\text{NEW}_{m+1}(x) = (\exists y)\, U_m(y, x).$$

We have constructed a layered network which is too large, since it is not guaranteed that t is reachable from each vertex in the network. Therefore, we perform a reachability analysis on the reversed network starting at t and eliminate all vertices and edges which are not found by this reachability analysis. Also for this purpose it is essential to work with an interleaved variable ordering. The resulting network is described by vertex and edge sets which we again denote by NEW_m and U_m. We have obtained the network containing exactly all vertices and edges lying on augmenting paths. The task is to find a maximal set of edge-disjoint s-t-paths in this network.

The idea of Malhotra, Pramodh Kumar, and Maheshwary (1978) is the construction of a *right-potent network*. The edge sets $U_m(x, y)$ are partitioned to the sets $S_m(x, y)$ of *selected* edges and the sets $R_m(x, y)$ of *remaining* edges. The partition is done in a way that for each vertex y there are at least as many selected edges starting at y as there are selected edges reaching y. Moreover, we want to have many selected edges. The right-potency property ensures that a flow reaching y via selected edges can leave y via selected edges.

For the construction of the right-potent network, we fix for each vertex x a complete ordering $\leq_x$ on the set of all vertices. All these orderings are represented by a priority function $\pi(x, y, z)$ which takes the value 1 iff $y <_x z$. A good priority function leads to the construction of many selected edges. Later, we present two possible priority functions and discuss their advantages. The right-potent network is constructed backward starting at t. Let t be the vertex of layer l. We select all edges from $U_{l-1}(x,y)$, i.e., $S_{l-1}(x,y) = U_{l-1}(x,y)$. In the general case, we know the set $S_m(y,z)$ of selected edges and the set $U_{m-1}(x,y)$. Now we are considering three layers and, therefore, variables x, y, and z for three vertices. Again, we use an interleaved variable ordering. In the first step, we represent by $P_m(x,y,z)$ all paths of length 2 from layer $m-1$ via layer m to layer $m+1$ using edges from $U_{m-1}(x,y)$ and $S_m(y,z)$, i.e.,

$$P_m(x,y,z) = U_{m-1}(x,y) \wedge S_m(y,z).$$

Then we describe by $P_m^*(x,z)$ whether such a path connects x and z, i.e.,

$$P_m^*(x,z) = (\exists y)\ P_m(x,y,z).$$

Our aim is to obtain a matching between the vertices of layer $m-1$ and layer $m+1$ based on the "edges" described by $P_m^*(x,z)$. For this purpose, we work with the chosen priority function. First, we choose for vertex x the first vertex z such that $P_m^*(x,z) = 1$, i.e.,

$$Q_m(x,z) = P_m^*(x,z) \wedge \left[\neg(\exists y)\ (P_m^*(x,y) \wedge \pi(x,y,z))\right].$$

It is still possible that z is reached by more than one edge. Hence, in a second step, we choose for vertex z the first vertex x such that $Q_m(x,z) = 1$, i.e.,

$$D_m(x,z) = Q_m(x,z) \wedge \left[\neg(\exists y)\ (Q_m(y,z) \wedge \pi(z,y,x))\right].$$

The function

$$T_m(x,y,z) = D_m(x,z) \wedge P_m(x,y,z)$$

describes edge-disjoint paths of length 2 whose second edges have been selected. We select all edges described by $(\exists z)T_m(x,y,z)$ and we try to select more edges after having removed the edges described by $(\exists z)T_m(x,y,z)$ from $U_m(x,y)$ and $(\exists x)T_m(x,y,z)$ from $S_{m+1}(y,z)$ (the edges are not really removed from $S_{m+1}(y,z)$, they are only removed for the computation of further edges to be included in $S_m(x,y)$). This process is continued until the D-set is empty. This construction ensures the right-potency property of the network of selected edges.

Finally, the network of selected edges is described by the sets $S_0, \ldots, S_{l-1}$. We compute edge-disjoint augmenting paths by selecting the edges from $S_0(x,y)$, i.e., we define $F_0(x,y) = S_0(x,y)$. The right-potency property ensures that there are edge-disjoint augmenting paths starting with the edges described by $F_0(x,y)$. If $F_{m-1}(x,y)$ has been computed, let

$$F_m(y,z) = (\exists x)\ F_{m-1}(x,y) \wedge T_m(x,y,z).$$

The function $F^+(x,y) = F_0(x,y) + \cdots + F_{l-1}(x,y)$ describes the edges on edge-disjoint augmenting paths. This set of augmenting paths is maximal with respect to the set of selected edges but not necessarily with respect to all edges of the layered network. Hence, we delete the edges described by $F^+(x,y)$ from the layered network. Afterwards, we delete all vertices no longer lying on a path from s to t. Then we try to select new edges but we do not start with empty sets. It is easy to see that all previously selected edges which are not removed can still serve as selected edges without disturbing the right-potency property. We repeat the whole process including an updating of $F^+(x,y)$ by adding the further augmenting paths. The process stops if a path from s to t does not remain in the layered network. The new flow $F_{\text{new}}(x,y)$ has the value 1 if (x,y) is a forward edge and $F^+(x,y)=1$ or if (y,x) is a backward edge and $F^+(x,y)=0$ or if (x,y) is not a forward edge, (y,x) is not a backward edge, and $F(x,y)=1$. This can be easily expressed by

$$F_{\text{new}}(x,y) = \left(F^*(x,y) \wedge F^+(x,y)\right) + \left(B(y,x) \wedge \overline{F}^+(x,y)\right) + \left(\overline{F}^*(x,y) \wedge \overline{B}(y,x) \wedge F(x,y)\right)$$

if $F^*(x,y)$ describes all forward edges and $B(y,x)$ all backward edges.

We repeat the whole process with the new flow $F(x,y) = F_{\text{new}}(x,y)$. We know from the theory on network flow algorithms that the shortest augmenting paths in this next phase are longer than in the previous phase.

We still have to discuss how to choose an appropriate priority function. The priority function should have a small OBDD size for the interleaved variable ordering $(x_{n-1}, y_{n-1}, z_{n-1}, \ldots, x_0, y_0, z_0)$. The datum proximity checks whether the binary number represented by y is smaller than the binary number represented by z (and it is independent of x). Its OBDD size is linear but the same vertices always have a high priority. This implies that we often select only a "few" edges. The relative proximity checks whether $\|y-x\| < \|z-x\|$, where

$$\|y-x\| = \sum_{0\le i\le n-1} |y_i - x_i| 2^i.$$

This priority function also has linear OBDD size, although its OBDD size is larger than the OBDD size for datum proximity. Nevertheless, experiments have shown that the relative proximity function leads to a faster algorithm. The reason is that it tends to select more edges.

This implicit network flow algorithm can serve as a prime example for an implicit graph algorithm. It is not the main purpose to beat explicit algorithms on graphs which can be represented explicitly. Implicit algorithms make it possible to solve problems which cannot be solved explicitly, since they cannot be represented explicitly in reasonable time and space. Experiments with the implicit network flow algorithm have shown that it has this property. Implicit representations save a lot of space. Algorithms on implicit representations are implicitly parallel, since vertices or edges are treated simultaneously if the OBDD representation has nodes which are used for the common representation of these vertices or edges. This also shows the drawback of the implicit network flow algorithm. It contains a lot of inherently sequential parts. The number of phases is only bounded by the number $|V|$ of vertices. It is only possible to perform a quite limited number of phases. If, e.g., $|V| \approx 10^{27}$, the number of phases has to be much smaller than $|V|$. Moreover, the layered network has a depth which equals the length of a shortest augmenting path. The layers of this network are computed sequentially. This implies that even a single phase with long shortest augmenting paths cannot be performed efficiently.

15.3 Counting Problems

For OBDDs, ZBDDs, and FBDDs, the number of satisfying inputs can be computed in linear time. These algorithms for the SAT-COUNT problem are based on simple graph traversals. Hence, counting problems can be solved by representing the set of all solutions by, e.g., an OBDD and by applying a SAT-COUNT algorithm.

Minato (1994, 1997) and Semba and Yajima (1994) have solved counting problems (like the n-queens problem for $n \leq 13$) with OBDDs and ZBDDs. They have not obtained new results and their BDD-based counting algorithms are either not essentially faster than backtracking algorithms or the results can also be obtained by known combinatorial formulas.

Löbbing and Wegener (1996) have shown that BDD techniques can support the more efficient solution of counting problems. They have demonstrated this for classical combinatorial chess problems. The moves of a knight on a chessboard (always the classical 8×8 chessboard with its usual partition into white and black squares) are not as easy to follow as for a castle or a bishop. Hence, *knight's tours*, i.e., Hamiltonian circuits on the *knight's graph* whose vertices represent the squares of the chessboard and whose edges represent moves of a knight (see Fig. 15.3.1) have fascinated mathematicians like Euler, Legendre, and Vandermonde. It has been known for a long time that a lot of different knight's tours exist but their exact number has been unknown. One might suspect that this number cannot be derived from a general formula on the number of knight's tours on $n \times n$ chessboards. The number of knight's tours (see below) is too large to allow an explicit enumeration. Hence, it seems to be necessary to take advantage of isomorphic situations and one has to recognize situations which cannot lead to knight's tours. OBDDs support these two objectives. Partial assignments which cannot be completed to a satisfying input are represented by a single 0-sink and partial assignments leading to isomorphic situations are represented by the same node of reduced OBDDs. Nevertheless, a direct application of OBDD methods to compute the number of knight's tours seems to be hopeless, since the OBDD size explodes.

First, we solve an easier problem and determine the number of coverings of the directed knight's graph by disjoint cycles. This problem can easily be reduced to the problem of computing the number α of one-to-one mappings from the white squares to the black squares where it is only allowed to map a white square w to a black square reachable by a knight's move from w. A knight's move always leads from a white square to a black one and vice versa. Because of symmetry, α is also the corresponding number of one-to-one mappings from the black to the white squares. A cycle covering is a one-to-one mapping on the set of squares of the chessboard where squares are mapped to squares reachable by a knight's move. Hence, each cycle covering consists of a one-to-one mapping from the white to the black squares and a one-to-one mapping from the black to the white squares. Each such pair of one-to-one mappings describes a cycle covering implying that the number of cycle coverings equals α^2. The number of possible destinations of a knight's move from a given square is 2, 3, 4, 6, or 8 and the possibilities can be described by 1, 2, or 3 Boolean variables. We work on variables describing moves starting at the white squares and we design a circuit testing whether each black square is reached at least once. Using a rowwise variable ordering, this circuit is translated into an OBDD representing the set of one-to-one mappings. Applying the SAT-count algorithm on the final OBDD, we obtain that $\alpha = 2\ 849\ 759\ 680$ and the number of directed cycle coverings $\alpha^2 = 8\ 121\ 130\ 233\ 753\ 702\ 400$. The final OBDD is of size 598 972. An approach with ZBDDs is a little faster and the final ZBDD has 406 660 nodes. The parameter α can also be computed by backtracking algorithms. Our most clever backtracking algorithm approach was by a factor of more than 6000 slower than the ZBDD approach.

It is much harder to check the property that the chessboard is covered by a single cycle. This is also known from an integer-programming approach to solve the TSP. The number of constraints to ensure that the graph is covered by disjoint cycles is small and these constraints lead to the relaxation called the assignment problem. This relaxation can be solved in polynomial time. The number of constraints to ensure that the graph is covered by a single cycle is much larger. Analogously, counting knight's tours is much harder than counting cycle coverings.

The solution of Löbbing and Wegener (1996) is based on divide-and-conquer, BDD techniques, and backtracking. The divide-and-conquer strategy is illustrated in Fig. 15.3.1.

The rows 1, 2, and 3 of the chessboard are denoted as low part L, the rows 4 and 5 are denoted as middle part M, and the rows 6, 7, and 8 as upper part U. We consider an "overlapping partition" of the chessboard into the lower "half board" LM consisting of L and M and the upper half board UM. The moves of a knight's tour are partitioned into moves belonging to LM and moves belonging to UM. A move starting at a square of L or reaching a square of L belongs to LM, similarly for U and UM. In order to obtain a unique partition, we define

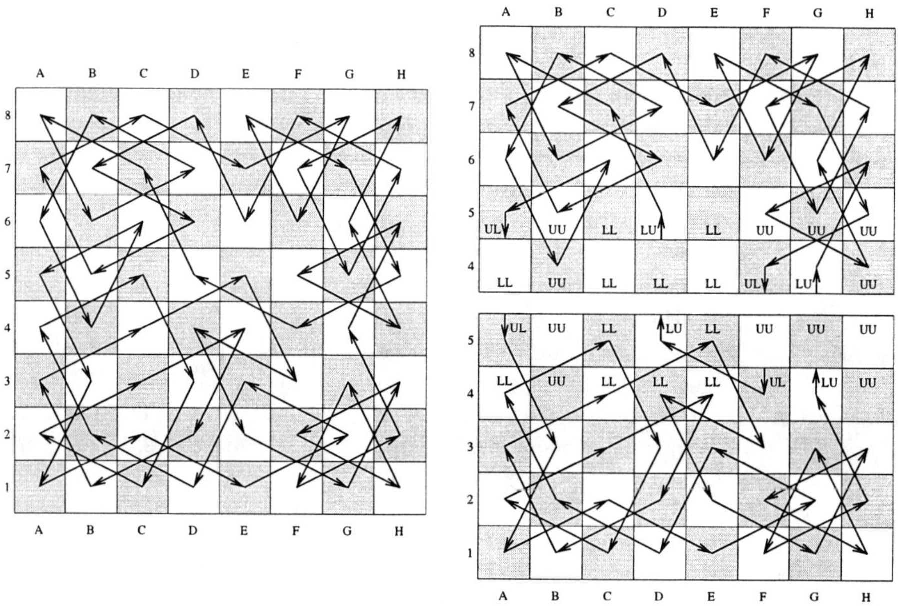

Figure 15.3.1: *A knight's tour on a chessboard and a partitioned chessboard.*

that a move from row 4 to row 5 belongs to LM and a move from row 5 to row 4 belongs to UM.

For a fixed knight's tour, we partition the set of squares of the chessboard into the sets LL, UL, LU, and UU. A square belongs to UL if it is reached by a move belonging to UM and left by a move belonging to LM, the other sets LL, LU, and UU are defined similarly. The squares of the rows 1, 2, and 3 always belong to LL while the squares of the rows 6, 7, and 8 belong to UU. We are left with 4^{16} classifications of the squares of the rows 4 and 5. It is not necessary to consider all these cases. We only consider the cases with at most 8 squares (of the rows 4 and 5) in LL. In order to correct this mistake we double the number of knight's tours obtained in those cases where UU contains at least 9 squares. Moreover, $1 \leq |UL| = |LU| \leq 8$, since we have to leave the half-boards and we reach a half-board on a knight's tour as often as we leave it. Altogether, our divide-and-conquer approach leads to

$$\binom{16}{1}^2 + \binom{16}{2}^2 + \cdots + \binom{16}{8}^2 = 383\,358\,644$$

cases. For each $i \in \{1, \ldots, 8\}$, we choose all pairs (A, B) of subsets of M of size i. Such a pair describes the following classification of the squares of M. Squares in $A \cap B$ belong to LL, squares in $A - A \cap B$ to UL, squares in $B - A \cap B$ to UL, and the remaining squares to UU.

Because of the huge number of cases, each case has to be handled very efficiently. Before describing the counting of knight's tours for a single case, we note that the divide-and-conquer strategy has some similarity with the concept of partitioned OBDDs.

In the following, we assume a fixed partition of M into LL, UL, LU, and UU, where $|\text{UL}| = |\text{LU}| \geq 1$ and $|\text{LL}| \leq 8$. We only consider the squares of the lower half-board LM and construct an OBDD which tests whether the moves starting at squares from LL $\cup$ UL (including the first three rows) reach different squares and whether they reach squares from LL $\cup$ LU. Each satisfying input describes a system of disjoint paths from squares of UL to squares of LU and perhaps some disjoint cycles. Now backtracking on the OBDD is used to determine the satisfying inputs describing cycle-free disjoint paths. Such a path system defines a function $f\colon \text{UL} \to \text{LU}$ where each source of a path is mapped to its terminal. Let $\#_1(\text{LL}, \text{UL}, \text{LU}, \text{UU}, f)$ be the number of cycle-free disjoint path systems describing the same function f. By symmetry, we also obtain the number $\#_2(\text{LL}, \text{UL}, \text{LU}, \text{UU}, g)$ of cycle-free disjoint path systems belonging to the function $g : \text{LU} \to \text{UL}$ mapping the sources of the paths in UM to their terminals. In our example,

$$f(\text{A5}) = \text{G4} \quad \text{and} \quad f(\text{F4}) = \text{D5}$$

and

$$g(\text{G4}) = \text{F4} \quad \text{and} \quad g(\text{D5}) = \text{A5}.$$

This leads to a valid pair (f, g) of functions, since (f, g) describes one cycle A5→G4→F4→D5→A5 on UL $\cup$ LU. The same function f together with g' defined by

$$g'(\text{G4}) = \text{A5} \quad \text{and} \quad g'(\text{D5}) = \text{F4}$$

leads to an invalid pair (f, g') describing two cycles A5→G4→A5 and F4→D5→F4. Fixing a valid pair, we obtain the number of corresponding knight's tours as the product of the numbers $\#_1(\text{LL}, \text{UL}, \text{LU}, \text{UU}, f)$ and $\#_2(\text{LL}, \text{UL}, \text{LU}, \text{UU}, g)$. We have to take the sum over all valid pairs (f, g) and over all cases (LL, UL, LU, UU) (the number of solutions for the cases where $|\text{UU}| \geq 9$ has to be doubled) in order to obtain the number of directed knight's tours. The number of undirected knight's tours is half the number of directed ones and equals 13 267 364 410 532.

This is an example where it is advantageous to work with MDDs (see Section 9.1). If there are, e.g., six possibilities to leave a square, it is better to work with a six-valued variable than with three Boolean variables.

15.4 Genetic Programming

Evolutionary and genetic algorithms are heuristics for the computation of good solutions for optimization problems. In Section 5.8, we have used these techniques for a heuristic solution of the variable-ordering problem. The automatic creation of computer programs by means of evolution is called *genetic programming.* One should not hope to obtain better algorithms for well-defined problems like sorting or the maximum network flow problem by genetic programming but many practical problems do not have such a clear structure.

A typical problem from machine learning is that some unknown procedure, e.g., nature, a program, or a black box, works in the background and we observe its input-output behavior on some inputs which perhaps are randomly chosen. We want to understand the unknown procedure. Since we only observe pairs of inputs and corresponding outputs, there is no hope to understand how the procedure really works but we may try to predict the output of the procedure for further inputs. It is easy to see that this is hopeless without further assumptions. Machine learning and genetic programming are based on the hypothesis that the unknown procedure is somewhat simple. We try to explain the observed data, denoted as training data, by a simple hypothesis which has to be taken from some given class called the concept class. Koza (1992) was the first to attack such problems with genetic programming.

We only investigate the case that the unknown function is a Boolean function $f\colon \{0,1\}^n \to \{0,1\}$ and that we know a set of training examples $(a, f(a))$, $a \in S \subseteq \{0,1\}^n$. The elements $a \in S$ are random elements from $\{0,1\}^n$ and S is small enough to observe $|S|$ experiments. In theory, $|S|$ is polynomially bounded with respect to n. In order to apply genetic programming, we have to design among others the following modules:

- An algorithm to create a population of random Boolean functions drawn from some subset of all Boolean functions,
- an algorithm to randomly mutate a Boolean function,
- an algorithm to compute the result of a random recombination of two or more Boolean functions,
- a fitness function which, for a Boolean function and a set of training examples, determines the value or fitness of this function with respect to the training data,
- an algorithm to choose deterministically or randomly the members of the next generation.

Moreover, we need a data structure which supports all these operations and leads to a succinct representation of many functions.

Since the aim is to produce a representation of a function $g\colon \{0,1\}^n \to \{0,1\}$ such that the chance that $g(b) = f(b)$ for a random input b is high, it is not necessary that g agree with f on the set S of training examples. This might even lead to bad generalizations g. Let us consider the example where the concept class consists of all polynomials $g\colon [0,1] \to \mathbb{R}$ whose degree is bounded by d and the unknown function is $f(x) = \sin x$. If we have $d+1$ examples, there is exactly one function in our concept class which is correct on the training examples. This is typically a polynomial with complicated coefficients which differs a lot from f. A polynomial of degree 2 is perhaps a much better generalization. If the number of training examples is larger than $d+1$, there is, with high probability, no polynomial of degree d which interpolates the training examples. In the case of Boolean functions, all representation types have the property that all Boolean functions can be represented. Our aim is to find a small-size representation of a function g which agrees with f on all (or at least most) elements $a \in S$. This aim is based on Occam's razor theorem (Blumer, Ehrenfeucht, Haussler, and Warmuth (1987)).

Theorem 15.4.1. *Let H be a set of Boolean functions $h\colon \{0,1\}^n \to \{0,1\}$ and let $f\colon \{0,1\}^n \to \{0,1\}$. Let S be a random subset of $\{0,1\}^n$ of size s. The probability that there exists a function $g \in H$ such that $g(a) = f(a)$ for all $a \in S$ but $Prob(g(b) = f(b)) \le \frac{1}{2} + \varepsilon$ for a random $b \in \{0,1\}^n$ is bounded above by $|H|(\frac{1}{2} + \varepsilon)^s$.*

For a given representation type, we choose H_m as the set of functions whose description length is smaller than m. Using canonical descriptions, this implies that H_m contains the functions with a small-size representation. If $m \ll s$, $2^m(\frac{1}{2} + \varepsilon)^s$ is close to 0, even for ε close to $\frac{1}{2}$. This result implies the following. If we are able to obtain a good *compression* of the training data, namely a representation with m bits instead of s, it is very unlikely that this function is very different from the unknown one.

Koza (1992) has discussed genetic programming for many types of problems and has worked with so-called S-expressions. In the case of Boolean functions, S-expressions are quite general formulas, i.e., the set of possible gate operations is large. Nevertheless, S-expressions are trees and, if a subformula is used twice, it has to be represented twice. Already Koza (1994) has recognized that graph representations can be much more succinct. He has introduced automatically defined functions (ADFs) to overcome this difficulty. For the special case of Boolean functions, Sakanashi, Higuchi, Iba, and Kakazu (1996) have applied DTs and ordered DTs because of the simplicity of the operation at the nodes. OBDDs combine the simplicity of a single operation at the inner nodes with the advantages of graph representations. Moreover, the efficiency of operations on OBDDs should support the design of genetic operators. Yanagiya (1995) and Droste (1997) were the first to design genetic-programming algorithms based on

OBDDs. It is no surprise that OBDDs outperform tree representations. The connection between genetic programming and Occam's razor theorem has been described by Droste (1998).

All these early approaches are based on some clever but nevertheless ad hoc ideas for the design of the genetic operations. Droste and Wiesmann (1998) have discussed the relations between the chosen representation type and the way genetic operators guide the search for a good representation. This has led to formal requirements for genetic operators. We do not go into the details of these general guidelines but we present genetic operators which fulfill the requirements of Droste and Wiesmann (1998).

The following genetic-programming system is based on π-OBDDs with a fixed variable ordering π. The user has to choose a size bound s such that no OBDD with more than s nodes is accepted as a member of the population, the population size μ, the number λ of children created in one generation, the number r of rounds or generations, and some numbers α and β where $0 < \alpha < 1$ and $\beta > 0$. In the beginning, μ random OBDDs of size at most s are created. Nobody has been able to describe an efficient implementation of this step. Therefore, we have to use an efficient algorithm which is a heuristic approximation of the idea. E.g., we may restrict the width of the OBDDs by $w = \lfloor s/n \rfloor$ and start with w nodes on each level and two sinks. Each edge leaving a node on level i leads to a random node on level $i + 1$. The first node on level 1 is chosen as the source and the reduction algorithm is applied to the resulting OBDD. Taking into account that the first and last levels of a reduced OBDD have a limited size, we can be a bit more careful and can increase w in an appropriate way.

The random mutation of an OBDD G is performed in the following way. The random variable M determines the number of inputs such that the value of the new OBDD G^* differs from the value of G. The probability that $M = k$ is set to $\alpha(1-\alpha)^k$ for $1 \le k \le 2^n$. With the remaining probability of $\alpha + (1-\alpha)^{2^n+1}$, $M = 0$. Typically, M is small and it is possible to choose M different random inputs and to create an OBDD G_{mut} which takes the value 1 on the chosen inputs. The result G^* of the mutation is the result of an EXOR-synthesis of G and G_{mut}. This mutation operator has the property that it may change the function g represented by G into each function g' but large changes are less probable than small changes.

Recombination is defined for two parents G_1 and G_2 representing g_1 and g_2, respectively. The child g_3 represented by G_3 should lie "between" g_1 and g_2, i.e., if $g_1(a) = g_2(a)$, also $g_1(a) = g_3(a)$, and both parents should have the same random influence, i.e., if $g_1(a) \ne g_2(a)$, $g_3(a)$ takes a random value. This is realized by the following algorithm whose worst-case runtime is exponential. Typical experiments with genetic experiments work with at most 20 variables and this allows exponential algorithms. The computation of G_3 is performed on an SBDD for G_1 and G_2 and resembles a synthesis algorithm *without* computed-table. We start with the pair (v_1, v_2) of sources of G_1 and G_2. There are two terminal cases. If $v_1 = v_2$, return v_1, since the corresponding subfunctions of g_1 and g_2 agree. If v_1 and v_2 both are sinks, return one of them with equal probability. In all other cases, we follow the recursive synthesis algorithm. Only the abandonment of the computed-table ensures that different inputs are handled independently. This algorithm may be described in a different but even less efficient way. Create an OBDD G_{ran} for a random Boolean function g_{ran} and compute an OBDD for $g_{\text{ran}} \wedge (g_1 \oplus g_2) + g_1 g_2$.This approach can be generalized to a larger number n of variables by choosing g_{ran} only as a "pseudorandom" Boolean function created in the same way as the members of the initial population. The children are produced independently by randomly choosing two parents and performing recombination and, afterwards, mutation.

The fitness function has two components. The aim is to obtain a small OBDD which works correctly on many of the training examples. Let G be an OBDD, $|G|$ its size, and $t(G)$ the number of training examples where G produces the same result as f. The fitness of G is defined by $t(G) - \beta|G|$ if $|G| \leq s$ and $-\infty$ otherwise.

The μ OBDDs with the highest fitness among the μ OBDDs of the last generation and the λ newly created OBDDs are chosen as members of the new generation. Ties are broken arbitrarily. The whole process is repeated for r generations and, finally, the fittest OBDD is presented as a result.

Experiments have shown that such a genetic-programming system is superior to genetic-programming systems not based on OBDDs. The most often used Boolean benchmark functions are the parity function and the multiplexer or direct storage access function. All OBDD-based experiments with the multiplexer MUX work with an optimal variable ordering where all control variables are tested before the data variables. We know that the multiplexer is almost ugly (Theorem 5.3.3), i.e., almost all variable orderings lead to exponential OBDD size. It is still possible that we find for many variable orderings good approximations of the multiplexer with small OBDD size, which would imply that we have a good chance of obtaining a small OBDD which agrees with the multiplexer on the random training set. Let us consider the equality test EQ_n testing whether $a \in \{0,1\}^n$ and $b \in \{0,1\}^n$ are identical. This function has linear OBDD size for interleaved variable orderings but is almost ugly. The function EQ_n takes the value 1 only on 2^n of the 2^{2n} inputs and we have a good chance that our training set contains only inputs mapped to 0. Such a training set can be interpolated by the 0-sink. Hence, it is a question whether the variable-ordering problem is important for OBDD-based genetic programming. In particular, it is interesting to investigate whether an OBDD-based genetic-programming system with a random variable ordering has a good chance of being successful for the multiplexer. Krause, Savický, and Wegener (1999) have answered this question negatively. From a complexity theoretical point of view, we have to prove lower bounds not only on the OBDD size of a function f but also on the OBDD size of all functions which are close to f. To make the notion of closeness precise, we define c-approximations of f.

Definition 15.4.2. A function $g \in B_n$ is a *c-approximation* of $f \in B_n$ if the probability that $f(a) = g(a)$ for a random input $a \in \{0,1\}^n$ is at least c.

Each function has a trivial 1/2-approximation, namely one of the two constants. Hence, we are interested in $(1/2 + \varepsilon)$-approximations for $\varepsilon > 0$. The result of Krause, Savický, and Wegener (1999) is the following.

Theorem 15.4.3. *Let $0 < \delta < \varepsilon$. For every large enough n, the following property holds for a fraction of at least $1 - n^{-2\varepsilon^2/\ln 2}$ of the variable orderings π for DSA_n. Each function which is a $(\frac{1}{2} + \varepsilon + n^{-(\varepsilon-\delta)/2})$-approximation of DSA_n has a π-OBDD size which is bounded below by e^{n^δ}.*

Combining this result with Occam's razor theorem, it is not too difficult to prove the following theorem.

Theorem 15.4.4. *The following statement holds for large n. If we have a random set of a polynomial number $m = m(n)$ of training examples for the direct storage access function DSA_n and a random variable ordering π with probability at least $1 - n^{-1/2}$, there is no π-OBDD of size $\frac{1}{10} m \log^{-1} m$ representing a function which equals DSA_n on the training set.*

Since the OBDD size of DSA_n equals $2n+1$, a good compression of the training examples should have almost linear size. The theorem states the following. If $m \gg n \log n$, we need a good variable ordering to obtain a good compression of the training examples. Hence, OBDD-based genetic-programming systems have to start with random variable orderings and also have to choose a good variable ordering in the evolutionary process. Such a system should be developed in the future. Here we prove Theorem 15.4.3, since it is a new type of lower bound result and since the proof uses methods from information theory in a nice way. It is easier to obtain an even better result for the inner product function IP_n but the result on DSA_n is of special interest, since DSA_n is the main example in genetic programming. The proof also uses one-round communication complexity. In order to obtain the proposed bound, Alice obtains the first $n' := \lfloor (1 - 2\varepsilon)n \rfloor$ of the $n + k$ variables with respect to π and Bob obtains the other variables. First, we prove that Alice gets with high probability not too many address variables. For DSA_n, we consider as usual the $k = \log n$ address variables $a = (a_{k-1}, \ldots, a_0)$ and the n data variables $x = (x_0, \ldots, x_{n-1})$.

Lemma 15.4.5. *With probability at least $1 - n^{-2\varepsilon^2/\ln 2}$, Alice obtains at most $(1 - \varepsilon)k$ address variables.*

Proof. The random variable ordering can be produced as follows. We take the k address variables and randomly choose for them one after another a free position among the $n+k$ possible positions. Then we continue in the same way with the data variables, i.e., the x-variables. During the first k steps of this process, there are always at most $(1-2\varepsilon)n$ free positions among the first $n' = \lfloor (1-2\varepsilon)n \rfloor$ positions and at least n open positions at all. Hence, the probability of each address variable to be given to Alice is at most $1-2\varepsilon$. We can upper bound the probability that Alice gets more than $(1-\varepsilon)k$ address variables by the probability of at least $(1-\varepsilon)k$ successes in k independent Bernoulli trials with success probability $1-2\varepsilon$.

The expected number of successes $\mathrm{E}(Z)$ equals $(1-2\varepsilon)k$. By Chernoff's bound, we obtain

$$\mathrm{Prob}\bigl(Z \geq (1-\varepsilon)k\bigr) = \mathrm{Prob}\bigl(Z \geq \mathrm{E}(Z) + \varepsilon k\bigr) \leq e^{-2\varepsilon^2 k} = n^{-2\varepsilon^2/\ln 2}. \qquad \square$$

In the following, we fix a variable ordering π where Alice gets at most $(1-\varepsilon)k$ address variables. She also gets at least $\lfloor (1-2\varepsilon)n \rfloor - k$ data variables. If Alice's address variables are fixed, there are at least n^{ε} data variables left which may describe the output. On the average, at least $(1-2\varepsilon)n^{\varepsilon} - o(1)$ of these variables are given to Alice. In order to enable Bob to compute the output exactly, Alice has to send him the value of her address variables and those data variables which can describe the output. If the information given from Alice is much smaller than this, Bob can compute the value of DSA_n only with a probability close to $1/2$. The information given from Alice to Bob is measured by the logarithm of the size of a π-OBDD computing the function DSA_n.

For a rigorous argument, let π be a variable ordering and let A (resp. B) be the corresponding set of address variables given to Alice (resp. Bob) and let X (resp. Y) be the corresponding set of data variables given to Alice (resp. Bob). Clearly, $|A \cup X| = n'$ and every computation in any π-OBDD reads first (some of) the variables in $A \cup X$ and then (some of) the variables in $B \cup Y$. Let g be a function represented by a π-OBDD G of size s. Because of the definition of c-approximations, we consider random inputs $(\tilde{a}, \tilde{b}, \tilde{x}, \tilde{y})$ where $\tilde{a}$ is a random setting of the variables in A, etc. In this situation, the following holds.

Lemma 15.4.6.

$$\mathit{Prob}\bigl(\mathit{DSA}(\tilde{a}, \tilde{b}, \tilde{x}, \tilde{y}) = g(\tilde{a}, \tilde{b}, \tilde{x}, \tilde{y})\bigr) \;\leq\; 1 - \frac{|X|}{2n} + \frac{1}{2n}\bigl(2 \cdot 2^{|A|}|X| \ln s\bigr)^{1/2}.$$

This lemma easily implies Theorem 15.4.3.

Proof of Theorem 15.4.3. Recall that $k = \log n$ and let $s < e^{n^{\delta}}$. For every variable ordering π, we have $(1-2\varepsilon)n - k - 1 \leq |X| \leq n$. Moreover, Lemma 15.4.5 implies that, with probability at least $1 - n^{-2\varepsilon^2}$, we have $|A| \leq (1-\varepsilon)k$. By substituting these estimates into the bound from Lemma 15.4.6, we conclude that the probability that DSA_n and g have the same value is at most $\frac{1}{2}+\varepsilon+\frac{1}{\sqrt{2}}n^{-(\varepsilon-\delta)/2}+\frac{1+\log n}{n}<\frac{1}{2}+\varepsilon+n^{-(\varepsilon-\delta)/2}$. This implies the theorem. □

To prove Lemma 15.4.6, we apply well-known inequalities on the entropy of random variables. The entropy $H(U)$ of a random variable U taking values $u\in\mathcal{U}$ is defined by

$$H(U)=-\sum_{u\in\mathcal{U}}\mathrm{Prob}(U=u)\log\mathrm{Prob}(U=u).$$

In the same way, the entropy $H(U|E)$ given some event E is defined. For a second random variable V, the conditional entropy $H(U|V)$ is the expected value of the random value $H(U|V=v)$. Moreover, let $H^*(z)=-z\log z-(1-z)\log(1-z)$ for $0<z<1$. Besides other well-known information theoretical inequalities, we use the following one whose easy proof is given by Krause, Savický, and Wegener (1999). If the random variables U and V take values in $\{0,1\}$,

$$H^*\big(\mathrm{Prob}(U=V)\big)\geq H(U|V).$$

Proof of Lemma 15.4.6. For each (a,b,y), let

$$q(a,b,y)=\mathrm{Prob}\big(\mathrm{DSA}(a,b,\tilde{x},y)=g(a,b,\tilde{x},y)\big)$$

for random assignments $\tilde{x}$ to the variables in X. The probability we are interested in is the average of all $q(a,b,y)$. Let $q(a,b)$ denote the average of $q(a,b,y)$ over all possible y and, similarly, let $q(a)$ denote the average of $q(a,b,y)$ over all possible b and y. Moreover, for each partial input a, let I_a be the set of partial inputs b such that the variable $x_{|(a,b)|}$ or $x_{a,b}$, for simplicity, is given to Alice. Since $H^*(z)\leq H(1/2)=1$ for all $z\in[0,1]$, the following claim implies that $q(a,b)$ has to be close to $1/2$ for many $b\in I_a$ if $|I_a|\gg\log s$.

Claim. *For every a,*

$$\sum_{b\in I_a}H^*\big(q(a,b)\big)\geq|I_a|-\log s.$$

Proof. Consider the π-OBDD of size s computing the function g. For every (a,x), let $h(a,x)$ be the first node where the computation path for (a,x) reaches a node testing a variable in $B\cup Y$ or a sink. Note that the sink reached by the computation path for (a,b,x,y) depends on (a,x) only via $h(a,x)$. This means there is a function $\Phi_{b,y}$ such that $g(a,b,x,y)=\Phi_{b,y}(h(a,x))$. Note that the size of the range of h is at most s.

If (a,b,y) is fixed and $b\in I_a$, DSA_n outputs $x_{a,b}$. Using the above-mentioned inequality between H and H^* and the fact that $H\big(U|f(V)\big)\geq H(U|V)$ for each function f, we conclude

$$\begin{aligned} H^*\big(q(a,b,y)\big) &= H^*\big(\mathrm{Prob}\big(\tilde{x}_{a,b} = \Phi_{b,y}(h(a,\tilde{x}))\big)\big) \\ &\geq H\big(\tilde{x}_{a,b}|\Phi_{b,y}(h(a,\tilde{x}))\big) \geq H\big(\tilde{x}_{a,b}|h(a,\tilde{x})\big). \end{aligned}$$

Now we use the fact $H(U_1|V) + \cdots + H(U_r|V) \geq H\big((U_1,\ldots,U_r)|V\big)$ for $\tilde{x}_{a,b}$, $b \in I_a$, and the vector $\tilde{x}_a$ of these random variables. This implies

$$\sum_{b\in I_a} H\big(\tilde{x}_{a,b}|h(a,\tilde{x})\big) \geq H\big(\tilde{x}_a|h(a,\tilde{x})\big).$$

In the next step, we apply the equalities $H(U|V) = H(U,V) - H(V)$ and $H\big(U,f(U)\big) = H(U)$ to obtain

$$H\big(\tilde{x}_a|h(a,\tilde{x})\big) = H\big(\tilde{x}_a, h(a,\tilde{x})\big) - H\big(h(a,\tilde{x})\big) = H\big(\tilde{x}_a\big) - H\big(h(a,\tilde{x})\big).$$

We have $H\big(h(a,\tilde{x})\big) \leq \log s$, since there are only s different possibilities for $h(a,\tilde{x})$. The random variables $\tilde{x}_{a,b}$, $b \in I_a$, are independent and take random values in $\{0,1\}$, i.e., $\tilde{x}_a$ is uniformly distributed over $\{0,1\}^{|I_a|}$ and $H(\tilde{x}_a) = |I_a|$. This implies

$$H\big(\tilde{x}_a|h(a,\tilde{x})\big) \geq |I_a| - \log s.$$

Putting all our considerations together, we obtain

$$\sum_{b\in I_a} H^*\big(q(a,b,y)\big) \geq |I_a| - \log s.$$

The function H^* is concave. Hence, this inequality implies the claim. □

We continue with the proof of Lemma 15.4.6. Let $\Delta(a,b) = q(a,b) - \frac{1}{2}$. Then we apply the inequality $H^*(\frac{1}{2}+t) \leq 1 - (2/\ln 2)t^2$ (estimate Taylor's expansion using the second derivative) to obtain

$$\sum_{b\in I_a} H^*\big(q(a,b)\big) = \sum_{b\in I_a} H^*\left(\frac{1}{2} + \Delta(a,b)\right) \leq |I_a| - (2/\ln 2)\sum_{b\in I_a} \Delta(a,b)^2.$$

Together with the claim, we get

$$\frac{1}{2}\ln s \geq \sum_{b\in I_a} \Delta(a,b)^2.$$

Using Cauchy's inequality, we obtain

$$\sum_{b\in I_a} |\Delta(a,b)| \leq \left(|I_a| \sum_{b\in I_a} \Delta(a,b)^2\right)^{1/2} \leq \left(\frac{1}{2}|I_a|\ln s\right)^{1/2}.$$

Recall that $q(a)$ is the average of all $q(a,b)$. Since b may take $2^{|B|}$ values, we get

$$\begin{aligned}
q(a) &= \frac{1}{2^{|B|}}\left(\sum_{b\notin I_a} q(a,b) + \sum_{b\in I_a} q(a,b)\right)\\
&\le \frac{1}{2^{|B|}}\left(2^{|B|} - \frac{1}{2}|I_a| + \sum_{b\in I_a}\Delta(a,b)\right)\\
&\le 1 - 2^{-|B|-1}\big(|I_a| - (2|I_a|\ln s)^{1/2}\big) = \psi(|I_a|),
\end{aligned}$$

where $\psi(t) := 1 - 2^{-|B|-1}\big(t - (2t\ln s)^{1/2}\big)$. The function ψ is concave. Let $a_1,\dots,a_m$, $m = 2^{|A|}$, be the possible values of a. Then

$$\frac{1}{m}\sum_{1\le i\le m} q(a_i) \le \frac{1}{m}\sum_{1\le i\le m}\psi(|I_{a_i}|) \le \psi\left(\frac{1}{m}\sum_{1\le i\le m}|I_{a_i}|\right) = \psi(|X|/2^{|A|}).$$

The last equality follows, since, by definition, the sum of all $|I_{a_i}|$ equals $|X|$. The left-hand side of the above inequality is the average of all $q(a)$ and this is the average of all $\mathrm{Prob}\big(\mathrm{DSA}(a,b,\tilde{x},y) = g(a,b,\tilde{x},y)\big)$ and, therefore, equal to $\mathrm{Prob}\big(\mathrm{DSA}(\tilde{a},\tilde{b},\tilde{x},\tilde{y}) = g(\tilde{a},\tilde{b},\tilde{x},\tilde{y})\big)$. We have proved that this probability is bounded above by

$$\psi(|X|/2^{|A|}) = 1 - \frac{|X|}{2\cdot 2^{|A|+|B|}} + \frac{1}{2\cdot 2^{|A|+|B|}}\left(2\cdot 2^{|A|}|X|\ln s\right)^{1/2}.$$

Since A and B are a partition of the $\log n$ address variables, $2^{|A|+|B|} = n$ and the lemma is proved. □

Bibliography

The following abbreviations are used.

DAC	Design Automation Conference
DATE	Design Automation and Test in Europe
ECCC	Electronic Colloquium in Computational Complexity
EDAC	European Design Automation Conference
EDTC	European Design and Test Conference
FCT	Fundamentals of Computation Theory
FMCAD	Formal Methods in Computer-Aided Design
FOCS	IEEE Conference on the Foundations of Computer Science
ICALP	International Colloquium on Automata, Languages, and Programming
ICCAD	IEEE/ACM International Conference on Computer Aided Design
ICCD	International Conference on Computer Design
ICEC	IEEE International Conference on Evolutionary Computation
ICVC	International Conference on Very Large Scale Integration and Computer-Aided Design
ISAAC	International Symposium on Algorithms and Computation
IWLS	International Workshop on Logic Synthesis
LNCS	Lecture Notes in Computer Science
MFCS	Mathematical Foundations of Computer Science
Reed-Muller	International Workshop on Applications of the Reed-Muller Expansion in Circuit Design
SASIMI	Synthesis and System Integration of Mixed Technologies
STACS	Symposium on Theoretical Aspects of Computer Science
STOC	ACM Symposium on Theory of Computing
SWAT	Scandinavian Workshop on Algorithm Theory

Abadir, M. S. and Reghbati, H. K. (1986). Functional test generation for digital circuits described using binary decision diagrams. IEEE Trans. on Computers 35, 375–379.

Ablayev, F. (1996). Lower bounds for one-way probabilistic communication complexity and their applications to space complexity. Theoretical Computer Science 157, 139–159.

Ablayev, F. (1997). Randomization and nondeterminism are incomparable for ordered read-once branching programs. (The printed title has the misprint "comparable.") ICALP '97, LNCS 1256, Springer-Verlag, Berlin, New York, 195–202.

Ablayev, F. and Karpinski, M. (1996). On the power of randomized ordered branching programs. ICALP '96, LNCS 1099, Springer-Verlag, Berlin, New York, 348–356.

Ablayev, F. and Karpinski, M. (1998). A lower bound for integer multiplication on randomized ordered read-once branching programs. ECCC Rep. No. 98–011.

Aborhey, S. (1988). Binary decision tree test functions. IEEE Trans. on Computers 37, 1461–1465.

Agrawal, M. and Thierauf, T. (1998). The satisfiability problem for probabilistic ordered branching programs. 13th IEEE Conf. on Computational Complexity, 81–90.

Ajtai, M. (1999). A non-linear time lower bound for Boolean branching programs. In: Proceedings 40th FOCS, 60–70.

Akers, S. B. (1978a). Binary decision diagrams. IEEE Trans. on Computers 27, 509–516.

Akers, S. B. (1978b). Functional testing with binary decision diagrams. 8th Int. Conf. on Fault-Tolerant Computing, IEEE Computer Society Press, Los Alamitos, CA, 75–82.

Alon, N. and Maass, W. (1988). Meanders and their applications in lower bound arguments. Journal of Computer and System Sciences 37, 118–129.

Ashar, P. and Cheong, M. (1994). Efficient breadth-first manipulation of binary decision diagrams. ICCAD '94, 622–627.

Ashar, P., Ghosh, A., and Devadas, S. (1992). Boolean satisfiability and equivalence checking using general binary decision diagrams. INTEGRATION, the VLSI Journal 13, 1–16.

Ashar, P. and Malik, S. (1995). Fast functional simulation using branching programs. ICCAD '95, 408–412.

Atkins, D. E. (1968). Higher-radix division using estimates of the divisor and partial remainder. IEEE Trans. on Computers 17, 925–934.

Aziz, A., Balarin, F., Cheng, S.-T., Hojati, R., Kam, T., Krishnan, S. C., Ranjan, R. K., Shiple, T. R., Singhal, V., Tasiran, S., Wang, H. Y., Brayton, R. K., and Sangiovanni-Vincentelli, A. (1994). HSIS: A BDD-based environment for formal verification. 31st DAC, ACM, New York, 454–459.

Babai, L., Hajnal, P., Szemerédi, E., and Turán, G. (1987). A lower bound for read-once-only branching programs. Journal of Computer and System Sciences 35, 153–162.

Babai, L., Nisan, N., and Szegedy, M. (1992). Multiparty protocols, pseudorandom generators for logspace, and time-space trade-offs. Journal of Computer and System Sciences 45, 204–232.

Babai, L., Pudlák, P., Rödl, V., and Szemerédi, E. (1990). Lower bounds to the complexity of symmetric Boolean functions. Theoretical Computer Science 74, 313–323.

Bäck, T. (1996) *Evolutionary Algorithms in Theory and Practice.* Oxford University Press, New York.

Bahar, R. I., Cho, H., Hachtel, G. D., Macii, E., and Somenzi, F. (1994). Timing analysis of combinational circuits using ADD's. EDAC '94, IEEE Computer Society Press, Los Alamitos, CA, 625–629.

Bahar, R. I., Frohm, E. A., Gaona, C. M., Hachtel, G. D., Macii, E., Pardo, A., and Somenzi, F. (1997). Algebraic decision diagrams and their applications. Formal Methods in System Design 10, 171–206.

Barrington, D. A. (1989). Bounded-width polynomial-size branching programs recognize exactly those languages in NC^1. Journal of Computer and System Sciences 38, 150–164.

Beame, P., Saks, M., and Thathachar, J. S. (1998). Time-space trade-offs for branching programs. 39th FOCS, 254–263.

Becker, B. (1992). Synthesis for testability: Binary decision diagrams. STACS '92, LNCS 577, Springer-Verlag, Berlin, New York, 501–512.

Becker, B., Drechsler, R., and Enders, R. (1997). On the computational power of bit-level and word-level decision diagrams. ASP Design Automation Conf., IEEE, Piscataway, NJ, 461–467.

Becker, B., Drechsler, R., and Theobald, M. (1997). On the expressive power of OKFDDs. Formal Methods in System Design 11, 5–21.

Becker, B., Drechsler, R., and Werchner, R. (1995). On the relation between BDDs and FDDs. Information and Computation 123, 185–197.

Bern, J., Meinel, C., and Slobodová, A. (1995). Efficient OBDD-based Boolean manipulation in CAD beyond current limits. 32nd DAC, ACM, New York, 408–413.

Bern, J., Meinel, C., and Slobodová, A. (1996). Some heuristics for generating tree-like FBDD types. IEEE Trans. on Computer-Aided Design of Integrated Circuits and Systems 15, 127–130.

Besson, T., Bouzouzou, H., Floricica, I., Saucier, G., and Roane, R. (1993). Input order for ROBDDs based on kernel analysis. EDAC '93, IEEE Computer Society Press, Los Alamitos, CA, 266–272.

Blum, M., Chandra, A. K., and Wegman, M. N. (1980). Equivalence of free Boolean graphs can be decided probabilistically in polynomial time. Information Processing Letters 10, 80–82.

Blumer, A., Ehrenfeucht, A., Haussler, D., and Warmuth, M. (1987). Occam's razor. Information Processing Letters 24, 377–380.

Bollig, B., Löbbing, M., Sauerhoff, M., and Wegener, I. (1996). Complexity theoretical aspects of OFDDs. In: *Representation of Discrete Functions* (Eds.: Sasao, T. and Fujita, M.), 249–268. Kluwer Academic Publishers, Norwell, MA.

Bollig, B., Löbbing, M., Sauerhoff, M., and Wegener, I. (1999). On the complexity of the hidden weighted bit function for various BDD models. RAIRO Theoretical Informatics and Applications 33, 103–115.

Bollig, B., Löbbing, M., and Wegener, I. (1995). Simulated annealing to improve variable orderings for OBDDs. IWLS '95, 5.1–5.10.

Bollig, B., Löbbing, M., and Wegener, I. (1996). On the effect of local changes in the variable ordering of ordered decision diagrams. Information Processing Letters 59, 233–239.

Bollig, B., Sauerhoff, M., Sieling, D., and Wegener, I. (1998). Hierarchy theorems for kOBDDs and kIBDDs. Theoretical Computer Science 205, 45–60.

Bollig, B. and Wegener, I. (1996a). Read-once projections and formal circuit verification with binary decision diagrams. STACS '96, LNCS 1046, Springer-Verlag, Berlin, New York, 491–502.

Bollig, B. and Wegener, I. (1996b). Improving the variable ordering of OBDDs is NP-complete. IEEE Trans. on Computers 45, 993–1002.

Bollig, B. and Wegener, I. (1997a). Complexity theoretical results on partitioned (nondeterministic) binary decision diagrams. MFCS '97, LNCS 1295, Springer-Verlag, Berlin, New York, 159–168. (Also: (1999) Theory of Computing Systems 32, 487–503.)

Bollig, B. and Wegener, I. (1997b). Partitioned BDDs vs. other BDD models. IWLS '97.

Bollig, B. and Wegener, I. (1998a). A very simple function that requires exponential size read-once branching programs. Information Processing Letters 66, 53–57.

Bollig, B. and Wegener, I. (1998b). Completeness and non-completeness results with respect to read-once projections. Information and Computation 143, 24–33.

Borodin, A., Razborov, A., and Smolensky, R. (1993). On lower bounds for read-k-times branching programs. Computational Complexity 3, 1–18.

Brace, K. S., Rudell, R. L., and Bryant, R. E. (1990). Efficient implementation of a BDD package. 27th DAC, IEEE, Piscataway, NJ, 40–45.

Brand, D. (1993). Verification of large synthesized designs. ICCAD '93, 534–537.

Brandman, Y., Orlitsky, A., and Hennessy, J. (1990). A spectral lower bound technique for the size of decision trees and two-level AND/OR circuits. IEEE Trans. on Computers 39, 282–287.

Breitbart, Y., Hunt III, H. B., and Rosenkrantz, D. (1993). The comparative complexity of binary decision diagrams representing Boolean functions. Tech. Rep., Univ. of Kentucky, Lexington, KY.

Breitbart, Y., Hunt III, H. B., and Rosenkrantz, D. (1995). On the size of binary decision diagrams representing Boolean functions. Theoretical Computer Science 145, 45–69.

Bryant, R. E. (1985). Symbolic manipulation of Boolean functions using a graphical representation. 22nd DAC, IEEE, Piscataway, NJ, 688–694.

Bryant, R. E. (1986). Graph-based algorithms for Boolean function manipulation. IEEE Trans. on Computers 35, 677–691.

Bryant, R. E. (1991). On the complexity of VLSI implementations and graph representations of Boolean functions with application to integer multiplication. IEEE Trans. on Computers 40, 205–213.

Bryant, R. E. (1992). Symbolic Boolean manipulation with ordered binary decision diagrams. ACM Computing Surveys 24, 293–318.

Bryant, R. E. (1996). Bit-level analysis of an SRT divider circuit. 33rd DAC, ACM, New York, 661–665.

Bryant, R. E. and Chen, Y.-A. (1995). Verification of arithmetic functions with binary moment diagrams. 32nd DAC, ACM, New York, 535–541.

Burch, J. R. (1991). Using BDDs to verify multipliers. 28th DAC, ACM, New York, 408–412.

Burch, J. R., Clarke, E. M., and Long, D. E. (1991). Representing circuits more efficiently in symbolic model checking. 28th DAC, ACM, New York, 403–407.

Burch, J. R., Clarke, E. M., Long, D. E., McMillan, K. L., and Dill, D. L. (1994). Symbolic model checking for sequential circuit verification. IEEE Trans. on Computer-Aided Design of Integrated Circuits and Systems 13, 401–424.

Burch, J. R., Clarke, E. M., McMillan, K. L., and Dill, D. L. (1990). Sequential circuit verification using symbolic model checking. 27th DAC, IEEE, Piscataway, NJ, 46–51.

Burch, J. R., Clarke, E. M., McMillan, K. L., Dill, D. L., and Hwang, L. J. (1992). Symbolic model checking: 10^{20} states and beyond. Information and Computation 98, 142–170.

Buss, S. R. (1992). The graph of multiplication is equivalent to counting. Information Processing Letters 41, 199–201.

Butler, K. M., Ross, D. E., Kapur, R., and Mercer, M. R. (1991). Heuristics to compute variable orderings for efficient manipulation of ordered binary decision diagrams. 28th DAC, ACM, New York, 417–420.

Büttner, W. and Simonis, H. (1987). Embedding Boolean expressions into logic programming. Journal of Symbolic Computation 4, 191–205.

Cabodi, G., Camurati, P., and Quer, S. (1994). Auxiliary variables for extending symbolic traversal techniques to data paths. 31st DAC, ACM, New York, 289–293.

Cabodi, G., Camurati, P., and Quer, S. (1996). Improved reachability analysis of large finite state machines. ICCAD '96, 354–360.

Calazans, N., Zhang, Q., Jacobi, R., Yernaux, B., and Trullemans, A.-M. (1992). Advanced ordering and manipulation techniques for binary decision diagrams. EDAC '92, IEEE Computer Society Press, Los Alamitos, CA, 452–457.

Chandra, A., Stockmeyer, L., and Vishkin, U. (1984). Constant depth reducibility. SIAM J. on Computing 13, 423–439.

Chang, S.-C., Cheng, D. J., and Marek-Sadowska, M. (1994). Minimizing ROBDD size of incompletely specified multiple output functions. EDAC '94, IEEE Computer Society Press, Los Alamitos, CA, 620–624.

Clarke, E., Fujita, M., and Zhao, X. (1995a). Hybrid decision diagrams – overcoming the limitations of MTBDDs and BMDs. ICCAD '95, 159–163.

Clarke, E., Fujita, M., and Zhao, X. (1995b). Applications of multi-terminal binary decision diagrams. Reed-Muller '95, Fujiki Printing Co., LTD, Iizuka, Japan, 21–27.

Clarke, E. M., German, S. M., and Zhao, X. (1999). Verifying the SRT division algorithm using theorem proving techniques. Formal Methods in System Design 14, 7–44.

Clarke, E., Khaira, M., and Zhao, X. (1996). Word level symbolic model checking – avoiding the Pentium FDIV error, 33rd DAC, ACM, New York, 645–648.

Clarke, E. M., McMillan, K. L., Zhao, X., Fujita, M., and Yang, J. (1997). Spectral transforms for large Boolean functions with applications to technology mapping. Formal Methods in System Design 10, 137–148.

Clarke, E. M. and Wing, J. M. (1996). Formal methods: State of the art and future directions. ACM Computing Surveys 28, 626–643.

Cleve, R. (1991). Towards optimal simulations of formulas by bounded-width programs. Computational Complexity 1, 91–105.

Cobham, A. (1966). The recognition problem for the set of perfect squares. 7th Symp. on Switching and Automata Theory, IEEE, Piscataway, NJ, 78–87.

Coppersmith, D. and Winograd, S. (1990). Matrix multiplication via arithmetic progressions. Journal of Symbolic Computation 9, 251–280.

Cormen, T. H., Leiserson, C. E., and Rivest, R. L. (1990). *An Introduction to Algorithms.* McGraw–Hill, New York.

Coudert, O. (1994). Two-level logic minimization: An overview. INTEGRATION, the VLSI Journal 17, 97–140.

Coudert, O. (1995). Doing two-level logic minimization 100 times faster. In: *Proceedings Sixth Annual ACM-SIAM Symposium on Discrete Algorithms* (SODA '95), SIAM, Philadelphia, 112–121.

Coudert, O., Berthet, C., and Madre, J. C. (1989a). Verification of synchronous sequential machines based on symbolic execution. Workshop on Automatic Verification Methods for Finite State Systems. LNCS 407, Springer-Verlag, Berlin, New York, 365–373.

Coudert, O., Berthet, C., and Madre, J. C. (1989b). Verification of sequential machines using Boolean functional vectors. IMEC-IFIP Workshop on Applied Formal Methods for Correct VLSI Design, IEEE Computer Society Press, Los Alamitos, CA, 111–128.

Coudert, O. and Madre, J. C. (1992). Implicit and incremental computation of primes and essential primes of Boolean functions. 29th DAC, IEEE Computer Society Press, Los Alamitos, CA, 36–39.

Coudert, O., Madre, J. C., and Fraisse, H. (1993). A new viewpoint on two-level logic minimization. 30th DAC, ACM, New York, 625–630.

Damm, C. and Meinel, C. (1992). Separating completely complexity classes related to polynomial size Ω-decision trees. Theoretical Computer Science 106, 351–360.

Devadas, S. (1993). Comparing two-level and ordered binary decision diagram representations of logic functions. IEEE Trans. on Computer-Aided Design of Integrated Circuits and Systems 12, 722-723.

Dias da Silva, J. A. and Hamidoune, Y. O. (1994). Cyclic spaces for Grassmann derivatives and additive theory. Bulletin of the London Math. Society 26, 140–146.

Drechsler, R., Becker, B., and Göckel, N. (1996). A genetic algorithm for variable ordering of OBDDs. IEEE Proc. on Computers and Digital Techniques 143(6), 364–368.

Drechsler, R., Becker, B., and Jahnke, A. (1998). On variable ordering and decomposition type choice in OKFDDs. IEEE Trans. on Computers 47, 1398–1403.

Drechsler, R., Becker, B., and Ruppertz, S. (1996). K*BMDs: A new data structure for verification. EDTC '96, IEEE Computer Society Press, Los Alamitos, CA, 2–8.

Drechsler, R., Drechsler, N., and Günther, W. (1998). Fast exact minimization of BDDs. 35th DAC, ACM, New York, 200–205.

Drechsler, R. and Göckel, N. (1997). Minimization of BDDs by evolutionary algorithms. IWLS '97.

Drechsler, R., Sarabi, A., Theobald, M., Becker, B., and Perkowski, M. A. (1994). Efficient representation and manipulation of switching functions based on ordered Kronecker functional decision diagrams. 31st DAC, ACM, New York, 415–419.

Droste, S. (1997). Efficient genetic programming for finding good generalizing Boolean functions. Genetic Programming '97, Morgan-Kaufmann, San Francisco, CA, 82–87.

Droste, S. (1998). Genetic programming with guaranteed quality. Genetic Programming '98, Morgan-Kaufmann, San Francisco, CA, 54–59.

Droste, S. and Wiesmann, D. (1998). On representation and genetic operators in evolutionary algorithms. Tech. Rep., Univ. Dortmund, Germany.

Dunne, P. E. (1985). Lower bounds on the complexity of 1-time only branching programs. FCT '85, LNCS 199, Springer-Verlag, Berlin, New York, 90–99.

Ďuriš, P., Hromkovič, J., Rolim, J. D. P., and Schnitger, G. (1997). Las Vegas versus determinism for one-way communication complexity, finite automata, and polynomial-time computations. STACS '97, LNCS 1200, Springer-Verlag, Berlin, New York, 117–128.

Ehrenfeucht, A. and Haussler, D. (1989). Learning decision trees from random examples. Information and Computation 82, 231–246.

Enders, R. (1995). Note on the complexity of binary moment diagram representations. Reed-Muller '95, Fujiki Printing Co., LTD, Iizuka, Japan, 191–197.

Enders, R., Filkorn, T., and Taubner, D. (1993). Generating BDDs for symbolic model checking in CCS. Distributed Computing 6, 155–164.

Feige, U. and Kilian, J. (1996). Zero knowledge and the chromatic number. 11th IEEE Conf. on Computational Complexity, 278–287.

Fogel, D. (1996). *Evolutionary Computation.* IEEE Press, Piscataway, NJ.

Fortune, S., Hopcroft, J., and Schmidt, E. M. (1978). The complexity of equivalence and containment for free single variable program schemes. ICALP '78, LNCS 62, Springer-Verlag, Berlin, New York, 227–240.

Freivalds, R. (1979). Fast probabilistic algorithms. FCT '79, LNCS 74, Springer-Verlag, Berlin, New York, 57–69.

Friedman, S. J. and Supowit, K. J. (1990). Finding the optimal variable ordering for binary decision diagrams. IEEE Trans. on Computers 39, 710–713.

Fujii, H., Ootomo, G., and Hori, C. (1993). Interleaving based variable ordering methods for ordered binary decision diagrams. ICCAD '93, 38–41.

Fujita, M., Fujisawa, H., and Kawato, N. (1988). Evaluation and improvements of Boolean comparison method based on binary decision diagrams. ICCAD '88, 2–5.

Fujita, M., Fujisawa, H., and Matsunaga, Y. (1993). Variable ordering algorithms for ordered binary decision diagrams and their evaluation. IEEE Trans. on Computer-Aided Design of Integrated Circuits and Systems 12, 6–12.

Fujita, M., Matsunaga, Y., and Kakuda, T. (1991). On variable ordering of binary decision diagrams for the application of multi-level logic synthesis. EDAC '91, IEEE Computer Society Press, Los Alamitos, CA, 50–54.

Fujita, M., McGeer, P. C., and Yang, J. (1997). Multi-terminal binary decision diagrams: An efficient data structure for matrix representation. Formal Methods in System Design 10, 149–169.

Gál, A. (1997). A simple function that requires exponential size read-once branching programs. Information Processing Letters 62, 13–16.

Garey, M. R. and Johnson, D. S. (1979). *Computers and Intractability: A Guide to the Theory of NP-Completeness.* W. H. Freeman, San Francisco, CA.

Gergov, J. (1994). Time-space tradeoffs for integer multiplication on various types of input oblivious sequential machines. Information Processing Letters 51, 265–269.

Gergov, J. and Meinel, C. (1994). Efficient Boolean manipulation with OBDD's can be extended to FBDD's. IEEE Trans. on Computers 43, 1197–1209.

Gergov, J. and Meinel, C. (1996). MOD-2-OBDDs – a data structure that generalizes EXOR-sum-of-products and ordered binary decision diagrams. Formal Methods in System Design 8, 273–282.

Goldberg, E. I., Kukimoto, Y., and Brayton, R. K. (1997). Canonical TBDD's and their application to combinational verification. IWLS '97.

Goldberg, E. I., Kukimoto, Y., and Brayton, R. K. (1998). Combinational verification based on high-level functional specifications. DATE '98, IEEE Computer Society Press, Los Alamitos, CA, 803–808.

Graham, R. L., Knuth, D. E., and Patashnik, O. (1994). *Concrete Mathematics.* Addison–Wesley, Reading, MA.

Gröpl, C., Prömel, H. J., and Srivastav, A. (1998). Size and structure of random OBDDs. STACS '98, LNCS 1373, Springer-Verlag, Berlin, New York, 105–115.

Hachtel, G. D., Macii, E., Pardo, A., and Somenzi, F. (1994a). Symbolic algorithms to calculate steady-state probabilities of a finite state machine. EDAC '94, IEEE Computer Society Press, Los Alamitos, CA, 214–218.

Hachtel, G. D., Macii, E., Pardo, A., and Somenzi, F. (1994b). Probabilistic analysis of large finite state machines. 31st DAC, ACM, New York, 270–275.

Hachtel, G. D. and Somenzi, F. (1996). *Logic Synthesis and Verification Algorithms.* Kluwer Academic Publishers, Norwell, MA.

Hachtel, G. D. and Somenzi, F. (1997). A symbolic algorithm for maximum flow in 0-1 networks. Formal Methods in System Design 10, 207–219.

Hamaguchi, K., Morita, A., and Yajima, S. (1995). Efficient construction of binary moment diagrams for verifying arithmetic circuits. ICCAD '95, 78–82.

Hardy, G. H. and Wright, F. M. (1979). *An Introduction to the Theory of Numbers.* Oxford University Press, New York.

Heap, M. (1993). On the exact ordered binary decision diagram size of totally symmetric functions. Journal of Electronic Testing: Theory and Applications 4, 191–195.

Heiman, R., Newman, I., and Wigderson, A. (1993). On read-once threshold formulae and their randomized decision tree complexity. Theoretical Computer Science 107, 63–76.

Heiman, R. and Wigderson, A. (1991). Randomized vs. deterministic decision tree complexity for read-once Boolean functions. Computational Complexity 1, 311–329.

Hirata, K., Shimozono, S., and Shinohara, A. (1996). On the hardness of approximating the minimum consistent OBDD problem. SWAT '96, LNCS 1097, Springer-Verlag, Berlin, New York, 112-123.

Hojati, R., Shiple, T. R., Brayton, R. K., and Kurshan, R. P. (1993). A unified approach to language containment and fair CTL model checking. 30th DAC, ACM, New York, 475–481.

Hopcroft, J. E. and Ullman, J. D. (1979). *Introduction to Automata Theory, Languages and Computation.* Addison–Wesley, Reading, MA.

Hosaka, K., Takenaga, Y., Kaneda, T., and Yajima, S. (1997). On the size of ordered binary decision diagrams representing threshold functions. Theoretical Computer Science 180, 47–60.

Hromkovič, J. (1997). *Communication Complexity and Parallel Computing.* Springer-Verlag, Berlin, New York.

Hu, A. J. and Dill, D. L. (1993). Reducing BDD size by exploiting functional dependencies. 30th DAC, ACM, New York, 266–271.

Hu, A. J., York, G., and Dill, D. L. (1994). New techniques for efficient verification with implicitly conjoined BDDs. 31st DAC, ACM, New York, 276–282.

Hyafil, L. and Rivest, R. L. (1976). Constructing optimal binary decision trees is NP-complete. Information Processing Letters 5, 15–17.

Immerman, N. (1988). Nondeterministic Space is closed under complementation. SIAM J. on Computing 17, 935–938.

Ishiura, N. (1992). Synthesis of multi-level logic circuits from binary decision diagrams. SASIMI '92, Seisei Insatsu, Osaka, 74–83.

Ishiura, N., Sawada, H., and Yajima, S. (1991). Minimization of binary decision diagrams based on exchanges of variables. ICCAD '91, 472–475.

Jacobi, R., Calazans, N., and Trullemans, C. (1991). Incremental reduction of binary decision diagrams. In: *Proceedings IEEE Internat. Symposium on Circuits and Systems*, Vol. 5, IEEE Press, Piscataway, NJ, 3174–3177.

Jain, J., Abadir, M., Bitner, J., Fussell, D. S., and Abraham, J. A. (1992). IBDDs: An efficient functional representation for digital circuits. EDAC '92, IEEE Computer Society Press, Los Alamitos, CA, 440–446.

Jain, J., Abraham, J. A., Bitner, J., and Fussell, D. S. (1992). Probabilistic verification of Boolean functions. Formal Methods in System Design 1, 61–115.

Jain, J., Adams, W., and Fujita, M. (1998). Sampling schemes for computing OBDD variable orderings. ICCAD '98, 631–635.

Jain, J., Bitner, J., Abadir, M., Abraham, J. A., and Fussell, D. S. (1997). Indexed BDDs: Algorithmic advances in techniques to represent and verify Boolean functions. IEEE Trans. on Computers 46, 1230–1245.

Jain, J., Bitner, J., Fussell, D. S., and Abraham, J. A. (1992). Functional partitioning for verification and related problems. Brown MIT VLSI Conf., MIT Press, Cambridge, MA, 210–226.

Jeong, S.-W., Kim, T.-S., and Somenzi, F. (1993). An efficient method for optimal BDD ordering computation. Int. Conf. on VLSI and CAD, ICVC '93, 252–256.

Jeong, S.-W., Plessier, B., Hachtel, G. D., and Somenzi, F. (1991). Variable ordering and selection for FSM traversal. ICCAD '91, 476–479.

Jóźwiak, L. and Mijland, H. (1992). On the use of OR-BDDs for test generation. 18th EUROMICRO Symp. on Microprocessing and Microprogramming 35, 159–166.

Jukna, S. (1987). Lower bounds on communication complexity. Math. Logic and Its Applications 5, 22–30.

Jukna, S. (1988). Entropy of contact circuits and lower bounds on their complexity. Theoretical Computer Science 57, 113–129.

Jukna, S. (1989). On the effect of null-chains on the complexity of contact schemes. FCT '89, LNCS 380, Springer-Verlag, Berlin, New York, 246–256.

Jukna, S. (1995). A note on read-k-times branching programs. RAIRO Theoretical Informatics and Applications 29, 75–83.

Jukna, S. (1999). Linear codes are hard for oblivious read-once parity branching programs. Information Processing Letters 69, 267–269.

Jukna, S. and Razborov, A. (1998). Neither reading few bits twice nor reading illegally helps much. Discrete Applied Mathematics 85, 223–238.

Jukna, S., Razborov, A., Savický, P., and Wegener, I. (1997). On P versus NP $\cap$ co-NP for decision trees and read-once branching programs. MFCS '97, LNCS 1295, Springer-Verlag, Berlin, New York, 319–326.

Jukna, S. and Žák, S. (1998). On branching programs with bounded uncertainty. ICALP '98, LNCS 1443, Springer-Verlag, Berlin, New York, 259–270.

Kalyanasundaram, B. and Schnitger, G. (1992). The probabilistic communication complexity of set intersection. SIAM J. Discrete Math. 5, 545–557.

Karpinski, M. and Mubarakzjanov, R. (1999). A note on Las Vegas OBDDs. ECCC Rep. No. 99-09.

Kebschull, U. and Rosenstiel, W. (1993). Efficient graph-based computation and manipulation of functional decision diagrams. EDAC '93, IEEE Computer Society Press, Los Alamitos, CA, 278–282.

Kebschull, U., Schubert, E., and Rosenstiel, W. (1992). Multilevel logic synthesis based on functional decision diagrams. EDAC '92, IEEE Computer Society Press, Los Alamitos, CA, 43–47.

Kimura, S. and Clarke, E. M. (1990). A parallel algorithm for constructing binary decision diagrams. ICCD '90, 220–223.

Kloss, V. (1966). Estimates of the complexity of solutions of systems of linear equations. Sov. Math. Doklady 7, 1537–1540.

Kolchin, V. F., Sevest'yanov, B. A., and Christyakov, V. P. (1978). *Random Allocations.* John Wiley (Halsted Press), New York.

Kovari, T., Sós, V., and Turán, P. (1954). On a problem of K. Zarankiewicz. Colloquium Mathematicum 3, 50–57.

Koza, J. (1992). *Genetic Programming. On the Programming of Computers by Means of Natural Selection.* MIT Press, Cambridge, MA.

Koza, J. (1994). *Genetic Programming II. Automatic Discovery of Reusable Programs.* MIT Press, Cambridge, MA.

Krajíček, J. (1995). *Bounded Arithmetic, Propositional Logic, and Complexity Theory.* Cambridge University Press, Cambridge, New York.

Krause, M. (1988). Exponential lower bounds on the complexity of local and real-time branching programs. Journal of Information Processing and Cybernetics (EIK) 24, 99–110.

Krause, M. (1991). Lower bounds for depth-restricted branching programs. Information and Computation 91, 1–14.

Krause, M. (1992). Separating $\oplus$L from L, NL, co-NL and AL(=P) for oblivious Turing machines of linear access time. RAIRO Theoretical Informatics and Applications 26, 507–522.

Krause, M., Meinel, C., and Waack, S. (1991). Separating the eraser Turing machine classes L_e, NL_e, co-NL_e and P_e. Theoretical Computer Science 86, 267–275.

Krause, M., Savický, P., and Wegener, I. (1999). Approximations by OBDDs and the variable ordering problem. ICALP '99, LNCS 1644, Springer-Verlag, Berlin, New York, 493–502.

Krause, M. and Waack, S. (1991). On oblivious branching programs of linear length. Information and Computation 94, 232–249.

Kremer, I., Nisan, N., and Ron, D. (1995). On randomized one-round communication complexity. 22nd STOC, 596–605.

Kriegel, K. and Waack, S. (1988). Lower bounds on the complexity of real-time branching programs. RAIRO Theoretical Informatics and Applications 22, 447–459.

Kukimoto, Y., Fujita, M., and Brayton, R. K. (1994). A redesign technique of combinational circuits based on gate reconnections. ICCAD '94, 632–637.

Kushilevitz, E. and Mansour, Y. (1991). Learning decision trees using the Fourier spectrum. 23rd STOC, 455–464.

Kushilevitz, E. and Nisan, N. (1997). *Communication Complexity.* Cambridge University Press, New York.

Lai, Y.-T., Pedram, M., and Vrudhula, S. B. K. (1993). BDD based decomposition of logic functions with application to FPGA synthesis. 30th DAC, ACM, New York, 642–647.

Lai, Y.-T., Pedram, M., and Vrudhula, S. B. K. (1994). EVBDD-based algorithms for integer linear programming, spectral transformation, and function decomposition. IEEE Trans. on Computer-Aided Design of Integrated Circuits and Systems 13, 959–975.

Lai, Y.-T., Pedram, M., and Vrudhula, S. B. K. (1996). Formal verification using edge-valued binary decision diagrams. IEEE Trans. on Computers 45, 247–255.

Lai, Y.-T. and Sastry, S. (1992). Edge-valued binary decision diagrams for multilevel hierarchical verification. 29th DAC, IEEE Computer Society Press, Los Alamitos, CA, 608–613.

Lee, C. Y. (1959). Representation of switching circuits by binary-decision programs. The Bell Systems Technical Journal 38, 985–999.

Liaw, H.-T. and Lin, C.-S. (1992). On the OBDD representation of general Boolean functions. IEEE Trans. on Computers 41, 661–664.

Lin, B. and Devadas, S. (1995). Synthesis of hazard-free multi-level logic under multiple-input changes from binary decision diagrams. IEEE Trans. on Computer-Aided Design of Integrated Circuits and Systems 14, 974–985.

Linial, N., Mansour, Y., and Nisan, N. (1993). Constant depth circuits, Fourier transform and learnability. Journal of the ACM 40, 607–620.

Löbbing, M., Sieling, D., and Wegener, I. (1998). Parity OBDDs cannot be handled efficiently enough. Information Processing Letters 67, 163–168.

Löbbing, M. and Wegener, I. (1996). The number of knight's tours equals 13, 267, 364, 410, 532 – counting with binary decision diagrams. The Electronic Journal of Combinatorics 3, #R5.

Long, D. E. (1993). bddlib—a binary decision diagram (BDD) package. Available online at http://www.cs.cmu.edu/~modelcheck/bdd.html

Lupanov, O. B. (1965). On the problem of realization of symmetric Boolean functions by contact schemes. Problems of Cybernetics 15, 85–99.

MacWilliams, F. J. and Sloane, N. J. A. (1977). *The Theory of Error-Correcting Codes.* Elsevier–North Holland, Amsterdam.

Mailhot, F. and de Micheli, G. (1993). Algorithms for technology mapping based on binary decision diagrams and on Boolean operations. IEEE Trans. on Computer-Aided Design of Integrated Circuits and Systems 12, 599–620.

Malhotra, V. M., Pramodh Kumar, M., and Maheshwary, S. N. (1978). An $O(|V|^3)$ algorithm for finding maximum flows in networks. Information Processing Letters 7, 277–278.

Malik, S., Wang, A. R., Brayton, R. K., and Sangiovanni-Vincentelli, A. (1988). Logic verification using binary decision diagrams in a logic synthesis environment. ICCAD '88, 6–9.

Manders, K. and Adleman, L. (1978). NP-complete decision problems for binary quadratics. Journal of Computer and System Sciences 16, 168–184.

Masek, W. (1976). A fast algorithm for the string editing problem and decision graph complexity. M.Sc. Thesis, MIT, Cambridge, MA.

Matsunaga, Y., McGeer, P. C., and Brayton, R. K. (1993). On computing the transitive closure of a state transition relation. 30th DAC, ACM, New York, 260–265.

Mayr, E. W., Prömel, H. J., and Steger, A. (1998). *Lectures on Proof Verification and Approximation Algorithms.* LNCS Tutorial. Springer-Verlag, Berlin, New York.

McFarland, M. C. (1993). Formal verification of sequential hardware: A tutorial. IEEE Trans. on Computer-Aided Design of Integrated Circuits and Systems 12, 633–654.

McMillan, K. L. (1994). *Symbolic Model Checking.* Kluwer Academic Publishers, Norwell, MA.

Meinel, C. (1988). The power of nondeterminism in polynomial-size bounded-width branching programs. Theoretical Computer Science 62, 319–325.

Meinel, C. (1990). Polynomial size Ω-branching programs and their computational power. Information and Computation 85, 163–182.

Meinel, C. and Slobodová, A. (1994). On the complexity of constructing optimal ordered binary decision diagrams. MFCS '94, LNCS 841, Springer-Verlag, Berlin, New York, 515–524.

Meinel, C., Somenzi, F., and Theobald, T. (1997). Linear sifting of decision diagrams. 34th DAC, ACM, New York, 202–207.

Meinel, C. and Theobald, T. (1996). Local encoding transformations for optimizing OBDD-representations of finite automata. Formal Methods in Computer-Aided Design, FMCAD '96, LNCS 1166, Springer-Verlag, Berlin, New York, 404–418.

Mercer, M. R., Kapur, R., and Ross, D. E. (1992). Functional approaches to generating orderings for efficient symbolic representations. 29th DAC, IEEE Computer Society Press, Los Alamitos, CA, 624–627.

Minato, S. (1993). Zero-suppressed BDDs for set manipulation in combinatorial problems. 30th DAC, ACM, New York, 272–277.

Minato, S. (1994). Calculation of unate cube set algebra using zero-suppressed BDDs. 31st DAC, ACM, New York, 420–424.

Minato, S. (1996). *Binary Decision Diagrams and Applications for VLSI CAD.* Kluwer Academic Publishers, Norwell, MA.

Minato, S. (1997). Arithmetic Boolean expression manipulator using BDDs. Formal Methods in System Design 10, 221–242.

Minato, S., Ishiura, N., and Yajima, S. (1990). Shared binary decision diagram with attributed edges for efficient Boolean function manipulation. 27th DAC, IEEE, Piscataway, NJ, 52–57.

Möller, D., Mohnke, J., and Weber, M. (1993). Detection of symmetry of Boolean functions represented by ROBDDs. ICCAD '93, 680–684.

Moret, B. M. (1982). Decision trees and diagrams. Computing Surveys 14, 593–623.

Motwani, R. and Raghavan, P. (1995). *Randomized Algorithms.* Cambridge University Press, Cambridge, New York.

Mukherjee, R., Jain, J., Takayama, K., Fujita, M., Abraham, J. A., and Fussell, D. S. (1997). FLOVER: Filtering oriented combinational verification approach. IWLS '97.

Narayan, A., Isles, A. J., Jain, J., Brayton, R. K., and Sangiovanni-Vincentelli, A. (1997). Reachability analysis using partitioned-ROBDDs. ICCAD '97, 388–393.

Narayan, A., Jain, J., Fujita, M., and Sangiovanni-Vincentelli, A. (1996). Partitioned ROBDDs – a compact, canonical and efficiently manipulable representation for Boolean functions. ICCAD '96, 547–554.

Nechiporuk, È. I. (1966). A Boolean function. Sov. Math. Doklady 7, 999–1000.

Nechiporuk, È. I. (1971). On a Boolean matrix. Systems Theory Research 21, 236–239.

Newman, I. (1991). Private vs. common random bits in communication complexity. Information Processing Letters 39, 67–71.

Nisan, N. and Wigderson, A. (1993). Rounds in communication complexity revisited. SIAM J. on Computing 22, 211–219.

Ochi, H., Ishiura, N., and Yajima, S. (1991). Breadth-first manipulation of SBDD of Boolean functions for vector processing. 28th DAC, ACM, New York, 413–416.

Ochi, H., Yasuoka, K., and Yajima, S. (1993). Breadth-first manipulation of very large binary-decision diagrams. ICCAD '93, 48–55.

Okol'nishnikova, E. A. (1993). On lower bounds for branching programs. Siberian Advances in Mathematics 3, 152–166.

Okol'nishnikova, E. A. (1997a). On the hierarchy of nondeterministic branching k-programs. FCT '97, LNCS 1279, Springer-Verlag, Berlin, New York, 376–387.

Okol'nishnikova, E. A. (1997b). On comparison between the sizes of read-k-times branching programs. In: *Operations Research and Discrete Analysis* (Ed.: A. D. Korshunov), 205–225. Kluwer Academic Publishers, Norwell, MA.

Panda, S. and Somenzi, F. (1995). Who are the variables in your neighborhood. ICCAD '95, 74–77.

Panda, S., Somenzi, F., and Plessier, B. F. (1994). Symmetry detection and dynamic variable ordering of decision diagrams. ICCAD '94, 628–631.

Picard, C. (1965). *Théorie des questionnaires.* Gauthier–Villars, Paris.

Pixley, C., Jeong, S.-W., and Hachtel, G. D. (1994). Exact calculation of synchronization sequences based on binary decision diagrams. IEEE Trans. on Computer-Aided Design of Integrated Circuits and Systems 13, 1024–1034.

Plessier, B., Hachtel, G. D., and Somenzi, F. (1994). Extended BDDs: Trading off canonicity for structure in verification algorithms. Formal Methods in System Design 4, 167–185.

Ponzio, S. (1995). A lower bound for integer multiplication with read-once branching programs. 27th STOC, 130–139.

Ranjan, R. K., Sanghari, J. K., Brayton, R. K., and Sangiovanni-Vincentelli, A. (1996). High performance BDD package based on exploiting memory hierarchy. 33rd DAC, ACM, New York, 635–640.

Ravi, K. and Somenzi, F. (1995). High-density reachability analysis. ICCAD '95, 154–158.

Razborov, A. A. (1991). Lower bounds for deterministic and nondeterministic branching programs. FCT '91, LNCS 529, Springer-Verlag, Berlin, New York, 47–60.

Razborov, A. A. (1992). On the distributional complexity of disjointness. Theoretical Computer Science 106, 385–390.

Razborov, A., Wigderson, A., and Yao, A. (1997). Read-once branching programs, rectangular proofs of the pigeonhole principle and the transversal calculus. 29th STOC, 739–748.

Rho, J.-K., Somenzi, F., and Pixley, C. (1993). Minimum length synchronizing sequences of finite state machine. 30th DAC, ACM, New York, 463–468.

Ross, D. E., Butler, K. M., Kapur, R., and Mercer, M. R. (1991a). Fast functional evaluation of candidate OBDD variable ordering. EDAC '91, IEEE Computer Society Press, Los Alamitos, CA, 4–10.

Ross, D. E., Butler, K. M., Kapur, R., and Mercer, M. R. (1991b). Exact ordered binary decision diagram size when representing classes of symmetric functions. Journal of Electronic Testing: Theory and Applications 2, 243–259.

Rosser, J. B. and Schoenfeld, L. (1962). Approximate formulas for some functions of prime numbers. Illinois Journal of Mathematics 6, 64–94.

Rudell, R. (1993). Dynamic variable ordering for ordered binary decision diagrams. ICCAD '93, 42–47.

Sakanashi, H., Higuchi, T., Iba, H., and Kakazu, K. (1996). An approach for genetic synthesizer of binary decision diagram. ICEC '96, 559–564.

Sauerhoff, M. (1998). Lower bounds for randomized read-k-times branching programs. STACS '98, LNCS 1373, Springer-Verlag, Berlin, New York, 105–115.

Sauerhoff, M. (1999a). Complexity theoretical results for randomized branching programs. Ph.D. Thesis. Univ. Dortmund. Shaker Verlag, Aachen, Germany.

Sauerhoff, M. (1999b). On the size of randomized OBDDs and read-once branching programs for k-stable functions. STACS '99, LNCS 1563, Springer-Verlag, Berlin, New York, 488–499.

Sauerhoff, M. and Wegener, I. (1996). On the complexity of minimizing the OBDD size for incompletely specified functions. IEEE Trans. on Computer-Aided Design of Integrated Circuits and Systems 15, 1435–1437.

Sauerhoff, M., Wegener, I., and Werchner, R. (1996). Optimal ordered binary decision diagrams for fan-out free circuits. SASIMI '96, Seisei Insatsu, Osaka, 197–204.

Sauerhoff, M., Wegener, I., and Werchner, R. (1999). Relating branching program size and formula size over the full binary basis. STACS '99, LNCS 1563, Springer-Verlag, Berlin, New York, 57–67.

Savický, P. (1998a). A probabilistic nonequivalence test for syntactic $(1, +k)$-branching programs. ECCC Rep. No. 98-051.

Savický, P. (1998b). On random orderings of variables for parity OBDDs. ECCC Rep. No. 98-068.

Savický, P. and Wegener, I. (1997). Efficient algorithms for the transformation between different types of binary decision diagrams. Acta Informatica 34, 245–256.

Savický, P. and Žák, S. (1996). A large lower bound for 1-branching programs. ECCC Rep. No. 96-030.

Savický, P. and Žák, S. (1997a). A lower bound on branching programs reading some bits twice. Theoretical Computer Science 172, 293–301.

Savický, P. and Žák, S. (1997b). A hierarchy for $(1, +k)$-branching programs with respect to k. MFCS' 97, LNCS 1295, Springer-Verlag, Berlin, New York, 478–487.

Savický, P. and Žák, S. (1998). A read-once lower bound and a $(1, +k)$-hierarchy for branching programs. Accepted for publication in Theoretical Computer Science.

Scholl, C., Becker, B., and Weis, T. M. (1998). Word-level decision diagrams, WLCDs and division. ICCAD '98, 672–677.

Scholl, C., Melchior, S., Hotz, G., and Molitor, P. (1997). Minimizing ROBDD sizes of incompletely specified Boolean functions by exploiting strong symmetries. EDTC '97, IEEE Computer Society Press, Los Alamitos, CA, 229–234.

Scholl, C., Möller, D., Molitor, P., and Drechsler, R. (1999). BDD minimization using symmetries. IEEE Trans. on Computer-Aided Design of Integrated Circuits and Systems 18, 81–100.

Schürfeld, U. (1983). New lower bounds on the formula size of Boolean functions. Acta Informatica 19, 183–194.

Schröer, O. and Wegener, I. (1998). The theory of zero-suppressed BDDs and the number of knight's tours. Formal Methods in System Design 13, 235–253.

Semba, I. and Yajima, S. (1994). Combinatorial algorithms using Boolean processing. Trans. of Information Processing Society of Japan 35, 1661–1673.

Shannon, C. E. (1949). The synthesis of two-terminal switching circuits. The Bell Systems Technical Journal 28, 59–98.

Shen, A., Devadas, S., and Ghosh, A. (1995). Probabilistic manipulation of Boolean functions using free Boolean diagrams. IEEE Trans. on Computer-Aided Design of Integrated Circuits and Systems 14, 87–95.

Shin, H. and Hachtel, G. D. (1997). Verification of combinational circuits using conjunctively decomposed implications. IWLS '97.

Shiple, T. R., Hojati, R., Sangiovanni-Vincentelli, A., and Brayton, R. K. (1994). Heuristic minimization of BDDs using don't cares. 31st DAC, ACM, New York, 225–231.

Sieling, D. (1995). Algorithmen und untere Schranken für verallgemeinerte OBDDs. Ph.D. Thesis. Univ. Dortmund. Shaker Verlag, Aachen, Germany.

Sieling, D. (1996). New lower bounds and hierarchy results for restricted branching programs. Journal of Computer and System Sciences 53, 79–87.

Sieling, D. (1998a). On the existence of polynomial time approximation schemes for OBDD minimization. STACS '98, LNCS 1373, Springer-Verlag, Berlin, New York, 205–215.

Sieling, D. (1998b). Variable orderings and the size of OBDDs for random partially symmetric Boolean functions. Random Structures and Algorithms 13, 49–70.

Sieling, D. (1998c). A separation of syntactic and nonsyntactic $(1, +k)$-branching programs. ECCC Rep. No. 98-045.

Sieling, D. (1999). The complexity of minimizing FBDDs. MFCS '99, LNCS 1692, Springer-Verlag, Berlin, New York, 251–261.

Sieling, D. and Wegener, I. (1993a). NC-algorithms for operations on binary decision diagrams. Parallel Processing Letters 3, 3–12.

Sieling, D. and Wegener, I. (1993b). Reduction of OBDDs in linear time. Information Processing Letters 48, 139–144.

Sieling, D. and Wegener, I. (1995a). Graph driven BDDs – a new data structure for Boolean functions. Theoretical Computer Science 141, 283–310.

Sieling, D. and Wegener, I. (1995b). New lower bounds and hierarchy results for restricted branching programs. In: *Workshop on Graph-Theoretic Concepts in Computer Science*, LNCS 903, Springer-Verlag, Berlin, New York, 359–370.

Sieling, D. and Wegener, I. (1998a). On the representation of partially symmetric Boolean functions by ordered multiple valued decision diagrams. Multiple-Valued Logic 4, 63–96.

Sieling, D. and Wegener, I. (1998b). A comparison of free BDDs and transformed BDDs. Tech. Rep. No. 697, Univ. Dortmund, Germany.

Simon, J. and Szegedy, M. (1993). A new lower bound theorem for read-only-once branching programs and its applications. DIMACS Series in Discrete Mathematics and Theoretical Computer Science 13, 183–193.

Sinha, R. K. and Thathachar, J. S. (1997). Efficient oblivious branching programs for threshold functions. Journal of Computer and System Sciences 55, 373–384.

Skyum, S. and Valiant, L. G. (1985). A complexity theory based on Boolean algebra. Journal of the ACM 32, 484–502.

Slobodová, A. and Meinel, C. (1998). Sample method of minimization of OBDDs. SOFSEM '98: Theory and Practice of Informatics, LNCS 1521, Springer-Verlag, Berlin, New York, 419–428.

Somenzi, F. (1998). CUDD: CU decision diagram package release 2.3.0. Tech. Rep., University of Colorado, Boulder, CO.

Stanion, T. and Sechen, C. (1994). Boolean division and factorization using binary decision diagrams. IEEE Trans. on Computer-Aided Design of Integrated Circuits and Systems 13, 1179–1184.

Supowit, K. J. and Friedman, S. J. (1986). A new method for verifying sequential circuits. 23rd DAC, IEEE, Piscataway, NJ, 200–207.

Szelepcsényi, R. (1988). The method of forced enumeration for nondeterministic automata. Acta Informatica 26, 279–284.

Tafertshofer, P. and Pedram, M. (1997). Factored edge-valued binary decision diagrams. Formal Methods in System Design 10, 243–270.

Takahashi, N., Ishiura, N., and Yajima, S. (1994). Fault simulation for multiple faults by Boolean function manipulation. IEEE Trans. on Computer-Aided Design of Integrated Circuits and Systems 13, 531–535.

Takenaga, Y. and Yajima, S. (1993). NP-completeness of minimum binary decision diagram identification. Tech. Rep. COMP 92-99, Institute of Electronics, Information and Communications Engineers, 57–62.

Tani, S., Hamaguchi, K., and Yajima, S. (1993). The complexity of the optimal variable ordering problem of a shared binary decision diagram. ISAAC '93, LNCS 762, Springer-Verlag, Berlin, New York, 389–398.

Tani, S. and Imai, H. (1994). A reordering operation for an ordered binary decision diagram and an extended framework for combinatorics of graphs. ISAAC '94, LNCS 834, Springer-Verlag, Berlin, New York, 575–583.

Thathachar, J. S. (1998a). On separating the read-k-times branching program hierarchy. 30th STOC, 653–662.

Thathachar, J. S. (1998b). On the limitations of ordered representations of functions. In: *Computer Aided Verification*, LNCS 1427, Springer-Verlag, Berlin, New York, 232–243.

Touati, H. J., Brayton, R. K., and Kurshan, R. (1995). Testing language containment for ω-automata using BDDs. Information and Computation 118, 101–109.

Touati, H. J., Savoj, H., Lin, B., Brayton, R. K., and Sangiovanni-Vincentelli, A. (1990). Implicit state enumeration of finite state machines using BDD's. ICCAD '90, 130–133.

Tsai, C.-C. and Marek-Sadowska, M. (1996). Generalized Reed-Muller forms as a tool to detect symmetries. IEEE Trans. on Computers 45, 33–40.

van Eijk, C. A. J. (1997). A BDD-based verification method for large synthesized circuits. INTEGRATION, the VLSI Journal 23, 131–149.

van Laarhoven, P. and Aarts, E. (1987). *Simulated Annealing. Theory and Applications.* Kluwer Academic Publishers, Norwell, MA.

Waack, S. (1997). On the descriptive and algorithmic power of parity ordered binary decision diagrams. STACS '97, LNCS 1200, Springer-Verlag, Berlin, New York, 201–212.

Wang, K. H., Hwang, T. T., and Chen, C. (1993). Restructuring binary decision diagrams based on functional equivalence. EDAC '93, IEEE Computer Society Press, Los Alamitos, CA, 261–265.

Wegener, I. (1984). Optimal decision trees and one-time-only branching programs for symmetric Boolean functions. Information and Control 62, 129–143.

Wegener, I. (1986). Time-space trade-offs for branching programs. Journal of Computer and System Sciences 32, 91–96.

Wegener, I. (1987). *The Complexity of Boolean Functions.* Teubner, Stuttgart, John Wiley, New York.

Wegener, I. (1988). On the complexity of branching programs and decision trees for clique functions. Journal of the ACM 35, 461–471.

Wegener, I. (1993). Optimal lower bounds on the depth of polynomial-size threshold circuits for some arithmetic functions. Information Processing Letters 46, 85–87.

Wegener, I. (1994a). Efficient data structures for Boolean functions. Discrete Mathematics 136, 347–372.

Wegener, I. (1994b). The size of reduced OBDD's and optimal read-once branching programs for almost all Boolean functions. IEEE Trans. on Computers 43, 1262–1269.

Werchner, R., Harich, T., Drechsler, R., and Becker, B. (1995). Satisfiability problems for ordered functional decision diagrams. Reed-Muller '95, Fujiki Printing Co. LTD, Iizuka, Japan, 206–212.

Wu, Y.-L. and Marek-Sadowska, M. (1993). Efficient ordered binary decision diagram minimization based on heuristics of cover pattern processing. EDAC '93, IEEE Computer Society Press, Los Alamitos, CA, 273–277.

Yanagiya, M. (1995). Efficient genetic programming based on binary decision diagrams. ICEC '95, 234–239.

Yao, A. C. (1983). Lower bounds by probabilistic arguments. 24th FOCS, 420–428.

Žák, S. (1984). An exponential lower bound for one-time-only branching programs. MFCS '84, LNCS 176, Springer-Verlag, Berlin, New York, 562–566.

Žák, S. (1995). A superpolynomial lower bound for $(1, +k(n))$-branching programs. MFCS '95, LNCS 969, Springer-Verlag, Berlin, New York, 319–325.

Žák, S. (1997). A subexponential lower bound for branching programs restricted with regard to some semantic aspects. ECCC Rep. No. 97-050.

Index